D0214488

Submarine Groundwater

Submarine Groundwater

Igor S. Zektser
Roald G. Dzhamalov

English Editor
Lorne G. Everett

Taylor & Francis
Taylor & Francis Group
Boca Raton   London   New York

CRC is an imprint of the Taylor & Francis Group,
an informa business

CRC Press
Taylor & Francis Group
6000 Broken Sound Parkway NW, Suite 300
Boca Raton, FL 33487-2742

© 2007 by Taylor & Francis Group, LLC
CRC Press is an imprint of Taylor & Francis Group, an Informa business

International Standard Book Number-10: 0-8493-3576-0 (Hardcover)
International Standard Book Number-13: 978-0-8493-3576-1 (Hardcover)

Library of Congress Cataloging-in-Publication Data

Zektser, I. S. (Igor' Semenovich)
   Submarine groundwater / by Igor S. Zekster, Lorne G. Everett, Roald G. Dzhamalov.
      p. cm.
   Includes bibliographical references and index.
   ISBN-13: 978-0-8493-3576-1 (alk. paper)
   ISBN-10: 0-8493-3576-0 (alk. paper)
   1. Groundwater flow. 2. Seawater. 3. Submarine geology. I. Everett, Lorne G.
II. Dzhamalov, R. G. (Roal'd Gamidovich) III. Title.

GB1197.7Z435 2006
551.49--dc22   2006012470

Visit the Taylor & Francis Web site at
http://www.taylorandfrancis.com

and the CRC Press Web site at
http://www.crcpress.com

Authors

Igor S. Zektser earned his doctorate in 1975 in hydrogeology science from the All Union Research Institute for Hydrogeology and Engineering Geology in the former Soviet Union. For the past 30 years, he has been head of the Laboratory of Regional Hydrogeology at the Water Problems Institute of the Russian Academy of Sciences.

Dr. Zektser's fields of specialization include quantitative assessment of groundwater resources, evolution of the contribution of groundwater to total water resources and water balance, assessment of the interrelation between ground and surface water and groundwater discharge into rivers, lakes, and seas, study of the main regularities of groundwater formation and distribution in various natural as well as disturbed conditions, and assessment of human impact on groundwater and groundwater vulnerability. He has published 12 monographs and more than 220 papers. Under his guidance as full professor, thirteen postgraduate candidates successfully earned their doctoral degrees.

Dr. Zektser is one of the original authors of the groundwater flow map of central and eastern Europe as well as the world hydrogeological map, and his research has earned him international recognition. He also served as expert and scientific leader for several projects of the International Hydrological Program, UNESCO, on groundwater resources assessment.

Dr. Zektser is president of the Russian National Committee of the International Association of Hydrological Sciences and is a member of the Russian Academy of Natural Sciences, the Russian Academy of Ecology, the New York Academy of Sciences, and the American Institute of Hydrology.

In 1991, Dr. Zektser was invited by the Environmental Protection Agency (EPA) to the Vadose Zone Monitoring Laboratory in the Institute for Crustal Studies at the University of California in Santa Barbara to serve as a visiting research professor. From 1997 to 1998, he worked as a Fulbright Scholar at the same laboratory at UCSB. From 2003 to 2004 he worked at the Bren School of Environmental Science and Management at UCSB, also as Fulbright Scholar. He has been a member of the Board of IGCP Scientific Council of UNESCO since 2003.

Dr. Zektser authored *Groundwater and the Environment, Applications for the Global Community* (CRC/Lewis Press, 2000). He is one of the authors and editor-in-chief of the international monograph *Groundwater Resources of the World and their Use* (UNESCO, 2004).

Roald G. Dzhamalov graduated from Moscow State University. He earned his doctorate in 1991 in hydrogeology science from St. Petersburg Mining Institute in the former Soviet Union. He has been head of the Environmental Hydrogeology Laboratory in the Water Problems Institute of the Russian Academy of Sciences for the last 20 years. He is also professor of the International University, Dubna, delivers lectures on hydrogeology, engineering geology, geocryology, landscape geochemistry, and is a supervisor of postgraduate students.

Dr. Dzhamalov's fields of specialization include studying and assessment of groundwater contribution to the water and salt balances of seas, evaluation of the groundwater contribution to total water resources and water balance, regional assessment and mapping of groundwater resources, studying of groundwater contamination and acidification under acid fallout and other sources, assessment of groundwater vulnerability, and studying of human impact on groundwater formation.

Dr. Dzhamalov has published 11 monographs and more than 120 papers. He is one of the principal authors and editors of groundwater flow maps for the USSR and other countries and regions. He is editor-in-chief of the World Map of Hydrogeological Conditions and Groundwater Flow, scale 1:10 000 000. He has also served as scientific leader and expert for several projects of the International Hydrological Program, UNESCO, and the Russian Basic Research Foundation on groundwater resources, its development and contamination. His research has gained him international recognition.

Dr. Dzhamalov is an academician of the Russian Academy of Natural Sciences and is a member of the American Institute of Hydrology, the International Association of Hydrological Sciences, and the International Association of Hydrogeologists. He has been awarded the Savarensky Prize from the Russian Academy of Sciences (2001) and a special diploma from the American Institute of Hydrology (1994).

English Editor

Lorne G. Everett is the 6th Chancellor of Lakehead University in Thunder Bay, Ontario, Canada, president of L. Everett and Associates, LLC, research professor (retired) from the Bren School of Environmental Science & Management at UCSB (Level VII), past director of the UC Vadose Zone Monitoring Laboratory, and senior vice president of Haley & Aldrich, Inc. The University of California describes full professor Level VII as "reserved for scholars of great distinction." Dr. Everett earned a Ph.D. in hydrology from the University of Arizona in Tucson, and is a member of the Russian Academy of Natural Sciences. He earned a Doctor of Science degree (Honoris Causa) from Lakehead University  in Canada for distinguished achievement in hydrology (1996). He received the Ivan A. Johnston Award for outstanding contributions to hydrogeology (1997), the Kapitsa Gold Medal — the highest award given by the Russian Academy — for original contributions to science (1999), the Medal of Excellence from the U.S. Navy, and the Award of Merit, the highest award given by ASTM International (2000), the C.V. Theis Award, the highest award given by the American Institute of Hydrology (AIH) for major contributions to groundwater hydrology (2002), and the Canadian Golden Jubilee Medal for "Significant Contributions to Canada" (2003). Dr. Everett is an internationally recognized expert who has conducted extensive research on subsurface characterization and remediation. He is chairman of the American Society for Testing and Materials (ASTM) Task Committee on Groundwater and Vadose Zone Monitoring (D18.21.02). He chaired the remediation session of the First USSR/USA Conference on Environmental Hydrogeology (Leningrad, 1990). Dr. Everett has published over 150 technical papers, holds several patents, developed 11 national ASTM Vadose Zone monitoring standards, and has authored several books, including *Vadose Zone Monitoring for Hazardous Waste Sites, and Subsurface Migration of Hazardous Waste.* His book *Handbook of Vadose Zone Characterization and Monitoring* is a best seller. His book *Groundwater Monitoring* was endorsed by the EPA as establishing "the state-of-the-art used by industry today" and is recommended by the World Health Organization for all developing countries.

Dr. Everett has made several presentations before the U.S. Congress, and participates on blue ribbon peer review panels for most U.S. Department of Energy installations. Dr. Everett is a member of the UC/LLNL Petroleum Hydrocarbon Panel, the DOE/EPA VOC Expert Committee, the Interagency DNAPL Consortium Science Advisory Board, and is a scientific advisor to the U.S. Navy's National

Hydrocarbon Test Site Program. Dr. Everett is a member of the DOE Executive Panel for both the Vadose Zone S&T Roadmap and the Long Term Stewardship Roadmap.

Dr. Everett is an expert witness with an established track record in over 55 court cases involving over $2 billion dollars.

Introduction

In recent decades, hydrologists and hydrogeologists from around the world have carried out investigations of water circulation, total water balance, and water resources of particular regions, countries, continents, and the planet as a whole. To a great degree, these investigations have progressed owing to activities implemented within the International Hydrologic Program of UNESCO. The problem of subsurface water exchange between land and sea has been the subject of numerous international symposia conducted in recent years by the International Association of Hydrological Sciences and International Association of Hydrogeologists.

Subsurface water exchange between land and sea is one of the important elements of natural water circulation. It includes two differently oriented and nonequipollent processes: groundwater discharge from land to sea and intrusion of seawaters to coasts.

Groundwater discharge to seas is formed from the zone of water-saturated rocks located within the draining influence areas of seas. It occurs constantly and everywhere, except in some regions of the Arctic and especially Antarctica that are chiefly occupied by permafrost. Seawater intrusion into shore and coasts or, as geologists say, intrusion of seawaters in natural conditions, has a local spreading. However, this process becomes more active in disturbed conditions, namely in areas where intensive groundwater extraction by wells located on the seacoasts creates conditions for drawing the seawaters upward.

Hydrologists and hydrogeologists have come to the necessity of studying the subsurface water exchange between land and sea both simultaneously and independently of each other. An impetus to development of these studies was given by the demands of the practice. Let us briefly discuss the basic problems and tasks facing the specialists that determined the main directions of investigations in the field of studying the subsurface water exchange between land and sea and submarine groundwater. For convenience, these problems are conventionally subdivided into the hydrological and hydrogeological.

Hydrological Aspects: Many recent publications devoted to investigations of the regional and global water balance indicate that the groundwater discharge to seas and oceans is the most poorly studied and difficult to determine element. Absence of basic quantitative data on groundwater discharge to seas and oceans often hampers the study of total water balance and water circulation in many regions, and of the Earth as a whole. Without data on groundwater discharge, the world water balance remained incomplete. Only in recent years were investigations carried out on the groundwater component of the world water balance, making it possible to obtain approximate differentiated estimates of the groundwater discharge from continents to seas and oceans. This research reveals the basic and general regularities in formation and distribution of the groundwater discharge to seas in differing natural conditions. This book briefly analyzes the results of these investigations.

Determining the role of groundwater in water and salt balances of seas, especially inland ones, is of practical importance. Groundwater discharge is the element most difficult to determine in present and prospective studies of the water and salt balances of seas and large lakes; nevertheless scientists answer a number of complicated questions, such as: What is this discharge in the quantitative respect? Does it exert a substantial influence on the water and salt balances of a water body? How will groundwater inflow change in the future? To what degree should the so-called groundwater component be taken into account when studying balances of salt and heat in a water body? Quantitative estimation of groundwater inflow is needed for developing water-managing measures aimed at the maintenance of optimal water, salt, temperature, and hydrobiological regimes of inland seas and large lakes. It is necessary to verify the water balance of the inland seas and large lakes, and more important, to predict its changes under influences of intensive human activity in water drainage areas. These factors are significant triggers in the development of works studying the interaction of groundwater and the land and sea.

Hydrogeological Aspects: Many groundwater-pumping wells are located on seacoasts, and the conditions of their work greatly depend on how groundwater and seawater interact. The objective of investigations into this interaction is to determine the most optimal yield of wells exploiting groundwater on the coasts. Intrusion of seawaters to aquifers presents a serious threat to the work of coastal well fields. Intensive exploitation of such well fields causes a change in water exchange within the system of "sea–groundwater." An important task facing the hydrogeologists is to determine the position of the dividing boundary of "fresh groundwater–salt sea-water" and hence to predict water quality at well fields located on seacoasts.

Groundwater discharge to seas is an important parameter of groundwater resources. In recent years many regions of the world have a deficit of water resources. In coastal regions the lack of water, especially of quality fresh water, can in many cases be considerably eliminated or even entirely compensated through using groundwater, which at present "uselessly" outflows to the sea. Some countries have had positive experiences using waters of large submarine springs being discharged to sea not far from the coast, as well as experiences of exploiting wells drilled in shelf and stripped fresh groundwater for water supply to populated seaside areas.

It is important to mention the results of widely carried out offshore geological investigations. The results of drilling works on sea and ocean bottoms have revealed considerable reserves of useful minerals such as oil, gas, coal, iron ores, manganese, and phosphorites. Thus, the task of researchers is to study the role of groundwater in formation of mineral deposits on sea bottoms. These studies are now only in the initial stage. Reliable conclusions on the origin of mineral deposits on the sea and ocean bottoms can be obtained by studying groundwater discharge to seas and the physical–chemical interaction processes of submarine groundwater, rocks, and sea-waters. The preliminary results of investigations in this field show that in the areas of groundwater discharge on the sea bottom, sharp changes occur in redox situation which can cause considerable phasal transformations.

We have named only the basic problems facing researchers of subsurface water exchange between land and sea. Nevertheless, those short descriptions show the scientific and practical interest to investigations in this direction.

This book analyzes and generalizes the results of multiyear investigations of submarine groundwater discharge to seas, gives the scientific-methodical principles of assessment of subsurface water exchange between land and sea, and demonstrates numerous concrete examples of such investigations carried out by scientists from various countries.

It should be noted that although the book is titled *Submarine Groundwater*, it also covers the results of assessment of groundwater discharge to large lakes. Therefore, strictly speaking, the book is devoted to subaqueous groundwater, and the term "subaqueous groundwater" includes "submarine groundwater" as well. However, taking into consideration that the methodical principles of study and assessment of groundwater discharge both to seas and large lakes are actually similar, and that the term "subaqueous groundwater" has no wide usage, the authors considered it reasonable to use the title *Submarine Groundwater*.

With the aim of describing the particular problems of studying submarine groundwater as completely and comprehensively as possible, the author involved the participation of a number of well-known scientists from the Commonwealth of Independent States, the United States, and Japan — all of whom have had experience in theoretical and especially regional investigations of groundwater discharge. Those chapters and sections were prepared by the following authors:

M. Khublaryan, A. Frolov, and I. Yushmanov (1.3); J. Cable, J. Martin, and M. Taniguchi (1.5); H. Loaiciga (2.1.1; 3.6); C. Langevin (2.1.2; 4.6); M. Liepnik and C. Baldwin (2.1.5); R. Coebett, J. Cable, and J. Martin (2.2.2); J. Clark and T. Stieglitz (2.2.3); V. Kiryukhin and A. Korotkov (3.1); A. Voronov and E. Viventsova (4.1.3); V. Veselov and V. Poryadin (4.3); G. Buachidze (4.4); B. Pisarsky (5.1); O. Podolny (5.2); A. Voronov, E. Viventsova, and M. Shabalina (5.5); N. Oberman (5.6); L. Everett (6.1–6.4); G. Koff (6.5).

The general scientific editor is I.S. Zektser; and the English editor is L.G. Everett.

Sections 1.1, 1.2, 1.4, (with participation of O. Karimova) 2.1.3, 2.1.4, 2.2.1, 2.2.4, 2.2.5, 3.2–3.5, (with participation of T. Safranova) 4.1, 4.2, (with participation of A. Meskheteli) 4.5, and 5.3 were prepared by I. S. Zekster and R. Dzhamalov under financial support of the Russian Fundamental Investigation Foundation (Grants No. 04-05 64566 and No. 05-05-65033). The general scientific editing was fulfilled by I.S. Zekster. L.G. Everett was the English editor. Authors and editors are grateful to Olga A. Karimova for preparing this book for publication.

The book is intended for a wide circle of readers — hydrologists, hydrogeologists, environmentalists, chemical geologists, and all those who deal with the problems of water circulation, water balance, water resources, and groundwater discharge.

Table of Contents

1 Groundwater–Seawater Interaction in the Coastal Zone

CONTENTS

1.1 PRESENT CONCEPT OF SUBSURFACE WATER EXCHANGE OF LAND AND SEA

The theory of subsurface water exchange between land and sea is closely connected with the general theory on groundwater flow and began its development as a branch of hydrogeology in the mid 20th century. Concrete investigations of submarine groundwater discharge and especially the intrusion of seawaters into the coastal shores had been carried out much earlier, for example, during the exploration of groundwater on coast areas. However, the study of water exchange between land and sea in the regional and global scales began relatively recently in conjunction with the neccessity to give a reliable assessment of the role of groundwater in the water and salt balances of particular seas as well as in global water circulation.

The investigations of the hydrogeological cycle, water balance, and water resources of particular regions, sea basins, continents, and the earth as a whole have been rather subdued because of the absence of sufficient quantitative data on the subsurface water exchange between land and sea. The problems of the last decades related to inland seas and lakes have also brought to the forefront the task of direct measurement and quantitative assessment of the role of submarine groundwater discharge in the water and salt balances of water bodies. The poor knowledge of submarine water and chemical discharge is partly due to the impossibility to measure directly these elements of water balance. Until recently they were determined by calculating the discrepancy in the water balance equation. As a result, final values included all the errors from the measurements of the rest of its terms. The prediction of changes in the water, salt, and hydrobiological regimes of some water bodies and the substantiation of measures on their keeping and protection requires a comprehensive study of the water exchange between land and sea.

The achievements of sea geology in geophysics, with the accumulation of unique data from deep drilling, raised a task to analyze and generalize these materials from the position of distribution and conditions of occurrence of submarine waters, their features, specificity of circulation, and to verify the parameters of their interaction with rocks, sea, and groundwaters of the land, as well as their influence on the biota. All this substantiates the necessity of investigating the role of submarine groundwater not only in the water balance of seas and oceans, but also in the geological processes occurring at their bottoms. In other words, an independent branch is being formed in general hydrogeology — *marine hydrogeology*. This book presents the scientific fundamentals of a large section of marine hydrogeology, dealing with the study of subsurface water exchange between land and sea. Special attention is paid to regional assessment and the identification of regularities in submarine groundwater discharge to seas and oceans, because these processes are manifested everywhere and can exert a significant influence on the water and salt balances of individual water bodies or their parts. The intrusion of seawaters into the coastal shores is also an element of subsurface water exchange between land and sea, but it has a limited character and is activated under the presence of human activity.

The current achievements of marine geology and geophysics make it possible to perform the hydrogeological zoning of the bottoms of seas and oceans, and to distinguish if hydrogeological structures have analogues on the continent. Of special interest in this respect are the most studied structures in shelf areas of seas and oceans where geostructural, hydrodynamic, and hydrochemical features of water exchange between land and sea are clearly manifested (Zektser et al. 1984; Korotkov et al. 1980). The factual data on the distribution and features of groundwater migration at different depths have provided the possibility of the quantitative assessment of groundwater flows in the covers of the earth crust and of their role in different geologic-hydrogeological processes (Kononov 1983; Zverev 1993; Dzhamalov et al. 1999; Shvartsev 1999).

As mentioned above, the new branch of *marine hydrogeology* has been formed at the intersection of two allied sciences. It has its own subject and objectives, aims, and tasks of investigations. A system of basic concepts and terms for this new science and its related definitions are given below. In the future, the proposed concept–terminology base will be improved in accordance with the development of the science itself. Some concepts were defined from terminology used for the study of groundwater flow on land (Zektser et al. 1984).

The general concepts and terms of marine hydrogeology are:

- *Subaqueous* or *submarine groundwater* is the water enclosed in rocks composing the bottom of large lakes, seas, and oceans.
- *Submarine groundwater flow* is the groundwater movement in rocks under the bottoms of lakes, seas, and oceans, occurring as the result of the general water circulation and geodynamic processes in the earth's crust.
- *Submarine ionic or chemical discharge* is the transfer of salts and chemical elements, dissolved in submarine groundwater, to lakes, seas, and oceans.

The general term *submarine groundwater flow* is defined as the water exchange between rocks, composing the bottom, and the marine basin. The *submarine groundwater discharge* implies an influx (discharge) of groundwater, generated on land, directly to sea. The reverse process is the *penetration (intrusion) of seawaters* into the shores and aquifers of the land under influence of different natural and artificial factors. All these processes in combination define the term *subsurface water exchange between land and sea.*

The quantitative characteristics of the subsurface water exchange between land and sea can be represented by, besides absolute values, specific indices such as *modulus* and the *linear flow rate* of groundwater discharge to the sea, as well a modulus of submarine groundwater discharge. The *modulus of groundwater discharge to the sea* is understood as losses of groundwater flow to the sea from a drainage area of 1 km², the discharge from which is directly to the sea. The *modulus of submarine groundwater discharge* is defined by the characteristics of groundwater flow rate from an area of 1 km² of aquifer discharge on the sea bottom. The *linear groundwater discharge to the sea* is defined as the losses of groundwater flow per a width unit of its front or the shoreline of the sea. The linear discharge can also characterize seawater intrusion into the shore.

By analogy with groundwater discharge to the sea, it is reasonable to introduce specific characteristics for the submarine chemical discharge. The term *chemical discharge* insufficiently reflects the essence of the physical-chemical processes of the transfer of dissolved salts with groundwater, but it is already used in the literature and it is unfeasible to replace it. In this case, the amount of dissolved salts or particular chemical elements (compounds) transferred with groundwater directly to the sea from 1 km² of drainage area will be termed as a *modulus*, and the amount from a width unit of groundwater flow front or from 1 km² of the shoreline will be defined as *linear losses of submarine chemical discharge.*

Groundwater is discharged to lakes, seas, and oceans in the form of:

- Juvenile waters during the degassing of the earth's mantle
- Sedimentation waters at the expense of their press-out during lithogenesis of marine sediments

- Subsurface component of the total river discharge (river low water runoff)
- Direct groundwater discharge to seas, not involving the river network

The juvenile water discharge represents a "subsurface" component of the water balance of seas and oceans. The quantitative assessment of the volume of juvenile waters discharged to the seas is rather difficult. As reported in the publication of Timofeev et al. (1988), the amount of juvenile waters at the present-day phase of the earth's evolution usually does not exceed 5% of the total hydrothermal discharge from volcanic areas. The isotopic composition of inert gases indicates that major volatile substances, including hydrogen, have been degassed at the initial phase of the planet's evolution (Verkhovsky et al. 1985). However, the complete degassing of the mantle has not yet occurred; and this is confirmed by the current emission of the mantle hydrogen and methane through rift zones (Kononov 1983). In the present geological epoch, juvenile gas- and water-containing fluids are associated with rift zones. The leading role belongs to the median-oceanic ridges. According to Vinogradov (1967), the annual amount of juvenile water, contributed from volcanoes, hot springs, and deep-seated faults, does not exceed 0.5–1.0 km^3 (an extremely small value for the current balance of the World Ocean).

The methodical procedure for calculating groundwater discharge drained by rivers, which in water-balance equations is included into the total river runoff, is well worked out. It should be taken into account that approximately one-third of the water discharge from rivers inflowing into the seas is generated at the expense of drained groundwater from the zone of intensive water exchange.

The waters in the underground part of the hydrosphere can be subdivided into (1) *groundwater of the land* that has an unsaturated zone in the upper part of the hydrogeological cross-section and is linked with the atmosphere, and (2) *subaqueous groundwater* spread under the bottom of large lakes and seas and hydraulically linked with waters of these water bodies. The most widely distributed type of underground water is subaqueous water (i.e., submarine waters occurring under the bottom of seas and oceans), the hydrodynamics and hydrochemistry of which are largely determined by their interconnection with seawaters. The submarine waters are sub-divided into *infiltration waters* generated on land at the expense of precipitation and surface runoff; *sedimentation waters* formed directly within the marine area because of accumulation of sediments and their subsequent diagenesis and katagenesis; and *juvenile waters* connected with the degassing of the mantle.

The infiltration waters discharging from the land are mainly spread in the shelf zone. Their current filtration recharge per unit time is much higher than the amount of elision recharge from sedimentation waters during sedimentary processes. There-fore, when conditions are favorable, the tongues of the confined filtration waters can intrude far into the sea area, reaching the continental slope and displacing the sedimentation waters. The results from the drilling of the sea bottom gives vivid evidence of the intrusion of fresh or low-mineralized groundwater formed on land (Scripps Institute of Oceanography). The constant interaction of infiltration waters with sedimentation waters and directly with seawaters leads, owing to the convective-diffusive processes and physical-chemical reactions, to the gradual equalization of the chemical composition and mineralization of submarine groundwater of different geneses. It is possible that at certain depths a transient zone exists — the spatial characteristics of which are, to a great degree, determined by hydrodynamic and physical-chemical gradients of the counter-moving submarine groundwater, and, depending on the geofiltrative properties of the water-bearing rocks, the groundwater of different geneses may have a by-layer occurrence.

Below, the specific features of the interaction between unconfined filtration water flow and seawaters are considered. Theoretically, if the filtration properties are homogenous, the shallow groundwater would be entirely discharged in the sea area and groundwater line (coastal zone). When a sufficiently thick shallow aquifer has a close hydraulic linkage with seawaters, usually in the coastal zone, a counter seawater flow is formed, which restricts the shallow groundwater current and causes it to wedge out and, hence, the discharge has a greater velocity. The velocity of the discharging shallow groundwater in this zone can be several times higher than the average velocity of the filtration flow above the shoreline (tidal zone). Such occurrences are true only for ideal homogenous aquifers. In nature the shallow aquifers are usually heterogeneous, whether for lithological composition or filtration qualities. Closer to the shoreline, the filtration properties of such sedimentary confined/unconfined aquifers are, as a rule, decreased due to prevailing clayey material and increased water pressure head. The entire submarine discharge of infiltration flow occurs not only in the narrow coastal or tidal zone but also at a distance of several kilometers inside the sea area. At the same time, the results of direct measurements show that a considerable part of the total shallow groundwater filtration flow is discharged in the coastal zone with a width of several hundreds of meters.

The shallow groundwater of the land, like confined groundwater, during submarine discharge transforms its composition and mineralization because of mixing and physical-chemical reactions with seawaters. However, because of the agricultural and industrial development of many coastal areas, the compositions of natural waters (including shallow groundwater) are subjected to significant changes. The nitrate concentrations in shallow groundwater during submarine discharge in some coastal areas of the United States, Australia, and the islands of Jamaica and Guam vary from 20–80 mg/l to 120–380 mg/l. Such high concentrations of biogenic elements in the submarine waters cause the anomalous growth of biomass in the seawater and the appearance of specific species of microorganisms and algae, which, in turn, can serve as indicators of submarine discharge of groundwater of a certain composition. In other words, the intensive human activity in the coastal drainage areas causes direct changes in shallow groundwater composition and exerts a significant influence on the environmental state of seawaters due to the transfer of biogenes, pesticides, heavy metals, and other toxic substances.

Submarine discharge of confined groundwater takes place through submarine springs usually associated with tectonic disturbances and areas with developing fissured and karst rocks, as well as through seepage via low-permeable covers of aquifers and sea-bottom sediments. It is important to analyze the principal possibility of a hydrodynamic interconnection between hydrogeological structures of the land and adjacent parts of oceans. It is acknowledged that the shelf and continental slope are parts of the land, which do not continue on the ocean bottom. A continental slope usually represents a tectonic flexure, within which the thickness of sedimentary rocks sharply reduces until the full wedging out and disappearance of some beds (Lisitsyn 1978). In other words, the hydrogeological structures of continents and oceanic depressions are dissociated tectonically, and the boundary between them runs along the continental slope. In this case, the submarine infiltrating waters, formed on the land, are wedged out mainly within the shelf and continental slope. The submarine groundwater discharge in the shelf and continental slope is considerably favored by the existing submarine canyons — i.e., deeply entrenched erosional valleys stripping not only Quaternary, but also more ancient, rocks.

The interaction between the infiltration and sedimentation waters in the shelf has not been fully studied until now. Some researchers consider that elision processes, i.e., the press-out of water from clayey sediments during the compaction and diagenesis of the latter, play a leading role in all the phases of the formation of artesian basins. However, the modeling of the gravitational consolidation of clayey rocks has shown that with real initial parameters the pressure distribution in clayey strata is controlled by the conditions at their upper and lower boundaries (Dyunin 2000). The determining influence on pore pressure distribution is exerted by the permeability of clays and the velocity of their sedimentation. In the actual conditions of seas and oceans, the sedimentation velocity of clayey rocks seldom exceeds 10^{-3} m/yr, and their permeability is usually over 10^7 m/day and can increase with depth due to the formation of macro- and meso-jointing during lithogenesis. In these cases, according to the data from the model, the pore pressure distribution in the clayey strata actually has a linear character and depends on a ratio of the pressures at the upper and lower boundaries of clays. The hydrostatic pressure of the clayey strata in the upper cross-section of marine sediments increases with depth. The pressure is almost always higher at the base than at the strata roof. Therefore, the submarine sedimentation waters, pressed out during clay consolidation, move predominantly from underneath upward into the source sea basin.

It should be noted that the upper part of the cross-section of the bottom sediments is mainly composed of sedimentary clays with anomalously high moisture and porosity (to 80%) and a low density (1.2–1.3 g/cm^3). The resultant changes in their physical properties with depth are chiefly caused by compressive (gravitational) consolidation. The most intensive reduction of the porosity and moisture of clayey rocks takes place within the first tens and hundreds of meters, and gradually the effect diminishes with larger depth. Thus, within the first 100 m of depth, the clays lose more pore water than within the next 100 m, and at a depth of 300–500 m the loss amounts to 70–80% of the initial moisture volume. After that the clay consolidation rates progressively slow down. Thus, a decrease in the porosity and moisture of clays with depth occurs in relation to exponential dependence at which the major pore sedimentation waters are pressed out at the initial steps of sediment consolidation and then discharged to the source basin. In general, the compression regime of rock consolidation depends on the lithological composition and on the ratio of pressure and temperature, which determines the physical-chemical and mineralogical processes of diagenesis and lithification of the sediment (Dhzamalov et al. 1999).

The vertical filtration of sedimentation waters provides their constant linkage with seawaters in this case. The water exchange between compacting sediments and the sea basin is accompanied by salt exchange. The invariable composition of the seawaters during recent geological epochs and salt exchange between sedimentation and seawaters may cause the mineralization and salt composition of these waters to be similar. This was confirmed by the drilling experiences from the research ship *Glomar Challenger*: boreholes, drilled into the ocean bottom, were stripped at depths to several hundreds of meters from the bottom surface; the water's total salt concentration and content of basic components were similar to those in modern seawater (Scripps Institution of Oceanography).

The specific feature of submarine sedimentation waters is the invariability of their chemical composition. If, in submarine groundwater of infiltration type, one can observe a change in the chemical composition and mineralization both in area and section, in the submarine waters of sedimentation type in the upper part of the geological cross-section, the regional hydrochemical zonality (in the sense widely known as hydrogeology) is actually absent.

The data from sea-bottom drilling indicates that the similarity of the chemical compositions of submarine and seawaters is most plainly apparent in areas with slow sedimentation rates. In the other deeper parts of the oceans, a substantial change is often observed in the concentrations of some chemical elements in the submarine sedimentation waters. Namely, the Ca concentration usually increases, whereas concentrations of Mg and K decrease with depth. According to McDuff and Gieskes (1976), the breakdown of basalts in the deep parts of oceanic depressions and, to a lesser degree, of dispersed volcanic material and carbonates in the sedimentary rocks are the main reasons for the increase of dissolved Ca and the decrease of Mg in the submarine waters.

Of greatest interest, in hydrogeochemical respect, are the areas with modern submarine volcanism in the zones of median oceanic ridges, where the major juvenile submarine waters are probably discharged. The magma in these regions is effused onto the ocean bottom or close to it, causing a powerful heat flux. The young oceanic crust is highly fissured because of the processes of cooling, compaction, and extension. The seawater saturates the fissured zone, cools, and destroys the magmatic body. The hydrothermal solutions of the rift zones are usually enriched with CO_2, He^3, H_2, metals, and other components. The temperature of these solutions can approach the magmatic solidification point (980°C), but close to the bottom surface it is usually around 10–30°C. At higher temperatures, the water has low values of density and viscosity, causing active convection and an increased ability to penetrate into rocks and dissolve and leach water-bearing rocks. The thermal gas- and water-containing fluids, enriched with different chemical elements, serve as the basic source of polymetals. The isotopic compositions of carbon and helium in the hydrothermal solutions often indicate the age of these elements. The fluids of the oceanic rift zones represent a mixture of chiefly juvenile seawaters; because of contact with the magma, the temperature in the latter increases and chemical composition changes.

The conceptual model of the water exchange between land and sea under consideration is based on the analysis of updated information and makes it possible to formulate the following conclusions which can be used as the methodical basis for creating regional mathematical models of the formation, distribution, and migration of submarine groundwater.

Submarine waters are subdivided into infiltration, sedimentation, and juvenile waters. The infiltration waters reside only within shelf areas and are entirely discharged to the continental slope. The sedimentation waters are generated everywhere on the bottom of seas and oceans, but prevail within the floor of the World Ocean. The juvenile waters are mainly found in the zone of median oceanic ridges and their role in the current water balance of the World Ocean is not significant.

The predominant type of submarine groundwater discharge is seepage from beneath. The constant interaction of sedimentation and seawaters due to the vertical water exchange makes their mineralization and salt composition similar (if there are no additional salt sources).

According to the conditions of the formation and distribution of different types of submarine waters, the ocean bottom is subdivided into shelf, floor, and median-oceanic ridges, which differ from each other by peculiar hydro-dynamic and hydrochemical regimes of interaction of submarine waters with groundwater of land, rocks, and seawaters.

Submarine waters serve as the basic agent and medium of the migration of chemical elements in the earth's crust. Because of the higher mineralization of groundwater compared with surface water, the influence of submarine chemical discharge on the salt balance of seas (especially of inland ones) can be significant. The migration of chemical elements with submarine waters is most boldly distinguished in the zone of median-oceanic ridges, where, due to the convection of seawater, the active leaching of young basalts and enrichment of thermal gas- and water-containing fluids with different components takes place.

1.2 GROUNDWATER DISCHARGE TO SEAS

Submarine groundwater discharge to the seas and oceans is the least studied element in the water and salt balance of the seas. There are two reasons for this. First, groundwater inflow is the only component of the water balance of the seas, which is not accessible for direct measurements, and thus usually no data are available for sufficiently accurate calculations of the underground component of water balance. At the same time, without data on groundwater discharge, the water balance of particular seas and oceans and the World Ocean as a whole remains incomplete.

The second reason is of a subjective nature. For many years hydrogeologists engaged in studying water balance have assumed that groundwater discharge plays an insignificant role in the water balance (compared with its other components) and therefore it can be determined using the average multi-annual water balance equation. In other words, in their opinion, groundwater discharge can be determined as a difference between the average annual values of precipitation, evaporation, and river runoff. Groundwater discharge calculated in this way entirely depends on how precisely the average values of precipitation, evaporation, and river runoff are esti-mated and include all the errors of these estimates, which in total often exceeds the real value of groundwater discharge directly to the seas.

By means of such calculations those scientists obtained, not the value of ground-water discharge, but rather groundwater discharge "mixed" with all the errors included in the assessment of the basic components of water balance. This led to false conclusions. The earlier (approximately before 1970) investigations of the water balance of the Caspian Sea can serve as the most vivid example in this respect, when the data on groundwater discharge of different researchers, who used the above-mentioned calculation technique, differed from each other by almost 150 times.

Such an approach seems to be incorrect. This is linked to the fact that ground-water discharge to the sea is usually small in comparison with other elements of water balance (precipitation, evaporation, river runoff) and can be determined only by direct hydrogeological methods.

In general, investigations of groundwater discharge to seas do not differ from investigations of groundwater discharge to lakes, therefore everything said above and below concerns the water balance of lakes as well.

A considerable stimulus to the organization and development of investigations of groundwater discharge to seas was given in recent years by the practical demands of solving the "problem of inland seas." The essence of this problem is that in many inland seas (primarily the Caspian and Aral Seas) and in large lakes, the water level undergoes considerable changes caused by natural factors and intensive human activities in water-catchment areas. The necessity to study the current and expected water and salt balances of these water bodies arose and, hence, priority was given to the assessment of the role of groundwater in the formation of these balances. It is necessary to investigate the influence of groundwater not only on water and salt balances of a water body, but also on its hydrochemical, thermal, and hydrobiological regimes. Today a considerable amount of experience has been gained in investigations aimed at the quantitative estimation of groundwater discharge to inland and coastal seas and large lakes. The basic goal of these investigations is to study the features and regularities of water and salt exchange of a water body with the land, as well as to substantiate predictions of changes in the underground component of the water balance under ever-increasing human activity.

It should be especially emphasized that groundwater formed on land and discharged in the coastal zone of seas and oceans in many cases exerts a great influence on hydrochemical, hydrogeological, and thermal regimes of seawater in the coastal zone, and also may influence the sedimentation processes.

The study of groundwater dynamics in artesian structures adjacent to seas shows that the groundwater flow in the upper hydrodynamic zone is usually directed toward the sea and generates submarine groundwater discharge. As mentioned above, submarine groundwater discharge occurs in the form of concentrated water springs along tectonic disturbances and in zones of developing fissured and karst rocks, and as a result of distributed seepage through low-permeable roofs of aquifers into marine bottom sediments. Large submarine springs represent the most vivid, but not basic type of groundwater discharge to seas. As the analysis of available data shows, groundwater discharges to the shores and sea bottom mainly as a result of seepage through low-permeable sea-bottom sediments. Groundwater discharge by seepage through low-permeable sediments has been determined for many artesian structures in areas with both natural and disturbed hydrodynamic regimes. Seepage processes often determine the groundwater dynamics of an artesian basin and play a significant or even determinant role in submarine groundwater discharge.

Groundwater seepage occurs in that part of the sea where piezometric levels of an aquifer are located above the seawater surface and where there is a pressure-head gradient for the ascending filtration of submarine waters. In the coastal zone of artesian structures, water-bearing systems with a direct ratio of hydrostatic pressure heads are formed. In this case, water successively seeps from underlying aquifers into overlying ones, then the total vertical groundwater flow is discharged directly to the sea. Thus, the entire filtration flow that is generated on land and submerged under the sea bottom must be discharged into the sea area. This process is greatly favored by an affluent of infiltrated groundwater in the form of sedimentation submarine waters located at certain depths under the sea bottom.

A considerable amount of experience in the regional assessment and mapping of groundwater discharge over extensive territories has been gained recently, and regularities of this discharge process have been revealed. This has enabled the determination of submarine groundwater discharge to lakes, seas, and the World Ocean, using the already tested approaches and techniques of the regional assessment of groundwater discharge from the land. Most large-scale investigations of submarine groundwater discharge to seas were carried out by Russian, French, American, Yugoslavian, Italian, and Australian specialists. At present, the assessment of the groundwater discharge has been carried out for the following inland and coastal seas: Caspian, Aral, Azov, Baltic, and Black Seas; the following large lakes: Baikal, Balkhash, Issyk-Kul, Sevan, Lagoda (Ladozhskoye Ozero), the Great Lakes (North America), and in certain coastal areas of the Atlantic and Pacific Oceans. It should be noted that during the above-mentioned investigations, the submarine groundwater discharge was determined directly by hydrogeological, hydrochemical, and isotope methods. Recent analogous investigations are carried out by taking into account special marine hydrogeological works. This circumstance seems rather important, because the reliability of the assessment of groundwater discharge to a sea depends heavily on the calculation method and extent of reliable initial data. Independent assessment of submarine discharge often makes it possible to verify the other components of water and salt balances of inland seas and large lakes, as well as to give more grounded predictions of possible changes in the groundwater exchange between land and sea due to the growing impact of human activity in drainage areas (Dhzamalov et al. 1977).

1.3 SEAWATER INTRUSION INTO FRESHWATER AQUIFERS*

Interaction of seawater with fresh groundwater of seaboard zones is a widespread natural phenomenon in coastal regions. This process involves, along with groundwater runoff from land to seas, intrusion of seawater into underground aquifers. In natural conditions, intrusion of seawater occurs only when density of the groundwater that

---

* This subchapter was written by Martin G. Khublaryan, Anatoliy P. Frolov, and Igor O. Yushmanov (Water Problems Institute, Russian Academy of Sciences, Russia, martina@aqua.laser.ru).

is hydraulically linked with it is lower than the density of this seawater. Differences alone in densities of sea and freshwaters and difference in pressures of rock layers determine the kind and velocity of seawater intrusion. Being much heavier, seawaters move along the layer base, with a dividing boundary between sea and freshwaters having a complicated form.

Circa the 19th to 20th centuries, reseachers of the European shoreline area, Baydon-Ghyben (1888–1889) and Herzberg (1901), established that seawaters can penetrate deeply into fresh underground aquifers and that the depth of a dividing border between sea salt and freshwaters is 40 times greater than the level of fresh shallow groundwater in an unconfined aquifer. From the hydrostatic equilibrium Equation $p_A = p_B$ in points A and B (Figure 1.3.1), where $p_A = \rho_s gH$ and $p_B = \rho_0 g(H+h)$ are the entire hydrostatic pressures, they obtained the so-called condition of Ghyben–Herzberg, $H = \rho_0 h/(\rho_s - \rho_0)$, where h is the freshwater level above sea level, H is the depth of the dividing border below sea level, and ρ_0, ρ_s are the densities of freshwater and saltwater, respectively. If we accept that $\rho_0 = 1$ g/cm^3 and $\rho_s = 1.025$ g/cm^3, then H = 40 h.

Seawater intrusion becomes more intensive with more intensive extraction of groundwater. Numerous field and laboratory investigations show that depth of seawater intrusion into coastal zones significantly depends on the hydrological characteristics (permeability) of water-bearing rocks that usually possess aniso-tropic properties and are nonhomogenous. Thus, the intrusion zone is negligibly short in thin low-permeable layers and long (reaching several kilometers) in thick well-permeable layers. Due to this, intrusion zones in multilayered systems are characterized by high variety. Multilayered systems consist of several water-bearing layers separated from each other by intermediate low-permeable beds, with tongues of saltwater intrusion appearing in each layer. However, sometimes when water seeps between neighboring layers through semipermeable interbeds, analyzing the hydrodynamics of the intrusion process is difficult. Depending on

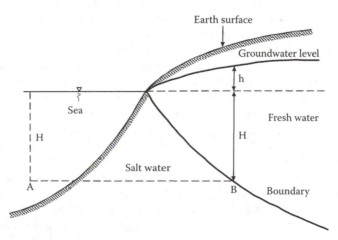

FIGURE 1.3.1 Distribution of freshwater and saltwater in an unconfined coastal aquifer.

the position of each layer, the flow rate of freshwater in it, and the thickness and permeability of water-bearing rocks, the depth of intrusion in some layers changes. A typical example of such a structure of an intrusion zone is Long Island, New York (Lusczynski and Swarzenski 1962). Researchers Schmorak and Mercado (1969) have revealed a multi-edged intrusion structure in a coastal aquifer of Israel. An attempt to analyze theoretically multi-edged intrusion zones was made by Todd and Huisman (1959), in which intermediate beds were supposed to be completely impermeable. The works of Collins and Gelhar (1971) and Podsechin and Frolov (1989) describe a hydrostatic approximation zone of seawater intrusion into a two-layered aquifer divided by a semipermeable interbed. In Volker and Rushton (1982), the finite element method is used to solve the 3-dimensional task with availability of pumping wells in layered media. Shamir and Dagan (1971) carried out an analysis of concentration distribution in complicated hydrogeological conditions where a well-conductive anisotropic aquifer is overlain by low-permeable sedimentary rocks with a dividing boundary between them having a sufficiently arbitrary shape.

Water-bearing layers can also contain geological formations (structures) with different permeability. In such cases, the filtration coefficient in Darcy's law (see below) changes in horizontal direction as well. Such layers are called horizontally heterogeneous.

With an increase of water extraction from underground aquifers located in coastal areas, submarine discharge of freshwater decreases. This results in formation of deep cones of depression that create hydraulic conditions favorable for intensification of seawater intrusion. Place, shape, and length of a dispersion zone depend on many factors, including the relationship between densities of salt and freshwaters, discharge or pressure head of freshwater, dispersion parameters of a water-bearing layer, etc.

In the contact zone of fresh groundwater and salt seawater, a transient area is formed where water salinity changes from mineralization of freshwater to salinity of seawater. Location and size of this transient zone depends on density of seawater, losses of subaqueous groundwater discharge, and other parameters. Creation of underground water-extracting fields in coastal areas and intensive pumping of fresh groundwater intensifies saltwater intrusion into fresh groundwater aquifers and causes difficult-to-remove contamination of underground water sources. Thus, for example, in Southern California since 1940, intensive exploitation of groundwater during drought periods caused many coastal aquifer's groundwater level to decline below sea level. Penetration of seawater into fresh groundwater aquifers alters their water quality, making it unusable for drinking water. Moreover, water-bearing rocks are saturated with salt and thus the deteriorated water quality in aquifers remains for many years, even after removal (stoppage) of intrusion and recovery of natural, freshwater discharge. Along with increased water extraction, other engineering actions on the coast, such as drainage of adjacent areas, creation of quarries, and so on, favor invasion of sea (salt) waters into fresh groundwater aquifers. This is seen, for example, in saltwater intrusion into a freshwater layer in the Silver Bluff, Miami, Florida drainage systems. This intrusion led to a groundwater level decline of a few feet, brought on by saltwater intrusion into the coast of 2–3 km (Brown and Parker 1945).

On the Baltic Sea coast, about 600 ths m³/day of fresh groundwater is extracted from artesian aquifers. This led to the formation of deep (40–60 m) cones of depression with a radius of up to 100 km, which stretched under the Baltic Sea area and created hydraulic conditions favorable for intensification of seawater intrusion (Gregorauskas et al. 1986, 1987; Frolov and Yushmanov 1998). In natural conditions, not disturbed by human activity, intrusion of saltwater is intensified not only by a difference in densities of salt and freshwaters, but also by tidal and piling-up phenomena.

The greatest experience in studying seawater intrusion into coastal water-bearing layers has been gained in the United States, Japan, Israel, and the Netherlands. In the countries of the former Soviet Union, the problem of seawater intrusion is important for some areas of the Baltic region, Kamchatka, the Black Sea, and Caspian Sea coasts, where phenomena of seawater leakage into coastal water-extracting facilities are observed (Goldberg 1982; Zektser 1983).

During the study of seawater penetration into fresh groundwater aquifers, attention is paid to the prediction of dynamics of water quality in well fields on sea coasts. Intrusion of seawater into underground aquifers is a complicated hydrodynamic process of joint movement of fresh and saltwaters differing by densities and other physical properties. In the zone of transition from fresh to saltwater, processes of dispersion and diffusive and convective mixing of liquids take place.

Two principally different approaches can be distinguished in the modeling of intrusion processes. In the first approach, freshwater and saltwater are accepted to be nonmixing liquids with a sharp dividing boundary between them; in the second, a smooth transition from fresh to saltwater is assumed. Modeling of intrusion in a regime of two nonmixing liquids is the subject of numerous investigations by both international and Russian scientists (Podsechin and Frolov 1989). The most general mathematical model of seawater intrusion into coastal nonhomogeneous freshwater layers under conditions of mixing liquids is presented by a system of equations for filtration and convective diffusion of dissolved salts (Khublaryan and Frolov 1988, 1994; Khublaryan et al. 1984, 1996, 2002; Reilly 1990; Senger and Fogg 1990).

The model given below of intrusion is based on the generalized filtration law for a liquid with density depending on the concentration of dissolved substances. It can describe confined and unconfined groundwater flows in nonhomogeneous and anisotropic water-bearing layers.

Accept the pressure head of freshwater by Equation 1.3.1

$$\psi(X,Y,T) = \frac{P(X,Y,T)}{\rho_0 g} + Y \tag{1.3.1}$$

where ρ_0 is the density of freshwater
g is the acceleration of gravity

X and Y are horizontal and vertical coordinates
$P(X,Y,T)$ is pressure in a point of layer with coordinates (X,Y)
T is time

The generalized filtration law of Darcy for an nonhomogeneous anisotropic layer and liquid with a variable density is expressed as

$$Q = -\frac{A}{\mu(\rho)}(\nabla P + \rho g) \qquad (1.3.2)$$

where Q is the vector of filtration velocity with components Q_X, Q_Y in directions X, Y (axis Y is directed vertically upward), $\mu(\rho)$ is liquid viscosity, ρ is the variable density of liquid

$$\nabla = \left(\frac{\partial}{\partial X}, \frac{\partial}{\partial Y}\right) \quad A = \begin{pmatrix} K_{xx} & K_{xy} \\ K_{yx} & K_{yy} \end{pmatrix} \qquad (1.3.3)$$

A is the tensor of permeability of water-bearing rocks, which in cases when the main axes of anisotropy coincide with the axes of coordinate system X, Y can be written as

$$A = \begin{pmatrix} K_x & 0 \\ 0 & K_y \end{pmatrix} \qquad (1.3.4)$$

and Equation (1.3.2) as

$$Q = -\frac{A\rho_0 g}{\mu(\rho)}(\nabla \psi + \rho_r j) \qquad (1.3.5)$$

where j is unit vector by axis Y, $\rho_r = (\rho - \rho_0)/\rho_0$.
An equation of continuity for an uncompressible liquid with a variable density in a nondeformable porous medium has the following form

$$\nabla(\rho Q) = 0 \qquad (1.3.6)$$

For relative concentration $C = (S-S_0)/(S_s-S_0)$, Equation 1.3.7 for keeping of a dissolved contaminant (sea salt) can be written as

$$\nabla(D\nabla C) - \nabla(VC) = \partial C / \partial T$$

$$D = \begin{pmatrix} D_{xx} & D_{xy} \\ D_{yx} & D_{yy} \end{pmatrix} \qquad (1.3.7)$$

where S is the salt content in a mixture of fresh and seawaters, S_s and S_0 are the salt contents in sea and freshwaters, respectively, $V = Q/n$, n is the effective porosity of water-bearing rocks, and D is the tensor of hydrodynamic dispersion.

Laboratory results reported in Reilly and Goodman (1985) make it possible to approximate a dependence of density and viscosity of mixed fresh and seawaters on the concentration of dissolved salts using the following linear relationships

$$\rho = \rho_0 (1 + \varepsilon C), \mu = \mu_0 (1 + 2,8 C), \varepsilon = (\rho_s - \rho_0)/\rho_0 \qquad (1.3.8)$$

where μ_0 is the viscosity of freshwater, and ρ_s is the density of seawater.

With the appearance of intrusion at the boundary of coastal groundwater flow with a saltwater body $(0 < y < 1, x = 0)$, there is a return point $(y = R)$, below which $(y < R)$ the flow is directed from sea to land $(Q_x > 0)$, and at $(y > R)$ coastal groundwater is discharged into the sea $(Q_x < 0)$.

Boundary conditions for a system of interrelated equations relative to pressure head and concentration are determined by concrete definition of a task.

On a free surface of unconfined flow with availability of infiltration recharge with a concentration C_{inf}, the boundary condition (for isotropic media) has the following form

$$D\frac{\partial C}{\partial N} = E_N (C - C_{\inf}), E_N = w/(1 + (\partial\phi/\partial x)^2)^{1/2} \qquad (1.3.9)$$

where D is the coefficient of diffusion, d/dN is a derivative along inside normal toward the boundary, E_N is the infiltration recharge per a unit of free surface length, $w(x,t)$ is the infiltration recharge per horizontal rock surface.

The initial condition of the task is initial distribution of concentration $C(x,y,0)=C_0(x,y)$, and in the case of unconfined water-bearing layers is the position of free surface $\phi(x,0) = \phi_0 (x)$.

Movement of the free surface is described by Equation 1.3.10

$$n\frac{\partial \Phi}{\partial t} = -k_y \left(\frac{\partial \psi}{\partial y} + \varepsilon C \right) + k_x \frac{\partial \psi}{\partial x} \frac{\partial \Phi}{\partial x} + w \qquad (1.3.10)$$

The defined marginal task can be realized using numerical methods.

The model was tested through comparing the modeling results (Frolov 1991) with the data (Kohout 1960b) from studying seawater intrusion in a Biscayne Bay coastal aquifer in the area of Cutler, Florida. This aquifer is composed of soluble limestone and loamy sandstones and has a thickness of up to 30 m. A sufficiently extended diffusion zone is observed where salinity changes from mineralization of freshwater to salinity of Atlantic saltwater. Here, an extensive recirculation zone of saltwater appears. Figure 1.3.2 demonstrates the results of finite-difference modeling of distribution of salt concentration in the circulation zone by the initial data of Kohout (1960b), as well as isohaline C = 0.5 obtained by the method of finite

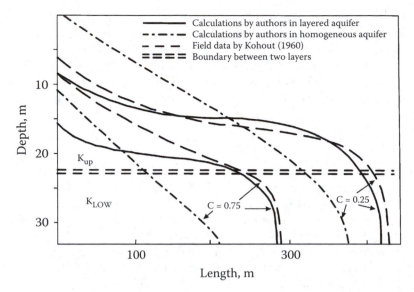

FIGURE 1.3.2 The results of finite-difference modeling of the distribution of salt concentration in the circulation zone by the initial data of Kohout (1960b) and obtained by the method of finite elements (Lee and Cheng 1974).

elements (Lee and Cheng 1974). The results of calculations carried out by different methods agreed well, especially at the layer base. The developed technique was used to calculate seawater intrusion into nonhomogeneous confined aquifers and to assess an influence of coefficients of filtration and dispersion on saltwater distribution zones.

Figure 1.3.3 (Khublaryan et al. 1996) shows a structure of sea salt concentration field for the case of filtration into a three-layered aquifer. One can see (Figure 1.3.3b) that when the difference in permeability between basic layer K_0 and intermediate layer K_n is great ($K_0/K_n > 1000$), the intrusion zone is divided into two parts. Penetration of seawater into the intermediate layer is actually not observed. At a lower difference ($K_0/K_n < 100$), one observes seepage of freshwater from the lower layer through a low-permeable layer into the upper part of an aquifer (Figure 1.3.3A).

It is convenient to model layers of variable thickness by means of specifying nonhorizontal dividing boundaries between zones with very different permeability (actually excluding mass transfer in low-permeable zones). Figure 1.3.4 (Khublaryan et al., 1996) demonstrates the calculated results of intrusion into a system of sloped layers in a confined/unconfined aquifer.

The work of Lal (1990) analyzes the proposal on how to prevent saltwater intrusion through creation of "underground dams" blocking seawater penetration to freshwater aquifers in coastal areas. Figure 1.3.5 (Khublaryan et al. 1996) shows the results of modeling a flow around the screen carried out on a hydrodynamic model.

Work of an injection (pumping) well can be modeled either by means of specifying a source (runoff) in the right-hand part of Equation (1.3.6) or using a boundary condition at the lower boundary of a water-bearing layer.

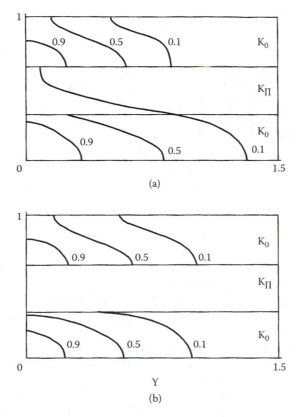

FIGURE 1.3.3 The structure of a sea salt concentration field for the case of filtration into a three-layered aquifer; (a) $K_0/K_n > 1000$; (b) $K_0/K_n < 100$.

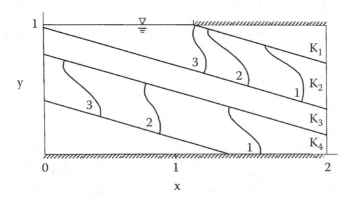

FIGURE 1.3.4 The calculated results of intrusion into a system of sloped layers in a confined/unconfined aquifer.

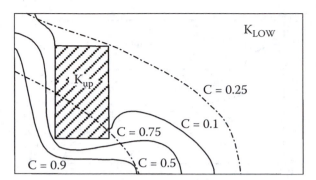

FIGURE 1.3.5 The results of modeling a flow around the screen carried out on a hydrodynamic model.

1.4   SUBMARINE SPRINGS

Submarine springs have attracted people's attention throughout the history of mankind. In ancient times, the interest was in groundwater springs on the sea bottom. In the first century BC, the Roman philosopher and poet Lucretius in his long poem, *On the Nature of Things*, described a submarine spring that manifested itself on the sea surface as "boiling" water.

It is not the objective of the authors here to represent as many submarine springs as possible. Only the best known and, hence, most studied examples of discharging submarine waters on the bottom of seas and oceans are given to provide the reader with a picture of the scale of this natural phenomenon.

One can find in special literature numerous cases on the practical usage of water from submarine springs, and from individual wells drilled in the sea bottom for water supply. Below are some particular examples.

Submarine springs are widespread on the submarine slopes of island systems with a steep mountainous relief (e.g., Hawaiian Islands, Philippine Islands, Greater Antilles, and Greater and Lesser Sunda Isles).

Korotkov et al. (1980) reported that there are zones of recharge areas at relatively low elevation, the submarine slopes of which have a great number of active submarine springs with some having high yields. Such areas include the Florida Peninsula (where hypsometric elevations are not higher than 100 m). Active submarine water discharge is provided there by strongly developed karstic Paleogene–Neogene limestone, which composes a major part of the peninsula, plenty of precipitation (1200–1400 mm/yr), and by the plain-type relief with extensive swamped territories that actually exclude runoff from the surface.

Similar conditions of water discharge are observed on the Yucatan Peninsula, the surface of which represents a lowland plain, only with a small area in the southeast occupied by the Maya Mountains. Limestone covers 100,000 km^2 (of total peninsula area ~180,000 km^2), over a shoreline of 1000 km. On sandy coastal islands, freshwater occurs in the form of lenses on the salt seawater. Hydrostatic levels of the freshwater are 1 m higher than sea level. One of the largest submarine springs in the world is located near Jamaica where a real freshwater "river" has been

FIGURE 1.4.1 Space image of the southern coast of Jamaica. Oval light spots indicate submarine springs (1974).

discovered with a flow rate of 43 m^3/sec. This spring was discovered 1600 m from the coast with water "breaking forth" to the sea surface from a depth of 256 m (Figure 1.4.1).

The Mediterranean Sea is especially rich with groundwater submarine springs. These springs are associated with fissures and karst canals in rocks. A high-yield submarine spring was discovered in the Aegean Sea near the southeastern coast of Greece. There are about 700 submarine springs on the coasts of the Adriatic Sea. In the Mediterranean, springs are located at deep depths, associated chiefly with zones of local tectonic disturbances in karst sediments (Cannes at a depth of 165 m; San Remo, 190 m; St. Martin Gulf, 700 m). In the Dinara coastal karst province, 32 individual and group submarine springs were discovered along 420 km of shoreline. Discharges of submarine springs in the Mediterranean Sea are often so large that they even form freshwater flows in the sea, e.g., at the mouth of the Rona River submarine springs discovered at the sea bottom form a freshwater stream among saltwaters. A similar freshwater "river" also flows in the Genovese Gulf.

The most detailed investigations of submarine springs using different methods were carried out at the southern coast of France between the cities of Marseilles and Cassis. The largest springs there are Port Miou and Bestuan. In 1964 the Geological and Mining Research Bureau and Water Society of Marseille established a specialized research organization to study these submarine springs and investigate the possibilities of exploiting them for water supply, as well as for the development of methodical techniques in the study of such types of springs. It was determined that the submarine springs are associated with karstic limestone of the Cretaceous age that forms a monocline dipping toward the sea. Karst hollows in this limestone are manifested at sea depths of 100 m. The divers who investigated the karst galleries of the Port Miou springs reached a depth of 45 m below sea level, passing through karst canals to a distance of over 1 km inside the continental landmass. The following investigations were done:

- Measuring points were equipped with flow meters, pressure gauges, and resistivity meters.
- Samples of water and soil were taken.
- Experiments with a fluorescent substance were carried out. This allowed determination of direction and velocity of a filtration flow.
- Geophysical experiments were fulfilled.

As a result of these investigations, it has been established that a water mass in the karst hollows is separated into two layers: a lower layer with seawater and an upper one with fresh groundwater of a lesser specific weight. It has been shown that with the velocity of salt seawater movement depthward, the karst gallery is inversely proportional to the pressure heads of discharging fresh groundwater. The pressure heads determine the discharge of fresh submarine water flow that moves toward the sea along the surface of much denser seawaters. The equilibrium of freshwater and saltwater depends on: (a) the pressure head gradient of freshwater flow and variations in the sea level, (b) the ratio of densities of freshwater and saltwater, and (c) the difference in their temperatures favoring the processes of diffusion.

The results obtained were used in the design and construction of a concrete dam on the path of setter intrusion through the main karst gallery. The dam was built in the deepest part of the karst gallery at approximately 500 m from its exit to the sea (Figure 1.4.2). The dam made it possible to constantly measure the discharging velocity of groundwater, control pressure losses along the karst gallery, prevent seawater invasion, and find optimal conditions for groundwater exploitation with the aim of water supply.

In the former Soviet Union, submarine springs were widely investigated on the Caucasian shelf of the Black Sea. There, near the Gantiadi Settlement, in the background of the blue seawater, a contrasting light spot with numerous fast-bursting bubbles create an illusionary effect of boiling water. This submarine spring with this effect has a yield of 0.3 m^3/sec. A larger spring, manifesting itself as "boiling" seawater near Gagry City, is clearly seen from the shore when the sea is calm and has a yield of 8 m^3/sec. "Hummocks of boiling" are located 100–150 m from the shore; these are manifestations on the sea surface of submarine karst springs located at depths of 5–10 m. Such springs are locally called "reprua."

Springs, caused by discharging fissured-veined waters, are usually associated with systems of large tectonic disturbances in intrusive and metamorphic rocks. They are manifested both onshore and offshore. In the Crimea on the northwestern slope of the Ayudag Mountains, three springs were detected at depths of 6–8 m. Submarine springs associated with dolomites and dolomitized limestone were revealed near the northwestern coast of Sicily, Italy; all these springs are at a depth of 10 m.

On the coast of Dalmatia near Shibenika, Croatia, the submarine springs are located at depths of 3 to 10 m. Their total yield equals 300 l/sec, with the largest spring among them having a yield of 90 l/sec.

Korotkov et al. (1980) described the hydrogeological conditions of the ground-water discharge on the Bahrein Islands in the Persian Gulf. In the coastal zone of these islands plenty of submarine springs discharge brackish water (with total mineralization ~4 g/l). The recharge area of this water is on the land mass of Saudi Arabia, and according to the opinion of the authors, the groundwater moves over 100 km under the sea bottom with a high pressure head for eventual discharge in the form of springs in the coastal zone of these islands.

A few years ago a Russian newspaper published a series of interesting articles about the Bahrein Islands, and in particular about submarine springs:

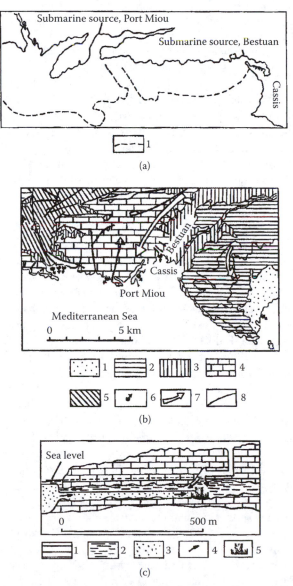

FIGURE 1.4.2 Groundwater discharge into the Mediterranean Sea near Cassis, France. A. Infrared thermal photograph of Port Miou and Bestuan submarine sources area: (1) boundary of fresh groundwater distribution in area. B. Schematic hydrogeological map of area: (1) alluvial sediments; (2) Cenomanian-Turonian limestone; (3) Upper Cretaceous rock of marine genesis; (4) limestone of Armenian facies; (5) carbonate rocks of Lower Cretaceous and Jurassic periods; (6) submarine sources; (7) groundwater flow direction; (8) fault. C. Scheme of karst gallery of Port Miou submarine spring: (1) limestone; (2) groundwater; (3) marine water; (4) water flow direction; (5) dam (Potie 1973).

There is always abundant fresh water on the Bahrein elder fields. Its reserves are replenished at any time at the expense of the submarine springs spouting in the mining areas. In some places, even "developed" springs can be found, that are equipped with hosepipes with float-corks on the sea surface: everyone may swim up and fill empty tanks with water chargeless! The name of Bahrein was not given to the archipelago accidentally: it is translated as "two seas" — a salt sea surrounding a sea of fresh water discharging as springs from the bottom of the ocean.

Numerous submarine springs have been found near the Atlantic coasts. Close to Florida, freshwater was detected 120 km from the coast. At the western coast of the Atlantic, submarine springs spout to the sea surface from depths of about 400 m. According to the data of American scientists, the entire Atlantic eastern coast has a subsurface runoff to the Atlantic Ocean and the Gulf of Mexico. The groundwater discharge into the ocean from only one area of Long Island, New York is estimated at 25 mln m^3/yr (slightly lower than 1 m^3/sec). Groundwater with a high pressure head was exposed by a well drilled into the ocean floor at 37 km from the coast in this part of the shelf (opposite the Delaware River mouth). In 1966 on the Blake Plateau southward of Savannah, Georgia, a 50-m deep depression filled with water having a temperature 2.5°C lower than the temperature of the surrounding seawater was detected on the sea bottom. The depression is located at 200 km from the coast at a depth of 510 m. This anomaly is connected with groundwater discharge.

Numerous submarine springs are often associated with subsurface canyons that usually represent submarine extensions of river mouths. For example, there is a submarine canyon over 1600 km long, about 700 km wide, and over 70 m deep which stretches from the mouth of the Ganges River to the Bay of Bengal. This canyon is associated with groundwater discharge.

Investigations of deep submarine springs started in 1964 when the depression "discovery" in the Red Sea was detected with a salt content of over 300°/$_{00}$. Later, similar brine depressions were found by the *Atlantis II*. Investigations by the bathyscaphe-submarine boat *Aluminaut* along the coast of Florida detected discharges of fresh and brackish groundwater 120 km from the coast at a depth of 510 m.

The use of groundwater of submarine springs for water supply has been widespread since ancient times. Numerous cases are described in the literature, when many centuries ago, people with the aid of different means (e.g., long bamboo tubes) used large submarine springs to obtain fresh drinking water, as well as for supplying ships with fresh water. Quantitative assessment of submarine discharge enables the discovery of additional water resources for water supply. The most representative example of the practical use of submarine springs is the construction of a special dam in the sea near the southeastern coast of Greece. The dam "blocked" discharge of submarine springs and, thus, created a freshwater lake inside the sea. The total yield of this submarine spring is over 1 mln m^3/day. Water from this "lake" is used for irrigation on the coastal territory.

Presently, new techniques and means for water extraction "underwater" are being adopted in a number of countries. In Japan a patent was obtained for a method of freshwater extraction from a submarine spring on the sea bottom. The authors of the patent have suggested separating freshwater of the spring from the seawater directly on the sea bottom. For this purpose, special equipment with transducers is installed above the spring. The transducers constantly measure the salt composition of the water. The operation of the equipment is completely automated. When the water salinity exceeds the permissible level, the water delivery to the user is automatically stopped, and the water is discharged to the sea until its salt content and composition reach the levels specified by the user. Italian specialists have suggested extracting waters of submarine springs using a special bell installed on the sea bottom, capping the spring. The ball is equipped with safety valves controlling water discharge and, if necessary, its composition.

With the increased development of technical means of drilling and testing of water wells on shelf, continent slopes, and on bottoms of seas and oceans, great prospects in the field of using submarine groundwater by offshore water wells are opened. Wells drilled on the shelf of Australia, off the Atlantic coast of the United States, on the land slope of the Gulf of Mexico, and in other places have exposed fresh submarine waters with low mineralization and a considerable pressure head. Drilling works in the Atlantic Ocean near the Florida coast have revealed freshwater 43 km away from the coast eastward of Jacksonville. A well drilled from shipboard at a depth of 250 m below sea level exposed waters with a mineralization of 0.7 g/l, and with a pressure head reaching 9 m above the sea level.

However, it should be kept in mind that the problem of practical usage of submarine waters directly in the sea is not simple. It is connected, first of all, with complicated water extraction (complicated equipment of water catchment) from discharging submarine springs on the sea bottom, the necessity and economic efficiency of such extraction, and the technical difficulties of well drilling in the sea.

In current scientific literature, there are still strong opinions on the possibility of using submarine groundwater as a source of inexhaustible resource. Therefore, it should be emphasized that conclusions on the possibilities of the practical use of submarine waters can be made only after carrying out special works on the assessment of exploitable reserves of such waters, including studies on the feasibility of their usage.

1.5 A REVIEW OF SUBMARINE GROUNDWATER DISCHARGE: BIOGEOCHEMISTRY OF LEAKY COASTLINES*

1.5.1 INTRODUCTION AND STATEMENT OF THE PROBLEM

Groundwater describes a large reservoir of poorly estimated and largely unavailable freshwater which resides in soil, bedrock fractures, and solution cavities deep within the Earth (Fetter 1994). Water supplies are generally obtained from surfacewaters and/or drawn from aquifers less than 500 m from the Earth's surface. Recharge to the water table and confined aquifers occurs by infiltration of precipitation, runoff, and stream waters (Figure 1.5.1.1). Less than 1% of the Earth's water is present as

* Section 1.5 of Chapter 1 was written Jaye E. Cable, Ph.D. (Louisiana State University, Baton Rouge, jcable@lsu.edu), Jonathan B. Martin, Ph.D. (University of Florida, Gainesville), and Makoto Taniguchi, Ph.D. (Research Institute for Humanity and Nature, Japan).

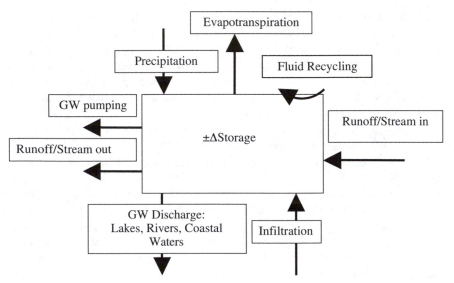

FIGURE 1.5.1.1 Water budget schematic for an aquifer is shown with major sources and sinks outlined. Not all inputs and outputs may occur for every aquifer. For example, fluid recycling is likely to be the most important in coastal aquifers where tides may temporarily propagate mixing into the aquifer.

groundwater at any one time, yet it plays a critical role in global freshwater storage (~98% of all freshwater is groundwater). Thus, understanding the movement and biogeochemistry of groundwater is important for managing resources. As precipitation and runoff (meteoric waters) infiltrate the land, dissolved elemental constituents, heavy metals, or sewage present in the water, as well as constituents leached from the surrounding substrate, may percolate into an aquifer. These materials (dissolved and particulate) will be transported with groundwater along topographic and potentiometric gradients. Eventually, groundwater may discharge into lakes, streams, or coastal water bodies.

The flow of water between continents and oceans is critical for global chemical fluxes. One of many important flow paths to the oceans is discharge of water from continents through coastal and continental shelf aquifers, commonly referred to as submarine groundwater discharge, or SGD (Bokuniewicz 1980). Evidence of groundwater discharge to the ocean and its significance to hydrologic and biogeochemical budgets have been well documented in recent decades (e.g., Valiela et al. 1978; Giblin and Gaines 1990; Valiela et al. 1990; Rutkowski et al. 1999; Taniguchi and Tase 1999; Krest et al. 2000). This input has been implicated in coastal eutrophication through the delivery of nutrient-laden groundwaters to coastal water bodies (e.g., Johannes 1980; Sewell 1982; Johannes and Hearn 1985; Lee and Olsen 1985; Valiela et al. 1990; Lapointe and Matzie 1996; Charette and Buesseler 2004). Groundwater inputs and their dissolved constituents are also considered responsible for local biological zonation in coastal water bodies by either preferentially excluding or encouraging flora and fauna growth (e.g., Kohout and Kolipinski 1967; Miller and Ullman 2004).

The flux of water from coastal nearshore and estuarine sediments has the potential to carry dissolved components to the coastal water column, thereby influencing its chemistry and nearshore ecosystems. Dissolved components could include contaminants derived from continental aquifer systems if much water flows from terrestrial aquifers, but the fluxes of continental contaminants will depend on the concentrations of solutes in the water as well as the volumes of terrestrial water discharging to the coastal zones. Attempts to quantify the water flux date back more than 100 years (Baydon-Ghyben 1888–1889; Herzberg 1901). Over the last decade, recent developments of a variety of techniques to measure SGD has revitalized the interest in the volume of terrestrial-aquifer water discharge and its potential source of dissolved components to the nearshore environment and influence on ecosystems (e.g., Taniguchi et al. 2003; Burnett et al. 2003). Because of the potential contaminant load that could arise from terrestrial sources (e.g., natural, agricultural, urban), studies that attempt to characterize groundwater-derived nutrient loads are of particular interest to coastal resource managers and communities wishing to control coastal eutrophication or preserve potable water supplies. Additionally, recirculating marine fluids that flush sediments due to physical and/or biological pumping in lakes and coastal water bodies provide no "new" water inputs, but they may provide an important biogeochemical transformation and transport mechanism. These meteoric and recirculating fluid inputs represent the total benthic water flux and both are important to benthic biogeochemical inputs.

Comprehensive reviews have been presented recently in the peer-reviewed literature discussing the volumetric importance of submarine groundwater discharge (Taniguchi et al. 2002; Burnett et al. 2003) and the ecological and biogeochemical role of groundwater discharge to coastal waters (Slomp and Van Cappellen 2004; Hancock et al. 2005). We will not attempt to duplicate these efforts in this chapter. Our goals here are to discuss the role of different sources of water as it relates to biogeochemical fate and transport into coastal waters and to evaluate the sources of groundwater nitrogen to the Gulf of Mexico and eastern coastlines of the United States.

1.5.2 ROLE OF FLUID COMPOSITION IN BIOGEOCHEMICAL ASSESSMENTS

One issue to address in evaluating SGD-derived biogeochemical inputs to coastal waters is how we define submarine groundwater discharge. Based on a review of recent literature that pertains to submarine groundwater discharge, described below in detail, differences among various techniques and studies suggest that meteoric water flowing to coastal zones through coastal and continental aquifers represent only a fraction of total discharge across the sediment–water interface. The remineralization rate of organic carbon in the sediment coupled with the total discharge of water across the sediment–water interface controls the fluxes of solutes. Submarine groundwater discharge is being used in the peer-reviewed literature today to refer to the total flux of water across the sediment–water interface — often without distinguishing sources of the fluids. For this reason, in reviewing previous studies, it is important to evaluate critically not only the magnitude of the fluxes but also the approach used to obtain these fluxes. The methods used may help identify the approximate fluid composition of SGD if one is not provided.

Groundwater discharge to coastal waters has been studied from many different perspectives, including those of hydrogeologists and oceanographers (e.g., Johannes 1980; Harvey et al. 1987; Pandit and El-Khazen 1990; Li et al. 1999; Moore 1999; Taniguchi et al. 2003). Consequently, the range in disciplinary expertise and techniques applied have lead to a wide range in estimates of groundwater discharge for some study sites. Within a single environment groundwater discharge estimates may vary seasonally, but otherwise should be similar regardless of the technique. However, between different environments, this range in estimates is expected due to differences in geology and climate. Geochemical methods for estimating groundwater discharge have utilized a variety of natural and artificial tracers, including but not limited to ^{222}Rn, Ra isotopes, ^{4}He, ^{131}I, SF_6, and Cl^- (Cable et al. 1996a,b; Rama and Moore 1996; Corbett et al. 1999, 2000a; Dillon et al. 1999; Top et al. 2001; Martin et al. 2002, 2004). Physical field measurements may employ seepage meters (Israelson and Reeve 1944; Lee 1977; Corbett et al., Section 2.2.2) or head measurements (Freeze and Cherry 1979; Harvey and Odum 1990). In addition, numerical modeling and water budget methods have been used to estimate groundwater inputs to coastal water bodies (Giblin and Gaines 1990; Pandit and El-Khazen 1990; Li et al. 1999).

Most studies today acknowledge that SGD is composed of multiple sources, but it has only been recent studies that have attempted to resolve the now obvious issue of how to quantify the relative magnitude of new inputs from continental aquifers and recirculating fluids as physical and biological mechanisms flush the upper 10–100 cm of sediments (e.g. Martin et al. 2004). Future studies of groundwater inputs and benthic nutrient loading clearly must be directed at resolving this critical question if we are to accurately assess and manage coastal groundwater resources. In this review we define two major components of SGD, one of which is derived from terrestrial aquifer flow and the other from recirculated seawater. To the extent possible based on reported data, water derived from terrestrial aquifers will be termed terrestrial SGD, while water recirculated through shallow sediments will be termed seawater recirculation or referred to as a mixing process within sediment pore waters (e.g., ventilation of bottom sediments). The total flux of water from the sediment to the water column will be more generally stated as benthic exchange or total benthic fluxes. These terms are used recognizing the fact that there will be no net flux of water if it circulates from the water column to the pore water and back to the water column. If the chemical composition of this water is altered by diagenetic reactions during recirculation, however, there will be a net flux of dissolved constituents to or from the overlying water column.

The flow path between the coastal aquifer and the coastal ocean has been described as a subterranean estuary because of similarities to physicochemical processes in surfacewater estuaries (Moore 1999), although significant differences are present because subterranean estuaries occur in a porous and chemically reactive medium. The flux of groundwater to the oceans may be large, perhaps up to 40% of local river fluxes (Moore 1996). Thus, submarine groundwater discharge is the focus of a broad range of research aimed at quantifying magnitudes of the water and chemical fluxes (Burnett et al. 2002), because of its potential importance to flow of nutrients and contaminants from continents to the oceans (Johannes 1980; Capone and Bautista 1985; Zimmermann et al. 1985; Simmons 1992; Millham and Howes 1994; Gallagher et al. 1996; Corbett et al. 2000b; Warnken et al. 2000; Tobias et al. 2001). Unlike other sites of water discharge to the oceans, such as at hydrothermal systems or cold seeps, submarine groundwater discharge provides only subtle thermal and chemical signals, thus complicating measurements of flow. Consequently, many techniques have been used to measure the flow, including seepage meters, tracers, mass balance calculations, and numerical flow models. Multiple techniques are often used to elucidate the most reproducible estimates of groundwater inputs (Millham and Howes 1994; Tobias et al. 2001) or to evaluate technique effectiveness (Giblin and Gaines 1990; Burnett et al. 2002). Discrepancies among estimates at a single location are often attributed to inaccuracies in one of the techniques. Another possible explanation for discrepancies may be the source of the water included as groundwater in various estimates (Cable et al. 2004; Martin et al. 2004).

Groundwater discharge often occurs as disperse seepage across the sediment–water interface in coastal water bodies, but it may also be delivered through less common features such as submarine springs (Kohout 1966). In areas where seepage is the primary delivery mechanism, recirculated seawater complicates quantification of terrestrial SGD (Figure 1.5.2.1). Recirculated seawater is driven by a complex set of environmental conditions that will vary depending on location,

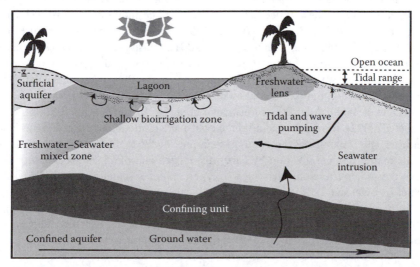

FIGURE 1.5.2.1 Cross-section schematic shows an open ocean, back-barrier lagoon, and coastal aquifer system. SGD is often quantified as the total flux from the sediments, but this water flux will be a sum of processes, such as bioirrigation, waves and tides pumping, convective mixing, and groundwater. Leaks from confined aquifers may occur if fractures, thinning, or discontinuity occurs in a confining unit. A subsurface freshwater–seawater mixing zone will appear near the shore where meteoric water discharges into surfacewaters. No scale is implied here; scales will vary greatly among coastal geologies. (Modified from Martin et al. 2005).

season, and local biology. Thus, no uniform fraction of recirculated seawater can be assigned to benthic water fluxes; each study site must be evaluated independently. As a consequence, in many submarine groundwater discharge studies worldwide, the source of the discharging fluids is not defined. Herein lies the most important problem. What is the biogeochemical significance of terrestrial SGD? Some studies consider all discharging fluids at the sediment–water interface to be groundwater, while others have included only that water originating as meteoric recharge to continental aquifers. This lack of defined or measured sources in some studies creates confusion about the biogeochemical significance of terrestrial groundwater discharge to coastal waters.

1.5.3 IMPORTANCE OF RECIRCULATING SEAWATER TO SUBMARINE GROUNDWATER

Mixing of marine water into the sediment pore spaces creates a shallow zone of recirculating pore water (<1 m) from the overlying water column, which may represent a significant fraction of the total groundwater input reported in the literature in some cases. Shallow advective mixing, or recirculated seawater, through permeable coastal sediments has been suggested to originate from at least three major categories of physical and biological mixing processes, but their controls, extent, and timing are unknown (Bokuniewicz 1992; Boudreau 1997; Burnett et al. 2002; Martin et al. 2004). These processes include wave and tidal pumping of water into the sediments (Riedl et al. 1972; Nielsen 1990; Shum 1992; Shum 1993; Li et al. 1999; Mu et al. 1999; Huettel and Webster 2001), density-driven flow (Rasmussen 1998), and passive or active flow through structures produced by burrowing organisms (Aller 1980; Smethie et al. 1981; Emerson et al. 1984; Aller and Aller 1992; Marinelli et al. 1994; Furukawa et al. 2000; Sandnes et al. 2000; Schluter et al. 2000; Warnken et al. 2000; Aller 2001; Furukawa et al. 2001; Meile et al. 2001; Timmermann et al. 2002). It is likely that all three contribute to mixing at the sediment–water interface. This mixing has been suggested previously to fuel microbial-mediated organic matter remineralization by carrying oxygen and oxidized species into pore waters that otherwise would have low oxygen contents (Huettel et al. 1996, 1998; Huettel and Webster 2001). The depths that water, and associated dissolved oxygen, may be pumped into the sediments depends on the mechanism driving the pumping. For example, the depth of tidal and wave pumping depend on the current speed and size of bedforms, but typically are only a few tens of centimeters (Shum 1992, 1993). The depth of density-driven flow depends on the density contrast between the overlying water and pore water (Rasmussen 1998). Density contrasts must be large, on the order of the difference between seawater and freshwater, to pump water to depths of 10–15 cm. Bioirrigation may drive water to depths greater than wave and tidal pumping or density-driven flow, and changes in concentrations of Cl⁻ and ^{222}Rn have recently been used to suggest that mixing occurs to depths of nearly 1 m over time scales of days to weeks (Cable et al. 2004; Martin et al. 2004, 2005).

Discrepancies among SGD estimates using several techniques in one location may shed some light on the magnitude of aquifer-derived groundwater versus other advective processes (Cable et al. 2004). In one case, seepage meters and salt and water budgets were used to estimate groundwater flow in Town Cove, Massachusetts (Giblin and Gaines 1990). They found the budgets both yielded estimates at least one order of magnitude less than the seepage meter measurements and concluded that seepage meters were overestimating the total input to the estuary. Five field techniques (three seepage meter types and two tracers) were used in a technique intercomparison study aimed at measuring groundwater inputs at the Florida State University Coastal and Marine Laboratory (Burnett et al. 2002). The traditional Lee-type seepage meter, a heat pulse automatic seepage meter, and an acoustic Doppler automatic seepage meter demonstrated very good agreement amongst themselves and with the tracers, radon and radium, but Burnett et al. (2002) also noted that a groundwater flow model for their study indicated groundwater inputs should be 8–10 times less than field measurements. When similar comparisons are made among the techniques used in the Banana River Lagoon, Florida (Cable et al. 2004), discrepancies in measured rates occur that are similar to those observed by these other studies. Seepage meters (3.6–6.9 cm/day) and the ^{222}Rn tracer (3.5 cm/day) were higher than a Cl$^-$ model (Martin et al. 2004) and a finite element model (Pandit and El Khazen 1990). In the Banana River Lagoon, differences between the Cl$^-$ model (0.01–0.02 cm/day) and groundwater model (0.06–0.2 cm/day) results are small, and these estimates more likely approximate groundwater discharge from distant sources (i.e., continental aquifers). Shallow mixing in the upper sediment column, where the seepage meter and Rn tracer would be most responsive, may represent greater than 90% of the total discharge if the Cl$^-$ and flow models are accurate.

Recirculation of seawater through shallow sediments is a well-known benthic boundary layer process (e.g., Boudreau and Jorgensen 2001). Its role in mass transfer from pore waters to coastal waters is studied by biologists, biogeochemists, and physicists, but it may affect estimates of groundwater discharge. Subtidal pumping was suggested as a mechanism for circulating seawater through sediments by Riedl et al. (1972) on the continental shelf of the southeastern United States. Extrapolating this water exchange to a global scale, they predicted ~90,000 km^3 of seawater may pass through porous sediments annually along the world's coastlines.

1.5.4   RELEVANCE OF FLUID COMPOSITION TO BENTHIC NUTRIENT LOADING

The difficulty with accurately measuring terrestrial groundwater discharge has prevented a global estimate of the magnitude of the flux. However, the lack of agreement on the composition or definition of SGD has contributed greatly to our poor understanding of terrestrial SGD fluxes. For example, "hydrologic flow" from passive margins has been crudely estimated on the basis of global hydrologic balances to be on the order of 100 km^3/yr or approximately 0.2% of river flow (COSODII 1987). In contrast, direct measurements made at various coastal settings around the world when extrapolated to the global coastline range from approximately 1% to 10% of river discharge (SCOR/LOICZ Working Group 112 1997; Taniguchi et al. 2002). These results point to two important characteristics of submarine groundwater discharge to the coastal ocean: (1) the magnitude of the flux is poorly constrained, but is potentially large, and (2) depending on the chemical composition of groundwater discharge, chemical fluxes could also be large. The magnitude of chemical fluxes depends on the origins of discharging waters as well as the diagenetic reactions that occur along the flow path, which will control their associated nutrient and contaminant concentrations. For example, if submarine groundwater discharge is composed entirely of meteoric water, the flux will be small compared to other fluxes such as rivers and hydrothermal systems, and except for reactive components, may be more dilute than seawater. On the other hand, if we include advective exchange between the overlying water column and the sediment–water interface in SGD estimates, the flux of water (and dissolved components) could be large relative to other sources.

Because coastal waters are often saturated with respect to oxygen, reactive components in this water could be regenerated organic matter or redox sensitive metals (Bokuniewicz 1992; Sandnes et al. 2000; Burnett et al. 2002). In a study of North Sea sandy sediments, de Beer et al. (2005) showed the magnitude of advective water transport and mineralization rates that may occur as pressure gradients pump overlying oxygenated seawaters into and out of the sediments. Piston flow velocities of about 2–6 μm/sec, or a corresponding pore water exchange rate of 160–500 l/m^2/day, were found in the shallow sediments (less than 15 cm depth). Aerobic degradation was found to be high considering the low organic matter content of these sandy sediments (~140 mmol/m^2/day) (de Beer et al. 2005). Coastal aquifer studies that investigate deeper reactions within the salinity transition (mixing) zone between infiltrated seawater and the seaward migration of fresh groundwater also find a chemically reactive environment. In studies of iron cycling and SGD in coastal salt marshes, iron reduction and aerobic respiration appear to account for most of the natural organic matter degradation (Snyder et al. 2004). This iron oxidation zone at the front between fresh groundwater and intruding saline waters has been referred to as the "iron curtain" (Charette and Sholkovitz 2002; Testa et al. 2002). Subsequent studies by Charette et al. (2005) have shown that permeable sediments associated with this mixing zone may act as a net sink for uranium and other biogeochemically active elements. In fact, these studies demonstrate that internal recycling of nutrients and other biogeochemical constituents within estuarine pore waters by downward-mixing oxygenated seawater is likely an important contributor to total benthic biogeochemical fluxes back to the water column.

In a study of the inland Upper Glacial aquifer of Long Island, New York, Slater and Capone (1987) measured denitrification rates of about 7.5 kg-N/ha/yr. Along the coast in the same study, denitrification rates were measured between 0.61–7.2 kg-N/ha/yr in the water table of Great South Bay, New York, where terrestrial groundwater discharge to the bay is known to exist. Up to 50% of the nitrate in these waters was lost to denitrification in marine sediments. Capone and Slater (1990) demonstrated that a water table's ability to remove or store nitrate was based on the amount of precipitation, and hence aquifer recharge, in previous months, which controlled the extent of terrestrial submarine groundwater discharge (Figure 1.5.4.1). During periods of low precipitation and low terrestrial SGD, denitrification and low nitrate inputs result in low nitrate concentrations in the sediments. During periods of moderate terrestrial SGD flow nitrate concentrations are dominated by denitrification, dilution, and diffusion. High terrestrial SGD flow rates will produce high nitrate concentrations as advection dominates over diffusion and denitrification.

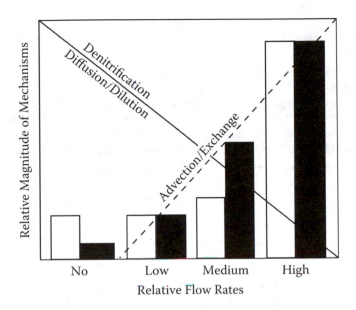

FIGURE 1.5.4.1 A conceptual model of mechanisms controlling pore water-derived nitrate concentrations in coastal waters. (Modified from D. Capone and J. Salter, 1990, *Biogeochemistry* 10: 277–288. With permission of Springer Science and Business Media). Advection is the rate of terrestrial submarine groundwater discharge. Upper (*white bar*) and lower (*black bar*) sediment horizons are separated to show how biogeochemical transformation rates, and thus pore water nitrate concentrations, will be affected by flow rates at different depths in the sediments.

Denitrification rates ranged from 6.1–39 kg-N/ha/yr in the Great South Bay coastal zone water table based on core measurements (Capone and Slater 1990). Similar estimates for volatilization of nitrogen are difficult to make and should be considered negligible in most cases in the groundwater system. Variations in time and space for denitrification or aquifer storage capacity are dependent on a combination of the quantity of precipitation recharging an aquifer and sediment hydraulic conductivity, which determines the ability of a substrate to transmit fluids. Slomp and Van Cappellen (2004) demonstrated in their review of controls on nutrient inputs via groundwater discharge that the residence time of the mixing zone, and ultimately, the available reaction time are critical to evaluating the nitrogen and phosphorus inputs.

In the Banana River Lagoon, Florida, estimated N, P, and Si fluxes to the sediment are greater than the estimated return of these solutes to the water column through lagoon water recirculation (Bhada 2007). They report the N flux to the sediment is only about 30% greater than the return flux to the water column, thus indicating a net burial of 70% N. In contrast, their P flux to the sediment may be twice the flux from the sediment via recirculation processes, while Si fluxes to the sediment are about 2–4 times greater than the flux from the sediment. Bhada et al. (2005) attribute these flux differences to and from the sediment to differences in the rates of remineralization of the elements, where Si would be the most refractory and has the greatest potential for burial. These results indicate that ventilation of the sediments (i.e., recirculation of seawater) is an important process and may limit the burial rate of nutrients into the sediment by enhancing their dissolved benthic flux to the water column.

A benthic nutrient source depends on the advective transport rate and the concentration of nutrients in the discharged water, both of which depend on the origins of the discharged water and water-solid reactions along its flow path (Bhada et al. 2005). In the following sections, nitrogen inputs to and transformations within aquifers are provided to demonstrate that measurements of aquifer nitrogen concentrations which occur in distal areas to the coastal aquifer zones are not necessarily representative of what may be discharged with terrestrial SGD.

1.5.5   SOURCES, FATE, AND TRANSPORT OF GROUNDWATER NITROGEN TO COASTAL WATERS

1.5.5.1   Nitrogen Storage and Accumulation

Groundwater nitrogen concentrations are given for four regions of the eastern United States to provide background for available N as it is transported through aquifers to the coast (Table 1.5.5.1). The four regions are defined as the Northeast (including Maine, New Hampshire, Massachusetts, Rhode Island, and Connecticut), the Mid-Atlantic (New York, New Jersey, Delaware, Maryland, and Virginia), the Southeast (North Carolina, South Carolina, Georgia, and the east coast of Florida), and the Gulf Coast (the gulf coast of Florida, Alabama, Mississippi, Louisiana, and Texas). These regions will be used consistently throughout this section to refer to specific coastal areas. Inorganic nitrogen as nitrate (NO_3^-) is the most common nitrogen species as compared to other species, such as ammonium (NH_4^+), nitrite (NO_2^-), and dissolved organic nitrogen (DON) concentrations. Nitrate concentrations have increased over time by as much as a factor of 3 in several aquifers along the eastern United States. For example, nitrate concentrations increased from 0–450 µM to 0–800 µM over a 12-year time span in a Cape Cod, Massachusetts, aquifer (Frimpter and Gay 1979; Smith et al. 1991). Likewise, in a shallow coastal plain aquifer of Georgia, nitrate concentrations increased from 71 µM in 1984 to 218 µM in 1992 (Lowrance et al. 1984; Lowrance 1992). Over 11 years, nitrate concentrations increased from 100–270 µM in the Upper Glacial aquifer of Long Island, New York (Ragone et al. 1976; Slater and Capone 1987). In most cases, nitrate concentrations in groundwater will be related to the size of the surrounding population and sewage waste percolation into aquifers. Urban and agricultural areas may supply the highest degree of nitrate contamination to aquifers. However, the increase in nitrate in the Georgia aquifer is likely a response to fertilizer applications since this area is not heavily populated.

1.5.5.2   Nitrogen Fixation and Nitrogen Fertilization

Nitrogen present in the atmosphere, sewage and industrial waste, and fertilizers can reach groundwaters and migrate along the aquifer flow path. Sources of nitrogen to groundwaters include nitrogen fixation and fertilization due to both residential and agricultural practices. Nitrogen fixation is negligible in groundwaters since net primary production is generally low in these waters. Substantial fertilization of

TABLE 1.5.5.1
Groundwater Nitrogen Concentrations (μM) for Several Coastal Aquifers of U.S. East and Gulf Coasts

Aquifer Type and Location	NO$_3^-$	NH$_4^+$	NO$_2^-$	DIN	DON	Reference
Northeast Region						
Unconfined; Falmouth, MA	1–693					Meade and Vaccaro (1971)[a]
Seeps/springs; Great Sippiwissett Marsh, Buzzards Bay, MA	10–100	2–12	0–3	12–115		Valiela et al. (1978)
Unconfined; Buttermilk/Buzzards Bay, MA	70	16		86		Valiela and Costa (1988)
Unconfined; Indian Heights Subbasin, Buzzards Bay, MA	89–1060	<0.1–1.5		91–1060		Weiskel and Howes (1991)
Unconfined; Indian Heights Subbasin, Buzzards Bay, MA	1292 ± 250[b]	488 ± 95		1780 ± 313	468 ± 378	Weiskel and Howes (1992)
Unconfined; Cape Cod, MA	0–450	0–65		0–515		Frimpter and Gay (1979)[a]
Unconfined sand and gravel aquifer; Cape Cod, MA	10–800					Smith et al. (1991)
Unconfined; Orleans, MA	0–393					Gaines et al. (1983)[a]
Seeps, interstitial fluids; Town Cove, MA	10–107					Giblin (1983)[a]
Unconfined/seeps; Cape Cod, MA	0–750					Giblin and Gaines (1990)
Seeps/springs (interstitial fluids), undeveloped; Salt Pond Bay, Eastham, MA	6 ± 7	2 ± 3				Portnoy et al. (1998)
Seeps/springs (interstitial fluids), moderately developed; Salt Pond Bay, Eastham, MA	109 ± 75	2 ± 1				Portnoy et al. (1998)
Seeps/springs (interstitial fluids), highly developed; Town Cove, Eastham, MA	203 ± 164	8 ± 14				Portnoy et al. (1998)
Unconfined; Childs River, Waquoit Bay, MA				133		Valiela et al. (1992)
Unconfined; Quashnet River, Waquoit Bay, MA				4		Valiela et al. (1992)
Unconfined; Sage Lot Pond, Waquoit Bay, MA				15		Valiela et al. (1992)

(Continued)

TABLE 1.5.5.1 (Continued)
Groundwater Nitrogen Concentrations (μM) for Several Coastal Aquifers of U.S. East and Gulf Coasts

Aquifer Type and Location	NO_3^-	NH_4^+	NO_2^-	DIN	DON	Reference
Riparian forest, unconfined; SPD, Kingston, RI	93–343					Nelson et al. (1995)
Wetland section of riparian zone, unconfined, PD; Kingston, RI	14–21					Jacinthe et al. (1998)
Wetland/upland transition near riparian forest, unconfined, MWD; Kingston, RI	7–36					Jacinthe et al. (1998)
Arkosic surficial/bedrock; Connecticut River Valley, CT	3–1069	1–136	1–2		0–1	Grady and Mullaney (1998)
Mid-Atlantic Region						
Upper Glacial unconfined (5 km inland); Long Island, NY	100					Ragone et al. (1976)
Long Island, NY	8–610					Bowman (1977)[a]
Upper Glacial unconfined (5 km inland); Long Island, NY	270 ± 10					Slater and Capone (1987)
Upper Glacial unconfined (interstitial fluids, coastal zone); Long Island, NY	21					Slater and Capone (1987)
Delmarva Peninsula, WD uplands; MD-DE-VA	140					Phillips et al. (1993)
Delmarva Peninsula, PD uplands; MD-DE-VA	70					Phillips et al. (1993)
Surficial confined region; Delmarva Peninsula, MD-DE-VA	6					Phillips et al. (1993)
Shallow clay aquifer, deciduous forest; Coastal Plain, MD	110–119					Peterjohn and Correll (1984)
Shallow clay aquifer, deciduous forest; Coastal Plain, MD	129					Jordan et al. (1993)
Columbia Aquifer, fertilized agricultural land; Choptank River, MD	520–2700					Weil et al. (1990)
Columbia Aquifer, manured agricultural land; Choptank River, MD	7–7430					Weil et al. (1990)

Description				Reference
Columbia Aquifer, forest; Choptank River, MD	0–7			Weil et al. (1990)
Coastal plain; Wye Narrows, MD	956–1163			Staver and Brinsfield (1996)
Brackish marsh, interstitial fluids; Rhode River, MD		14–19	43–64	Jordan and Correll (1985)
Southern coastal plain, water table, agricultural land; VA	195–603	1.4–7.4	0.1–72.6	Simmons et al. (1992)
Southern coastal plain, water table, wetlands; VA	0.3–6.8	1.9–13.0	0.1–1.8	Simmons et al. (1992)
Southern coastal plain, water table, urban area, VA	0–4.8	65–1749	0–0.6	Simmons et al. (1992)
Coastal plain, water table, agricultural land near Cherrystone Inlet, VA	492 ± 8	4.8 ± 0.2	0.6 ± 0.2	Reay et al. (1992)
Coastal plain, water table, forested land near Cherrystone Inlet, VA	44 ± 19	9.2 ± 1.4	1.8 ± 0.3	Reay et al. (1992)
Unconfined (interstitial fluids); Phillips Creek, Delmarva Peninsula, VA	1071			Gallagher et al. (1996)
Unconfined (interstitial fluids), Rappahannock River, VA	1028			Gallagher et al. (1996)
Unconfined (interstitial fluids); Cherrystone Inlet, Delmarva Peninsula, VA	1106			Gallagher et al. (1996)
Unconfined (interstitial fluids); Piankatank River, VA	928			Gallagher et al. (1996)
Unconfined, coastal plain; VA	561–1467	1–5	0–10	Gallagher et al. (1996)
Southeast Region				
Coastal plain, water table; NC	1–2247	1–296		Gilliam et al. (1974)
Shallow aquifer, forest, coastal plain; NC	123			Jacobs and Gilliam. (1985)
North Inlet, interstitial fluids, wetlands; SC		2–400		Agosta (1985)
Clambank Creek, North Inlet, interstitial fluids, wetlands; SC	0.40 ± 0.03[b]	40 ± 5		Whiting and Childers (1989)
Bly Creek, North Inlet, interstitial fluids, wetlands; SC	0.29 ± 0.03[b]	30 ± 6		Whiting and Childers (1989)
Shallow aquifer, deciduous forest; coastal plain, GA	71			Lowrance et al. (1984)
Shallow aquifer, pine and deciduous forest, coastal plain; GA	218			Lowrance et al. (1992)

(Continued)

TABLE 1.5.5.1 (Continued)
Groundwater Nitrogen Concentrations (μM) for Several Coastal Aquifers of U.S. East and Gulf Coasts

Aquifer Type and Location	NO_3^-	NH_4^+	NO_2^-	DIN	DON	Reference
Upper Floridan Aquifer; Suwannee River Basin, FL	1280–1430[c]					Katz et al. (1999)
Alligator Reef, interstitial fluids; The Keys, FL	8.49 ± 4.78					Simmons and Netherton (1986)
French Reef, interstitial fluids; The Keys, FL	5.92 ± 2.21					Simmons and Netherton (1986)
Porous limestone; The Keys, FL	0.05–2890[b]	0.77–2750		470–1035		Lapointe et al. (1990)
Gulf Coast Region						
Floridan Aquifer; Franklin and Wakulla Counties, FL	0.25 ± 0.05[b]	9.9 ± 1.3				Bugna et al. (1996)
North Florida Springs; Franklin and Wakulla Counties, FL	26 ± 4[b]	0.34±0.15				Bugna et al. (1996)
Unconfined, interstitial fluids; Turkey Point, Apalachee Bay, FL	43 ± 12	35.8 ± 0.1	0.28 ± 0.14			Rutkowski et al. (1999)
Unconfined, interstitial fluids; St. Joseph's Bay, FL	20 ± 5	6 ± 5	0.17 ± 0.15			Rutkowski et al. (1999)
Water table, moderately developed area (99% NH_4^+); Little St. George Island, FL				24 ± 19		Corbett et al. (2002)

[a]As reported in Valiela et al. (1990).

[b]This nitrogen concentration represents nitrate plus nitrite.

[c]Sources of nitrogen to groundwater are from fertilizer applications (49%) and farm animal wastes (45%).

DIN, dissolved inorganic nitrogen; DON, dissolved organic nitrogen; PD, poorly drained; SPD, somewhat poorly drained; MWD, moderately well drained; WD, well drained.

residential gardens and lawns may occur in more populated areas, but it is difficult at times to quantify this nitrogen flux to aquifers. Flipse et al. (1984) estimated nitrogen inputs to a Long Island, New York, aquifer based on residential irrigation waters to be about 0.83 kg-N/ha/yr (Table 1.5.5.2). In contrast, Flipse et al. (1984) also found that anthropogenic fertilizer application to farms on Long Island contributed about 17 kg-N/ha/yr to groundwater. In this case, 20 times more nitrogen entered the Long Island aquifer due to agricultural practices. In a comprehensive study of Rhode Island salt ponds, Lee and Olsen (1985) found that the combined effect of septic systems, lawn irrigation, and pets contributed as much as 93.4 kg-N/ha/yr to groundwaters flowing toward Green Hill Pond. Agriculture was responsible for a total of 33.8 kg-N/ha/yr to groundwaters entering the salt ponds. In addition, precipitation averaged 12% of the total inorganic nitrogen input to groundwaters feeding these ponds with a flux range of 3.3–12.6 kg-N/ha/yr. Both manure and anthropogenic fertilizers have been applied to farmland in Maryland. Weil et al. (1990) estimated that up to 174 kg-N/ha/yr from anthropogenic and up to 377 kg-N/ha/yr from manure fertilizer may have entered the Columbia aquifer in Maryland as a result of this application. Anthropogenic fertilizers used in the Apalachicola River watershed have contributed as much as 2.3 kg-N/ha/yr NH_4^+ and 13 kg-N/ha/yr total nitrogen to Georgia groundwaters and 3.5 kg-N/ha/yr NH_4^+ and 7.6 kg-N/ha/yr total nitrogen to Alabama groundwaters (Fu and Winchester 1994).

1.5.5.3 Denitrification and Volatilization

As the groundwater moves through an aquifer, nitrogen may be through denitrification or volatilization. Denitrification measurements made in a few aquifers of the northeastern United States demonstrate the variability of this process and its dependence on soil type and flow rate (Table 1.5.5.3). Smith et al. (1991) measured denitrification rates of $2.3 \cdot 10^{-7}$ kg-N/ha/yr in a Cape Cod, Massachusetts, aquifer, while denitrification in a Kingston, Rhode Island, aquifer beneath a university farm was found to be 6 kg-N/ha/yr (Groffman et al. 1996). Total nitrate removal rates in the Kingston, Rhode Island, aquifer were measured as high as 120 kg-N/ha/yr, suggesting that poorly drained aquifer soils must have a huge storage capacity for nitrate (Nelson et al. 1995). In a Rhode Island forest, denitrification rates ranged from 0.092 kg-N/ha/yr in moderately well-drained riparian soils to 13 kg-N/ha/yr in poorly drained riparian soils (Jacinthe et al. 1998). These studies each indicate that nitrate may be removed more efficiently by aquifers with lower transmissivity. Total nitrate removal must include removal by all mechanisms, physical, chemical, and biological, that occur in aquifer soils. Physical loss mechanisms may include groundwater discharge to surfacewaters or apparent loss due to a dilution effect. This dilution can be caused by upwelling of deeper low nitrate groundwater or by infiltration of low nitrate precipitation and runoff (Starr et al. 1996). Biological/chemical loss terms are usually microbially mediated, such as denitrification or immobilization into biomass. This removal may vary based on the hydraulic conductivity of the substrate, concentration gradients in the aquifer, the magnitude of nitrogen inputs, and the volume of recharge to the aquifer. In riparian forests of Virginia, Correll et al. (1992) found that 85% of nitrate was removed from the

TABLE 1.5.5.2
Inputs of Nitrogen (kg- N/ha/yr) to Groundwaters Are Given for Areas along the U.S. East and Gulf Coasts

Description and Location	Natural Other[a]	Fertilizer	Manure	Anthropogenic Other[b]	N Form	Reference
Northeast Region						
Septic systems; Childs River, Waquoit Bay, MA	1.8			18.9	DIN	Valiela et al. (1992)
Septic systems; Quashnet River, Waquoit Bay, MA	1			4.5	DIN	
Septic systems; Sage Lot Pond, Waquoit Bay, MA	0.2			5.4	DIN	
Indian Heights Subbasin; Buzzards Bay, MA	0.004 ± 0.001			58 ± 20[c]	TDN	Weiskel and Howes (1991)
Residential; Ninigret Pond, RI	5.4	2.71		22.6	DIN	Lee and Olsen (1985)
Residential; Green Hill Pond, RI	12.6	17.2		93.4	DIN	
Residential; Potter Pond, RI	8.9	12.4		60.4	DIN	
Residential; Point Judith Pond, RI	3.3	1.5		29.4	DIN	
Mid- Atlantic Region						
Irrigation; Twelve Pines, Suffolk County, NY		17		0.83	TN	Flipse et al. (1984)
Agricultural land; Columbia Aquifer, MD		83–119	71–167		TN	Weil et al. (1990)
Recharge; Gott's Marsh, Upper Patuxent Estuary, MD	0.5–3.1				TDN	Heinle and Flemer (1976)
	0.1–0.7				NO_3^-	
	0–5				NH_4^+	
	2–8				DON	

Description and Location			Species	Reference
Recharge, wetlands; South Chesapeake Bay, VA	1.2		NO_2^-	Simmons et al. (1992)
Recharge, urban area; South Chesapeake Bay, VA	2.2		NO_3^-	
Recharge, urban area; South Chesapeake Bay, VA	0.4		NO_2^-	
Southeast Region				
Farmland (1,530,000 ha); Apalachicola River, GA	13		TN	Fu and Winchester (1994)
	2.3		NH_4^+	
Gulf Coast Region				
Sewage; Apalachicola River Estuary, FL		11	TN	Fu and Winchester (1994)
Farmland (1,340,000 ha); Apalachicola River, AL	7.6		TN	
	3.5		NH_4^+	

[a] Natural sources of N to groundwater are precipitation unless otherwise stated in Description and Location.

[b] Anthropogenic N sources to groundwater are specified in the Description and Location (residential, e.g., lawn irrigation, pets, and septics).

[c] This value represents septic effluent (48 ± 17 kg-N/ha/yr) and residential lawn fertilization (8 ± 3 kg-N/ha/yr).

TABLE 1.5.5.3
Denitrification or Other *In Situ* Removal Mechanisms (kg- N/ha/yr)
Measured in Several Coastal Aquifers of the U.S. East and Gulf Coasts

Watershed/Estuary	Soil Type	Denitrification	*In Situ* Removal	Reference
Northeast Region				
Unconfined aquifer; Cape Cod, MA	Sand and gravel	$2.33 \cdot 10^{-7}$		Smith et al. (1991)
Riparian forest; Kingston, RI	Glaciofluvial and loamy sands	6		Groffman et al. (1996)
Riparian forest; Kingston, RI	Glaciofluvial and loamy sands		120	Nelson et al. (1995)
Mid- Atlantic Region				
Marine sediments; Great South Bay, NY	Sand and clay glacial outwash	3.9–19.4		Slater and Capone (1987)

shallow water table aquifer before groundwater entered nearby streams. They suggested that denitrification was the most important mechanism for nitrate loss in groundwaters of this area. Subsequent review here summarizes some of the SGD-derived nitrogen fluxes presented in the literature without delineation of fluid source unless such information was provided in the cited material.

1.5.5.4 Summary of SGD- Derived Nitrogen Fluxes

In the Northeast region of the United States, a substantial quantity of nitrogen is delivered to marine waters via SGD. As such, some of the most comprehensive studies of this input to coastal bays and estuaries have been performed in this region. The highest fluxes of nitrogen from groundwater discharge are found in Massachusetts. For example, in an ongoing (more than 20 years) study of Great Sippewisset Marsh, Valiela and others have measured groundwater nitrate and ammonium fluxes as high as 16,288 and 568 kg-N/ha/yr, respectively (Tables 1.5.5.4 and 1.5.5.5). Nitrite fluxes ranged from 0–363 kg-N/ha/yr into this same marsh (Valiela et al 1978; Valiela et al. 1990). Nitrate enters Buttermilk Bay via groundwater at rates of 5454–39,211 kg-N/ha/yr, while 3–16 kg-N/ha/yr NH_4^+ was found discharging into these same waters (Valiela and Costa 1988; Weiskel and Howes 1991). Dissolved inorganic ditrogen (DIN) fluxes from groundwater discharge to Waquoit Bay range from 5–21 kg-N/ha/yr (Valiela et al. 1992). In more recent studies of coastal aquifer dynamics in Waquoit Bay, DIN fluxes were found to be about 228 kg-N/ha/yr at the head of the bay (Abraham et al. 2003; Talbot et al. 2003). Talbot et al. (2003) report that 80% of this flux is derived from terrestrial (fresh) SGD, while the remaining 20% enters the bay from the salinity transition zone of the subterranean estuary. Giblin and Gaines (1990) found a flux of 409 kg-N/ha/yr SGD-NO_3^- discharging into Town Cove, Massachusetts. SGD measurements of NH_4^+ and DIN into Narragansett Bay, Rhode Island, and bordering coastal lagoons were

TABLE 1.5.5.4

Summary of Groundwater- Derived Nitrogen Fluxes (kg- N/ha/yr) to Coastal Waters along the U.S. East and Gulf Coasts

Watershed/Estuary	SGD- N Flux	SGD- NO$_3$ Flux	SGD- NH$_4^+$ Flux	Reference
Northeast Region				
Great Sippewisset Marsh, Buzzards Bay, MA	147–1412	123–1228	25–147	Valiela et al. (1978)
Buttermilk/Buzzards Bay, MA	0.22	0.18	0.04	Valiela and Costa (1988)
Indian Heights Subbasin, Buttermilk/Buzzards Bay, MA	31 ± 3			Weiskel and Howes (1991)
Town Cove, Cape Cod, MA	94	85–254		Giblin and Gaines (1990)
Childs River, Waquoit Bay, MA	17			Valiela et al. (1992)
Quashnet River, Waquoit Bay, MA	0.5			Valiela et al. (1992)
Sage Lot Pond, Waquoit Bay, MA	2			Valiela et al. (1992)
Waquoit Bay, MA	228[d]			Talbot et al. (2003)
Nauset Marsh estuary (99% NO$_3$), Eastham, MA	1228–3684			Portnoy et al. (1998)
Conanicut Point, benthic fluxes, Narragansett Bay, RI			138 ± 4[a]	McCaffrey et al. (1980)
Mid- Atlantic Region				
Great South Bay, NY	60	10.2–61.4		Capone and Slater (1990)
Coastal Plain (TN), Wye Narrows, MD	1–44			Staver and Brinsfield (1996)
Gott's Marsh, DIN, Upper Patuxent estuary, MD	1–44	0–14	1–10	Heinle and Flemer (1976)
Gott's Marsh, DON, Upper Patuxent estuary, MD	1–22			Heinle and Flemer (1976)
Rhode River estuary, Chesapeake Bay, MD-VA	16		14	Jordan and Correll (1985)
Urban, Elizabeth River, Chesapeake Bay, VA	230			Charette and Buesseler (2004)
Southern coastal plain, agricultural land; VA		25–1251	0.86–27.02	Simmons et al. (1992)
Southern coastal plain, wetland; VA		–0.245–7.246	4.79–22.11	Simmons et al. (1992)

(Continued)

TABLE 1.5.5.4 (Continued)
Summary of Groundwater- Derived Nitrogen Fluxes (kg- N/ha/yr) to Coastal Waters along the U.S. East and Gulf Coasts

Watershed/Estuary	SGD- N Flux	SGD- NO₃ Flux	SGD- NH₄⁺ Flux	Reference
Southern coastal plain, wetland; VA		$-0.245-7.246$	$4.79-22.11$	Simmons et al. (1992)
Southern coastal plain, urban; VA		$-2.33-0.982$	$16.33-2589$	Simmons et al. (1992)
Coastal plain, agricultural land near Cherrystone Inlet, VA		$1-1850$	$0.3-133.9$	Reay et al. (1992)
Coastal plain, forest near Cherrystone Inlet, VA		$3-31$	$1-101$	Reay et al. (1992)
Phillips Creek, Delmarva Peninsula, VA		263		Gallagher et al. (1996)
Rappahannock River, VA		252		Gallagher et al. (1996)
Cherrystone Inlet, Delmarva Peninsula, VA		272		Gallagher et al. (1996)
Piankatank River, VA		228		Gallagher et al. (1996)
Coastal Plain, unconfined; VA		3045 ± 1977	215 ± 399	Gallagher et al. (1996)
Southeast Region				
Neuse River subestuary, Pamlico Sound, NC		$3.94 \pm 3.94*$	$275 \pm 63*$	Fisher et al. (1982)
South River subestuary, Pamlico Sound, NC		$1.74 \pm 1.15*$	$139 \pm 23*$	Fisher et al. (1982)
Belle Baruch Marine Lab, North Inlet, SC			$100-228^{a,c}$	Krest et al. (2000)
Banana River Lagoon, Merritt Island, FL	$3.3-3.8$			Bhada et al. (2005, in press)
Taverneir Basin, Florida Keys, Florida Bay, FL	$38-126$			Corbett et al. (1999)
Gulf Coast Region				
Turkey Point, Apalachee Bay, FL	$36 \pm 10^{a,b}$			Rutkowski et al. (1999)

[a]Reported as including recirculated seawater.

[b]SGD-N flux reflects 11 ±3 kg-N/ha/yr of new N from groundwater and 25 ± 7 kg-N/ha/yr of pore water recycled N.

[c]Approximately 40%–75% of this SGD flux is recirculated seawater based on an independent salt and water budget (Morris 1995).

[d]Talbot et al. (2003) found that 80% of benthic DIN delivered to Waquoit Bay was via terrestrial SGD. Flux was calculated by us from area provided in Abraham et al. (2003).

DIN, dissolved inorganic nitrogen; DON, dissolved organic nitrogen; SGD, submarine groundwater discharge.

TABLE 1.5.5.5
Summary of Nitrogen Inputs and Outputs (kg- N/ha/yr) for Groundwaters of the U.S. East and Gulf Coasts

Region	N Inputs				N Outputs	
	Anthropogenic Fertilizer	Manure Fertilizer	Aquifer Recharge	Irrigation Waters	Denitrification	Aquifer Discharge
Northeast						
≤ 1985	33.8			236		388–17492
1986–1998					139	6446–40211
Mid- Atlantic						
≤ 1985	17			0.83		1015–1762
1986–1998	110–174	160–377	3.8		14.2–53.7	5261–6004
Southeast						
≤ 1985						733
1986–1998	15.3					2974–3149
Gulf Coast						
≤ 1985						
1986–1998	11.1			11		47
Total						
≤ 1985	50.8			237		2136–19987
1986–1998	136–200	160–377	3.8	11	153–193	14728–49411

177 kg-N/ha/yr and up to 109 kg-N/ha/yr, respectively (McCaffrey et al. 1980; and Lee and Olsen 1985). Within the Northeast the magnitude of nitrogen entering surfacewaters via SGD has increased by as much as 17 times between 1985 and 1998 (see Table 1.5.5.5).

In the Mid-Atlantic, SGD-derived nitrogen was found entering coastal waters of New York, Maryland, and Virginia (Table 1.5.5.4 and Table 1.5.5.5). Nitrate in SGD discharged to Great South Bay, New York, at 60–299 kg-N/ha/yr in 1985 and at 172–317 kg-N/ha/yr in 1990 (Capone and Bautista 1985; Capone and Slater 1990). An increase in this flux by a factor of about 3 occurred over a 5-year period. Capone and Bautista (1985) also found NH_4^+ discharging into Great South Bay at 89–449 kg-N/ha/yr. Total dissolved nitrogen fluxes from Gott's Marsh to Patuxent River estuary, Maryland, were 9.4 kg-N/ha/yr, of which 5.1 kg-N/ha/yr was estimated to be organic nitrogen (Heinle and Flemer 1976). Staver and Brinsfield (1996) found 60 kg-N/ha/yr NO_3^- discharging into the Wye River estuary of Maryland. Another well documented area for SGD is located within the many tributaries and subestuaries entering Chesapeake Bay, Virginia. The highest nitrogen flux derived from SGD (1816 kg-N/ha/yr) for this region is nitrate, and it was found in southern Chesapeake Bay near agricultural land (Simmons et al. 1992). Next to an urban area in southern

Chesapeake Bay, ammonium was the most abundant nitrogen species in SGD with a flux of 1134 kg-N/ha/yr (Simmons et al. 1992). SGD-derived nitrogen has increased by a factor of up to 5 from 1985 to 1998 in the Mid-Atlantic region (Table 1.5.5.5).

Groundwater discharge to coastal waters of the Southeast has been estimated in several areas, but this region and the northern Gulf of Mexico coast have not received nearly the attention the other regions have (Table 1.5.5.4 and Table 1.5.5.5). Benthic flux chamber measurements of nitrogen demonstrated that both NH_4^+ and NO_3^- were being advected into North Carolina's South River (SR) and Neuse River (NR) subestuaries. Ammonium ranged from 178 (SR) to 353 (NR) kg-N/ha/yr, while nitrate ranged from 8 (SR) to 17 (NR) kg-N/ha/yr (Fisher et al. 1982). Using nearshore seepage meter experiments, Whiting and Childers (1989) found NH_4^+ and $NO_3^- + NO_2^-$ fluxes into North Inlet, South Carolina, ranging from 25–39 kg-N/ha/yr and 0.35–0.42 kg-N/ha/yr, respectively. Groundwater discharge to coastal waters has been estimated off South Carolina to be 350 $m^3 \cdot sec^{-1}$ (Moore 1996). Using pore fluid nitrate concentrations of approximately 0.40 μM from a 32-km^2 area of North Inlet, South Carolina (see Table 1.5.5.1), the subsurface fluid flux of nitrate to South Carolina coastal waters may be as high as 85 kg-N/ha/yr. Total nitrogen entering the Little River, Georgia, watershed via SGD is estimated at 12.5 kg-N/ha/yr (Lowrance et al. 1983). Groundwater seepage velocities measured in the Indian River lagoon, Florida, ranged from 6.6–8.9 cm/day (Zimmermann et al. 1985). Although nitrogen species were not investigated for this study, it was demonstrated that groundwater flow was an important dissolved reactive phosphate source to the lagoon waters. If nitrogen is also present in the aquifer upstream of the lagoon, groundwater discharge may deliver N as well to these surfacewaters. SGD-derived nitrogen is a serious concern in the Florida Keys, due to the porous limestone aquifer underlying the islands and the presence of active sewage injection wells. DIN and total nitrogen fluxes of 2813 and 250 kg-N/ha/yr, respectively, have been measured off the Keys (Lapointe et al. 1990; Rutkowski et al. 1999). A factor of 4 increase has occurred for nitrogen entering coastal waters of the Southeast via SGD since 1985 (Table 1.5.5.5).

The Gulf Coast of the United States represents the area where the least investigations of SGD have been performed (Table 1.5.5.4 and Table 1.5.5.5). Only the northeastern Gulf of Mexico has estimates of groundwater discharge and dissolved constituents. SGD into the northeastern Gulf of Mexico is estimated to be as high as 710 $m^3 \cdot sec^{-1}$ (Cable et al. 1996b). If surficial aquifer concentrations of nitrate and ammonium from Turkey Point, Florida (see Table 1.5.5.1) are used, the flux of nitrogen may be as high as 962 kg-N/ha/yr NO_3^- and 234 kg-N/ha/yr NH_4^+ into this 620-km^2 area of the Gulf of Mexico. Nearshore seepage meter measurements of SGD reveal total nitrogen fluxes of 11–36 kg-N/ha/yr into a smaller (10 km^2) portion of the Gulf of Mexico (Rutkowski et al. 1999). Estimates of SGD-derived nitrogen were not found prior to 1985, so it is not possible to determine if a change in this flux has occurred over time. Progress is being made in several different regions of the United States to fill in gaps in submarine groundwater discharge information, but a paucity of data was revealed in this review for Louisiana, Alabama, South Carolina, North Carolina, and Delaware.

TABLE 1.5.5.6
Submarine Groundwater Discharge Estimates of Nitrogen (kg- N/ha/yr) for Various Coastal Waters Worldwide

Watershed/Estuary	Geology	Study Area (km²)	SGD Flux as NO_3	Reference
Marmion Lagoon, Perth, Australia	Limestone	25.3	0.056	Johannes and Hearn (1985)
Peel Inlet, Harvey Estuary, Australia	Limestone	NR[a]	34[b]	Sewell (1982)
Mariana Islands, Guam	Uplifted carbonate island	256	69	Matson (1993)
Lake Biwa, Japan	Volcanic	674	227–1136	Taniguchi and Tase 1999
Laguna de Celestun, Yucatan, Mexico	Karst, limestone aquifer	28	2037	Herrera-Silveira (1994); Herrera-Silveira and Comin (1995)

[a]NR, not reported.
[b]Submarine groundwater discharge flux is reported as nitrogen, not nitrate.

Estimates of SGD-derived nitrogen in most areas of the world are less common than the northeastern United States. We present a few areas by way of example, but for a more comprehensive review of worldwide SGD-N estimates, refer to Slomp and Van Cappellen (2004). In Western Australia, studies of the Marmion Lagoon and Peel-Harvey Estuary revealed groundwater nitrogen fluxes to coastal waters ranging from 0.056–34 kg-N/ha/yr (Table 1.5.5.6). SGD-N fluxes were also relatively small (69 kg-N/ha/yr) in the Mariana Islands, Guam (Matson 1993). In contrast, Lake Biwa, Japan, is located in an urban area and resides in a porous volcanic substrate. Fluxes of nitrogen as SGD ranged from 227–1136 kg-N/ha/yr (Taniguchi and Tase 1999). The Yucatan Peninsula, Mexico, is unique in that no surfacewater drainage enters the sea; the carbonate landscape is so porous that almost all precipitation infiltrates the limestone to recharge the aquifers. Groundwater resource management has become a serious concern for Yucatan residents as the waste from urban areas also infiltrates these aquifers. SGD-N fluxes from Laguna de Celestún alone were found to be about 2037 kg-N/ha/yr (Herrera-Silveira 1994; Herrera-Silveira and Comin 1995).

1.6   CONCLUSIONS

The magnitude of nutrients and other pollutants that terrestrial SGD carries will be critical for estimates of external loading of nutrients and pollutants to coastal bays and estuaries, because this source could represent new material. However, it will be important to separate the magnitude of the new material (nutrients) carried from the surface through the terrestrial aquifers to these coastal waters from the amount of nutrients regenerated from the marine sediments as the water flows through the sediment. Similar to nutrients derived from benthic processes of recirculation of water through the sediments, the magnitude of the water flux associated with terrestrial SGD will also be an important variable, but may be difficult to measure without first identifying the alongshore extent of seaward edge of the seepage face around the water body.

2 Methods of Study and Quantitative Assessment of Groundwater Discharge into Seas and Lakes — Advantages and Limitations

CONTENTS

2.1   METHODS OF LAND STUDYING IN THE COASTAL ZONE

2.1.1   HYDRODYNAMIC METHODS TO CALCULATE SUBMARINE GROUNDWATER DISCHARGE*

Submarine groundwater discharge (SGD) bypasses river networks and reaches the ocean directly as seepage through its bottom sediments and formations. It is important in the water and chemical balances of the coastal zone (Zektser and Loaiciga 1993; Church 1996; Zektser 2000; Loaiciga and Zektser 2001). Submarine groundwater discharge and its associated chemical fluxes in coastal karst aquifers may be larger than those associated with river discharge, in particular during low stream flow (Moore 1996). Thus grew the interest in developing

---

* Section 2.1.1 was written by Hugo Loaiciga, Ph.D. (University of California, Santa Barbara, hugo@geol.ucsb.edu).

methods to estimate SGD accurately (Paulsen et al. 2001). Its estimation, however, poses unique challenges because it often involves the deployment of measurement equipment in coastal waters as well as on land. The highly dynamic nature of groundwater–seawater interactions in the coastal zone hampers the accurate quantification of SGD (Gordon et al. 1996). Relatively short time series and the sparsity of hydrogeological data in most coastal regions constitutes another serious impediment to the accurate estimation of SGD. This chapter presents a review of the hydrodynamic and numerical methods to estimate submarine groundwater discharge. The former included flow nets, piezometers, and seepage meters. The latter consists of numerical simulation with or without coupling of hydraulic and density-driven phenomena governing SGD.

Flow Nets. Flow nets have been used to analyze regional flow problems and to estimate submarine groundwater discharge to oceans and lakes (Zektser and Loaiciga 1993). It is a simple yet powerful method to obtain preliminary estimates of the latter. Figure 2.1.1.1 depicts the elements that enter the flow-net approach.

The total SGD to the ocean ($q = q_i + q_2 + q_3 + q_4$, $L^3\ t^{-1}$) is given by the flow-net equation:

$$q = \frac{Kp\Delta hb}{n} \tag{2.1.1.1}$$

in which K is hydraulic conductivity ($L\ t^{-1}$); p is the number of stream tubes in the flow net (p = 4 in Figure 2.1.1.1); Δh is the change in hydraulic head between the two boundary equipotential contours ($\Delta h = 6 - 0 = 6$ m in Figure 2.1.1.1, L), in which the hydraulic head represents a freshwater equivalent, that is, after it is corrected for salinity (which affects the groundwater density; see the section on numerical simulation below); b is the thickness of the coastal aquifer perpendicular to the plane of Figure 2.1.1.1 (L); and n is the number of equipotential head drops

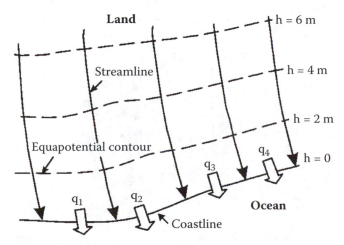

FIGURE 2.1.1.1 Plan view of a flow net in the coastal zone: $p = 4$; $\Delta h = 6$ m; $n = 3$ (see text).

in the flow net (n = 3 in Figure 2.1.1.1, between 6 m and 4 m, 4 m and 2 m, and 2 m and 0 m).

The construction of the coastal flow net requires groundwater level data, which may be sparse in many regions. In addition, the flow-net Equation (2.1.1.1) is based on simplifying hydrogeologic assumptions that may be too restrictive in many coastal zones (e.g., the assumption of homogeneous-isotropic hydraulic conductivity). It also assumes steady-state submarine discharge.

Piezometers. The estimation of submarine groundwater discharge by the piezometer method relies on the measurement of the gradient of hydraulic head and the application of Darcy's law. Consider Figure 2.1.1.2, which shows the measurement approach for a particular deployment of the piezometer system. Nested piezometers (in Figure 2.1.1.2 each nested piezometer consists of a cluster of two side-by-side piezometers) are located along a transect of the coastal zone to estimate the submarine groundwater discharge, q_v (for simplicity we assume herein that it is a vertical flux), which in this case flows toward the ocean along a coastal zone of width L):

$$q_v = \sum_{j=1}^{r} K_j w L \frac{\Delta h_j}{\Delta z_j} \qquad (2.1.1.2)$$

in which r is the number of nested piezometers; K_j is the hydraulic conductivity of the sediments where the piezometers are opened to groundwater flow; L is the width of the coastal zone perpendicular to the plane of Figure 2.1.1.2, along which submarine discharge flows toward the ocean; w is the width of the zone of influence of the nested piezometers in the plane of Figure 2.1.1.2; Δh_j is the vertical hydraulic

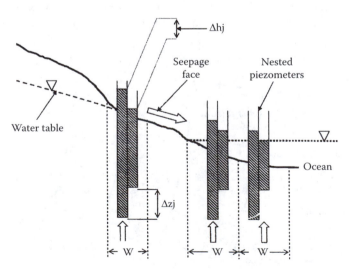

FIGURE 2.1.1.2 Nested piezometers deployed to measure submarine groundwater discharge as seepage.

head difference at the j-th nested piezometer (adjusted for groundwater salinity variations); Δz_j is the vertical distance between the piezometers' screen intervals.

Equation 2.1.1.2 assumes vertical seepage. If the horizontal flow component is not negligible, it may be calculated by applying Darcy's law along horizontal streamlines driven by a horizontal hydraulic gradient. Having the vertical (q_v) and horizontal (q_h) components of the groundwater discharge, the total flux is computed by $q = \sqrt{q_h^2 + q_v^2}$.

The piezometer method requires the determination of the hydraulic conductivity of coastal-zone sediments, which may prove an arduous undertaking (Landon et al. 2001). Other complications arise from the deployment, anchoring, and maintenance of, and access to, the piezometers, which may become onerous except in coastal environments with minimal surf. The cyclical nature of tides may cause reversal of seepage direction in parts of the coastal zone, which may introduce analytical difficulties in the estimation of the net submarine discharge (Shih et al. 2000).

Seepage Meters. The classic seepage meter is illustrated in Figure 2.1.1.3 (see Lee 1977; Fellows and Brezonik 1980; Isiorho and Mayer 1999; Landon et al. 2001 for applications that involve seepage meters). The collector box is inserted into the ocean sediments as shown in Figure 2.1.1.3. Previous researchers have used a cylindrical collector box (Lee [1977] used the cutoff end of a 55-gallon steel drum as a collector box), to which a short tube is attached leading to a plastic bag where the submarine discharge accumulates. The average submarine groundwater discharge $\bar{q}$ (L t^{-1}) during a period of time (t) is given by the following equation:

$$\bar{q} = \frac{V}{A_x t} \tag{2.1.1.3}$$

In Equation (2.1.1.3) V is the volume of groundwater captured inside the plastic bag during the time t; A_x is the cross-sectional area of the collector box perpendicular to the direction of the discharge $\bar{q}$. If the (horizontal) area A of ocean floor

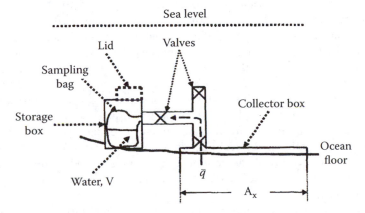

FIGURE 2.1.1.3 Schematic of a seepage meter; q is the average seepage measured in the meter.

within which $\bar{q}$ applies is A, the total volume of submarine discharge (in $L^3\,t^{-1}$) equals $q = \bar{q}A$.

This method yields an average discharge over the time period (t), instead of a temporal record of the actual discharge. Care must be taken to avoid energy losses between the collector box and the plastic bag that might yield an underestimate of the average discharge. Recently, Paulsen et al. (2001) introduced a digital seepage meter that measures the velocity of submarine discharge in the tube that connects the collector box to the sampling bag at small intervals, thus providing a nearly continuous record of the submarine groundwater discharge at the measurement location. Notice also that this method provides a point measurement of submarine groundwater discharge. An estimate of the regional submarine discharge requires multiple measurements of the kind illustrated in Figure 2.1.1.3 spread throughout the coastal zone.

2.1.1.1 Numerical Simulation

Decoupled Density-Hydraulic Simulation of Submarine Groundwater Discharge. One version of this method uses a freshwater-equivalent hydraulic head, and the effect of groundwater salinity must be input externally to the simulation. In other words, the effect of groundwater salinity variation in the mixing zone on SGD is not explicitly simulated. This approach is based on the solution of the 3-dimensional equation of groundwater flow, which is given by the following expression (the buoyancy effect exerted by differences in groundwater density is neglected in Eq. 2.1.1.4; see Langevin and Gao, 2006):

$$\frac{\partial K \frac{\partial h}{\partial x}}{\partial x} + \frac{\partial K \frac{\partial h}{\partial y}}{\partial y} + \frac{\partial K \frac{\partial h}{\partial z}}{\partial z} \pm N = S \frac{\partial h}{\partial t} \qquad (2.1.1.4)$$

in which K is hydraulic conductivity ($L\,t^{-1}$); h(x,y,z,t) is hydraulic head (L); N is the external water input (such as recharge, $L^3\,L^{-2}\,t^{-1}$, positive) or output (such as groundwater pumping, $L^3\,L^{-2}\,t^{-1}$; negative); S is the storage coefficient ($L^3\,L^{-2}\,t^{-1}$); x, y, z are Cartesian orthogonal coordinates; and t is the time-independent variable. The hydraulic head (h) is a freshwater equivalent. That is, it equals the elevation head (z) at any location plus the pressure head (h_p) corrected for salinity, in which the correction factor equals the ratio ρ_s/ρ_f, where ρ_s and ρ_f are the actual density of groundwater and the density of fresh groundwater (1 gm/cm^{-3}, approximately), respectively. In other words, h = z + (ρ_s/ρ_f) hp. Equation (2.1.1.4) is supplemented by initial hydraulic head conditions and boundary conditions that render h a well-defined problem. Typical boundary conditions are no-flow values at the bottom and lateral sides of the aquifer and a constant-head value in the freshwater–seawater mixing zone. If solved as a steady-state problem, its right-hand side term is equal to zero.

Equation (2.1.1.4) is discretized and expressed in finite-difference form using a stable solution algorithm (see Waterloo Hydrogeologic, Inc. 2000). Figure 2.1.1.4

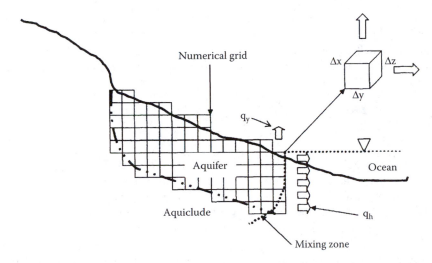

FIGURE 2.1.1.4 Cross-sectional view of a finite-difference grid used to simulate coastal groundwater flow.

shows a typical cross-section of a finite-difference grid in a coastal aquifer. Upon solution of Equation (2.1.1.4) for the hydraulic heads in the downstream boundary of the aquifer (named mixing zone in Figure 2.1.1.4), the submarine groundwater discharge equals the summation of the Darcian fluxes in each of the finite-difference cells that make up that boundary.

The horizontal component of the submarine groundwater discharge equals:

$$q_v = \sum_{i=1}^{n} K_i^h \Delta z_i \Delta x_i \left(\frac{h_i - d_i}{\Delta y_{id}} \right) \tag{2.1.1.5}$$

in which n is the total number of boundary cells in the finite-difference grid that have horizontal discharge to the ocean (see Figure 2.1.1.4); K_i^h is the horizontal hydraulic conductivity of the i-th cell (i = 1, 2, ..., n); Δz_i is the vertical dimension of the i-th cell; Δx_i is the horizontal dimension of the i-th cell (perpendicular to the plane of Figure 2.1.1.4); Δy_{id} is the horizontal distance between the cell centers where the hydraulic heads h and d; are calculated; d_j is the constant-head hydraulic head at the downstream boundary. Likewise, the vertical component of the submarine groundwater discharge is equal to:

$$q_v = \sum_{j=1}^{m} K_j^v \Delta y_j \Delta x_j \left(\frac{h_j - d_j}{\Delta z_{jd}} \right) \tag{2.1.1.6}$$

in which m is the number of top boundary cells where there is upward vertical discharge to the coastal zone (see Figure 2.1.1.4); K_j^v is the horizontal hydraulic conductivity of the j-th cell (j = 1, 2, ..., m); Δy_j is the horizontal dimension of the j-th cell (in the same plane of Figure 2.1.1.4); Δx_j is the horizontal dimension of the j-th cell (perpendicular to the plane of Figure 2.1.1.4); Δz_{jd} is the vertical distance between the cell centers where the hydraulic heads h_j and d_j are calculated; d_j is the constant-head hydraulic head at the downstream boundary. The magnitude of the total SGD is

$$q = \sqrt{q_h^2 + q_v^2} \qquad (2.1.1.7)$$

This method does not couple salinity (and, thus, density) of groundwater with hydraulic phenomena. Salinity-related corrections to hydraulic head must be input externally by assigning increasing density to groundwater as the ocean floor is approached. In addition, the numerical simulation method requires detailed knowledge of the aquifer's properties in the coastal zone with which to calibrate and validate the numerical model. If a validated model is achieved, on the other hand, it provides a powerful tool to predict SGD under a variety of scenarios. These include the effect of groundwater pumping in the coastal zone, climatic fluctuations, changes in sea level, and the like.

Coupled Density-Hydraulic Simulation of Submarine Groundwater Discharge. In this case the groundwater density $\rho(p,c)$ is a function of fluid pressure (p) and solute concentration (c), where the solute can be a single dissolved constituent (i.e., chloride) or an indicator of inorganic dissolved substances such as the total dissolved solids (TDS). Generally, the effect of high-solute concentration dominates pressure-related effects in $\rho(p,c)$. The unknown field variables are p and c, which vary with space and time. One must then solve for p and using two coupled differential equations, one describing the conservation of groundwater mass (or continuity equation) and the other solute conservation (known as the equation of hydrodynamic dispersion). The continuity equation for groundwater is as follows (with the aquifer porosity [n] independent of solute concentration):

$$-\left\{\frac{\partial(\rho(p,c)q_x)}{\partial x} + \frac{\partial(\rho(p,c)q_y)}{\partial y} + \frac{\partial(\rho(p,c)q_z)}{\partial z}\right\} \pm N = \frac{\partial(\rho(p,c)n)}{\partial t} \qquad (2.1.1.8)$$

in which the Cartesian components q_x, q_y, q_z of the groundwater flux vector q along the unit vectors i, j, k, respectively, are obtained from Darcy's law (g, k, and μ denote the Earth's gravitational acceleration, aquifer permeability, and the dynamic viscosity of groundwater, respectively):

$$\overline{q} = -\frac{k}{\mu}\left[\frac{\partial p}{\partial x}\overline{i} + \frac{\partial p}{\partial y}\overline{j} + \left(\frac{\partial p}{\partial z} + \rho g\right)\overline{k}\right] \qquad (2.1.1.9)$$

The equation of hydrodynamic (solute) dispersion is as follows (in which D is the coefficient of hydrodynamic dispersion, G is a solute source [positive if it is an input to groundwater, negative otherwise], ∇ is the gradient operator applied to the solute concentration (c), and $D \nabla$ denotes the vectorial dot product):

$$-\left(\frac{\partial(cq_x)}{\partial x} + \frac{\partial(cq_y)}{\partial y} + \frac{\partial(cq_z)}{\partial z}\right) + n\left(\frac{\partial(D \cdot \nabla_c)}{\partial x} + \frac{\partial(D \cdot \nabla_c)}{\partial y} + \frac{\partial(D \cdot \nabla_c)}{\partial z}\right) + G = \frac{\partial(nc)}{\partial t} \qquad (2.1.1.10)$$

The solution of Equations (2.1.1.7) through (2.1.1.9) requires an empirical (or constitutive) relationship expressing the dependence of groundwater density on solute concentration and fluid pressure, $\rho(p,c)$, plus initial and boundary conditions for pressure and concentration. It has been assumed above that isothermal conditions prevail in the aquifer. The resulting set of equations must be solved numerically, and specialized codes have been developed for that purpose (Voss 1984). Upon solution for the pressure and concentration fields, the total SGD is approximated by summing up the component of the groundwater flow vector perpendicular to the ocean floor (q) over the area of ocean floor (q_n) of interest (W).

Mathematically, the submarine groundwater discharge (q) is expressed as follows:

$$q = \int_W q_\eta dw \qquad (2.1.1.11)$$

The coupled density-hydraulic simulation of SGD is a challenging numerical and hydrogeologic problem. Accurate and high-resolution data are indispensable to calibrate and test the numerical simulation model. The numerical solution method necessitates stable and convergent algorithms. If successfully developed and tested, however, the numerical model for SGD becomes a very powerful prediction tool.

All the estimation methods presented above provide estimates of submarine groundwater discharge that are hindered by uncertainty. That uncertainty arises from model parameters, from the complexity of the coastal environment, and from approximate measurement of variables such as hydraulic head and groundwater salinity. These limitations call for sensitivity analysis designed to reveal the variability inherent to the estimation of SGD.

Another important consideration is the scale at which the presented methods best capture the process of SGD. The seepage meter and piezometer approaches are best suited to problems with flow paths of tens to hundreds of meters in length, or flow areas at the scale of hectares or less. The flow-net approach can be applied along the coastline for as long as the hydraulic data are available and the model assumptions hold. A typical scale of linear distance along the coastline within which the flow-net assumptions apply is the kilometer. The numerical simulation method is flexible and can handle complex hydrogeology as well as 3-dimensional flow patterns in coastal areas of variable size.

Cost is another key issue. The flow-net method requires the least amount of data, although the study area must have a well-characterized hydraulic-head map.

Numerical simulation requires detailed data and substantial modeling sophistication. The piezometer and the seepage meter methods rely on a dense network of piezometers and collector boxes in the coastal zone, respectively, some of which are subject to tidal effects and/or installed in ocean sediments. This poses measurement challenges and accessibility limitations.

2.1.2 NUMERICAL MODELING OF SUBMARINE GROUNDWATER DISCHARGE*

Submarine groundwater discharge can be defined as the mass transfer of groundwater across the sea floor. SGD may consist of freshwater, brackish water, or seawater depending on the location and time of year. SGD has been the subject of increasing concern because of the possibility that groundwater discharge may be partially responsible for nutrient loading (Uchiyama et al. 2000; Masterson and Walter 2001) or pollutant contamination (Johannes 1980; Li et al. 1999) to coastal marine estuaries. Consequently, new methods for quantifying SGD are continuously evolving in an effort to improve our understanding of terrestrial inputs to the marine environment.

Numerical modeling is a sophisticated way of distributing aquifer recharge, in space and time, to the various outflow boundaries. Use of numerical models is appealing because a simulation provides spatially and temporally detailed estimates of SGD rates. A well-calibrated numerical model is a scientifically defensible tool for estimating SGD within an entire study area by interpolating and extrapolating field measurements in space and time. Numerical models can also be used to predict future SGD rates or rates for other hydrologic conditions (e.g., predevelopment) provided a reasonable set of hydrologic stresses can be assigned for those conditions. Another advantage for using numerical models is that they provide a framework for integrating and organizing different sources and types of field data, and for developing new conceptual ideas about system dynamics. Preliminary simulations with a numerical model can provide guidelines for designing a field study of SGD, including sampling frequencies and locations. A sensitivity analysis with a numerical model can be used to identify the processes and hydrogeologic parameters that have the largest effect on SGD, which often correspond with the parameters that require further study.

Numerical models are beginning to emerge as viable tools for understanding and quantifying SGD. This is evident in the literature where the number of reported studies using numerical models specifically for SGD continues to grow (Robinson 1996; Uchiyama et al. 2000; Langevin 2001, 2003; Prieto 2001; Kaleris et al. 2002; Destouni and Prieto 2003; Oberdorfer 2003; Smith and Nield 2003; Smith and Zawadzki 2003). Several of these papers are from a special issue of *Biogeochemistry* dedicated to SGD (Burnett et al. 2003b). The journal, *Groundwater*, has also released a special issue on coastal groundwater in November 2004. Future use of numerical models for understanding and quantifying SGD is expected to increase as field

---

* Section 2.1.2 was written by Christian Langevin, Ph.D. (Center for Water and Restoration Studies, Miami, Florida, langvin@usgs.gov).

methods for measuring SGD become more reliable, computer programs for simulating groundwater flow to the coast improve, and as computer processing speeds increase.

This section provides an overview of numerical modeling methods and their applicability to studies of SGD. The general approaches for representing SGD are presented, followed by a brief description of the modeling approach. Computer codes used for numerical studies of SGD are also presented. The section concludes with some challenges for developing numerical models of SGD. Numerical modeling of groundwater flow is a complex task, and modeling coastal groundwater flow provides unique challenges. It is beyond the scope of this short section to describe all the details for developing and calibrating a numerical model of SGD. Instead, this chapter provides an overview, and interested readers are referred to additional references where appropriate.

2.1.2.1   Approaches for Representing Coastal Groundwater Flow

Several different approaches can be used to represent groundwater flow in coastal environments. Most groundwater flow models are based on an assumption of constant fluid density, which is valid for freshwater environments where spatial and temporal variations in fluid density are minimal. The constant-density assumption may also be applied in certain coastal areas if the interface between fresh and saline groundwater is highly dispersed, and density gradients are insignificant (Essaid 1990). In most coastal aquifers, however, the constant-density assumption is not valid, and the location and shape of the interface between fresh and saline groundwater affects the groundwater flow pattern and distribution of SGD. If density variations cannot be neglected, then either a sharp or dispersed interface approach can be used. With the sharp interface approach, freshwater and seawater are treated as immiscible fluids, whereas the dispersed interface approach allows the freshwater and seawater to mix. One advantage of the sharp interface approach is that simulations run relatively quickly on a computer, because a solution to the solute transport equation is not required, and aquifer layers can be represented with a single model layer. The disadvantage of the sharp interface approach is that salinity concentrations are not calculated within the aquifer and SGD consists only of freshwater. Thus, if the interface is truly dispersed at the scale of the investigation, then there may be difficulties in interpreting the results of the sharp interface model and comparing those results with field measurements. Most numerical models of SGD use the dispersed interface approach because the capability to represent the dispersive mixing process is often necessary. The disadvantage of the dispersed interface approach is that numerical simulations may require long computer runtimes. A single aquifer layer may require many model layers to accurately capture the flow and transport patterns. As a result, dispersed interface models often have a much higher level of resolution (and more computations) than constant-density or sharp interface models.

Figure 2.1.2.1 shows examples of generalized flow paths that would result from numerical models based on the three different approaches. In this example, the

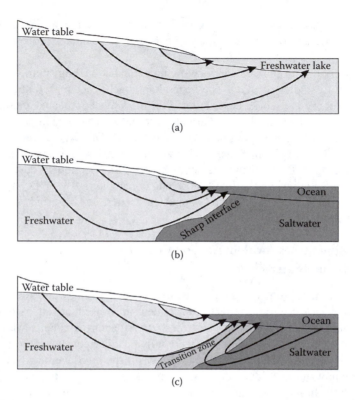

FIGURE 2.1.2.1 Generalized groundwater flow paths for the three numerical modeling approaches. A. Constant-density groundwater flow. B. Variable-density groundwater flow with a sharp interface. C. Variable-density groundwater flow and solute transport with a dispersed interface.

hydrogeologic setting is a shallow unconfined coastal aquifer; the setting is the same for each case, with the exception that discharge is to a freshwater lake for the constant-density approach (Figure 2.1.2.1A). For the two variable-density approaches (Figure 2.1.2.1B,C), SGD is into an ocean. With the constant-density approach, groundwater generally will discharge much farther offshore than for the two variable-density approaches (Figure 2.1.2.1B,C). With the variable-density approaches, the presence of dense seawater acts as a barrier to freshwater flow and causes most of the groundwater to discharge near the shoreline. With the sharp interface approach, nearly all the flow is within the freshwater part of the aquifer, and SGD consists only of fresh groundwater. Groundwater flow in the saltwater part of the aquifer occurs only in response to interface movement. Groundwater flow paths for the dispersed interface approach (Figure 2.1.2.1C) are quite different from the other two (Figure 2.1.2.1A,B). With the dispersed interface approach, a circu-lation pattern is generated as brackish groundwater is discharged through the sea floor. The flux of saline groundwater upward through the sea floor draws seawater into the interface. The combination of gravity-driven freshwater flow and mixing causes seawater to be pumped through the aquifer. This groundwater flow pattern was first observed by Kohout (1960a) in a saltwater intrusion study near Miami, Florida, and can generally be reproduced with a numerical model based on the dispersed interface approach.

Another approach for representing the interface is that of Bakker (2003), which is similar to the sharp interface approach, but allows one to input a dispersed interface as an initial condition. The code was tested by comparing results with those from two other codes (Bakker et al. 2004). Mixing between freshwater and seawater (hydrodynamic dispersion) is not represented during the simulation, but individual density isosurfaces are tracked and allowed to move in response to the flow field. The main advantage of this approach is that computer runtimes are relatively fast because an aquifer layer can be represented with a single model layer. The limitation is that the interface configuration is required as input, rather than being calculated by the model. There currently are no reported applications of this method for the study of SGD, but the approach is applicable, particularly at the regional scale.

2.1.2.2 Computer Programs

Numerous computer codes have been developed to simulate variable-density groundwater flow with both the sharp and dispersed interface approaches. Many more computer codes are available for simulating constant-density groundwater flow. Development of variable-density codes was motivated primarily by the need to simulate saltwater intrusion, a growing problem in many coastal regions where aquifer withdrawals have substantially lowered water levels. Most of these computer codes are directly applicable to SGD and other variable-density groundwater problems, including aquifer storage and recovery and deep-well wastewater injection. In general, the computer codes can be divided into two main types, depending on whether the finite-difference or finite-element method is used to solve the governing equations. The finite-difference method is conceptually straightforward and models tend to run relatively quickly, but a rectilinear grid is required, which poses complications for irregularly shaped domains. Finite-element models are very flexible in terms of the mesh design, but some find the method to be more conceptually difficult, and simulation times can be relatively long. Each method has advantages and disadvantages, and it is up to the investigator to select a code based on the problem at hand. Zheng and Bennett (2002) present some of the practical considerations (evaluation of the code's assumptions and solution options, level of documentation, costs, and reliability) that should be used when selecting a groundwater code. For a list of variable-density computer codes, readers are referred to Kolditz et al. (1998) and Sorek and Pinder (1999). Langevin et al. (2004) provide a description of four variable-density groundwater codes based on the MODFLOW program (McDonald and Harbaugh 1988; Harbaugh et al. 2000).

For numerical modeling studies of SGD, several different codes have been used. Smith and Zawadski (2003) used the finite-element FEFLOW code (Diersch, 1996), which simulates a dispersed interface, to run the model to steady-state conditions. Smith and Nield (2003) applied the constant-density, finite-difference MODFLOW code (McDonald and Harbaugh 1988) to their study of SGD in Cockburn Sound, Western Australia. Destouni and Prieto (2003) used the finite-element SUTRA code

(Voss 1984) to simulate representative SGD conditions for three coastal aquifers of the Mediterranean Sea. Uchiyama et al. (2000) developed a finite-difference code to simulate SGD and associated nutrient transport at Hasaki Beach, Japan. Langevin (2001, 2003) used the finite-difference SEAWAT code (Guo and Langevin 2002) to simulate SGD to Biscayne Bay, Florida. Kaleris et al. (2002) developed a constant-density MODFLOW model to simulate regional-scale SGD, and used the finite-difference, variable-density SWIFT code (Reeves et al. 1986) to simulate SGD through permeable sea bottom features referred to as a pockmarks.

2.1.2.3   Development of a Numerical Model

The numerical simulation of SGD requires that a coastal aquifer be divided, or discretized, into many small cells, or elements. Each cell is then assigned representative values for hydrogeologic properties, such as hydraulic conductivity, specific storage, and in some cases, porosity and dispersivity. Recharge, groundwater withdrawals, and other aquifer stresses must also be represented by carefully assigning internal and external boundary conditions. In addition, spatial and temporal scaling issues should be considered as part of an SGD modeling study. Will the model be 2-dimensional or 3-dimensional? Are regional SGD estimates required, or will the model represent a small field area? How will the transient nature of the system be handled? Will the model represent tidal fluctuations, or can steady-state conditions be assumed? A preferred modeling strategy is to start simple and build complexity as necessary. Even if the project requires a fully 3-dimensional, transient model, a 2-dimensional, steady-state model can provide useful information about required grid resolution and other numerical concerns. Transient simulations with a 2-dimensional model can also yield insight into problems that may be encountered with a transient, 3-dimensional model. Many successful modeling studies maintain two parallel modeling tracks. A simple model runs quickly, and can be run many times to identify the simulations that should be performed with the larger, more complicated model. In addition, the spatial and temporal resolution used for the model will determine how the model results can be compared with field data. SGD measurements from a seepage meter employed during a single tidal cycle will have questionable value for a steady-state model that represents average annual conditions.

An example of a model developed to simulate SGD is the 2-dimensional cross-section model of the Biscayne aquifer near Miami, Florida (Langevin 2001) shown in Figure 2.1.2.2. This model is based on the dispersed interface approach, and thus both the flow and transport equations are solved as part of the simulation. Each cell in the cross-section model is 200 m in the horizontal by 2 m in the vertical. Lateral inflow of freshwater is assigned to the left boundary. Fresh recharge is applied at the top, and the Biscayne Bay estuary is represented with a constant-head and constant-concentration condition.

Once the input data for the numerical model are prepared, a computer is used to solve a set of algebraic equations that approximate the governing equations of groundwater flow. In some cases, the transport of seawater salts is also simulated to estimate the spatial and temporal distribution of aquifer salinity and the salinity

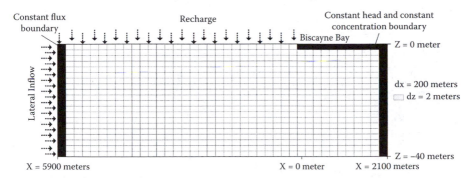

FIGURE 2.1.2.2 Model grid and boundary conditions used for a numerical simulation of submarine groundwater discharge to Biscayne Bay, Florida. (Modified from Langevin 2001.)

of SGD. Simulation of aquifer salinity requires that the solute transport equation is solved, in addition to the groundwater flow equation. The numerical solution results in groundwater heads at each cell, groundwater fluxes across each cell face, and if the solute transport equation is solved, concentrations at each cell. As an example of output from a numerical model, Figure 2.1.2.3 shows contours of simulated salinity for the Biscayne Bay cross-section model (Figure 2.1.2.2). Figure 2.1.2.3 also shows salinity values from screened or open-hole intervals of monitoring wells. A difference between measured and simulated values is most likely an indication that one or more the boundary conditions, hydrogeologic parameters, or hydrogeologic processes are not accurately represented by the model.

SGD estimates from numerical models contain a high level of uncertainty. This uncertainty can be reduced by calibrating the model to field data (e.g., aquifer heads and salinities). Field measurements of SGD are particularly useful for model calibration. If the model can reproduce SGD measurements at several locations for various conditions, confidence is gained in the accuracy of the simulation. Thus, one might expect the model to produce reasonable SGD estimates for other locations and conditions. Regardless of the calibration, a numerical model will never provide an exact replication of the field system, and thus uncertainty in model results is unavoidable.

If the hydrogeologic properties and the hydrologic stresses assigned to the model are reasonably accurate, the calculated groundwater fluxes will provide a reasonable estimate of groundwater fluxes within the aquifer. Estimates of SGD can be obtained from the numerical model by evaluating or summing the groundwater fluxes to the ocean boundary. Continuing with the Biscayne Bay cross-section model, Figure 2.1.2.4 shows the estimated brackish SGD to the estuary. All of the discharge is within the first two model cells (about 300 m of the shoreline). Farther offshore, results from the numerical model indicate downward flow from Biscayne Bay into the aquifer — a pattern expected for coastal groundwater flow.

The brief introduction to modeling SGD presented here is a simplified description of the process. In practice, substantial effort is required to ensure that the numerical model is an accurate representation of the system. Fortunately, guidelines have been established to help with the development of a conceptual model and the

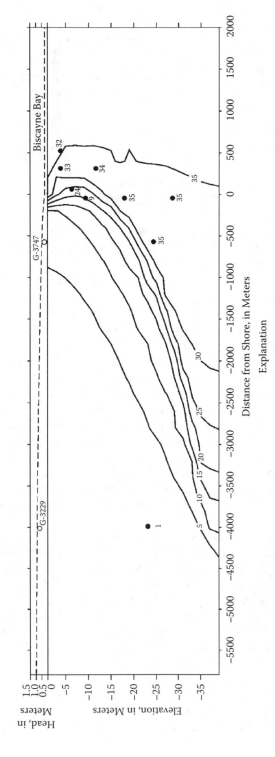

FIGURE 2.1.2.3 Comparison of simulated and measured salinity values for the numerical model of submarine groundwater discharge to Biscayne Bay, Florida. (Modified from Langevin 2001.)

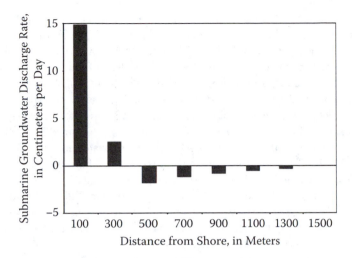

FIGURE 2.1.2.4 Simulated estimates of brackish submarine groundwater discharge to Biscayne Bay, Florida. Positive values indicate groundwater flow into Biscayne Bay. Negative values indicate flow from Biscayne Bay downward into the underlying aquifer.

translation of the conceptual model into a numerical model (e.g., Anderson and Woessner 1992; Hill 1998; Reilly 2001).

2.1.2.4 Challenges for Simulating Submarine Groundwater Discharge

Most modelers acknowledge that the freshwater component of SGD is simply the difference between aquifer recharge and losses to other sinks, such as evapotranspiration, groundwater withdrawals, and river baseflow (Li et al. 1999; Langevin 2001; Kaleris et al. 2002; Destouni and Prieto 2003; Oberdorfer 2003; Smith and Nield 2003). A significant problem with simulating coastal groundwater flow is that in many cases the SGD is within the uncertainty range of the aquifer's sources and sinks. Nearly all the documented SGD modeling efforts report large uncertainties in simulated SGD rates due to a lack of field data. For example, Smith and Zawadski (2003) could not define the local hydrostratigraphy at their study site in Florida because borehole data were not available. Smith and Nield (2003) indicated that the assigned recharge rate, a key water-balance component for estimating SGD, contained large uncertainties. Langevin (2001) had detailed groundwater salinity measurements at coastal monitoring wells, owing to a previous study by Kohout (1960a), but no reliable measurements of SGD were available for model calibration. Oberdorfer (2003) noted that hydraulic head and conductivity data are rarely collected offshore and thus, the existing models have never been calibrated to represent the recirculation of seawater in the coastal mixing zone.

The lack of important field data is largely related to the historical focus of coastal groundwater studies, which has been saltwater intrusion. In Florida, for example, a primary water resource issue has been locating and monitoring the position of the saltwater interface. Where the interface moved inland, groundwater monitoring wells were abandoned once the salinity exceeded drinking water standards. In studies of SGD, groundwater salinity measurements are useful for model calibration, even if the salinity values are relatively high. Fortunately, much of the data collected for saltwater intrusion studies are useful for a study of SGD, but additional field data at the coastline and offshore will be required to calibrate models of SGD.

In addition to challenges posed by a lack of field data, the type of field data used for model calibration may pose unique challenges. Rates of SGD simulated by a numerical model may not be comparable with rates determined by field measurements if one includes the quantity of recirculated seawater and the other does not (Oberdorfer 2003). With the constant-density and sharp interface modeling approaches, for example, only the freshwater portion of SGD is represented by the model. This means that model results cannot be compared directly with measurements from seepage meters, which include the fresh and recirculated seawater components. However, if the dispersed interface approach is used, the simulated rate of SGD includes both the freshwater and recirculated seawater components. In general, this concern can easily be avoided by ensuring that model results can be compared with field measurements, or by adjusting field measurements in such a way that they can be compared with model results.

Another challenge with using a numerical model to estimate SGD is the issue of scale. Numerical models of SGD are often developed at the regional scale, whereas calibration data and aquifer information are collected at the local scale. Scaling issues have important consequences for some of the most basic modeling assumptions. For example, most models are based on the assumption that the aquifer can be treated as a porous medium, as opposed to a fractured or karst system, and that Darcy's law is valid (an underlying assumption in most modeling codes). In many areas, submarine groundwater springs are thought to be at the intersection of a karst conduit, or fracture trace, with the sea bottom. These permeable features can sometimes be explicitly included in a numerical model if the geometry of the feature is known and if the feature is relatively large compared to the model cell size. If the features are relatively small, their effects on flow can be roughly incorporated by adjusting the bulk hydraulic properties of the model cell; however, this approach can lead to errors in SGD estimates, particularly if solute transport is represented. It is not surprising that carbonate aquifers with preferential flow paths can pose unique challenges for the development of a numerical groundwater model.

Use of a numerical model to estimate SGD is can be very complex, time consuming, and plagued with potential pitfalls. Most computer codes are fairly primitive in their ability to detect user errors. This often leads to simulations where the computer solves a problem that does not correspond with the problem intended by the modeler. Fortunately, graphical user interfaces, which can help in detecting user errors, are available for many of the modeling codes. In general, the design of a numerical model requires a strong quantitative background in hydrogeology or engineering, a firm understanding of the governing flow and transport equations and the numerical methods used to solve them, and the ability to use programming languages, spreadsheets, or other computer programs to manipulate large datasets.

As a final caution, numerical models are best used as one method in a toolbox of methods for understanding and estimating SGD. Numerical models have the potential to provide useful and reliable results, but they require substantial and reliable field data for calibration. The general thinking among variable-density modelers is that the existing computer codes adequately represent the physics that govern the process of SGD. The true challenge lies in the ability to develop a numerical model that accurately represents field conditions. Numerical modeling of SGD is a relatively new application of variable-density modeling, and the number of reported studies is limited. However, the number of studies is expected to rapidly increase as the importance of SGD continues to receive widespread recognition.

2.1.3 COMPLEX HYDROLOGIC-HYDROGEOLOGICAL METHOD USING SPECIFIC DISCHARGE DATA

The quantitative estimation of groundwater discharge to seas should be carried out on the basis of reliable and tested methods for the study and assessment of ground-water formation and migration, especially in coastal zones. Of special value among these methods is the integrated hydrologic-hydrogeological method of subdividing a river hydrograph for a multiyear period that has been successfully used for regional estimation of the groundwater discharge in the former USSR, countries of Central and Eastern Europe, and the world as a whole (Kudelin 1960; GUGK, 1974; GUGK 1983; Hydroscience Press 1999). Using various modifications of this method, the multiyear average moduli of groundwater discharge are determined for basic water-bearing systems (aquifers), drained by rivers in the coastal zones of the seas and oceans. Then, by means of analogy these moduli are applied to geologic-hydrogeo-logically similar coastal areas drained directly by the sea, without participation of rivers. It should be kept in mind that the integrated hydrologic-hydrogeological method was developed for the regional estimation of groundwater discharge to rivers from the zone of intensive water exchange. Therefore, its use to estimate SGD must be combined with analogous methods.

For subdividing river hydrographs, it is necessary to use the entire period of observations for the natural water discharge within a measuring range in order to avoid possible errors and to obtain the most stable and reliable values of absolute and specific characteristics of groundwater discharge. When the observation row is sufficiently long (>20 years), it becomes possible to determine multiyear average subsurface discharge, and sometimes the estimation of its standard level is also possible. In this case, it is necessary to treat all the obtained hydrometric data, which can be rather time consuming. At the same time, according to Lvovich, in order to calculate multiyear average values of groundwater discharge, it is sufficient to take data from a 4-year time frame: 2 years with mean water levels, 1 dry year, and 1 wet year. Years with mean water levels are selected not only according to the annual average discharge, but they must be also typical according to the intra-annual distribution of groundwater discharge. The dry and wet years must correspond approximately to the years with the 75%–80%-th and 20%–25%-th water provision (Lvovich 1974).

The essence of hydrologic-hydrogeological methods lies in that groundwater discharge in the zone of intensive water exchange in areas with a constant river network is chiefly formed under the draining influence of river systems. The determination of a groundwater component in the total river runoff gives the possibility to estimate a value of regional groundwater discharge. The hydrologic-hydrogeological approach is based on the genetic subdivision of the river runoff hydrograph. Namely, it is subdivided into two parts. The first includes the discharge from sloped surfaces, inside topsoils, and precipitation falling onto water surfaces in river channels. All these elements form the direct discharge. Subsurface recharge or base discharge is considered as the second part of the hydrograph.

At present, a vast number of scientifically based methods and engineering techniques are used for the genetic subdivision of hydrographs. And the majority of scientists proceed from the opinion that stable mean water level is formed only at the expense of groundwater discharge (except rivers having predominantly natural and artificial regulation). The necessary condition for determination of stable mean water level is the absence of summer rains and winter thawing weather during a period exceeding the time during which a high-water wave passes along the measuring hydrometric range.

The basic difference of the existing methods concerns the hydrograph subdivision during floods and spring high water. The approaches used can be conventionally divided into three groups: (1) methods that do not take into account the effect of shore regulation during flooding; (2) techniques reducing the effect of shore regulation to a slight decrease in subsurface recharge of the river; and (3) those that accept stoppage of subsurface discharge to the river caused by the shore regulation process.

The practice of the genetic subdivision of hydrographs shows the necessity of taking into account interaction between surface and groundwater, i.e., a degree of their hydraulic linkage.

The literature reports semicalculating methods to subdivide hydrographs of the total river discharge with the use of different constants characterizing the features of river basins (Linsley et al. 1962; De Wist 1969; Freeze and Cherry 1979).

The simplest way is to divide a river discharge hydrograph by a horizontal line AD (Figure 2.1.3.1). However, such a method usually leads to overestimation of a flood period at the expense of the following mean water level. Because of this, a sloping line AD′ is used. The point where this line is crossed with the line of flood going down (D′) is obtained by calculating a number of days (N) beginning from the flood peak to its stop. The value N depends on the drainage area parameters: size, slope, surface roughness, as well as amount of precipitation or snow melt. Tentative N-values for drainage areas of different sizes are given in the publication by Linsley et al. (1962). There is also an approximate formula: $N = F^{0.2}$, where F is the area of the drainage basin in square miles (Linsley et al. 1962; De Wist 1969).

More often the hydrograph subdivision is performed by a line ABD′ (Figure 2.1.3.1), where the AB-line corresponds to decreasing groundwater discharge observed before a rainfall or thawing weather to a point located above the flood peak. The line BD′ connects point B with the discharge in N-days after the flood peak.

Also, extrapolation of the postflood groundwater recharge depletion curve is used in hydrograph separation. For this purpose, a period of time is considered distant from the moment of rain stop where there is no doubt that the river runoff

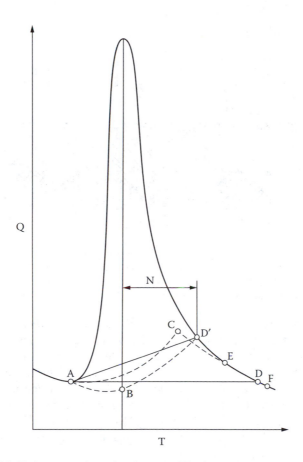

FIGURE 2.1.3.1 Various procedures for river runoff hydrograph separation (De Wist, 1965).

is formed at the expense of groundwater recharge. This curve, which is extrapolated by graphical or analytical means, gives the point of surface runoff finish (E) and the groundwater recharge curve (ACEF; Figure 2.1.3.1). For this, a segment of stable mean water level is considered, which follows directly after the flood. This hydrograph segment graphically reflects the dynamics of groundwater recharge depletion and is extrapolated for the flood period to point C (Figure 2.1.3.1) corresponding to the peak of groundwater discharge. However, the time when groundwater discharge starts and the shape of its rise (line AC) are taken arbitrarily.

For rivers with a flood regime, one often uses the scheme of hydrograph subdivision suggested by Hatterman and Fridrich (Keller 1965). The line dividing surface and subsurface discharges is represented by a lower enveloping curve passing across the low points of the hydrograph (similarly to line AD; Figure 2.1.3.1).

A rather simple method for assessment of groundwater recharge of rivers suggested by Vundt (Keller 1965) is worthy of attention. The method is based on multiyear observations of minimum river runoff with a distinct stable groundwater discharge (artesian discharge). Then by an average value of the minimum monthly average values, the shallow water discharge is determined (Figure 2.1.3.2).

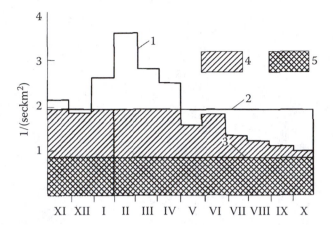

FIGURE 2.1.3.2 Specific river discharge values as indices of groundwater flow (Keller 1965): (1) Average monthly minimal discharge values. (2) Average annual minimal discharge values. (3) Average long-term minimal discharge values. (4) Unconfined groundwater flow. (5) Sustained groundwater (artesian) flow.

To obtain a more accurate assessment of the subsurface component in the river runoff, the method of Vundt can be added by plotting a single hydrograph of subsurface discharge (Linsley et al. 1962). Single hydrographs are widely used in hydrologic calculations of flood runoff. However, the complexity of groundwater discharge during floods and its dependence both on the parameters of drainage areas and the hydrogeological properties of water-bearing rocks do not allow the implementation of typical single hydrographs. Plotting a single hydrograph requires detailed study of hydrologic-hydrogeological features of a concrete basin, which is rather time consuming.

Based on hydrodynamics methods, it has been suggested to divide the hydrograph during the flood period if the preflood (q_s) and postflood (q_f) river runoffs are provided exclusively by groundwater recharge (Verigin 1963).

The above illustrated methods for dividing river runoff hydrographs are approximate. Their lack of precision is connected with a too formal approach to such a complicated natural process like groundwater discharge in drainage basins. The errors of these methods are related to the intention to determine subsurface recharge of rivers exclusively by data on river runoff without properly studying the hydrogeological conditions of the basin. This is their main disadvantage, because hydrometric data do not reflect the complexity of the geologic-hydrogeological structure of a river basin and, more importantly, does not enable strict scientific substantiation for the above-mentioned methods of hydrograph subdivision.

To obtain reliable data on subsurface recharge of rivers, it is necessary to carry out an integrated study of surface and subsurface discharges within a drainage area with substantiation on the type and degree of their interaction. The investigations of our researchers show that the processes of shore regulation during floods on plain-located rivers lead to a considerable decrease and sometimes even to stoppage of subsurface recharge of the rivers, which should be taken into account in the separation of hydrographs (Kudelin 1960; Walton 1970; Lvovich 1974). Russian specialists have worked out an integrated hydrologic-hydrogeological method to separate the hydrograph.

The basic advantage of this method is that it takes into account a type and degree of the interrelation between ground and surfacewaters in a river basin, which are determined through analyzing the existing geologic-hydrogeological information. On the basis of data from literature publications and of field investigations, the typical schematic draining maps are compiled for different areas of a river basin (Figure 2.1.3.3). These maps reflect drained aquifers and their lithological compositions, as well as levels of groundwater and riverwater in different seasons of the year. The type of hydraulic linkage of a river with an aquifer, depending on the ratio of ground and riverwater levels, determines the regime of groundwater discharge to the river and, also, on the different hydrograph dividing schemes described below.

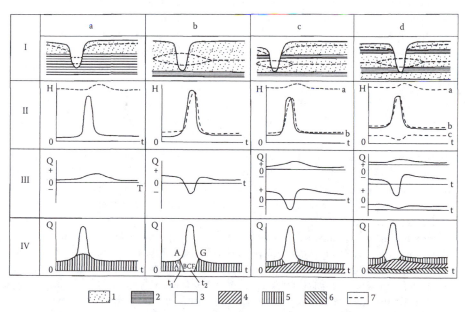

FIGURE 2.1.3.3 Analysis of stream hydrograph separation under different hydrogeological conditions in a stream basin (Kudelin 1960). I – Type of groundwater flow regime and its relation to stream stages: (a) unconfined groundwater flow regime not connected hydraulically with stream; (b) unconfined groundwater flow regime connected hydraulically with stream; (c) mixed groundwater flow regime; (d) mixed flow regime with unconfined and confined groundwater. II – Typical fluctuations in stream stage and groundwater levels near stream bank: a,b,c – hydrographs of groundwater discharge from upper, middle and lower aquifers. III – Hydrograph of groundwater discharge to stream. IV – Scheme for the analysis of stream hydrograph: (1) water-bearing rocks; (2) impervious rocks; (3) surface runoff; (4) groundwater discharge from aquifers not hydraulically connected with stream; (5) groundwater discharge from aquifers hydraulically connected with stream; (6) artesian flow; (7) groundwater level.

The separation of a river runoff hydrograph, while taking into account shore regulation, is widely used to calculate groundwater discharge from aquifers hydraulically
connected with rivers (Figure 2.1.3.3,b). The technique is based on distinguishing
a subsurface component in the river discharge during spring high water and autumn
flood periods, using data on river runoff rate and velocity of the pursuit flood wave.
During summer and winter, the mean water level periods of groundwater discharge
is accepted to be equal to the river level.

The subsurface recharge of the river is stopped (point A) at the ascending phase
of high water (flood). However, through the measuring river section, the groundwater, drained by the river in its upper reaches before flooding, continues transient
penetration. The time interval (t), after which this penetration stops, is determined
by the velocity of the flood wave. The groundwater discharge through the measuring
section will gradually decrease along line AB. Point G corresponds to the end of
the flood in the measuring section and to the start of groundwater discharge. As in
the river upstream, where the flood ends usually first (point C), the groundwater,
drained there, reaches the measuring section before the end of the flood. In this case,
the groundwater discharge will increase along line FG.

Subsurface recharge from aquifers not hydraulically connected with the river
(see the draining scheme in Figure 2.1.3.3,a) is determined on the basis of spring
surveying and the investigations of the dynamics of reference springs, with the use
of the following formula:

$$Q = qk_1 + \ldots + qk_i + \ldots + qk_n \qquad (2.1.3.1)$$

where q is the river mean water runoff; k_1, ..., k_n are the coefficients (monthly) of
spring discharge dynamics characterizing groundwater discharge to the river, and
are determined based on observations over the reference spring discharge during a
year. However, for many poorly studied regions, the data on the coefficients of spring
discharge dynamics are absent. At the same time, it is known that at the expense of
replenishment of groundwater reserves during the flood, at the end of the latter a
general increase of groundwater discharge is observed with a dynamics coefficient
of 1.5–3.0. Therefore, for such regions a scheme of hydrograph separation is applicable, as suggested by Voskresensky, and which reflects a certain increase in groundwater discharge by the end of the flood. The comparison of this modified method
with more exact calculations, taking into account the coefficients of spring dynamics,
showed a difference in the obtained moduli of groundwater discharge, equal only
to 10–15% (Popov 1968).

Hydrograph separation of total river runoff at the mixed recharge of the river,
from an aquifer connected and not connected hydraulically, is fulfilled in two steps.
In the first step, the groundwater discharge from aquifers hydraulically connected
with the river is determined by the above-described method. After that, the lower
part of the hydrograph is overlapped by the discharge from aquifers not connected
hydraulically with the river (Figure 2.1.3.3,c).

The more prominent difficulties appear when the hydrograph of total river runoff
is divided at river recharge jointly from shallow and artesian aquifers (see the

draining scheme in Figure 2.1.3.3,d). The artesian recharge of the river can be determined by the regime of ascending springs. The hidden artesian water discharge through the river channel is determined by means of analytical calculations or modeling with the use of additional hydrogeological information.

The use of hydrograph separation of the total river runoff is chiefly used for relatively small river basins under natural conditions. A reference measuring area of drainage basins should be limited, depending on completeness with which the groundwater is drained and on the type of a passing flood wave. The analysis of the conditions of the groundwater discharge within the Russian Platform has shown that its amount increases, as a rule, with an extension of the river basin area approximately to 1000 km². Then the groundwater discharge moduli are stabilized and do not depend on further increase of the drainage area. On the other hand, the upper limit of the measuring area for rivers on plains should be restricted to 50 ths km². Before this limit, the flood wave is not split yet, which makes it difficult to divide the hydrograph (Popov 1968).

The selection of the size of the reference-measuring drainage area depends on the results of the analysis of the geologic-structural and hydrogeological conditions of groundwater discharge formation, which determines the character of the linkage of the groundwater with riverwaters and the degree to which a territory is drained. The regularities, established in formation of the groundwater discharge within the former USSR, indicate that the plain-type territories of platforms with erosion-induced troughs of up to 60 m deep are characterized by the low draining ability of groundwater. Therefore, the typical measuring area of drainage basins can here reach the first tens of thousands of square kilometers. Shields and folded regions with erosion-induced troughs at a depth of 100–150 m and deeper belong to mean- and highly drained territories where measuring areas of drainage basins are reduced to a few thousands or even hundreds of square kilometers.

On regulated rivers, in cases of the availability of long observation rows, the separation of hydrographs should be made with the use of river discharges measured before their artificial regulation. In regions where networks of stationary hydrometric posts are not dense or observation rows are not numerous enough to determine groundwater discharge with the required precision, during stable mean water periods, special hydrometric investigations are to be carried out. As a result, groundwater discharge is estimated within an ending measuring range of small drainage basins having a homogenous hydrogeological structure.

Special attention should be paid to the influence of lakes and swamps in river basins on groundwater discharge formation. In the estimation of groundwater discharge by means of hydrograph separation, the influence of lakes can be neglected if the coefficient of the total lakes' area is not more than 15%. In cases of higher coefficients, the regulating influence of lakes should be removed by carrying out special hydrometric surveying. Lakes actually do not affect the dynamics of groundwater discharge. This is due to the fact that the hydrologic regime of a water stream is under influence only of the upper (to 60 m) active layer of up- and downriver lakes, which actively accumulates and returns the water during flood periods. The mean-level water discharge from swamped drainage areas is formed at the expense of groundwater, which must be taken into account in the separation of hydrographs (Popov 1968).

In some cases, groundwater discharge can be determined by the calculation of a change in river mean-level water runoff within an area between two hydrometric posts. The calculated change of river runoff in an area having no water inflow (or with the deduction of a sum of inflows) for the period of stable mean water level will characterize groundwater discharge from drained aquifers or groundwater recharge from absorbing river runoff. Measuring hydrometric ranges (posts) should be selected so that a difference in river runoffs within the first and second ranges would exceed the total error in the measurements of the river runoffs.

The direct hydrometric measurements in river basins make it possible to establish the type and degree of the interrelation between ground and surfacewaters in particular areas of the river, to reveal the influence zones of specific conditions of groundwater discharge formation (e.g., karst, faults, icings, etc.), and to assess the effect of naturally and artificially regulated river runoff upon the subsurface recharge of rivers. The latter seems to be especially important, because hydrotechnical facilities on rivers significantly hamper or exclude the possibility of using methods of hydrograph separation.

Periodic hydrometric measurements or seasonal surveying should be carried out in the period of stable mean water level, when possible flood runoff amounts to only 10–15% of the total discharge in this period. The measurements at all temporal posts should be performed within a short time, in order to minimize the influence of possible changes in meteorological conditions. The areas for hydrometric measurements should be selected taking into consideration the hydrologic-hydrogeological features of a river basin. These features should (1) reflect the different conditions of groundwater and surfacewater interrelations along the river length; (2) include recharge and discharge zones of groundwater; and (3) characterize quantitatively the influence of different natural and artificial factors on subsurface recharge of a river.

Single measurements during mean water level periods give a picture of minimum or almost minimal groundwater discharge; therefore, they must be reduced to annual or multiyear average values, taking into account intra-year dynamics. For this purpose, hydrometric areas of temporal surveying should include measuring posts with a multiyear row of observations. The temporal measuring ranges (posts) should be located in a river basin, depending on the requirements of measurement accuracy. Possible errors in measurements depend on the ratio of drainage areas versus discharges in temporal ranges. When the ratio of discharges in neighboring ranges increases to 50, the error in the calculated values decreases to 5% (Popov 1968).

The interpretation of the periodic measurements of river runoff becomes much more reliable and exact when they are combined with observations of groundwater dynamics in a drainage basin, as well as with hydrochemical and isotopic determinations of natural waters. Hydrochemical and isotopic methods for estimating groundwater inflow to a river have been developed (Zektser 1977; Freeze and Cherry 1979). They are based on the comparison of the total mineralization of surfacewater, groundwater, and riverwater (i.e., the concentration of some ion or isotope) during different seasons of the year. However, hydrochemical processes occurring during groundwater discharge to a river are usually not considered, which makes the calculation a little conventional.

The method is based on the use of two basic equations:

$$CQ = CgwQgw + C'gwQ'gw + CawQaw + CswQsw \qquad (2.1.3.2)$$

$$CQ = Qgw + Q'gw + Qaw + Qsw$$

where C, Cgw, C'gw, Caw, and Csw are the values of total mineralization or concentration of a particular ion (isotope), respectively, in riverwaters, groundwater of aquifers not hydraulically connected with the river, groundwater of aquifers hydraulically connected with the river, artesian, and surfacewaters (slope runoff); Q, Qgw, Q'gw, Qaw, and Qsw are, respectively, the values of the total river runoff, groundwater discharge from aquifers not connected hydraulically with the river, aquifers hydraulically connected with the river, artesian, and surfacewaters.

The equations given make it possible to obtain the formulae for determining total groundwater discharge, as well as any subsurface aquifer participating in river recharge.

$$Qgw = Q\frac{C - Csw}{Cgw - Csw} \qquad (2.1.3.3)$$

While using the hydrochemical method, one should use conservative ions that do not participate in chemical reactions in riverwater and do not change their concentration.

In principle, the same method, but only with the use of natural isotopes, can be used to assess seepage of artesian waters to a shallow aquifer when studying total groundwater discharge to rivers (Valdes 1985).

Subsurface recharge of rivers can be also determined by means of comparison of electrical conductivity of groundwater, riverwater, and surfacewater, because it characterizes their mineralization (Visocky 1970).

Use of the above-described methods of genetic subdivision of river hydrographs and their modifications is reasonable for coastal territories with a well-developed river network. In this case only the SGD from the upper hydrodynamic zone that is most actively drained by rivers and seas is estimated. The major groundwater discharge to seas is formed from the coastal areas just adjacent to this zone of intensive water exchange and having predominantly fresh groundwater. At the same time, methods of hydrograph separation should be used in combination with other methods of regional assessment of groundwater discharge — hydrogeodynamic or multiyear average water balance, depending on the degree of knowledge on the coastal regions and availability of initial data.

2.1.4   MEAN PERENNIAL WATER-BALANCE METHOD

Assessment of groundwater discharge is made in many cases on the basis of *in situ* measurements and water-balance methods, which represent, in the general form, experimental methods of studying groundwater discharge. Experimental methods include lysimetry and the method of preparing a general water balance of territories when all balance components (precipitation, evaporation, and river runoff) are determined by direct independent measurements. The basic experimental method to study and assess the processes of groundwater discharge formation in the zone of intensive water exchange is the use of lysimeters. Lysimetric installations enable carrying out direct observations over evaporation from shallow water surfaces and infiltration of precipitation into shallow groundwater, both in natural conditions and during land irrigation. For this purpose, in a vessel (container) filled with natural soil, installed into the earth with the top open to the surface, a certain water level is artificially maintained. By a value of its rise or drop, one determines (taking into account the soil yield) an amount of infiltrated precipitation to this level or, in contrast, an amount of evaporation from this level.

There are many lysimeter designs: one with variable level maintained to correspond to natural conditions; one with an open surface located in the zone of aeration; one equipped with automated devices and transducers for temperature and moisture; ones for hydraulics, weight measuring, etc. In the practice of studying water balance, the most widely used lysimeters in many countries have an area of about 1 m² and a height (or depth from the Earth's surface) of 3–5 m, with water level maintained at a depth of 0.5–3.0 m. By creating various conditions on the lysimeter's surface and in the surrounding environment, one can investigate changes in elements of shallow water balance under the influence of reclamation, different types of vegetation or different states of topsoil, slope of the Earth's surface, etc.

It is known that the amount of infiltration recharge of groundwater depends mainly on two factors: depth of water level and lithology of the zone of aeration. Both evaporation and infiltration decrease with depth. Quantitative excess of one of these processes over a certain time period enables their summation to determine atmospheric recharge of shallow groundwater. However, as multiyear investigations in the Moscow region have shown, even at the maximum identity of conditions and designs of lysimeters, scattering in the results is 10–20%, increasing sometimes to ≥50%. Achieving satisfactory results requires at least double repetition for each experimental variant during a number of years. Only after doing this and having revealed and removed erroneous data is it possible to apply the quantitative regularities determined by the lysimeters to similar natural conditions and, through this, determine atmospheric recharge for a larger territory.

Unique experimental investigations of the shallow groundwater balance have been carried out in the central regions of Russia using lysimeters of the Shchemilovsky experimental polygon of VSEGINGEO (All-Russian Scientific Research Institute for Hydrogeology and Engineering Geology). The specific feature of the Russian central region is the wide spreading of glacial sediments composed of boulder loams and sands, which creates different conditions for groundwater outflow. The territories are low to well drained. In climatic respect, they are referred to as zones of moderate recharge. Typically, the shallow groundwater has a seasonal and chiefly spring and autumn recharge. The temperature conditions are distributed during the year so that they provide steady freezing of the unsaturated zone in winter. The observations were carried out with two or three multiple repeated independent events discontinuously over two decades. The data on groundwater annual average recharge with depths in different areas with different compositions of monoliths are presented in Table 2.1.4.1.

As can be seen from Table 2.1.4.1, the annual recharge at all depths in sands and loamy sands is higher than in loams; lower in open sites than in forest (despite the fact that the precipitation reaching the Earth's surface in forest is lower by 15%); with depth of the groundwater level the recharge amount at first increases, but then reduces; the intra-annual recharge is irregular and reaches the highest value during snow melt. Quantitative recharge values can highly change in different conditions, but the general regularities remain constant.

The specific feature of the areas of the Lower Don River and North Caucasus is their association with the alluvial terraces of the rivers Don, Terek, Kuban, and others, which determines the peculiar conditions of groundwater outflow. In climatic respects, the territory has scarce recharge. The shallow groundwater is characterized by year-round and chiefly winter recharge. The temperature conditions are such that they provide an absence of steady freezing of the unsaturated zone in winter.

The balance components were investigated in several balance-controlled sites of the North Caucasian water-balance station. In a cold period of humid years at a shallow groundwater level depth of 0.8–1.3 m, the infiltration of precipitation in loamy rocks is near to the total precipitation, i.e., the infiltration coefficient is near to 1 and is often equal to 0.95. At a groundwater depth of 1.5–2.3 m the infiltration coefficient reduces to 0.2–0.7. In the years that are, by meteorological conditions, close to the standard at a groundwater depth of 0.9–2.3 m, it amounts to 0.6–0.36, and in dry years –0.1–0.16.

Analysis of the published data shows that at the present lysimeter methods for evaluating groundwater-balance components are widely used in many regions. They are actually the only method of direct determination of recharge and

TABLE 2.1.4.1

Multiyear Average Values of Groundwater Atmospheric Recharge in the Central Part of the Moscow Artesian Basin

Depth of Water Level (m)	Open Area				Forest	
	Sands and Sandy Loams		Loamy Rocks		Sandy Loams	
	Water Layer (mm)	Precipitation (%)	Water Layer (mm)	Precipitation (%)	Water Layer (mm)	Precipitation (%)
1.0	117.7	18.0	10.2	1.6	335.7	62.9
2.0	252.4	40.5	101.4	16.2	458.0	85.9

discharge of groundwater. At the same time, lysimeter-aided investigations require multiyear observations, special equipment, and large expenses for labor-consuming Earth works. As with any method of investigation, this one needs to be controlled and compared with other methods. Among the latter, recommendations include hydrodynamic methods and methods of moisture-transfer assessment in the unsaturated zone — the combination of which considerably increases the reliability and completeness of the evaluation of the groundwater balance.

In water-balance investigations for more fundamental interpretations and extrapolation of data of direct measurements of groundwater recharge (lysimeter data, stationary observations, etc.) as a preliminary step, it is necessary to make zoning of the territory under study according to the conditions of groundwater discharge. Within the distinguished zones (with prevailing groundwater recharge/discharge), it is reasonable to estimate infiltration or losses of groundwater flow in key areas, the natural features of which would reflect all the geologic-hydrogeological and physical-geographic features typical of the given region. Stationary observations in discharge- and balance-controlled sites are of special importance in this respect.

Obtained values of infiltration recharge enable estimation of groundwater discharge only in cases of sufficiently simple hydrogeological conditions, when infiltration is the only source of groundwater formation. In more complicated conditions when, alongside infiltration recharge, vertical seepage from underlying aquifers and lateral inflow/outflow take place, the estimation of particular components of groundwater discharge and balance requires the use of hydrodynamic methods, and first of all, the use of modeling.

Regional estimation of recharge or discharge of relatively deep aquifers located below the zone drained by the local river network can be done through compiling and solving a mean perennial water-balance equation for a river basin or part of a basin. The method, suggested in 1960 (Kudelin 1960), is based on a mean perennial water-balance equation compiled and solved for the groundwater recharge/discharge areas.

For this purpose, at first, the structural-hydrogeological analysis of a territory is carried out to define areas with prevailing groundwater recharge/discharge. Multiyear deep groundwater discharge is determined by Equation (2.1.4.1):

$$\pm W = X - Y - Z, \quad 0 \qquad (2.1.4.1)$$

where X is the multiyear average precipitation, mm/yr; Y is the multiyear average river runoff, mm/yr; Z is the multiyear average evaporation, mm/yr; $\pm W$ is the multiyear average recharge/discharge of aquifers, mm/yr.

The main condition for the possible usage of this method for estimating infiltration recharge of deep groundwater is to know all the basic elements of the water-balance equation (X, Y, Z) determined by methods independent from each other.

The mean perennial water-balance method can be used for estimating groundwater discharge to the sea from deep artesian aquifers having a clearly expressed recharge area. The mean perennial water-balance equation is compiled for the groundwater recharge area on land; and then by a difference among precipitation, evaporation, and river runoff, the infiltration recharge of groundwater discharged to the sea is determined. The method is attractively simple; however, its use is restricted by several circumstances. First, it is suitable for calculating groundwater discharge to the sea only from those aquifers that are surely isolated by proof-beds from over- and underlying aquifers, i.e., there must be certainty that the water discharge from aquifers under estimation reaches the sea, but is not lost through seepage to other layers. Second, the given method can be reliably used only in those cases when deep infiltration (and, hence, also groundwater discharge to the sea) to be estimated exceeds the total error in calculation of the rest of the water-balance elements. Taking into consideration that the accuracy of regional determination of basic elements of the balance — precipitation and especially evaporation — remains presently not high, the second above-mentioned restriction of the water-balance method for calculating groundwater discharge to the sea becomes very important, which often excludes the possibility of its practical usage.

Multiyear average precipitation values are usually calculated based on data of multiyear observations of precipitation at meteorological stations located within territories under estimation. Here, it is desirable that the meteorological stations be placed regularly over the recharge areas to sufficiently and fully characterize their climatic conditions. Then, on the basis of these calculated values, maps of contour lines are compiled on appropriate scales, taking into account the terrain relief. These maps enable determination of multiyear average precipitation amounts for each water-balance area, which are verified by the data of the meteorological stations located within this area.

Multiyear average values of river runoff in the water-balance areas are calculated using the runoff data of the existing hydrometric posts in the recharge area. On the basis of these values, the maps of moduli or of a river runoff layer are compiled. It should be noted that real river runoff values are often mistaken due to an influence of different water-pumping facilities or dams. In such cases, it is reasonable to use the runoff observation data recorded before construction of the water-pumping systems within the drainage areas under estimation.

In situ observations of evaporation are seldom or briefly carried out or are entirely absent in many regions. In this connection, multiyear average evaporation from the land surface in groundwater recharge areas is usually calculated with the aid of different analytical relationships or computer programs, taking into account concrete natural conditions and available initial information. For such calculations, data on radiation balance, temperature, moisture contents of the air and underlying surface, precipitation, and other factors are used. It should be noted that calculated evaporation values include all the errors of the determined initial parameters; therefore, analysis of the latter should be carefully done. For this purpose, all available data of direct evaporation measurements within territories under study and in regions with similar natural and climatic conditions are involved.

The data of directly measured and calculated water-balance elements, as well as the compiled maps of precipitation, river runoff, and evaporation for the recharge areas, make it possible to perform water-balance calculations for each distinguished area. The water-balance residual value, obtained as a result of calculations, represents the summed-up groundwater recharge within each water-balance territory, including errors in the estimated basic elements of water balance. As already mentioned above, in the case when this residual value exceeds the summed-up error of all the estimated elements in the water-balance equation, the obtained recharge values of aquifers can be used for regional estimation of their groundwater discharge. But in order to remove significant errors, the results of fulfilled water-balance calculations requires thorough hydrogeological and hydrological analysis and control.

One of the first works devoted to regional quantitative estimation of groundwater discharge to the sea, based on the water-balance method, was carried out by Dzhamalov (1973) for the Tersko-Kumsky artesian basin. The Tersko-Kumsky artesian basin is located in the East Fore-Caucasian area and covers ~75 ths km². The groundwater of the basin is the basic and often the only source of water supply and irrigation for the settlements, industrial enterprises, pastures, and dry areas.

The natural boundaries of the Tersko-Kumsky artesian basin are: in the south, the piedmonts of Caucasus; in the west, the watershed part of Stavropolskaya Upland; in the north, the Karpinsky Arch; in the east the basin enters within the water area of the Caspian Sea. By relief type, the territory is subdivided into two parts: in the south, the area of piedmonts and frontal ridges; and in north, the Caspian Lowland.

The East Fore-Caucasian territory under consideration is occupied by the river basins of Terek, Kuma, Sulak, and Kalaus. The specific feature of the region is the exclusively irregular distribution of surfacewaters. The piedmont area is characterized by the availability of numerous mountainous streams joining into furious rivers. The plain part of the region is poor in surfacewaters. The major rivers represent typical steppe small streams, usually dried out in summer.

The basic climatic element determining the size of groundwater recharge is precipitation. Distribution of the precipitation shows certain regularity. The highest amount of rain (600–900 mm/yr) falls in the area of the piedmonts and on the Stavropolskaya Upland. Toward the north and east the annual average precipitation gradually decreases; in the central areas it seldom exceeds 300 mm. In the piedmonts the zonality of precipitation distribution is distinctly vertical. There, with an increase of hypsometric elevations from 300–400 m to 900 m, the amount of precipitation increases from 500 mm/yr to 800–900 mm/yr, i.e., the precipitation build-up is equal on average to 60–70 mm/yr per every 100 m of height.

The freshwater and low-mineralized groundwater of the Tersko-Kumsky artesian basin is associated chiefly with the aquifers of Quaternary, Pliocene, and Sarmatian sediments.

The Tersko-Kumsky artesian basin has pronounced recharge areas in the piedmonts of the Major Caucasus and on the eastern slopes of the Stavropolskaya Upland. There exist a great amount of precipitation, coarse-grained sediments, and a well-developed hydrographic network — all this favors the formation of a powerful flow of fresh groundwater recharging the artesian aquifers.

The recharge areas are covered by a rather dense network of meteorological and hydrometric posts and stations possessing a multiyear row of observations. This circumstance enables estimation of the groundwater discharge in the artesian basin using the mean perennial water-balance method.

The calculations carried out for each distinguished water-balance area within the groundwater recharge areas resulted in obtaining the multiyear average values of deep infiltration. It was established that the most intensive recharge of artesian waters occurs within the Kabardinskaya sloped plain where deep infiltration reaches 74–90 mm/yr. This is the area of piedmonts characterized by abundant precipitation and high permeability of alluvial sandy-pebble soils and bedrocks underlying them, contributing to favorable conditions for groundwater recharge, which amounts to $566 \cdot 10^6$ m³/yr.

Low values of deep infiltration equal to 17–34 mm/yr are obtained for the eastern slopes of the Stavropolskaya Upland. There, losses for infiltration are not high because of unfavorable climatic conditions and the occurrence of covering low-permeable loessial loams.

For the entire Tersko-Kumsky artesian basin, the total multiyear average atmospheric recharge for an area of 35,560 km² is $1408 \cdot 10^6$ m³/yr, or 1.4 km³/yr. This water mass forms a deep groundwater discharge within the Quaternary and Neogene rocks of the basin and is distributed between aquifers of these sediments. Intensity of artesian aquifer recharge can be characterized by the recharge modulus that represents a recharge value from a unit of recharge area and is equal for the given artesian basin to 1.3 l/sec·km².

The average modulus of artesian discharge for the entire basin (total area = 75,000 km²) is equal to 0.6 l/sec·km². The total discharge, provided by recharge, amounts on average to 8600 m³/day per 1 km of the flow front at its total length of 450 km.

The maps of piezometric surfaces of the Quaternary and the Apsheronsky district's water-bearing systems show that the groundwater flow in the Tersko-Kumsky and Tersko-Sulaksky interfluves is directed toward the Caspian Sea. Gradual decline of the piezometric levels of the water-bearing systems toward the Caspian Sea indicates the existence of constant groundwater discharge to the sea.

According to the calculations carried out, the total groundwater discharge to the sea from the Quaternary water-bearing system is equal to about 2.8 mln m³/yr at a flow width of 255 km along the calculated contour line. The total groundwater discharge to the sea from the Apsheronsky water-bearing system is equal to 5.7 mln m³/yr at a flow width of 275 km along the calculated contour line. Considerable discharge reaching 2.3 mln m³/yr at a flow width of 83 km is also observed in the Tersko-Sulaksky interfluve. This is caused by the considerable slopes of the piezometric levels (to 0.001) in the coastal zone, and by a higher water conductivity of the rocks in the Apsheronskaya water-bearing system (to 100 m²/day).

In general, the groundwater discharge to the Caspian Sea from the Tersko-Kumsky artesian basin is not great mainly because of groundwater natural discharge via seepage within the land. The groundwater discharge to the Caspian Sea from the entire coastal zone is discussed in detail in Chapter 4.2.

2.1.5   USE OF GEOSPATIAL TECHNOLOGIES IN MAPPING
AND ANALYSIS OF SUBMARINE GROUNDWATER*

The term geospatial technologies refers to a rapidly evolving suite of technologies centered around geographic information systems (GIS), while encompassing global positioning systems (GPS), subsurface visualization, and several technologies for gathering, processing, and classifying aerial photography converted to a digital format, and satellite imagery that is gathered in a digital format. What all these technologies have in common is that they are digital and all have a geographic frame of reference. With GIS, GPS, and aerial imagery, this frame of reference is typically the surface of the earth (which can include the sea surface or scaled representation seabed in the form of bathymetric data). Subsurface visualization frequently incorporates surficial geographic data, but also has data on locations and characteristics of subsurface features such as strata, faces, and properties of geologic units (e.g., porosity, transmissivity), existence of petroleum reservoirs, confined and unconfined aquifers, and aquitards and aquifuges (Raper 1989). In this section the authors intend to describe each of these geospatial technologies to a limited extent, and provide examples of how the technologies are used to map and analyze data related to groundwater systems, marine systems, and submarine data. Examples of the uses of these technologies in the mapping and analysis of submarine groundwater, although still infrequent and nascent at present, will be summarized. The intention of the authors is to provide insight into geospatial technologies; how they are revolutionizing mapping and management of earth and earth resource science-related data; and how these evolving tools can be applied in future investigations of submarine groundwater.

2.1.5.1   GIS in Mapping and Spatial Analysis

Geographic information systems is a term that refers to an evolving suite of computer programs designed to work together as part of a "package." These programs can store, display, analyze, and generate products such as maps or reports or help serve data over the Internet from specially structured sets of spatial data, now increasingly referred to a *geodatabases*. Geodatabases are composed of geographic data (features such as rivers, political boundaries, geologic mapping units, aquifers, seas and oceans, springs, wells, etc.) linked to database tables containing descriptive attribute data (flow of the river, name of the country and its population, mineralogy of the geologic formation, transmissivity of the aquifer, average salinity of the sea, average depth of the ocean, name of the spring, screened interval of the well, etc.) (Burroughs 1986). GIS data are stored in a series of co-registered layers. Typically data are divided into layers by thematic content; thus aquifers would be in one layer, soils data in another, geologic formations in yet another (or stored in multiple layers differentiated by geologic or depth criteria). Rivers, springs, recharge areas, lakes, seas, wells, political boundaries, groundwater model grid lines, and man-made infrastructures (e.g., oil wells, roads, etc.) might all be in separate layers. Some GIS

---

* Section 2.1.5 was written by Mark Liepnik, Ph.D. and Christopher Baldwin, Ph.D. (Sam Houston State
  University, Huntsville, Texas).

systems such as the Port of Rotterdam have as many as 500 layers (ArcNews 2003). A typical groundwater GIS would have less than 20 layers (many of which are listed above). In addition to an explicit link between a geographic feature and a table of attribute data, GIS contains tools for managing coordinate systems, projections, datums, ellipsoids, scale information, and other aspects of cartographic representation. GIS contains many spatial analysis tools. These tools typically include spatial measurement tools (distance, perimeter, area, and in special cases, volume estimating tools), buffer-zone generation tools, map overlay tools, and spatial join tools. There are also tools for extracting information from the attribute data to help select and identify features of interest (Figure 2.1.5.1).

GIS software has existed since the mid 1960s (Foresman 1993), but commercial off-the-shelf software dates to the early 1970s, and recently most users have standardized programs on a very limited number of fully configured multiple-module GIS software packages, of which the *ArcGIS* software from the Environmental Systems Research Institute in Redlands, California is the world leader (www.esri.com). Other software in use by Earth scientists include products like *Geomedia* from Intergraph Corporation in Huntsville, Alabama, and of course specialized *quasi*-GIS software such as *Rockbase, ERDAS, Vulcan, ERMapper* (the latter two packages from Australia), and subsurface visualization software such as Landmark Graphics' *ZGraphics* and Dynamic Graphics's *Earthvision,* finds widespread use for some aspect of subsurface visualization, remote-sensing image analysis, and mapping. It is also worth noting that groundwater and surfacewater modeling software has been closely integrated with GIS, for example the *ArcHydro*

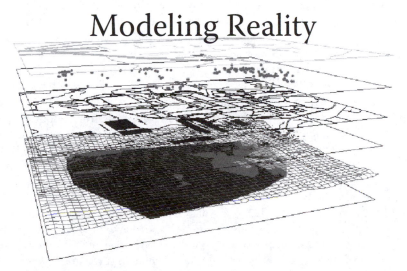

FIGURE 2.1.5.1 Illustration of the layer structure in geographic information systems (GIS). Some layers such as roads, buildings, and wells would be vector format with line, polygon, and point feature types, respectively. The photography, digital elevation model, and groundwater table would be in a raster format. This image was created using ESRI's *ArcGIS* software.

software from ESRI and the *Visual MODFLOW* package from Scientific Software Group in Sandy, Utah.

2.1.5.2   Other Geospatial Technologies

A number of other technologies frequently function synergistically with GIS. An exhaustive list would include global positioning systems (GPS), digital aerial photography and satellite remotely sensed imagery processing software, computer-aided design (CAD) software, laser total station surveying technology, and most recently, LIDAR (light detection and ranging) technology, and 3-dimensional sub-surface mapping and visualization software. Of these technologies, CAD, laser surveying, and LIDAR are of tangential importance in most subsurface mapping and analysis applications and will not be discussed further; while GPS and the processing, classification, and analysis of aerial photography (typically in digital raster format) and remotely sensed satellite imagery are of greater significance.

GPS refers to global positioning systems. It is a technology that relies on a group of satellites in a high orbit (typically 11,000 miles) that are in radio frequency contact with receivers at sea or on the ground. They are able to perform internal computations to determine a receiver's location based on signal time lags and broadcast data from three or more satellites that happen to be in view (Sweet 2003). Receiving line-of-sight signals from additional satellites typically improve the accuracy of the locational estimate. Generally, autonomous GPS receivers can accurately determine the X,Y location of a receiver getting signals from four satellites to within meters to decimeters and elevation to within meters. A host of factors affect and can materially improve the accuracy of positional fixes, and it is possible by using methods such as differential correction and utilization of broadcast base-station corrections to obtain positional accuracies better than 1 centimeter (Leick 2004). Of note to the readers of this book: GPS is frequently the locational method of choice while at sea; and due to the lack of topographic obstructions at sea, GPS often performs better than it does for determination of location on land, which may encounter obstructions (such as buildings and topo-graphic features, and tall trees can block or reflect signals from satellites, partic-ularly those low to the horizon.) The two GPS systems in use today are the United States Defense Department *Navstar* system (24 satellites) and the Russian Defense Ministry *GLONASS* system (15–18 satellites). Recently, the European Union (EU) announced its intention to launch the *Galileo Project*, which envisions a system of 30 satellites. The Russian Republic is expected to cooperate in this system and initial U.S. government attempts to dissuade the EU from developing *Galileo* have been discontinued in favor of grudging support. *Galileo* will significantly enhance GPS accuracy, and this is especially true when used in concert with the increasing number of broadcast base station corrections which adjust for local factors affect-ing positional accuracy estimates (Hesseldahl 2004). The significance of GPS for the current discussion is that GPS is the standard method for providing locational data to all field investigations, particularly those at sea, and is readily incorporated into GIS. Typically, mapping grade GPS units such as those manufactured by Trimble, Garmin, Ashtech, Wild, etc. are capable of storing locational data as points with annotation, as line features such as the track of a ship, or as area features such as the perimeter of a large area of subsurface seeps that are then brought into a GIS as point, line (arc), or polygon (area) features with the corresponding annotation providing attributes stored in linked database tables.

Increasingly, subsurface visualization software is playing an important role in many investigations involving the Earth's subsurface, but this highly specialized and costly software while extremely powerful is not yet universally used nor is it entirely compatible with other spatial analysis and mapping technologies. Where it impinges most directly on the current discussion is that it is most widely used in oil and gas reservoir visualization, and the presence of subsurface offshore groundwater can have a bearing on oil and gas production, although the groundwater aquifers overlying or interbedded with oil and gas reservoirs are of limited direct interest with respect to oil production (Park 2004).

2.1.5.3   Use of GIS in Groundwater Mapping and Modeling

Groundwater resource mapping, analysis and management applications of GIS and other geospatial technologies date back to the early 1980s (Leipnik and Zektser 1994). Large water resource management agencies in the United States, such as the U.S. Bureau of Reclamation and the California Department of Water Resources as well as researchers affiliated with the U.S. Geological Survey (USGS), have been conducting investigations of groundwater occurrence, behavior, and resources using GIS for over 25 years. These investigations typically take the form of using a GIS as a preprocessor and postprocessor for large, finite difference groundwater flow models such as the *MODFLOW* model (Scientific Software Group 2003). A GIS would be used to store input parameters such as topographic contours, thickness of the vadose zone, aquifers, aquitards, and aquifuges and depth to bedrock, boundary conditions, transmissivity of various geologic formations, and location of recharge areas, discharge areas, and wells and springs. Output in the form of pressure heads, water table surface elevations, and contours for cones of depression induced by pumping wells generated by the model would be displayed in the GIS in the context of surface features which might include layers of data on roads, streams and lakes, and jurisdictional boundaries. This last feature, while not physiographic data, is very important nonetheless as regulation of groundwater resources often depends on the location of groundwater resources in relation to national, state, and municipal boundaries. A typical application of *MODFLOW* would be to assess the impact of pumping a well field on regional groundwater behavior. So, for example, were a well field developed in a valley with an alluvial fill and a series of contact- or fault-controlled springs, would extraction of groundwater lower the pressure head (and/or water table) sufficiently to cause diminution or loss of flow in the springs (Johnson 2003)?

Other groundwater applications of GIS include aquifer vulnerability assessment and well head protection applications. In the former case, a method referred to as DRASTIC has been evolved by the U.S. Environmental Protection Agency explicitly based on the GIS capabilities of map overlay (EPA 1992). In DRASTIC, information on key parameters related to the vulnerability of aquifers to anthropogenic contamination are stored in co-registered layers in a GIS. These layers contain information on *d*epth to the permanent water table, net *r*echarge, *a*quifer characters, *s*oil characteristics, *t*opographic factors, *i*mpact of the vadose zone, and hydraulic *c*onductivity. Quantitative scores are assigned to polygonal areas for each of these criteria and through use of Boolean algebra in a map overlay function using GIS software such as *Arc/Info* (now typically *ArcGIS*) would be used. A summation of scores resulting in a series of quantitative vulnerability estimates for multiple areas within a region covered would be generated by this stepwise map overlay. The resulting vulnerability "map" could be used to assess vulnerability of aquifers and limit detrimental uses such as hazardous and soild waste disposal facilities sitting in those areas deemed most vulnerable. These assessments can be made using either vector- or raster-based GIS, such as *ERDAS* along with classified remotely sensed imagery (Napolitano 2002).

Well head protection applications of GIS use the buffer zone generation function of GIS along with spatial joints rather than the more typical map overlay (Hendricks 1992). Well head protection usually takes the location of a public water supply well and generates a buffer zone around that well, the buffer zone polygon (typically a circle with a radius of say 1000 m can then be compared to locations of features stored in other GIS layers that may potentially pose a source of concern). Potential sources of pollution could include underground storage tanks, hazardous and solid waste disposal facilities, mines, factories, septic tanks, and sewage treatment plants. Occasionally, other features such as locations where endangered species are present or where subsidence is occurring might be included in this proximity-based spatial analysis.

Groundwater-related GIS projects have occasionally involved assessment of interaction with marine environments. These assessments have typically involved the assessment of saltwater intrusion (Heath 1989). Saltwater intrusion typically occurs when wells situated in transmissive coastal plain alluvial sediments are over-drafted; as a cone of depression forms in the shallow lower-density freshwater an up-coning and intrusion of the underlying saline water occurs. Because of the differences in specific gravity, the cone of impression in the underlying saline water is significantly greater than the cone of depression in the overlying freshwater (Freeze and Cherry 1989). This phenomenon is termed saltwater intrusion, but might be better thought of as an upwelling rather than drawing inland of a front of saline water toward a well. In any event, this process is of concern in various areas, specifically the Mid-Atlantic coastal plain of the United States and parts of Southern California, and undoubtedly in other coastal areas where groundwater overdraft is occurring. In a sense this is an anthropogenic inverse process to the natural phenomena of groundwater discharge from seeps in the ocean floor.

GIS has been used along with finite difference and finite element groundwater flow models to understand the behavior of saltwater intrusion and to prevent its effects. An example of such studies includes work by the Orange County, California Water District to inject treated waste water in a line of large-diameter wells along the coast to impose a hydrodynamic barrier of nonpotable "fresh" water between the marine subsurface waters and inland areas with active well fields with deep cones of depression induced in even deeper alluvial basins (EPA 1999). A different example of use of GIS in indirect study of marine groundwater involves modeling of subsidence in the Venice, Italy region. The long-term pumping of wells on the mainland of Italy has caused subsidence, and given rising sea levels and the limited "freeboard" of historic Venice, any further subsidence is a threat. GIS-based modeling has helped to persuade the authorities to discontinue use of wells on the mainland (Strozzi 2003). Analogously, in the Houston, Texas region subsidence due to groundwater overdraft has induced several meters of subsidence from now-discontinued wells. Unfortunately rebound even after the aquifers have fully repressurized is only a small fraction of the essentially irreversible subsidence. Thus both saltwater intrusion and coastal subsidence induced by groundwater overdraft are areas where submarine groundwater comes into play and where GIS has been used along with groundwater flow models to good effect (Kasmarek and Robinson 2003).

2.1.5.4 Study of Inland Seas with GIS

Another application of GIS with implications for the study of submarine groundwater has been the assessment of inland seas. Examples of inland seas include the saline Caspian Sea, the Great Salt Lake (Utah), the Dead Sea, and the Mono Lake (California), as well as the brackish Aral Sea and Salton Sea (California), and the fresher waters of the various Rift Valley lakes in Africa, the chain of rapidly receding lakes in Nevada (Walker, Pyramid, Topaz, etc.), and last but not least the fresh waters of Lake Baikal in Siberia (which contains approximately 19% of all the fresh liquid surfacewater on Earth) (International Geological Congress 1984). These "inland seas" have fascinated hydrologists for theoretical and practical reasons. Most notably these inland seas are of interest because of their unusual water chemistry, ecological significance, frequently varying water levels, and (usually) increasing salinity. Many detailed studies of these lakes have been performed, and for the past 20 years GIS has played a central role in these investigations. The book *Salton Sea Atlas* is the best exemple of this type of study (Redlands Institute 2001). Other studies using GIS have involved the Great Salt Lake, Mono Lake, and the Aral Sea (Ressl 2001). In many of these studies the role of subsurface groundwater discharges to these bodies of water has been recognized, although typically only after the study of water balance has indicated discrepancies between surfacewater inflows, potential evapotranspiration, and observed water level changes (or rather, in many cases, lack of change). Many inland seas must have significant groundwater contributions to their water balance since surfacewater inflows are often inadequate to explain the observed volume of water and its persistence in the face of shrinking inflows due to the common practice of upstream diversions, mostly for agriculture, although in the United States also for municipal consumption in arid states such as California and Nevada (Anthoney 1998).

In some areas such as Mono Lake in southeastern California, visible clues that groundwater discharges to these inland seas were occurring was demonstrated by the presence of "Tufa Towers." These sometimes 20-m high, calcium carbonate-dominated mounds are evidence of springs locally changing the water chemistry on the lake bed. Falling lake levels have left them easily observable, in contrast to most seepage of groundwater into inland seas which is imperceptible. In other places such as the bed of the now totally diverted and largely desiccated Owens Lake in California, meadows and springs in formerly underwater areas show that springs were present before the lake dried up as a result of the total diversion of the Owens River for water supply to Los Angeles. In other inland seas that have not retreated — the Caspian Sea and the Great Salt Lake have actually seen rising water levels — the impact of groundwater inflows is less visible. Thus it is usually inferred from water-balance studies; and since potential evaporation is problematic, so too is groundwater contribution, which is the last term inferred in the water-balance equation. GIS can help to better understand the role of groundwater by mapping areas where aquifers intersect the lake bed or where sublacustrine springs are present.

2.1.5.5 Undersea GIS

The book *Undersea with GIS* (Wright 2002) summarizes the current advanced state of applications of GIS and related technologies in the submarine realm. Interest in using GIS in such applications is understandable, given the great expansion in bathymetric data now available along with the growing hunt for subsurface resources and extension of territorial waters into areas of the deep continental shelf. There are a wide variety of submarine applications for GIS, but the following sampling gives the flavor of the available data, techniques, and products. Perhaps the most important application involves not submarine groundwater per se, but where the liquid filling the porous submarine formations is oil (Figure 2.1.5.2). Offshore sedimentary basins offer one of the few areas for potential super giant oil fields and involve the investment of tens or even hundreds of billions of dollars annually on the part of the world's energy companies. GIS plays a role in all aspects of subsurface oil and gas exploration and production, from helping to identify lease boundaries using GIS and GPS to analyzing subsurface seismic and drill ship data to studying seafloor structures for platform and pipeline installation and integration of GIS with subsurface visualization for targeting directional drilling. Given the magnitude of the investments involved, use of state-of-the-art integration of a vast array of technologies of which the geospatial ones play a keystone role is understandable (Low and Riemersma 2004).

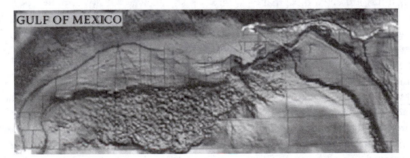

FIGURE 2.1.5.2 The use of geographic information systems in offshore oil and gas exploration and production is illustrated by this digital elevation model of bathymetric data with an added layer of lease boundaries for the upper Gulf of Mexico.

Although offshore oil and gas production is undoubtedly the most common and most sophisticated example of submarine GIS, there are many other applications. One is literally *submarine* in character since it involves use of bathymetric data along with GPS and sonar to guide submarines. Both research submersibles and military ballistic missile and hunter-killer submarines use geospatial technologies extensively. Given the secrecy surrounding military applications, the best-known examples of submersible use of geospatial technologies has been in regard to the use of submersible vehicles tethered to surface ships and side-looking sonar in the discovery of lost wrecks, such as the *Titanic*, *Bismarck*, and *Central America*. A related area of undersea GIS application has been in underwater archeology, where a coordinate grid is often imposed to track locations of artifacts; this may be coupled with mosaics of undersea photographic imagery, subsurface sonar data, and other information inside a GIS (Cooper et al. 2002).

In addition to assisting in the discovery of lost ships and other examples of underwater archaeology, GIS has been used in the study of submarine landslides and canyons (e.g., Monterey canyon) and the study of ancient drowned shorelines (e.g., in the Black Sea region) (Wong et al. 1999). In this situation, bathymetric data and information about submarine sediments and their location and thickness are critcally important. In a more regional context, GIS along with seismic data and geologic models has recently been applied to the study of offshore earth-quake faults such as that responsible for the tsunami in the Indian Ocean (Rennie 2005).

2.1.5.6   Investigations of Submarine Groundwater Discharge

The discharge of freshwater to the oceans from both unconfined and confined aquifers, loosely referred to as SGD, has been the subject of intense study for some time and gains in importance as greater levels of quantification and modeling take place related to this subject (for a comprehensive recent review see Taniguchi et al. 2002). Both individual researcher/aquifer and multinational/multiaquifer studies are currently active and are in process of being synthesized and method-ologically correlated under the rubric of, for example, the SCOR Working Group 112 Program ("Magnitude of Submarine Groundwater Discharge and Its Influence on Coastal Oceanographic Processes") and the linked LOICZ (Land-Ocean Inter-action in the Coastal Zone, a program of the International Geosphere Biosphere Program) (e.g., Burnett 1999; Kontar and Zektser 1999). These authoritative groups have resolved that three linked topics should form the focus of their work and comprises (i) modeling and calculation that is dependent on quantified field data sets; (ii) direct measurement of water fluxes; and (iii) typology and global-ization wherein regional scale knowledge of geological, hydrological, and ocean-ographic parameters that influence or control SGDs may be known and may be extrapolated with confidence to all relevant coastal sites (Figure 2.1.5.3). It is this latter area of interest that makes use of a developing set of remote-sensing techniques.

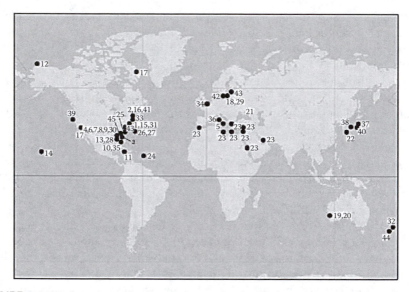

FIGURE 2.1.5.3 Locations of submarine groundwater discharge research sites around the world. Numbers correspond to: (1, 15, 31) Chesapeake Bay, MD; (2) Great South Bay, NY; (3) Crescent, FL; (4–9, 30) NE Gulf of Mexico; (10, 35) Florida Keys and Florida Bay; (11) Discovery Bay, Jamaica; (12) North Slope, AK; (13) West Florida; (14) Hawaii; (16) Cape Cod, MA; (17) Baffin Island, Canada; (18, 29) Puck Bay, Poland; (19, 20) Perth, Australia; (21) Black Sea, Crimea; (22) Beppu, Japan; (23) Mediterranean Sea; (24) Barbados, WI; (25) East coast Florida; (26) Coastal Georgia; (27) Coastal South Carolina; (28) West coast Florida; (32) Hawke Bay, New Zealand; (34) Plymouth, U.K.; (35) Wilmington, NC; (36) Capo Palinauto, Italy; (37) Tokyo Bay, Japan; (38) South of Osaka Bay, Japan; (39) Coastal Oregon; (40) Sagami Bay, Japan; (41) Coastal bays of Northeast U.S.; (42) Laholm Bay, Sweden; (43) Saltmarsh estuaries, South Carolina; (44) Golden Bay, New Zealand; (45) Indian River lagoon, Florida.

2.1.5.7   Future Directions for Research

In a field as rapidly evolving as geospatial technologies it is easy to make predictions about future technological and research directions, have those predictions prove true, and then be passé before a book describing them appears in print. With that caveat in mind we will try to make a few predictions about possible future directions for valuable research involving both undersea groundwater and GIS and related technologies.

Three areas of possible application of geospatial technologies are of significant potential. One is trying to take the local and limited regional groundwater flux studies (seeps) that have been performed and using the capability of GIS to manipulate global databases to estimate global groundwater contributions to sea levels and water chemistry and circulation patterns. At the moment we appear to be significantly undercounting this important component of the total freshwater flux to the oceans. A second potential area for research would involve the identification of offshore groundwater aquifers and their potential to provide beneficial water resources in arid areas. Many of these aquifers may contain saline or brackish (often over briny) water, but in areas like the Persian Gulf they may still represent untapped resources that are of an order less saline than ocean waters currently used to provide potable water in selected areas, typically through flash distillation using otherwise flared natural gas. As liquification and re-injection of natural gas becomes more common as development in areas such as Arabia continues to mine existing onshore aquifers, the potential of offshore aquifers should not be ignored. And last, as oil and gas production in offshore basins continues to grow and concerns over the environmental consequences to the marine environment of such development also continues to grow apace, greater interest in the location and behavior of subsea aquifers that may generate brines or may be contaminated with hydrocarbons that then seep out of the seabed may arise from the oil and gas industry. In all these situations geospatial technologies will prove not only helpful but essential to a more integrated and more precise and truly global understanding of the roles of submarine groundwater.

2.2 METHODS OF MARINE HYDROGEOLOGICAL INVESTIGATIONS

2.2.1 GENERAL REMARKS

As mentioned above, groundwater discharge into the sea causes different anomalies in seawater and sea bottom sediments. Studying the anomalies makes it possible to reveal and often delineate the sources of groundwater discharge and to give its quantitative assessment. Therefore, marine hydrogeologic studies include a wide complex of visual, remote sensing, geophysical, geochemical, and other works aimed at studying different anomalies directly in the sea (marinewater and bottom sediments), caused by groundwater discharge. Thus, marine hydrogeologic studies (more exactly "hydrogeologic studies in the sea") are aimed at studying hydrogeological interaction between groundwater and marinewater and underground exchange between the land and sea.

The main methods for marine investigations aimed at revealing, studying, quantitative assessing, and mapping submarine groundwater are given in Figure 2.2.1.1. These methods can be subdivided into (1) remote and visual (cosmic survey, aerial survey, visual observations; (2) methods of directly inspecting submarine springs (investigation by bathyspheres, divers, indicator methods, and flow metering); (3) methods of studying the near-the-bottom water layer (determining anomalies in chemical, gas, and isotope composition, electric conductivity, and temperature); and (4) methods of studying bottom sediments, including filtration properties through the use of flow meters of different construction, seismic-acoustic, thermal-electric profiling of the bottom–water boundary, studying pore water chemistry, isotope and gas composition, etc. It is not possible to describe each method in detail (the latter is given in the special literature); therefore, only a short description and examples of applying certain of the most widely used methods for studying SGD, which allows groundwater inflow into the sea to be quantitatively assessed, will be presented here.

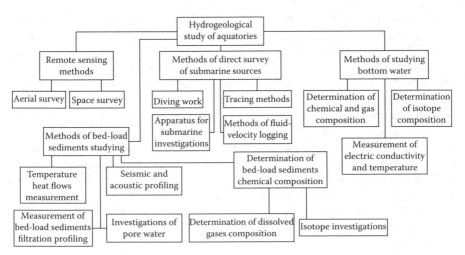

FIGURE 2.2.1.1 Methods of hydrogeologic investigations of water area of seas and lakes.

The best perspective for studying SGD is to use remote sensing methods, primarily multispectral and infrared surveys of the sea's surface from a plane or spacecraft. Multispectral surveys use highly sensitive radiation in different spectrums. The survey is based on the differences in the spectral reflections of sunlight by different objects. This method makes it possible to get maximum light contrast when interpreting elements of the sea surface using corresponding light filters.

The possibility of using remote sensing methods for studying groundwater discharge into the sea is based on the fact that submarine groundwater springs cause changes in the main marinewater parameters (color, transparency, temperature, water surface, structure) as recorded on aerial photographs. Large-scale aerial photographs make it possible to study peculiarities of geologic and geomorphologic structure of the coastal sea and land in detail, and to single out faults and intensively fissured zones, from which large submarine springs are usually confined. A change of marinewater color or transparency caused by submarine springs is fixed on aerial photographs by more or less essential changes of a photo tone (shade) in a certain site that indicates a submarine spring. In some areas, groundwater springs are so abundant and their areas so wide (for example, in the Al-Marj area of the Mediterranean Sea, where submarine springs are observed along the shore at a distance of 200 km) that anomalies caused by them are fixed even on satellite photographs.

The first U.S. satellite launched with the aim of studying resources of the Earth carried a multispectral scanning device for registering electromagnetic radiation in four wavelength ranges: green and two closer to infrared. These ranges were not chosen at random, but were selected for different purposes. In particular, marinewater is transparent in the "green" range, and therefore, bottom relief, turbidity, and different admixtures can be seen in high-altitude photographs. Thus, large karst springs, where the water usually contains air particles or bubbles, can be detected. Similar springs can be observed in aerial photographs of the coast of Jamaica, where upward karst water flows from turbid ovals on the sea surface. Multispectral surveys were used by the USGS to study submarine springs in Jamaica, Sicily, and Hawaii.

Infrared surveys are based on measuring the intensity of different natural sur-faces' thermal radiation, including the sea surface, in the infrared spectrum of electromagnetic waves. Using present-day infrared radiometers, it is possible to measure temperature changes on the sea surface with the accuracy of a 10th to even a 100th part of a degree by artificial satellites above the Earth.

The infrared survey is very effective for detecting the center of SGD. The temperatures of groundwater and marinewater usually differ. Submarine groundwater discharge is observed by typical contrasts in the form of tails ("trains") in photo-graphs obtained by an infrared survey. It should be noted that the use of infrared surveys for mapping the centers of groundwater discharge in seas is possible only if the temperature anomalies caused by submarine discharge reach the sea surface and differ from the surrounding marinewater temperature by a value exceeding the radiometer's sensitivity. Positive or negative temperature anomalies, formed due to groundwater rising to the surface, are clearly indicated in thermal photographs by a corresponding change in the tone.

Infrared surveys were carried out when studying water resources in Long Island, New York, and, in particular, for determining the places of groundwater discharge in Long Island Sound. Here, the summer temperature of marinewater is about 20°C and the groundwater is 10–15°C. Due to this range, submarine discharge is distinctly observed along the shoreline by dark, hazy areas in the summer photographs.

Aside from detecting submarine springs, infrared surveys allow for a number of other important data to be obtained. Thus, submarine spring depth is approximately calculated, based on the area of the surface spot. The temperature of a spring can be determined by thermal photographs. Observations of both separate sites and large water areas are carried out by repeated surveys (particularly satellite imagery which does not require additional costs, is highly periodic, and covers wide territories).

Investigation of groundwater discharge by complex remote-sensing methods was carried out at Big Quill Lake, a saline lake in Saskatchewan, Canada. Nine centers of groundwater discharge with a total area of 4 km^2 have been discovered on the lake bottom by remote soundings, including aerial exploration by a reflected-wave method, cosmic surveys in four ranges from the *Landsat 1* satellite, and infrared aerial surveys. The combined use of remote-sensing methods in this area made it possible to map with great accuracy the subaqueous submarine discharge and to delineate the anomalies caused by groundwater discharge and flows in the lake.

Since ancient times, methods of determining subaqueous submarine discharge have been known. Vigorous concentrated submarine springs cause different changes on the sea surface. Water domes have been noted where there is "boiling" of the sea surface, which causes a change of marinewater color because of the gas bubbles. A change of color in the area of a submarine spring discharge can be caused by turbidity and gas bubbles. Gas bubbles at the sea surface are caused by discharging water of a different origin. In karst areas, siphon springs can entrap air as they flow through the land. Discharge along the deep faults can contain water carrying meth-ane, hydrogen sulfide, and carbon dioxide. Certain thermal springs carry gases of magmatic origin.

A change of marine water color can be caused by different chemical reactions occurring under groundwater discharge on the sea bottom. A spring was discovered by a change of water color on fumarole fields of the Bartnu-Vukhu volcano on the Indonesian coast. The water there is red because of the oxidation of iron brought by thermal submarine springs from considerable depth, while springs located at a shallow depth can be seen mainly in the coastal sea. Even effective springs can remain unnoticed if the depth of their discharge exceeds some tens of meters. Karst-spring discharge is subjected to fundamental annual fluctuations. In a dry season, their concentration decreases considerably and they become less visible.

When submarine springs are studied, immediate subaqueous observations using light diving equipment are very important. When hydrogeologic subaqueous surveys are conducted, such methods as morphometric description and application of different colorings and indicators are widely used. These methods include radioactive, chemical, and thermal sampling by different current and flow meters, etc. of marine-water along cross-sections and profiles, and filming, photographing, and measuring of a submarine spring discharge and its hydrodynamic parameters. However, sub-aqueous hydrogeological investigations have some essential restrictions. The main drawback is the maximum permissible depth of diving in light diving equipment (40 m) when using compressed air (with an increase in diving depth, the time of a diver's stay under water decreases).

Investigations with light diving equipment were made along the Black Sea shelf in the area of the Gagry group of submarine springs with the purpose of studying the mechanism of groundwater discharge through karst springs on the sea bottom. A detailed study of submarine springs using diving equipment was made near the southern coast of France off Marseille. Divers took samples of water and rock, and made morphological observations of submarine springs. They managed to move far along the karst channel and take subaqueous photographs. Near the coast of Greece in the Argolicos Gulf, 400 m from shore, divers managed to obtain water samples out of the submarine spring Aqualos at a depth of 72 m below sea level. The spring, a narrow fracture, was described, and its discharge measured.

Description of hydrogeological investigations using diving equipment will not be complete without further experience in measuring submarine spring debit and head by piezometers and flow meters of different construction. By determining the head of groundwater, discharge through bottom sediments was made by special piezometers in Utah Lake, where the locations of groundwater discharge into the lake water were discovered by an infrared survey from a plane at 1100 m. These sites closely coincide with anomalies of sources of sodium, magnesium, and potassium. Visual observations showed a triangular ice-free space with a base of 5 km along a coastal line in the discharge area. The triangular apex is at a distance of 3 km from the shore. Further, visual observations detected a great number of turbid swirlings 3 m in diameter, which are caused by groundwater discharge on the bottom. At the site of a real groundwater discharge, discovered on the lake bottom by infrared survey and aerovisual observations, a needle filter, equipped with a piezometer, is forced to a depth of 9 m. It should be noted that in wells bored on the coast near the water line, groundwater level is above the level of the lake; and in piezometers in the lake water area where the anomaly takes place, the level reaches a height of 65 m above lake level. This allows researchers to affirm that a discharge of confined groundwater is occurring out of the deep aquifers. When investigating Utah Lake, a combined set of methods for studying subaqueous groundwater discharge was used, including infrared surveys from a plane, aerovisual observation, hydrochemical investigation of lacustrine water, and direct measurements of groundwater head on the lake bottom.

Flow meters can be used for assessing groundwater flow filtration through bottom sediments. Flow meter construction consists of enclosing a part of the bottom by a cylinder of a certain diameter, thus isolating a circle through which the sub-aqueoous discharge is measured. The open part of the cylinder is forced into the bottom sediments. The closed part of the cylinder, which has a small outlet, is connected to flow meters of different construction. By measuring the volume of water passing through a flow meter at a time unit from a certain circle of the bottom, a module of groundwater vertical discharge is obtained. By equipping this module with simple electronic devices, it is possible to make long-term regime observations. Using similar flow meters, subaqueous groundwater discharges were studied in Lake Michigan, glacial Lake Sally (Minnesota), the Gulf of Mexico (near Florida), and Lake Tanpo (New Zealand).

Studying groundwater springs at considerable depth is possible by using bathy-scaphes, which make underwater photography and other observations possible over long time periods at any depth. Investigations made in the U.S. *Aluminaut* bathy-scaphe of the Florida shelf have revealed subaqueous freshwater and saltwater discharges and permitted their conditions to be determined. Investigations of the formation of hot brines in the Red Sea bottom and deep areas of Lake Baikal were carried out by Russian scientists in the *Pisis* bathyscaphe.

From the arsenal of methods for studying groundwater submarine discharge, different indicators are used. The method discussed here makes it possible to deter-mine the location of spring discharge conditions and, in some cases, to determine submarine spring loss by dilution of the different indicators in seawater in a spring recharge area or karst hollow. Color-indicator materials, such as fluorescein, colored spores, different isotopes, and fresh spring water, were used to dilute the marinewater.

To map and determine the discharge of subaqueous groundwater springs, radio-active indicators are used. Serious research into submarine groundwater discharge into the seas by isotope methods is being carried out in the United States at present (Moore 1996; Cable et al. 1996a; Kontar and Burnett 1998). The technique of using seepage meters (Cable et al. 1997a,b) for determining groundwater filtration rates through marine bottom sediments (Cherkoner and Hensel, 1986) was successively improved by American specialists.

The measurement of electric conductivity is widely used as a method to study anomalies in marinewater composition and properties. This method is based on the relationship between dissolved solids content (salinization), chemical compo-sition, and the water's electrical resistance. Marinewater-specific resistance along profiles at different depths is measured by resistance meters used to detect ground-water springs in the sea. Temperature is registered simultaneously with measuring resistance based on the calculated water resistivity and salinity. Maps of marine-water salinity distribution are compiled, and the locations of submarine springs are determined by isoline configuration. The clearest results are obtained for highly concentrated karst springs.

In the detection of submarine springs, and, in some cases, for preliminary assessment of their water loss, anomalies of marinewater chemical composition are studied along profiles. These investigations are based on the essential difference between dissolved solids content and chemical composition of marinewater and groundwater.

Samples of marinewater for chemical analysis are usually taken along parallel profiles near the bottom, at the water table (surface), and in the intermediate depth. Similar methods were used by Buachidze and Meliva (1967) when they studied karst submarine springs near the Caucasus coast of the Black Sea. As a result, 24 hydro-chemical anomalies interrelated with submarine springs were discovered. The maximum depth at which the anomalous chloride ion content registered as different from the normal seawater by 5 g/l, was 400 m.

Similar surveys were carried out by different researchers in the Mediterranean Sea, along the Florida coast, and in other areas. It is very convenient to use automatic analyzers onboard a research ship that will allow the marine water to be continuously determined along a profile.

In recent years, the analysis of isotope composition of marine water has been used for studying submarine groundwater discharge. Changes in concentration of tritium, deuterium, radiocarbon, and oxygen in submarine springs enable opinions on their origin, recharge sources, water flow rate, and water-exchange terms. In recent years, more than one attempt has been made to estimate quantitatively SGD, using different geophysical methods and in particular those based on measurement of electrokinetic and temperature effects caused by groundwater filtration into the sea through bottom sediments.

An attempt to evaluate the velocity of submarine groundwater vertical filtration by means of measuring the filtration electrical field in the Sea of Japan and Barents Sea was undertaken by Korotaev et al. (1980). The essence of the method follows that during movement, groundwater, filtrating through capillaries in bottom sediments, carries with itself electrical charges. The movement of the charges is a convective current that creates intensity in the electromagnetic field. Comparing the intensity of the electromagnetic field with the filtration velocity, the authors determined groundwater flow through the bottom sediments. However, any parameter in natural conditions can be measured until a limit of the relationship between power of useful signal and noise is exceeded. The measurements cease to be representative when noise exceeds the useful signal. Seawater is a good conductor and moving in the magnetic field of the Earth, induces currents and fields that create noise in studying the electrokinetic effects of groundwater filtration through bottom sediments. Wave mixing is one type of such groundwater movement. As reported in Leibo (1977), generation of a electromagnetic field during wave mixing can be caused by the migration of suspended particles, movement of liquids in the upper layer of unconsolidated bottom sediments due to wave-induced changes in the pressure, and by water transport along the bottom. In this case, electrical fields can appear with an intensity of about 0.5 mB/m. In Leibo (1977), the maximum measured potential is 1.3 mB per each 19 m, i.e., the field intensity is equal to 0.068 mB/m. Thus, in the case presented the noise can be 7 times higher than the useful signal.

Noises are limited not only to hydrodynamic sources. Electrical fields of the ionosphere, the intensity of which considerably exceeds that of electrical fields induced by SGD, adds to the problem. Thus, evaluation of SGD during slipage through bottom sediments using electrical field induced by the filtration flow is not a recommended method. Lyalko et al. (1978) have suggested using temperature data to evaluate SGD. The approach is based on analyzing the balance of heat flux in the near-bottom layer through the sea bottom. This was the first attempt to use thermometric data for quantitative estimation of SGD.

As mentioned above, directly measuring submarine groundwater in the sea cannot be accomplished by traditional hydrogeological methods and equipment used on land. Therefore, in studying the interaction between groundwater and seawater it is necessary to use indirect indications of SGD, namely, to obtain more complete information on: the physical and chemical fields in bottom sediments and near-bottom seawater, the changes under the action of groundwater bottom discharge, and the geological cross-section of bottom sediments. Section 2.2.3 discusses concrete examples showing the experience of studying anomalies in the composition and properties of near-bottom seawater and sea bottom caused by SGD.

2.2.2   Direct Measurements of Submarine Groundwater Discharge Using Seepage Meters*

2.2.2.1   Introduction and Historical Perspective

The transport of groundwater into coastal zones is a significant process in the geochemical, nutrient, and carbon budgets of many marine nearshore waters (Valiela et al. 1978; Valiela and Teal 1979; Capone and Bautista 1985; Lapointe and O'Connell 1989; Capone and Slater 1990; Valiela et al. 1990; Simmons 1992; Moore and Church 1996). The quantification of SGD has been approached by several different methods, including tracer studies (Moore 1996; Rama and Moore 1996; Cable et al. 1996a,b; Moore and Shaw 1998; Corbett et al. 1999; Hussain et al. 1999; Corbett et al. 2000a; Charette et al. 2003), hydrogeologic methods (e.g., mass balance approaches) (Corbett et al. 1997), and numerical modeling (Chan and Mohsen 1992; Li and Jiao 2003; Wilson 2003). However, there still remains only one direct approach to measure the rate and quality (i.e., nutrients, dissolved organic carbon, etc.) of fluid flux in and/or out of the sediments — the seepage meter. Seepage meters were initially introduced as a viable field measurement tool for estimating fluid discharge from canal sediments by Israelson and Reeve (1944). Lee (1977) improved this device and demonstrated its utility in lakes and coastal marine systems. Since 1977, seepage meters have been used in marshes, lagoons, bays, rivers, and lakes with varying degrees of success (e.g., Shaw and Prepas 1989; Belanger and Walker 1990; Shaw and Prepas 1990; Belanger and Montgomery 1992; Reay et al. 1992; Simmons 1992; Libelo and MacIntyre 1994; Cable et al. 1996b, 1997a,b; Gallagher

---

* Section 2.2.2 was written by Reide Corbett, Ph.D. (East Carolina University, Greenville, North Carolina corbettd@mail.ecu), Jaye E. Cable, Ph.D. (Louisiana State University, Baton Rouge), and Jonathan B. Martin, Ph.D. (University of Florida, Gainesville).

et al. 1996; Corbett et al. 1999; Martin et al. 2000; Chanton et al. 2003; and many others).

The seepage meter is still an invaluable tool in the arsenal of methods for evaluating the importance of groundwater discharge to water and nutrient budgets in coastal systems. Many modifications and improvements have been made since the design developed by Lee (1977) for coastal systems. In addition, studies over the past two decades have helped constrain the environments best suited for the use of the instrument. This review summarizes the current knowledge of seepage meters to date, including the different types of meters actively used, advantages and disadvantages to these different designs, and environmental factors affecting the use of seepage meters.

2.2.2.2   Review of Seepage Meter Styles

Since Lee's design (1977), researchers have sought to improve approaches for direct measurements of seepage discharge from sediments. Automated meters have been designed that use several different technologies, including heat pulse (Taniguchi and Fuko 1993; Krupa et al. 1998), acoustic (Paulsen et al. 2001), dye dilution (Sholk-ovitz et al. 2003), and an electromagnetic flow meter (Rosenberry and Morin 2004) to improve measurement accuracy, increase sampling frequency, and reduce labor. The following is a short summary of the different designs and their advantages and potential disadvantages (Table 2.2.2.1).

TABLE 2.2.2.1
Types of Meters, Costs, Benefits, Drawbacks, and Inventors

Meter Type (Inventor)	Approximate Price	Benefits	Drawbacks
Lee-Type (Lee 1971)	<$100	Low cost, ease of use	Labor intensive
Heat Pulse (Taniguchi and Fuko 1993; Krupa et al. 1998)	~$1,100	Autonomous, off-the-shelf technology, easily integrated to different collection devices	Complex system, not commercially available
Ultrasonic (Paulsen et al. 2001)	~$5000	Autonomous, low power, well-developed technology, bi-directional flow, high sampling frequency	Not submersible; if submersible then low detection
Dye Dilution (Sholkovitz et al., 2003)	~$18,000	Autonomous, off-the-shelf equipment, bi-directional flow, high sampling frequency	Cost; complicated construction and initial standardization
Electromagnetic (Rosenberry and Morin, 2004)	$10,000	Autonomous, off-the-shelf electronics, bi-directional flow, high sampling frequency	Cost; power/data storage requirements, detection limits

2.2.2.3   Lee-Type Seepage Meter

The Lee-type meter is a simple and inexpensive method to measure seepage directly at the sediment–water interface. The meter consists of the top or bottom section (cut ~15 cm from either end) of a standard 55-gallon drum with an open port placed near the rim. The meter is placed open-end down in the sediments, typically at a slight angle with the port upslope (Figure 2.2.2.1a). A plastic bag is attached to the port to measure the change in water volume over a known time and area. The type of plastic collection bag used with this meter has varied in size from a latex condom (250 ml) (Isiorho and Meyer 1999) to a 565-L (30-cm radius, 2-m length) plastic bag (Shinn et al. 2002). However, most studies utilizing the Lee-type seepage meter have used 4-L plastic bags (Shaw and Prepas 1989, 1990; Cable et al. 1997a,b; Corbett et al. 1999). This measurement technique is an attractive tool because of its simplicity and ease of use. However, this technique is fairly labor intensive, as bags must be manually added and removed and volumes measured. The amount of time the collection bags must be left on the instrument in order to collect enough water for an accurate measurement is dependent on the seepage rate, so some preliminary work may be required. Therefore, this time-integrated measurement may prevent interpretation of temporal variations in the seepage rate.

2.2.2.4   Heat-Pulse Meter

Taniguchi and Fuko (1993) designed a seepage meter that uses a string of thermistors in a column positioned above an inverted funnel covering a known area of sediment (Figure 2.2.2.1b). The string of thermistors has more recently been fitted to a traditional Lee-type meter. The basis of the method is the measurement of travel time of a heat pulse generated within the column by a nichrome wire induction heater. Since heat is a conservative property, the travel time is a function of the advective velocity of the water flowing through the column. Thus, once the system is calibrated, measurements of seepage flow can be made automatically on a near-continuous basis. In addition, this meter can measure both outflow (positive seepage flux) and inflow (negative seepage flux) at a fairly high frequency (minutes) over several days.

2.2.2.5   Krupa Seep Meter

The automated Krupa seep meter system measures both seepage (heat-pulse meter) and the water quality of SGD and surfacewater. The system consists of two parts: the seepage meters and an onshore steel shelter (Figure 2.2.2.1c). Bi-directional flow and volume between the aquifer and surfacewater (seepage and recharge) are measured with heat-pulse flow meters (Krupa et al. 1998). Water quality sensors measure the temperature, specific conductance, salinity, pH, and dissolved oxygen of the water both inside and outside the dome. This sensor also measures the depth of the water above the sensor to record tidal fluctuations. One dome also has a surfacewater velocity sensor to measure the

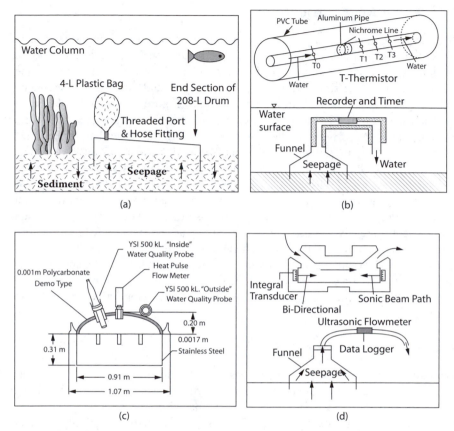

FIGURE 2.2.2.1 Variations in manual and automatic seepage meters: (a) Lee-type seepage meter; (b) heat-pulse meter (modified from Taniguchi and Fuko 1993); (c) Krupa Seep meter (image courtesy of Steve Krupa); (d) ultrasonic meter (modified from Paulsen et al. 2001); (e) dye-dilution meter (modified from Sholkovitz et al. 2003); (f) electromagnetic seepage meter. (Photo courtesy of Don Rosenberry.)

velocity of the surfacewater passing over the dome (perpendicular to the beach) and a photosynthetically active radiation sensor to measure the amount of light near the sediment–water interface.

The onshore shelter contains a weather station as well as computers to store data and power for the sensors. Meteorological parameters recorded are wind speed, gusts, and direction, barometric pressure, air temperature, rain, and photosynthetically active radiation on the land. All data from the meters and sensors are conveyed via cables to two onshore data loggers. The domes were designed to facilitate water quality and isotope sampling from inside and outside the dome. Each dome has a port to attach tubing. The tubing is used with a peristaltic pump to purge and obtain water samples from each meter. All told, this design is fairly cumbersome, but you get a thorough assessment of the seepage site.

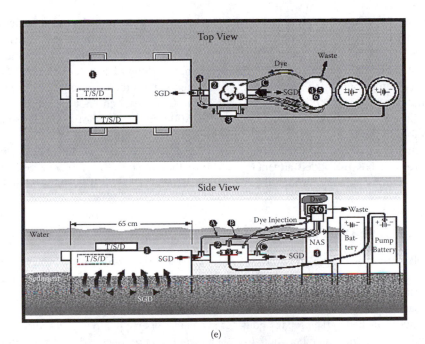

(e)

(f)

FIGURE 2.2.2.1 *(Continued)*

2.2.2.6 Ultrasonic Meter

Paulsen et al. (2001) adapted an ultrasonic flow meter, typically used in many engineering and material science applications that require the measurement of low flow rates, to attach to a steel funnel that is inserted into the sediments. Tubing attached to the inverted funnel directs the captured SGD through an ultrasonic flow tube, modified from the Contrlotron 1010 model, to be used beneath the sea surface

(Figure 2.2.2.1d). As Paulsen et al. (2001) describes, two piezoelectric transducers mounted on opposite ends of the flow tube generate and monitor bursts of ultrasonic signals (~400 sec^{-1}). The propagation time of the sound is dependent on temperature, salinity, and the speed of the fluid through which it flows. The calculation of water velocity in or out of the sediments is then dependent on the difference in upstream and downstream travel time of sound within the flow meter and not the temperature or salinity. The ultrasonic meter can resolve SGD on the order of 1 cm/day, comparable to the heat-pulse meter. The instrument can be programmed to make measurements at frequencies from once per second to once per day. In addition, this instrument has the added advantage of using the sound velocity to constrain the water salinity flowing through the instrument. However, this is a fairly complicated and expensive instrument.

2.2.2.7 Dye-Dilution Meter

Sholkovitz et al. (2003) designed an automated seepage meter that utilizes a dye-dilution method to measure both inflow and outflow of water across the sediment–water interface (Figure 2.2.2.1e). The instrument has four major components, including (1) a seepage housing similar to the one used by Paulsen et al. (2001); (2) a dye-mixing chamber attached to the seepage housing; (3) a submersible pump to recirculate water within the mixing chamber; and (4) a modified, battery-operated, *in situ* nutrient analyzer to inject dye and measure absorbance. The instrument injects dye (Alizarol Purple SS, maximum absorption at 588 nm) into the dye-mixing chamber and measures the absorbance of the dye solution at three locations (the dye-mixing chamber and the inlet and outlet of the mixing chamber) over time. The rate at which the dye solution is diluted by the inflow or outflow of water in the dye-mixing chamber is proportional to the flow rate of water across the sediment water interface.

The device can provide flow rate data on an hourly time scale over periods of days to weeks to months and accommodate a huge range of flow rates by simply using mixing chambers of different volumes. This instrument allows operation in both autonomous and manual modes, and like the other meters, can measure both inflow and outflow of seepage. Although still a complicated system relative to the Lee-type meter, Sholkovitz et al. (2003) suggest that the underlying principles and electronics are less complex than the heat-pulse and acoustic methods.

2.2.2.8 Electromagnetic Seepage Meter

Rosenberry and Morin (2004) have recently adapted a commercially available flow velocity meter designed for borehole measurements (groundwater wells) to a standard seepage meter to evaluate SGD. The electromagnetic flow meter (EFM) calculates the velocity of a fluid by measuring the voltage that is induced as it passes through an electromagnetic field. The flow meter uses Faraday's law of electromagnetic induction to measure the process of flow. These meters provide extremely accurate readings with no moving parts.

This is one of the simplest autonomous meters developed to date that simply requires off-the-shelf "upgrades" to the Lee-type meter. The EFM can be attached to the end of a 3-m section of 5.1-cm inside diameter flexible tubing. This flexible tubing is attached to the exit port of the Lee-type seepage meter or some other collection device with a hose barb. A more recent design incorporates the EFM into the exit port of the meter itself (Figure 2.2.2.1f). The limitations of this instrument are the power requirements, data collection and storage, and the minimum measurable flow velocities. The manufacturers of the EFMs used by Rosenberry and Morin (2004) include the Century Geophysical Corp. (Tulsa, Oklahoma) and Quantum Engineering Corp. (Loudon, Tennessee). The Century Geophysical EFM requires a 64-volt power supply, draws 250 mA of current, and provides a 0–5-volt DC output signal directly to a computer. Power and data acquisition for this system were performed by a geophysical logging truck. The Quantum Engineering EFM was connected to a digital data logger and power was provided by an AC inverter connected to a 12-volt deep-cycle battery (battery required charging after ~12 hours) (Rosenberry and Morin 2004). Minimum velocities of these two EFMs ranged from approximately 5–17 cm day when attached to the traditional Lee-type meter. These rates are much greater than those obtained by the other autonomous meters. The use of off-the-shelf instrumentation, although fairly expensive (~$15,000), certainly creates a simple system that can be assembled quickly, but also leads to several of the disadvantages described. Further work in this area may provide a more viable method for coastal deployment.

2.2.2.9   Applications and Environmental Calibrations

Assessments of the mechanisms controlling seepage meter measurements are critical to understanding the sources and relative magnitudes of pore water being advected from sediments. Field experiments are often needed with each new application to ascertain potential impacts on the measurements. These impacts include hydrostatic pressure differences between the flexible bag and the rigid underlying sediments (e.g., Shaw and Prepas 1989; Cable et al. 1997a), bioirrigation/bioturbation effects (e.g., Aller 1980; Martin et al. 2004, 2005), and the Bernoulli effect (Huettel et al. 1996; Shinn et al. 2002) induced by physical drivers (winds, waves, and currents) moving water across the sediments. In addition, temporal and spatial variability in seepage are often present due to variations in sediment type, tides, population densities of resident burrowing organisms, and physical forcing by winds, waves, and currents. Seepage meters are the only direct approach to measuring the water flux across the sediment–water interface. Therefore, a better understanding of the environmental variables that create concerns about the accuracy of these devices is imperative.

2.2.2.10   Anomalous Short-Term Influx

Time-series measurements using empty bags with the traditional Lee-type seepage meter show a decrease in seepage rate over a short period of time (Figure 2.2.2.2) (Shaw and Prepas 1989; Cable et al. 1997a). Seepage rates tended to reach a more

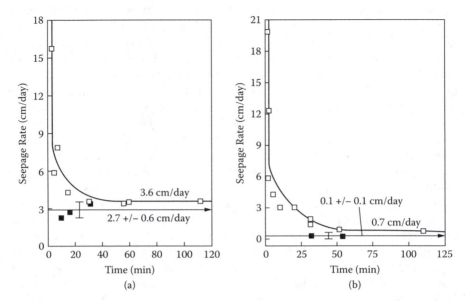

FIGURE 2.2.2.2 Results from time-series seepage measurements with empty (open symbols) and 1000-ml prefilled bags (closed symbols) in the norteast Gulf of Mexico, September 24, 1991, from experimental (a) and control (b) Lee-type seepage meters to demonstrate seepage meter response relative to collection time interval. Individual prefilled bag measurements are shown together with the mean (arrow). Control meters are placed inside a small children's plastic swimming pool filled with sediments. (Modified from Cable et al. 1997a.)

steady state after ~30 min. This rapid initial influx of water into the sampling bag was first documented by Shaw and Prepas (1987). They suggested that this "anomalous, short-term influx" of water was a result of the process used to manufacture the bags. The bags are collapsed from the original partially expanded state prior to attaching to the seepage meter. Once attached to the meter, the expansion of the bag to the original state creates a pressure differential between the inside and outside of the bag, drawing water into the bag. Shaw and Prepas (1989) and Cable et al. (1997a) demonstrated that prefilling the collection bags can prevent this anomalous influx of water associated with the bag's mechanical properties. Cable et al. (1997a) suggest that the most dependable seepage measurements with the Lee-type meter are obtained by using a 1000-ml prefilled collection bag and to take numerous measurements throughout a day at any one location.

2.2.2.11   Tidal Influences on SGD

Tidal variations in coastal systems influence the nearshore water table and therefore have a direct impact on the discharge of groundwater into coastal waters. Field and modeling studies in coastal aquifers show that tides can influence spatial and temporal groundwater discharge processes as well as nearshore groundwater salt concentration (Robinson and Gallagher 1999; Li and Jiao 2003; Chanton et al. 2003). Tidal fluctuations change the water table height as a result of mechanical loading of the aquifer at the oceanic extension, propagation and attenuation of the pressure wave created by the loading, and the flow of groundwater within the aquifer (Enright 1990). For example, as sea level increases, the aquifer bears a greater load creating a pressure gradient in the immediate vicinity of the loading. This pressure gradient (wave) is propagated inland and is attenuated since the matrix is bearing a portion of the load (Figure 2.2.2.3). The tidal loading and pressure gradient results in a change in the nearshore water table. In fact, Corbett et al. (2000b) showed that groundwater flow in an unconfined coastal aquifer (St. George Island, Florida) can change directions during a 12-hour period due to changing surfacewater levels.

The change in hydraulic gradient associated with the change in sea level dtue to tides or other processes has been shown to drive SGD velocities and direction in many different environments (Taniguchi 2002; Chanton et al. 2003; Charette and Sholkovitz 2004; Lambert and Burnett 2004). Tidal experiments conducted in the Florida Keys using Lee-type seepage meters, hourly measurements, and prefilled 4-L collection bags is just one example of the dependence of SGD on tidal height (Figures 2.2.2.4 and 2.2.2.5). Due to the highly porous underlying geology, changes in tidal height on the Atlantic side of these islands propagate quickly o the Florida Bay side. In fact, this is one of many locations that reverse flow was observed in the field (Figure 2.2.2.4) (Chanton et al. 1995; Corbett and Cable 2003). Measurements made at Hammer Point, Key Largo (Florida Bay side of the island), in February and March 1995, show a strong relationship to tidal stage in the Atlantic Ocean. When Atlantic tides were low, negative seepage rates in Florida Bay were observed and when Atlantic tides were high, positive seepage rates were observed. Measurements conducted later that year showed similar results, with decreasing seepage rates observed during the falling Atlantic tide (Chanton and Burnett 1995). Time-series seepage measurements were also made at the bayside well cluster location described by Shinn et al. (2002). These measurements responded to the Atlantic head pressure in a direct fashion, although negative rates were not observed. In this environment, there is a clear direct correlation between the tidal height and seepage rate, as maximum discharge rates were measured during the highest Atlantic tide level

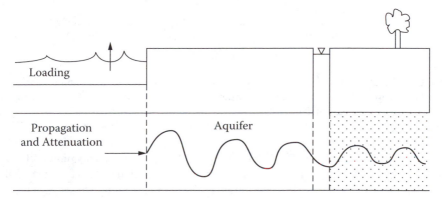

FIGURE 2.2.2.3 Propagation of pressure wave into adjacent coastal aquifer associated with the change in surfacewater level. Notice the attenuation of the pressure wave as it moves further inland.

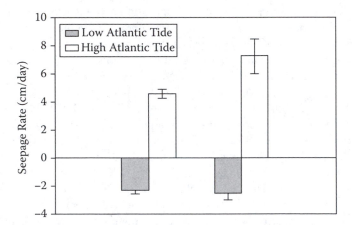

FIGURE 2.2.2.4 Florida Bay seepage rates determined at Hammer Point during a low Atlantic tide (*filled bars*) and high Atlantic tide (*open bars*). Each bar represents the mean and standard error of six individual seepage meters measured over the same time interval. (Modified from Corbett and Cable 2003.)

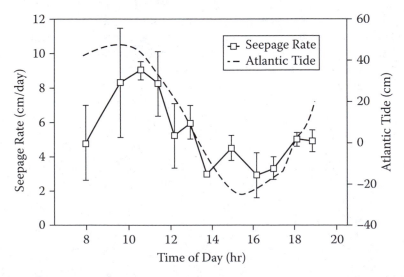

FIGURE 2.2.2.5 Tidal experiment conducted at the bayside well cluster (Shinn et al 2002) during August 1996. Seepage rates are mean and standard error of two individual meters (fiberglass dome-shaped meters described by Shinn et al. 2002). Relationship between Florida Bay seepage and Atlantic tide is easily identified. (Modified from Corbett and Cable 2003.)

(Figure 2.2.2.5) (Chanton et al. 2003; Corbett and Cable 2003). This direct relationship is unique to island environments because of the propagation of the hydraulic gradient through the island. In systems that measure SGD directly via seepage meters in the same locations as the change in water level, an indirect relationship between tidal stage and SGD is typically observed (Taniguchi et al. 2002; Chanton et al. 2003; Charette and Sholkovitz 2004; Lambert and Burnett 2004). This relationship is not surprising because the hydraulic gradient across the submarine aquifer will be steepest during low tide. In many cases, there has also been a time lag between the peak tides and the peak seepage rates, which are a function of the inland wave propagation.

2.2.2.12 Wind, Waves, Currents, and the Bernoulli Effect

Seepage meters have drawn criticism due to the possible influence of Bernoulli-type flow induced around seepage meters. Shum (1992, 1993) demonstrated that water movement over a seabed can provide significant penetration of water into sediments if microtopography, such as ripples, are present. Other variables that influence circulation of water into sediments include sediment permeability, ripple length and amplitude, and the wavelength, height, and period of water waves. Recent work related to effects of bottom topography on advective flow of pore waters demonstrated a pressure gradient will occur as currents move across the top of the seabed (e.g., Huettel and Gust 1992; Huettel et al. 1996). Shinn et al. (2002) argued that the Bernoulli effect on seepage measurements would be greater than any groundwater signal, and subsequently dubbed the effect, "Bernoulli's Revenge." Shinn et al. (2002) stated that the positive profile of seepage meters could create a Bernoulli effect that probably accounted for most of the water measured in the seepage collection bags, based on their observations in several locations of Florida Bay. In the simplest representation, bottom currents create a differential pressure between surrounding sediments and the seepage chamber top due to an increase in flow velocity above the chamber (Figure 2.2.2.6). According to Huettel and others working in sandy ripple beds, this differential pressure can drive vertical pore water velocities as high as 3 cm/hour (i.e., 72 cm/day) in 2–3 cm-high sand mounds when the bottom water current velocities approach 10 cm/sec (Huettel et al. 1996). While topography is certainly important in the rate of pore water advection through sediments, recent work conducted in the Indian River lagoon, Florida (Martin et al. 2005) suggests that it is unlikely the seepage meter acts as a significant enhancer of advective flow.

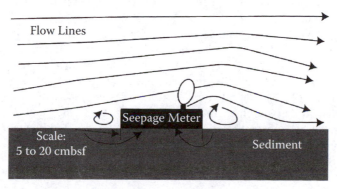

Flow Lines

Seepage Meter

Scale:
5 to 20 cmbsf

Sediment

FIGURE 2.2.2.6 Conceptual diagram of pressure gradients resulting from changes in bottom topography within the benthic boundary layer currents, where a seepage meter represents a macrotopographical feature. (Modified from Huettel et al. 1996.)

Huettel et al.'s (1996) work was performed in a wave tank where they could perform pore water velocity sensitivity analyses based on changing current velocities and mound heights. The vertical velocities they measured from sand mounds in a test tank are on order of magnitude greater than typical field-measured seepage results found in the Indian River lagoon (Cable et al. 2004). Therefore, a field program was initiated in the Indian River lagoon to evaluate the effects of physical forces that may induce a pressure gradient, and thus, the Bernoulli-type flow artifact in seepage meters. Average seepage rates (mean of duplicate meters) for a given time of day were compared to instantaneous wind speeds and current velocities at each station (Figure 2.2.2.7). These physical processes each may affect the pressure gradients surrounding seepage meters. Wind was used as an indirect measure of the turbulence in the water column and how pressure differentials may be affected. Currents are the only direct measure of artifacts associated with Bernoulli-type flow. Based on the collected data, Martin et al. (2005) concluded that there was no relationship between wind speeds, current velocities, or wave heights to the measured seepage rates at each station. While such artifacts may exist and may be more important in different settings, the magnitude of any effects is apparently insufficient to dominate the results in most cases.

2.2.2.13 Spatial and Temporal Variations

Variations in direct groundwater discharge observations are a function of scale for both time and space. Over small distances, groundwater discharge rates may vary due to sediment heterogeneity (Zimmerman 1991), variations in sediment thickness (Cherkauer and Nader 1989), or varying density of macrofauna (Zimmerman et al. 1985). On larger spatial scales, numerical and analytical modeling has shown an exponential decrease of seepage rates with increasing distance offshore (McBride and Pfannkuch 1975; Pfannkuch and Winter 1984; Winter and Pfannkuch 1984; Fukuo and Kaihotsu 1988). Although many field studies show this same exponential trend (Bokuniewicz 1980; Connor and Belanger 1981; Belanger and Walker 1990; Rosenberry 1990; Schafran and Driscoll 1993; Rosenberry 2000), other field areas are dominated by local hydrogeologic heterogeneity rather than the larger-scale topographically controlled seepage variability (Woessner and Sullivan 1984; Belanger and Mikutel 1985; Cherkauer and Nader 1989; Belanger and Walker 1990; Cable et al. 1997a). To evaluate the potential of an exponential decrease in seepage rate and to quantify the integrated estimate of offshore discharge, Taniguchi et al. (2003) suggested using more than five locations of measurement perpendicular to the shoreline. In some coastal systems, the geologic conditions might be such that substantial groundwater discharge occurs kilometers offshore, such as springs or seeps. These springs occur nearshore (Cable et al. 1996b) or offshore (Swarzenski et al. 2001) and are often a direct conduit of unaltered groundwater into surfacewaters.

The rate of SGD can also vary at any one location on different time scales from seconds to hours to weeks to months. Wave action can cause pore water to oscillate in its flow direction due to variations in pressure gradients with periods of seconds

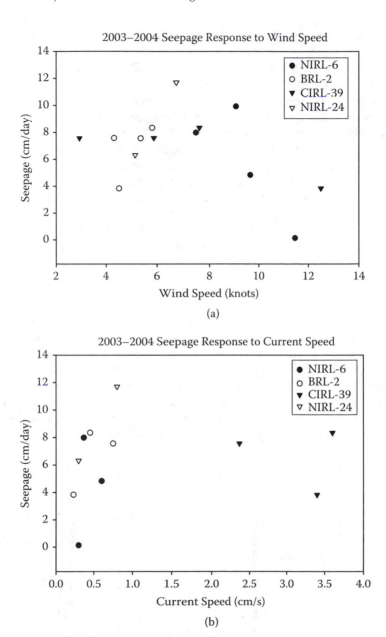

FIGURE 2.2.2.7 Seepage rates were compared to wind speed (**A**) and current velocity (**B**) in Indian River lagoon, Florida. (Data from Martin et al. 2005.)

to minutes (Webb and Theadore 1968, 1972), whereas tides act on time scales of hours to weeks (see above discussion) and can set up equally persistent discharge patterns. Long-term patterns can also be established by other large-scale sea level variations and by changes in the onshore hydraulic gradients, as due, for example, to variations in recharge (precipitation). Several studies have recognized a significant relationship between precipitation and SGD. Lewis (1987) showed that seepage rates along the west coast of Barbados, West Indies were almost twice as high during the wet season (November 1985) relative to measurements collected at the same sites in the dry season (May 1985). This was attributed to an increased aquifer recharge and therefore groundwater flow during the wet season. Similarly, Cable et al. (1997a) compared the seepage rate as measured by a Lee-type seepage meter in the northeast Gulf of Mexico to water table elevation observed at a nearby (15 km) USGS monitoring well. When the daily water table height and the average seepage rates were plotted vs. time (Figure 2.2.2.8), similar patterns were observed. When precipitation was lower, the measured seepage rates and water table heights were also low. As the water table elevation increased throughout the summer, a complementary rise in seepage rates was observed. All these processes conspire to produce SGD that is both spatially and temporally dynamic on a variety of scales and must be taken into account in order to provide an accurate assessment of groundwater discharge in all coastal settings.

2.2.2.14 Coastal Zones that can Utilize Seepage Meters

Taniguchi et al. (2002) summarized the recent studies that have attempted to estimate the magnitude of SGD or indicate if SGD in the area studied was significant by means of seepage meters, piezometers, and/or geochemical/geophysical tracers on a worldwide basis. This thorough review of the available literature demonstrates the widespread utility of the seepage meter, as this method of direct measurement was used in more than 40% of the reported studies (N = 45). The studies are primarily focused on the east coast of the United States, Europe, Japan, and Oceania, with fewer studies on the west coast of the United States and Hawaii, and no quantitative

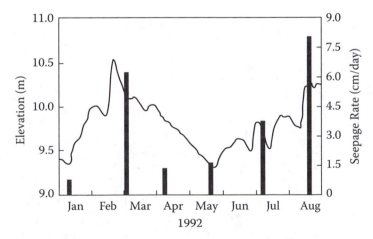

FIGURE 2.2.2.8 Daily water table elevations (*solid line*) above mean sea level (*m*) measured at a U.S. Geological Survey monitoring well in Sopchoppy, Florida; and the average seepage rate (cm/day) (*clased bars*) measured nearshore at Turkey Point, Florida in the northeast Gulf of Mexico. (Modified from Cable et al. 1997b.)

data available for South America, Africa, India, or China (Taniguchi et al. 2002). The coastal environments include estuaries, coastal bays, continental shelves, salt marshes, and coral reefs. The sediments of these environments range from mud and sand, silty-sand, coarse sand, to limestone.

Based on the variation of the environmental setting of the studies utilizing seepage meters, it may be concluded that this method could be used in most coastal permeable sediment environments worldwide. In fact, Shinn et al. (2002) showed how a modified seepage meter could be used in nearshore, hard bottom environments by cementing the meter to the bottom. However, at least one environmental factor that should be considered before using these instruments in any field setting is the current speed.

As previously mentioned, many investigators have recognized the potential increase in the variability of field measurements in stream and tidal environments where currents may be significant (Libelo and MacIntyre 1994; Shinn et al. 2002; Murdoch and Kelly 2003). Libelo and MacIntyre (1994) showed that the hydraulic head within the bag of a conventional seepage meter decreased as a function of flume-water velocity. A decrease in the bag's hydraulic head relative to the surrounding water would create an erroneously high seepage rate. It should be noted that at flow velocities less than 20 cm/sec, the difference in hydraulic head could not be accurately measured. Therefore, it may not be possible to obtain accurate direct measurements of seepage in environments with current velocities approaching 20 cm/sec without various corrections (Murdoch and Kelly 2003). With this limitation, it may not be possible to use seepage meters in many stream or riverine environments. However, Libelo and MacIntyre (1994) recommended covering the collection bag to isolate it from pressure gradients resulting from water movement in order to provide more accurate measurements in these sorts of settings.

2.2.2.15 Conclusions and Recommendations for Users

The seepage meter has been used in many different environments, including marshes, lagoons, bays, rivers, lakes, and coral reefs with varying degrees of success by many researchers since their introduction in 1944. Since the description of the conventional design, e.g., top or bottom section of a 55-gallon drum with an open port placed near the rim to attach a plastic collection bag (Lee 1977), researchers have sought to improve these direct measurements of seepage discharge from sediments through more accurate, less labor-intensive, and autonomous designs. However, seepage meters, whether conventional or newly designed, are not without fault. Most field measurement devices have limitations, and precautions should always be considered when implementing a field program that includes seepage meters, especially in environments with high current velocities (>20 cm/sec). However, when steps are taken to minimize error and control experiments are used, seepage meters make a very cost-effective technique for evaluating SGD from sediments. Though the technique does not identify the source of fluid discharge, other techniques (e.g., geochemical/geophysical tracers) used in conjunction with the seepage meter can be quite handy in evaluating advective sources.

2.2.3 Isotope and Tracers Techniques*

2.2.3.1 Introduction

Submarine groundwater discharge is recognized as an important source of water, nutrients, and other dissolved constituents to the coastal ocean (e.g., Johannes 1980; Church 1996). It occurs both as diffuse seepage and as focused flow at vents or cold seeps, depending on its hydrogeological setting. Globally, the total freshwater flux from SGD is estimated to be on the order of 10% of the freshwater flux from rivers (Zektser and Loaiciga 1993; Burnett et al. 2003a), although in some settings submarine discharge can exceed river discharge, particularly during times of low flow (e.g., Moore 1996). In areas where the flux of groundwater is focused into vents or regions of strong seepage, diverse ecosystems can be found (Paull et al. 1984; Bussmann et al. 1999; Rutkowski et al. 1999).

Given the spatial scales of the discharge zones, location and quantification of the groundwater flow to the coastal ocean can be challenging. A common approach for quantifying these discharge rates is to use geochemical tracers that are naturally enriched in groundwater relative to seawater and have well-understood chemistries within the marine environment. Most tracer studies of SGD have used naturally occurring radionuclides from the uranium and thorium decay series (e.g., Cable et al. 1996a; Moore 1996; Moore and Shaw 1998; Hussain et al. 1999). The flux of groundwater is determined by closing the tracer's mass balance within the marine environment, which requires knowledge of the end member (groundwater and seawater) concentrations, the flushing rate of the coastal water, surfacewater fluxes, and the removal pathways of the tracer (e.g., radioactive decay, gas exchange with the atmosphere, etc). In some settings, deliberate tracer experiments may be the best approach, especially when hydraulic connections between source areas and the coastal ocean need to be established.

2.2.3.2 Uranium and Thorium Series Nuclides

Uranium and thorium series radionuclides occur naturally in Earth material and include isotopes of uranium (^{238}U, ^{235}U, ^{234}U), thorium (^{234}Th, ^{232}Th, ^{230}Th), radium (^{228}Ra, ^{226}Ra, ^{224}Ra, ^{223}Ra), and radon (^{222}Rn, ^{220}Rn, ^{219}Rn). The decay sequence and half-lives ($t_{1/2}$) are shown in Figure 2.2.3.1. A by-product of the decay series is stable helium (as ^{4}He), which is produced by the α-decays. Because their mobility differs in aqueous environments, primarily due to their affinity to remain adsorbed on particle surfaces, daughter nuclides are often in disequilibrium with their parents (e.g., Ivanovich and Harmon 1992). Additionally, the activities (total number of disintegrations per unit time and volume) of each isotope can vary by order of magnitudes between different physico-chemical environments, such as those found in the saturated ground and the coastal ocean.

While uranium, radium, and radon are relatively mobile in aqueous systems, thorium is not; it adsorbs strongly to particles. This behavior of thorium is the basis

---

* Section 2.2.3 was written by Jordan F. Clark, Ph.D. (University of California, Santa Barbara jfclark@geol.ucsb.edu) and Thomas Stieglitz, Ph.D. (James Cook University, Queensland, Australia).

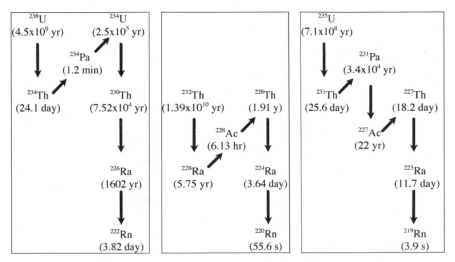

FIGURE 2.2.3.1 The ^{238}U, ^{232}Th, and ^{235}U decay series to radium (data from Ivanovich and Harmon 1992). Below each nuclide is its half-life. The vertical *arrows* are α-decays and the *arrows* up and to the right are β-decays.

for the uranium and thorium series methods for determining SGD. The mean life of thorium within the ocean is very short and depends on location. In the open ocean, its mean life is on the order of a half year while closer to the coast it is much shorter, typically less than a month (Cochran 1992). Thorium is effectively scavenged from the water column and transported to the sediments. As a result, the *in situ* (water column) production of its daughter radium isotopes is very low. The principle sources of radium to the ocean are external and include river discharge, shallow sediment fluxes, and SGD. The half-life of the longest-lived radium isotope, ^{226}Ra ($t_{1/2}$ = 1602 yr), is comparable to the mixing time of the ocean, while the half-lives of the other isotopes (^{223}Ra, ^{224}Ra, and ^{228}Ra) are much shorter (Figure 2.2.3.1). The half-lives of ^{223}Ra and ^{224}Ra are on the order of the mixing time of coastal embayments and the continental shelf while the half-life of ^{228}Ra is intermediate between ^{226}Ra and the short-lived isotopes. Away from the external sources, the activities of the short-lived radium isotopes are very low because of dilution and radioactive decay.

The dissolved thorium activity in groundwater is extremely low because of adsorption to the aquifer material (Krishnaswami et al. 1982; Osmond and Cowart 1992). However, because much of the thorium is adsorbed on the pore walls, the *in situ* production of radium is relatively high. Compared with seawater, the activity of the radium isotopes in groundwater is much higher, often by two or three orders of magnitude (Table 2.2.3.1). Small groundwater fluxes to the ocean can be distinguished with radium isotopes because of this concentration contrast.

Isotopes of radon, a noble gas, have very short half-lives (<4 days) and they reach secular equilibrium (after five half-lives) with their parent radium isotopes very quickly (<20 days). Thus, their activity in water is controlled primarily by the total activity (absorbed and dissolved) of their parent isotopes. The large difference in the total radium activity found between groundwater and seawater leads to a large

TABLE 2.2.3.1
Ground and Ocean Water Tracer Activities (Concentrations)

Nuclide	Groundwater		Open Ocean Surfacewater	
	Activity	Ref.	Activity	Ref.
^{226}Ra	1 to 4000 mBq/l	a	2 mBq/l	b
^{228}Ra	1 to 3000 mBq/l	a	0.2 mBq/l	b
^{222}Rn	2 to 200 Bq/l	a	2 mBq/l	b
^{4}He	40 to 200 µcc STP kg^{-1}	c	38 µcc STP kg^1	e
	>500 µcc STP kg^{-1}	d		

[a]King et al. (1982); Krishnaswami et al. (1982); Cecil and Green (2000); Tricca et al. (2001).
[b]Cochran (1992).
[c]Solomon et al. (1996); for shallow young groundwater.
[d]Bottomley et al. (1984); Stute et al. (1992); Clark et al. (1997); for >10,000-year-old groundwater.
[e]Weiss (1971); equilibrium concentration at 15°C and 34 PSU.

variation in the activity of radon isotopes in these two aqueous environments. Due to its nonreactive nature it does not have an affinity to adsorb to particles, and is readily dissolved in water. The activity of dissolved ^{222}Rn in groundwater, which is typically between 2 and 200 mBq/l, is much higher than the other members of the uranium-thorium decay series (Osmond and Cowart 1992; Cecil and Green 2000). Similarly, groundwater activities are also significantly higher than those of seawater, which is typically less than 2 mBq/l in the open ocean. Because of its 3.8-day half-life, its enrichment in groundwater relative to seawater, and the relatively simple analytical techniques (see below), ^{222}Rn is a particularly useful tracer for SGD (e.g., Cable et al. 1996a; Hussain et al. 1999; Top et al. 2001; Burnett and Dulaiova 2003; Stieglitz 2004). It has also been used to investigate groundwater discharge into rivers (e.g., Ellins et al. 1990; Cecil and Green 2000; Cook et al. 2003).

Groundwater discharge rates into the ocean can be determined with isotopes of radium and radon using a mass balance approach. To close the mass balance, sources and sinks within the study area must be well known. The principal loss mechanisms are radioactive decay, advection, and dispersion by coastal currents, and for radon only, air-sea gas exchange. A greater fraction of the short-lived nuclides is lost by radioactive decay than the long-lived ones because the time scales of the other loss mechanisms are relatively long. Sources of radium and radon to the coastal ocean include *in situ* production, continental runoff, diffusion from and irrigation of marine sediments, groundwater discharge, and advection of seawater into the coastal environment. Tidal pumping of animal burrows, such as those associated with tropical mangroves (Stieglitz et al. 2000), may also be an important source of these nuclides in some settings. The flux due to seawater advection is generally miniscule and can be neglected. While the *in situ* production rate can be determined very well, the fluxes from surfacewater and shallow sediment are more difficult to estimate and often require ancillary measurements. Finally, to convert groundwater radium or radon flux to water flux, the source groundwater activity must be known. This is generally determined through direct measurements of groundwater samples collected from terrestrial wells. Multiple wells need to be sampled because the natural variability in groundwater typically exceeds a factor of 3, especially if more than one aquifer is responsible for the SGD.

2.2.3.3 Helium

The isotopes of helium (^{3}He, ^{4}He) are important tracers for crustal fluids. Their isotope ratios vary significantly between different reservoirs in the geosphere. The ^{3}He/^{4}He isotope ratio of the atmosphere ($\sim$1.4$\cdot$10^{-6}) lies between that of the deep crust ($<$5$\cdot$10^{-8}) and of the upper mantle ($\sim$10^{-5}). Furthermore, their concentration in water varies by many orders of magnitude. The helium concentration of surfacewater (including seawater) is maintained near its solubility equilibrium value ($\sim$40 μcc STP kg^{-1}) (Weiss 1971) by gas exchange at the air–water interface. In waters that are unable to exchange with the atmosphere, such as groundwater below the water table and the ocean below the thermocline, the isotopic composition and concentration of helium can change as a result of the production of either ^{3}He by the decay of tritium or ^{4}He (and ^{3}He through resulting nuclear reactions) from α-decay of uranium and thorium series nuclides. Additionally, the discharge of helium-rich fluids can change the composition and concentration of dissolved helium in the receiving water.

Measurable changes from equilibrium values in the isotopic composition and concentration of helium are usually apparent in shallow groundwater after only a few years of isolation from the atmosphere (e.g., Schlosser et al. 1989; Solomon et al. 1996). However, these changes are small. Similarly, changes in the helium composition of pore water are small. Typically, helium concentrations in groundwater do not exceed equilibrium concentrations by more than two orders of magnitude until it has been isolated for tens of thousands of years (e.g., Bottomley et al. 1984; Stute et al. 1992; Clark et al. 1997). Thus, small fluxes of relatively old groundwater can be easily traced with helium. It is much more difficult to identify small fluxes of relatively young groundwater. For this reason, the helium method has been used more frequently for quantifying the discharge of groundwater and hydrothermal fluids into lakes (e.g., Sano et al. 1990; Aeschbach-Hertig et al. 1996; Clark and Hudson 2001) than into the coastal ocean. A notable exception is Top et al. (2001), who used helium to quantify SGD in Florida Bay.

2.2.3.4 Deliberate Tracer Experiments

Deliberate tracers are synthetic compounds that have small natural sources and well-understood chemistries. They are used in controlled experiments, where they are introduced at known times and quantities and monitored at known points. Thus, they are primarily used to determine hydraulic connections and rates of advection and dispersion in aqueous systems. In studies of SDG, their primary use is to determine travel times between point sources and coastal waters (e.g., Dillon et al. 1999, 2000).

For a tracer to work in studies of SGD, it must be conservative, not retarded in porous media, and measurable after significant dilution (typically >10,000:1). Sulfur hexafluoride (SF_6), a synthetic gas used primarily by the electrical industry, is one of the few tracers to meet these criteria. It is a nontoxic (Lester and Greenberg 1950), insoluble gas with a Henry's law coefficient on the order of 150 (Wilson and Mackay 1993). Its atmospheric mixing ratio, which is increasing at a rate of about 7% per year, was about 4 pptv (part per trillion by volume) in the late 1990s (Christophorou et al. 1997). At this time, equilibrium concentrations in the surface ocean were about $1 \cdot 10^{-15}$ mol/l (Law et al. 1994). The concentration of SF_6 at equilibrium with the pure gas is on the 10^{-4} mol/l, more than 10 orders of magnitude greater than surface oceanwater concentration. The detection limit of dissolved SF_6 is about $1 \cdot 10^{-17}$ mol/l using the most sensitive purge and trap gas chromatography (Law et al. 1994).

SF_6 has been used successfully as a tracer in ocean-scale mixing experiments (Ledwell et al. 1993), demonstrating the volume of water that can be economically "tagged" with this tracer. Laboratory and field experiments have shown that its movement in porous media is not slowed (retarded) relative to groundwater flow (Wilson and Mackay 1993; Nelson and Brusseau 1996; Gamlin et al. 2001). Additionally, it has been used in a number of large-scale groundwater studies where more than $1 \cdot 10^6$ m^3 of water has been tagged and its movement followed for more than 4 years (e.g., Clark et al. 2004). SF_6 differs from ionic and dye tracers in that it is a gas. Hence, it is lost from solution via gas exchange across the air–water interface.

Dillon et al. (1999, 2000) used SF_6 to establish hydraulic connections and travel times between terrestrial point sources (septic tank and a shallow waste water injection well) and the coastal ocean in south Florida. These authors injected to the groundwater system small volumes of water (<250 L) that had been saturated with pure SF_6, taking advantage of the large difference between coastal water in equilibrium with the atmosphere and tagged water in equilibrium with the pure gas. The injection points were within 25 m of the coastline. Tracer concentrations were monitored in nearby wells and the coastal ocean to determine travel times over the period of a few months. These experiments clearly show the utility of SF_6 to establish hydraulic connections and travel times between terrestrial point sources and the coastal ocean, provided that the tracer is detected away from the injection point. Experiments that end without detecting the tracer in coastal waters do not necessarily indicate that no hydraulic connections exist. Longer and greater than expected travels times and dilutions can also explain why tracer was not observed. Hence, the limiting factoring for SF_6 experiments is often what is a reasonable monitoring period and frequency of sampling.

2.2.3.5 Experimental Techniques

Radium Isotopes. Dissolved radium is typically extracted from large-volume water samples (60–1000 L) using MnO_2-coated acrylic fibers, which strongly adsorb and collect the dissolved radium (Moore 1976). The removal efficiency of the fibers can be determined by passing the water through two sets in series and determining residual activity on the downstream fibers. More recently, other extraction agents and methods have been tested, including the use of a specially fabricated manganese resin (Moon et al. 2003).

Activities of the short-lived radium isotopes, [223]Ra and [224]Ra, are measured in the field or in the laboratory shortly after sample collection. The most common method is the detection of the α-decay of their respective daughter nuclides, [219]Rn and [220]Rn, with photomultiplier tubes, and their identification with a delayed coincidence circuit (Moore and Arnold 1996). The MnO_2 fibers are partially dried, and placed in a closed-loop air circulation system connected to photomultiplier tubes. The respective daughters are flushed from the fiber into the tubes where counters discriminate the decay of the [224]Ra daughters ([220]Rn and [216]Po) from the [223]Ra daughters ([219]Rn and [215]Po) by electronically gating the registered counts with very high detection efficiencies. The measurement efficiency is affected by fiber surface conditions and water content, and is at an optimum when the water film covering the MnO_2 fiber has a thickness on the order of the recoil distance of the radon daughters (e.g., Sun and Torgersen 1998).

Activities of the long-lived radium isotopes, [226]Ra and [228]Ra, are determined by leaching the radium from the fibers with HCl after completion of the measurements of the short-lived isotopes, and subsequent α-particle spectrometry (e.g., Hancock and Martin 1991), or by γ-ray spectrometry on leached and $BaSO_4$ co-precipitated radium (Moore 1984). Samples need to be aged to allow for equilibration of radium with its daughters.

More recently, Kim et al. (2001) experimented with the use of a commercial radon-in-air monitor to detect [224]Ra and [226]Ra from the α-decay of their radon daughters. This instrumental setup is not commonly used to date to detect radium isotopes; however, it is used in various SGD studies to detect [222]Rn. The instrument is described below.

Radon. Because radon is a gas and is lost from solution across the air–water interface, water samples for radon analysis need to be collected in a fashion that prevents gas loss. Traditionally, up to 5 L of water are collected in evacuated glass bottles. Radon is extracted from these samples by sparging with helium and is collected by cryogenic trapping on charcoal. After heating, the radon from the traps is transferred to α-scintillation cells for counting (Mathieu et al. 1988). *In situ* production of [222]Rn in the water sample through decay of the parent [226]Ra is usually small, and can be determined by aging the sample for five half-lives of [222]Rn (~3 weeks) to establish equilibrium between the parent and daughter isotopes and recounting [222]Rn.

An alternative method determines the [222]Rn activity by liquid scintillation (Cook et al, 2003; Theodorsson and Gudjonsson 2003). This method requires adding an oil-based scintillant to water samples of known volume and equilibrating the [222]Rn between the water–gas–scintillant phases. The solubility of radon in the scintillant is much higher than in water and, thus, radon strongly partitions into this phase.

Recently, a third method has been developed that provides a significant simplification of [222]Rn data collection, utilizing commercial radon-in-air monitors (Lane-Smith et al. 2002). Seawater is pumped directly through an air-water equilibrator. The [222]Rn-enriched air is circulated in a closed air-loop, which passes through the monitor, to establish equilibrium between the circulating air and continuously pumped seawater. The monitors count α-decays of [222]Rn daughters, and discriminate different decays in energy-specific windows. With this method, continuous measurements of ^{222}Rn can be made. Applications include time-series recording of ^{222}Rn activity in one location (Burnett and Dulaiova 2003), or mapping from a moving platform the spatial distribution of ^{222}Rn in coastal waters to detect groundwater entry points (Stieglitz 2004).

Helium. Helium samples (between 10 and 40 ml) are collected in copper tubes sealed with steel pinch-off clamps. The copper tubes are flushed thoroughly to ensure that no air is sealed along with the water sample. In the laboratory, these tubes are attached to high-vacuum inlet systems that lead to noble gas mass spectrometers. Helium is extracted from the water and separated from other gases by a series of cold traps and titanium getters (e.g., Beyer et al. 1989). Typically, neon is analyzed along with helium and its concentration used for the excess air correction. In the ocean, excess air forms as a result of the dissolution of bubbles forced below the water surface by wave action. The mass spectrometer is calibrated with equilibrated water samples and known quantities of air.

Sulfur Hexafluoride. SF_6 is typically analyzed on a gas chromatograph equipped with an electron capture detector. It is separated from other gases with a molecular sieve 5A column held at room temperature (Wanninkhof et al. 1987; Law et al. 1994). During most studies, discrete water samples are collected and the SF_6 is extracted from the water using either a head space (Wanninkhof et al. 1987) or purge and trap method (Law et al. 1994). Clark et al. (2004) have shown that samples can be collected and stored in evacuated septa bottles (Vacutainers) for more than 6 months providing an inexpensive sample container that allows many samples to be collected in the field for analysis later in the laboratory. Recently, Ho et al. (2002) developed an automated field SF_6 measurement system that can be used on moving platforms (small boats). With their system, water is continuously pumped through a gas-stripping chamber (a membrane contractor) and the extracted gas is periodically (every 2 minutes) sent to a gas chromatograph for SF_6 quantification. The extraction efficiency is determined using dissolved oxygen meters up- and downstream of the stripping chamber. Underway measurements greatly improve the chances of determining coastal discharge points of water tagged with SF_6 during deliberate tracer experiments.

2.2.3.6 Discussion

Fluxes of SGD into the ocean are estimated primarily by (1) direct volumetric measurements with seepage meters, (2) calculations using numerical models of flow in coastal aquifers, or (3) geochemical tracer mass balance approaches. Although in many cases comparable, the three methods will not necessarily yield the same results when applied to a particular site, due to the different spatial and temporal scales of their respective measurements and the nature and origin of the SGD measured with each method.

Advantages of direct seepage meter measurements are that they are relatively simple, typically include conductivity recorders, and can be inexpensive. However, such devices provide point measurements, and a relatively large number of measurements are required to adequately account for local heterogeneities of flow (e.g., Stieglitz 2004). This is particularly so where fractured aquifers connect to the ocean and where the magnitude of SGD is small and spatially spread out. The conductivity data collected simultaneously is critical for determining if the SGD is recirculated seawater or continental groundwater.

In contrast, geochemical tracer techniques mentioned here temporally and spatially average SGD fluxes in an area. The measured flux is the sum of the different sources contributing to SGD and includes aquifer discharge and the flux of recirculated seawater. In principle the geochemical methods can distinguish SGD from different aquifer systems by using the multitracer approach (Top et al. 2001; Moore 2003). SGD has been investigated in a variety of coastal settings including salt marshes, estuaries, and the continental shelf, using geochemical techniques. In most cases, the studies have used radium and radon isotopes. Case studies using these techniques can be found elsewhere in this volume.

In addition to SGD studies, the radium quartet has been used to quantify coastal ocean mixing rates and shelf waters ages (Moore 2000a,b; Rengarajan et al. 2002; Burnett and Dulaiova 2003). The half-lives of the short-lived radium isotopes are of a similar scale as coastal mixing rates. Hence, variations in their activities offshore record the flushing time of tracers derived from SGD as well as any nutrients or contaminants dissolved in the source water. Knowledge of the flushing rate is critical for closing the tracer's mass balance when quantifying the groundwater flux.

Quantification of the sediments fluxes due to diffusion and irrigation are also needed for the mass balance calculation to determine the fraction of SGD that is meteoric. Rates of water-sediment exchange have been determined in a number of settings, using both radium and radon isotopes (Hammond et al. 1977; Berelson et al. 1982; Hartman and Hammond 1984; Hancock and Murray 1996). These studies typically quantified the exchange with benthic chambers.

The geochemical techniques provide complementary information, although few studies have employed the multitracer approach. Notable exceptions include Top et al. (2001) and Moore (2003). Top et al. (2001) examined SGD into Florida Bay, using both radon and helium. The fluxes calculated with each tracer differed, suggesting that SGD is occurring in a two-layered aquifer system beneath the bay. In a similar fashion, Moore (2003) demonstrated the utility of the suite of radium isotopes to distinguish between different aquifers discharging into the coastal ocean by using a three-end member model (one young aquifer, one old aquifer, one ocean) to explain the radium distribution in the northeast Gulf of Mexico.

The multitracer approach often leads to improved models of groundwater discharge. Each radionuclide effectively carries a clock that records when the nuclide entered and how quickly it is lost from the marine system. Because of their different half-lives, the relative importance of the sources and sinks differs for each tracer. Furthermore, the relative abundance of each tracer within groundwater varies according to the aquifer material, aquifer properties (porosity, hydraulic conductivity, etc.), and groundwater age (residence time). For these reasons, studies of SGD benefit from using multiple tracers as illustrated by a few studies.

2.2.4 STUDY OF ANOMALIES IN THE NEAR-BOTTOM LAYER OF SEAWATER AND SEDIMENTS

Anomalies in sea sediments are studied by different geophysical methods including the study of changes in thermal fluxes, seismo-acoustic profiling, thermo-electrical profiling at the "seawater–sea bottom" boundary, as well as by determination of the hydraulic capacity of sea sediments and the analysis of chemical, isotopic, and gas compositions of pore solutions in the sediments.

A complex geophysical method for studying anomalies in the bottom layer of seawater and sea bottom sediments (referred to in the literature as "marine hydrogeological complex") was developed and used in Russia. It includes continuous seismo-acoustic sounding conducted into the boards of a research ship and the simultaneous and continuous measurement of temperature and salt content in the near-bottom water layer by means of special equipment (Zektser et al. 1984, 1986). The essence of the method lies in determining the geological structure of the sea bottom along a continuous sounding profile with the aid of seismo-acoustic signals. The depth of seismo-acoustic sounding can reach 1000 m. Temperature and electrical conductivity (salt content) of seawater at the "bottom–seawater" boundary are registered along the same profile (Figure 2.2.4.1). This method, the principle of which represents marine hydrogeological surveying, enables detection and delineation of anomalies in the physical and chemical fields on the sea bottom, caused by different types of groundwater submarine discharge (i.e., concentrated karst springs, water discharge through tectonic and weakened zones and mud volcanoes, seepage through bottom sediments, etc.). Most distinctive anomalies, caused by groundwater submarine discharge, are manifested by the distribution of salt content and temperature at the "bottom sediments–near-bottom seawater layer" boundary. However, it is not necessarily so that these anomalies even if taken in combination would be connected with groundwater discharge. They may be caused by other reasons (e.g., piling up and out phenomena, currents, homorganic and bioorganic processes, etc.). To interpret anomalous values of these parameters properly, it is necessary to know the geological cross-section of sea-bottom sediments, which can be obtained by analysis of seismo-acoustic profiles.

Equipment for marine hydrogeological surveying include a seismo-acoustic profilograph; an on-bottom towed probe with transducers to measure temperature, electrical conductivity, and pressure; and registering devices. The seismo-acoustic profilograph provides information on the geological structure and often on the lithological composition of sea sediments within the first few hundred meters of depth. Design and operation of the profilograph are given below.

High-voltage capacitors accumulate electrical energy that is transmitted via a commutator, controlled by a synchronization system, to the excitation source. When discharging, the energy originates on the source as an acoustic wave (signal), which, reflecting from rock layers with different coefficients of reflection and spreading, comes to the piezoelectric receiver (receiving coiled cable with transducers installed). After amplification, the signal is transmitted to the self-recorder. At present, there are several ways of exciting oscillations in water for seismo-acoustic profiling. In our case, a group excitation source with a single energy transmission path to elementary

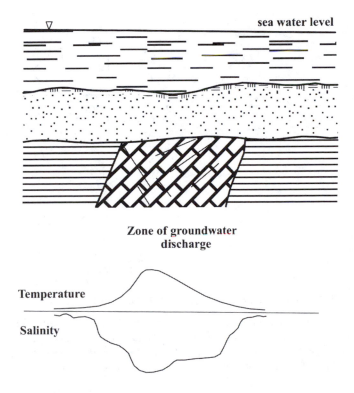

FIGURE 2.2.4.1 Salinity and temperature in seawater connected with groundwater discharge.

sources was used. This source is characterized by simple design and high resolution (Kalinin et al. 1975).

The excitation block includes a cable with sources as a load consists of a high-voltage rectifier and a communication system that provides discharge of reservoir capacitors in certain moments of time. A system of piezoelectric PDS-24 elements connected in parallel into the so-called receiving coiled cable serve as an acoustic receiver. An acoustic signal received by the coiled cable and transformed into an electrical one enters into the inlet of the rectifier. To obtain reliable information, it is important to have a proper hanger system for the group source and receiving coiled cable. Position of bearing arms relative to the ship and sea surface and hanging process are shown in Figure 2.2.4.2. The group source and receiving coiled cable were submerged with the use of 5-kg weights; additional weights were applied to stabilize the cable position. Optimal towing velocity is 5–6 km/hr. Such velocity enables reaching an optimal correlation between time spent for the given profile and picture quality on the profilogram, and the possibility to work without the receiving cable breaking off. The next component of the marine hydrogeological complex is the bottom-towed probe containing measuring transducers of electrical conductivity, temperature, and pressure.

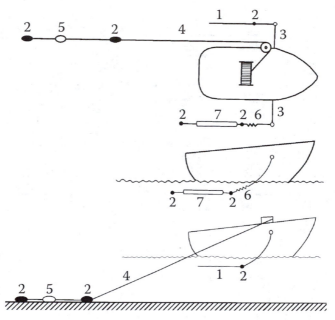

FIGURE 2.2.4.2 Equipment for sea hydrogeological surveying: (1) source of seismo-acoustic signal; (2) seismo-acoustic receiving coiled cable with transducers installed; (3) rubber shock absorber; (4) outboard arm; (5) probe with transducers of temperature, electrical conductivity, and pressure; (6) cable-wire; (7) different sinking and stabilizing weights.

The probe is made of brass in an ellipsoidal shape with transducers built into its rear part. As the sea works were carried out at a depth of no deeper than 500 m, the probe was hermetically insulated using polytetrafluorethylene (Teflon) liners, epoxy resin, and special sealing paste. One of the measuring elements installed on the probe is the transducer of water electrical conductivity. It represents a two- or four-electrode measuring unit, depending on conditions of surveying. The two-electrode unit is a platinum electrode to which AC voltage is applied and, through the surrounding space, is closed on the probe housing. The drawbacks of such a measuring unit are interferences caused by electrochemical processes occurring at the platinum–electrolyte boundary. Ion dipoles, orienting one of their poles to the metal, create on the platinum surface a double electrical layer that can be supposed to be a certain capacitor introducing errors into measurements (Grilikhes and Filanovskiy 1980). However, the simplicity of design and specificity of the construction that excludes clogging by silt and algae at the water–bottom boundary make two-electrode units useful for measuring seawater electrical conductivity.

The four-electrode measuring unit enables obtaining data on the salt content with a higher accuracy. The unit represents a dielectric tube with electrodes pressed-in inside in the form of stainless-steel rings. Current flowing between the external electrodes creates on the internal electrodes a potential difference proportional to the electrical conductivity of water. A disadvantage of such a system lies in that when measuring in a near-bottom layer of water, the four-electrode unit can be easily clogged by liquid silt.

Temperature-sensitive elements in the probe is represented by thermistor MT-52 placed into a metal protective shell. The properties of the thermistor itself and construction of the shell make it possible to obtain a time constant (time lag) equal to about a tenth of a second which is important when working in towing regime. Depth of the probe towing is controlled using a typical set of pressure transducers able to work at different depths. Of great importance is proper suspension of the probe and fixing of stabilizing weights (see Figure 2.2.4.2). A port electronic block receiving and transforming signals that come from the transducers through the cable enables continuous measurement of electrical conductivity and temperature with an accuracy of 0.01‰ and 0.05°C, respectively.

Before profiling, the measuring transducers are calibrated in the laboratory. Based on the established dependence of output voltage on electrical conductivity, temperature, and pressure, appropriate calibrating curves are plotted. The parameters to be measured are determined by these curves. Registering devices include a graph plotter (N306) and multichannel recorders (KSP-type) with a velocity and duration of recording similar to the recorder of the seismo-acoustic profilograph.

Another possible way to treat information coming from the bottom-tugged probe is the use of a computer directly on shipboard. Signals from measuring transducers installed on the probe are sequentially set via the simplest commutator into an analog-to-digital converter. After being encoded and converted, the signal is input into the computer where it is processed by a preset program. Results obtained can be tape, printer, or perforator recorded. The computer abilities enable conjoint usage of a graph plotter to immediately compile maps of salt content and temperature for the region under study. Thus, the sea hydrogeological complex method enables us to carry out marine hydrogeological surveying with high productivity and at low expenses, and to detect and map groundwater discharge on the sea bottom.

The given equipment and methodical procedure of marine hydrogeological investigations were tested in an experiment site located on the Black Sea shelf near Gudauta, Caucasus. The site lies on structural lowland of the fundament composed chiefly of rocks of the Jurassic, Cretaceous, and Paleogene–Neogene ages and is expressed as a bank in the sea-bottom relief. High tectonic dislocations, karst phenomena, disturbances in the top undersea sediment layers by erosion processes — all promote the intensive submarine discharge of groundwater in this area. The area stretches for 40 km along the shoreline from the cape of Pitsunda to the Gumista River and for 20 km offshore restricted by an isobath of 200 m.

The area is an extension of the Bzybsky artesian basin opened toward the sea. The basin is characterized by alternating high-permeable karst rocks and waterproof marl and clayey sediments. The main aquifer systems include the sediments of the Lower and Upper Cretaceous–Paleogene periods with karst, fissure-karst, and fissure-stratum types of groundwater circulation.

The Lower Cretaceous aquifer system is mostly spread in the interfluve of the Bzyb and Gumista Rivers where the experiment area is located. The system is composed of limestone, dolomites, and sandstone 500–600 m thick. Piezometric surface slopes are directed toward the sea and amount to 0.06–0.016. Absolute groundwater pressure heads vary from 5–180 m, and water conductivity in the zone of fissure-stratum circulates 60–70 m^2/day reaching in some cases 210 m^2/day. Water of this aquifer system has a mineralization of 0.6–1.7 g/l and contains chiefly gases of atmospheric origin and methane.

Groundwater discharge from the system occurs in river valleys along the shoreline, and also in karst submarine springs. Large karst springs on land (e.g., Reprua, Tsivtskali, Gagripshsky, and other springs) form rivers and lakes. For example, a spring in the upper reaches of the Chernaya River has a yield of 1400 l/sec. The temperature of the spring water is 9–11°C; mineralization is 0.25 g/l. The Reprua submarine springs have the same properties discharge at a depth of 5–10 m below sea level, and are located 2.5 km northwest from the city of Gagra. The temperature of the spring water is 9.5°C.

The aquifer system in the Upper Cretaceous–Paleogene sediments consists of limestone with interbeds of marl and clays from 250–630 m thick. The slope of the piezometric surface, directed toward the sea, ranges from 0.05–0.03. Water conductivity varies from 10–130 m^2/day. As the aquifer inclines from northeast to southwest, and with the increasing distance from the recharge areas and closer to the sea, the fissure-karst freshwaters in the zone of intensive water exchange are gradually replaced by stratum-fissured waters of the chloride-sodium type with high mineralization. This trend is disturbed in tectonically weakened zones. Groundwater is discharged in the form of ascending springs on land and on the sea bottom. The yields of some of these springs reach several tens and even hundreds of liters per second. Large ascending springs are located near the settlements of Akhali-Afoni, Anukhva, and others. Their yields reach 20 l/sec; water temperature is 10–15°C; mineralization is not more than 0.5 g/l. The spring in the settlement of Besleti has a yield of 200 l/sec.

The sufficient hydrogeological knowledge of the Bzybsky artesian basin and the presence of two pronounced water-bearing systems with groundwater flow directed toward the sea make this shelf area ideal for testing the methodical procedure of marine hydrogeological investigations. An integrated profiling by a rectangular grid was carried out in the area. The profiles were set along and perpendicular to the shore (with an interval of 2–4 km). To illustrate this, we can examine three complex profiles obtained during field work in 1991. Two profiles (nos. 10 and 14) were obtained for the area of New Afon; the third (no. 5) was located at the traverse of the Bambora Cape. The distance between profiles no. 10 and no. 14 was 2.5 km.

After the electrical conductivity was recalculated into the salt content and the temperature and salt contents were checked in the reference points, the curves for distribution of these parameters were plotted, and the related geological cross-section was drawn with the aid of a seismo-acoustic profilograph (Figure 2.2.4.3).

Profiles nos. 10 and 14, 1.5 km from shore, distinctly showed the areas of decreased salt contents. The curves of temperature did not have contrasting cycles and showed a smooth rise of temperature from 11–8.0°C with distance from the shore and with depth. On no. 14 the salt content sharply dropped to 16.5‰ and then rose smoothly to an average level of 19.9‰. The temperature curve had a slight decrease bias, relative to minimal salt content, toward the sea. The anomalous values of salt content and temperature on the seismic diagram corresponded to a vent

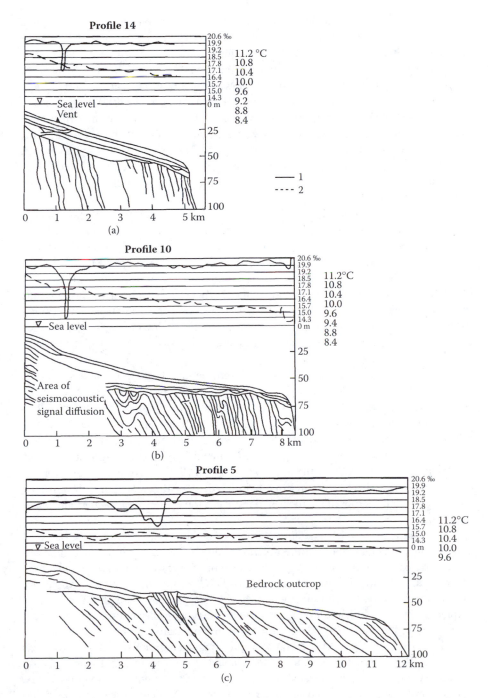

FIGURE 2.2.4.3 Complex profiling in the Gudauta experiment site: (1) the curve of salt content distribution; (2) the curve of temperature distribution.

registered by the seismo-acoustic profilograph. The gryphon mouth lay at a depth of 19 m and was associated with edging layers overlain by thin modern sea sands and sandy silts. The vent is shown on the diagram as a narrow cone of about 3 m high above the sea bottom. Reflection of the signal occurred because of a large density of the submarine karst spring jet possessing a high-pressure head. Ordinary mushroom-shaped spreading of the vent jet (Korotkov et al. 1980), which does not reach the sea surface, is not seen on the seismic diagram, because the densities of the vent jet and the surrounding seawater are equalized. This is the reason why the vent has a cone-shaped reflection on the diagram. The vent mouth is not shown on the seismo-acoustic profile. It may be because the vent mouth is filled with clastic rock material and its diameter is less than that of the resolution of the seismo-acoustic profiling instrument.

As can be seen in no. 10, a distinct zone of signal spreading with sharp boundaries vividly coincide with the boundaries of anomalous salt content equal to 14.3‰. Such spreading of the signal can be explained by the ascending flow of the gas-saturated karst groundwater. Unfortunately, it is impossible to observe the geological structure in the zone of signal spreading, which is of the greatest interest.

In our opinion, the site under study in the Bzybsky artesian basin has a submarine discharge of the karst groundwater. Having a considerable pressure head, this groundwater rises upward along a karst hollow or a tectonic dislocation and is discharged through porous overlying Quaternary sediments, causing a spreading of the seismo-acoustic signal.

The anomalous decrease of the salt content on profile no. 5 coincides with the exposed Paleogene bedrocks on the sea bottom, registered on the seismo-acoustic profile (see Figure 2.2.4.3). A slight salt decrease in the beginning of the profile is explained by the desalinating effect of the White River. Two areas can be distinguished in the anomalous site of 3.5 km long, located 4 km apart from the shore. The first, which precedes the basic anomaly, lies in the zone of edged modern sea sediments. The salt content there decreases from 19.2‰ (averaged value for the profile) to 17.6‰. The second area of maximum desalination reaching 15.8‰ exactly coincides with the exposed bedrocks and indicates the presence of fresh groundwater discharge.

The relatively small temperature anomaly on profile no. 14 and the absence of anomalous temperatures on profile no. 10 are due to the fact that the investigations have been carried out in early March when temperatures of karst and seawaters in the coastal zone are almost equivalent at 10–11°C. Profile no. 5 shows a slight rise of the temperature which coincides with the salt content anomaly. This can be explained by the discharge of deeper groundwater from the aquifers of the Upper Cretaceous–Paleogene period, which have high temperatures.

Based on the results of the complex profiling technique, six anomalous areas caused by SGD were distinguished at the Gudauta experiment area (Figure 2.2.4.4), with salt content anomalies located deep of the bottom sediments–seawater boundary. At the same time anomalies in salinity on the bottom sediments–seawater boundary caused by SGD are noted at the depth of 10–50 m, mainly within the coastal zone by a strip of 2 km in width. In some areas its width reaches 6 km. The works, carried out at the stage of detailed study, included hydrochemical, filtration, and isotopic

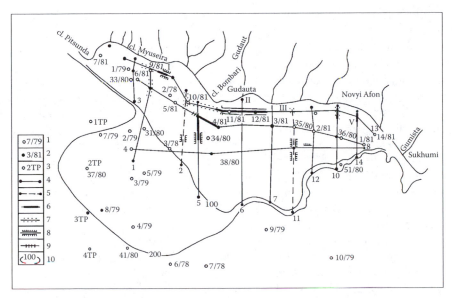

FIGURE 2.2.4.4 Schematic map of sea hydrogeological surveying in the Gudauta, Caucasus experiment site: (1) sea complex station (numerator = number of station, denominator = year of work execution; (2) station which registered an anomaly; (3) station for measuring heat flux; (4) line of complex profiling (seismo-acoustics, temperature, electrical conductivity); (5) line of seismo-acoustic profiling; (6) area of decreased salt content at the boundary of bottom sediments (seawater, detected by complex profiling); (7) zone of spreading seismo-acoustic signal; (8) exposure of bedrocks on the sea bottom; (9) zone of tectonic disturbances; (10) isobaths.

testing of bottom sediments and determination of gas composition in the near-bottom layer of seawater. Sediment samples were taken by one-pass ground tubes.

Coefficient of filtration in the samples was measured in the laboratory on shipboard with the aid of a filtration tube from the Russian Ministry of Natural Resources organization, SPETSGEO. Because of the low hydraulic capacity of sea sediments, one measurement took 2 or 3 days. The measurements show that the filtration coefficients in the coastal zone vary within a wide range from $4 \cdot 10^{-1}$ to $1 \cdot 10^{-2}$ m/day for fine-grained clayey sands with interbeds of silt and shelly rocks, and from $2 \cdot 10^{-3}$ to $1 \cdot 10^{-6}$ m/day for silts. The decrease of the coefficients is observed with increasing distance from the shore because of an increasing clay ratio in the sediments.

The highest filtration coefficient, registered at stations 11/81 and 12/81 and equal to 0.06 and 0.4 m/day, respectively, is related to the location of anomalous salt concentration (Figure 2.2.4.4), detected an isobath of 20 m during the profiling. The filtration coefficient at station 4/81 located at the boundary of the remotest salt content anomaly VI is relatively high (0.007 m/day) compared with the other stations. The high-filtration coefficients coinciding with the places of low-salt concentrations confirm the presence of SGD.

After taking sediment samples, the pore water was intermittently pressed out with the help of a hydraulic press or centrifuge to determine their hydraulic capacity. Chlorion in pore water was measured on shipboard. The six-component chemical analysis and determination of tritium content were carried out in a stationary laboratory.

How mineralization, chemical, and isotopic compositions of pore waters are distributed in the plane and cross-section considerably depends on the intensity of the exchange between groundwater and surfacewater. Unlike seawater, pore waters in the near-bottom layer are not subjected to external influences, therefore their chemical composition is rather stable. The most vivid indication of SGD is reflected in the vertical distribution of chlorion in the pore waters of sea sediments. It acts as a good indicator of its susceptibility to changes in the physical-chemical environment, and its good solubility. Gradients of chlorine concentration in pore waters of sea sediments are formed both because of seepage of fresh or salt groundwater and at the expense of sedimentation waters reflecting salt content changes in the sea in the past. Here, diffusive processes considerably smooth gradients of chlorine content, formed by sedimentation waters.

Investigations of the pore solutions in the Black Sea in the area of the Gagra-Gantiadi settlement have established that the chlorine concentration decreases near the submarinewater springs in the upper layer of bottom sediments and also with decreasing depth. At the same time, near the sources of groundwater discharge on the sea bottom some elements such as zirconium, vanadium, chromium, barium, nickel, and others have an increased concentration (Brusilovskiy 1971). Over 30 testing stations were installed at the experiment site of Gudauta where chemical six-component analyses of pore waters were carried out (Table 2.2.4.1). The results (Figure 2.2.4.4) show an increase of the chlorion in the pore waters with distance from the shore and with depth. At the same time, data on the major stations show a decrease of mineralization and chlorine content in the pore waters with depth of bottom sediments. This gives evidence of the penetration of fresh groundwater through bottom sediments. Thus, a decrease of groundwater discharge with distance from the shore leads to an increase of chlorine in pore waters. Influence of groundwater discharge on chlorion content in pore waters of sea sediments is mentioned by several researchers (Shishkina 1980; Manheim 1967).

At the experiment site, we studied the heat flux and composition of gases, dissolved in the near-bottom layer of seawater, jointly with experts from the Hydrogeology and Engineering Geology Sector of the Georgian SSR Academy of Sciences under the leadership of Buachidze. As is known, one of the factors that misrepresents the natural temperature field of the Earth's crust is the ascending flow of groundwater and particularly its filtration through sea bottom sediments. Even weak ascending or descending movement of groundwater can be a reason for considerable temperature anomalies (Ogilvi 1959), which provides the basis to use data on heat flux and temperature in sea sediments for studying the conditions of SGD and giving quantitative assessment. The heat flux was measured by a common technique using a standard thermogradientograph PTG-ZMTB with a 1.5-m-long probe with a 1-m base between the temperature transducers (Lyubimova et al. 1974). Stations for measuring heat flux were located on the steep southwestern slope of the Gudauta bank at a depth of 180–340 m.

The measured values of heat flux considerably differ from each other (by 2.0 or 2.5 times) even at a short distance. At the same time, anomalous high values of

TABLE 2.1.4.1
Concentrations of Chlorion and Tritium in Pore and Seawaters in the Gudauta Experimental Area

Station Number	Testing Interval of the Rock Column (cm)	Chlorine Concentration (g/l)			Sea Depth at Station (m)	Temperature in Near-Bottom Layer (°C)	Depth of Sampling (m)	
		Pore Water	Near-Bottom Layer of Seawater	Surface Layer of Seawater			Tritium Concentration in Seawater (TE)[a]	Tritium Concentration in Pore Waters (TE)
2/81	30–35	10.10	9.35	9.09	50	16.5	0/50 ± 4	
	70–75	9.94					20/45 ± 3	40 ± 6
	130–135	9.86					50/58 ± 6	
3/81	25–30	10.0	9.67	9.24	50	14.2	0/37 ± 4	
	60–70	9.9					20/65 ± 5	62 ± 7
	100–110	9.94						
	140–150	9.9					50/11 ± 3	
	185–195	10.36					50/11 ± 3	
4/81	35–40	10.13	9.60	9.13	52	9.5		9 ± 2
	75–80	10.17						
	135–140	10.23						
5/81	35–40	10.42	9.44	9.17	52	13.8	0/39 ± 3	
	120–125	10.05					20/55 ± 5	29 ± 3
	160–165	10.1					52/54 ± 4	
6/81	30–35	10.14	9.65	9.79	50	14.5	0/73 ± 4	
	80–85	9.98					20/63 - 4	32 ± 2
	130–135	10.33					50/44 ± 5	
7/81	35–40	10.78	9.98	9.61	49	15.2	0/44 ± 4	
	75–80	10.67					48/45 ± 4	

Date	Depth		9.42	9.36	Cl	Temp	Ratio	Age
8/81	150–155	10.4	9.42	9.36	26	21.2	49/1 ± 5	39 ± 3
	—						26/69 ± 4	
	85–90	4.27						
	155–160	2.09						
11/81	25–30	9.5	9.40	9.25	24	23.6	0/45 ± 5	
	—	8.4					20/104 ± 6	
							24/104 ± 5	59 ± 5
12/81	20–25	10.1	9.42	9.25	24	23.5	0/55 ± 4	
	70–75	8.81					20/48 ± 4	
							24/51 ± 6	
13/81	30–35	9.71	9.02	8.94	22	25.2	0/68 ± 4	
							20/38 ± 4	
							22/45 ± 4	
6/78	15–25	10.56	—	—	200			
	65–75							
7/78	15–25	10.71	—	—	500			
	70–85	10.56						
10/79	15–25	10.7	—		610			
	70–90	10.25						
8/79	10–20	9.93	—	—	200			
	75–85	9.62						
4/79	15–25	9.93	—		180			
	70–85	9.62						
3/79	15–25	9.6°	—	—	142			
	65–75	10.56						

(Continued)

TABLE 2.1.4.1 (Continued)
Concentrations of Chlorion and Tritium in Pore and Seawaters in the Gudauta Experimental Area

Station Number	Testing Interval of the Rock Column (cm)	Chlorine Concentration (g/l)			Sea Depth at Station (m)	Temperature in Near-Bottom Layer (°C)	Depth of Sampling (m)	Tritium Concentration in Seawater (TE)	Tritium Concentration in Pore Waters (TE)
		Pore Water	Near-Bottom Layer of Seawater	Surface Layer of Seawater					
5/79	15–30	9.62		—	136				
	70–85	9.47							
7/79	15–25	9.93	—	—	164				
	70–80	9.47							
2/79	15–25	9.62	—		128				
	65–75	9.16							
3/78	20–30	8.54	—						
	60–70	9.16							
1/79	15–25	8.69	—		54				
	70–80	8.69							
2/78	—	—	—						
	55–70	9.16							
31/80	25–35	9.78			105				
	80–100	9.31							
	200–238	8.85							
33/80	20–40	9.00	—		54				
	100–140	8.54							
	300–350	8.07							

Sample	Depth							
34/80	20–40	9.00	10.8	10.0	69	7.2		
	100–120	8.69						
	200–220	8.69						
35/80	20–35	8.85	10.8	10.2	56	7.9		21.9 ± 8
	80–100	8.69						
	160–180	8.69						
36/80	20–40	8.69	10.8	10.2	55	8.7	0/74.8 ± 9 25/65.2 ± 8	34.2 ± 7
	100–130	8.54						
	240–270	8.23					55/36.6 ± 7	
37/80	10–40	10.24	11.8	10.2	168		0/88 ± 6 168/23 ± 4	16.6 ± 10
	100–140	9.78						
	250–290	8.85						
38/80	5–30	9.16			67			50.6 ± 13
	40–60	8.23						
41/80	10–40	9.94	11.8	10.2	200	8.8	16/49 ± 10	
	100–140	9.31					200/16 ± 8	
	250–290	8.69						
51/80	10–55	9.70			200	7.8	0/40 ± 9	
	180–230	9.47					200/26 ± 8	

1 TE corresponds to the concentration at which one atom of tritium (3HI) is per 1018 atoms of ¹H; 1 TE is equivalent to specific activity of 7.2 decays/ (mln liters of water), 31 TE is equivalent to specific activity of 7.2 decays/(mln liters of water), 3.24 pKi/l or 0.12 Bk/l.

heat flow in the Gaudskaya sandbank are probably caused by submarine deep thermal water discharge along fractures that are typical to this region and can be determined by the use of some geophysical methods.

Studying gas composition of seawater in the near-bottom layer is important for determination of the origin, age, and discharge areas of submarine groundwater. For studying gases dissolved in the near-bottom layer of seawater and in pore waters, we have developed a technique for the acquirement and degassing of samples. Seawater samples are taken from the near-bottom layer with the use of a hermetic probe with an internal pressure of to $2.9-10^7$ Pa. For taking water samples at a certain distance from the bottom, the probe is equipped with a fixing reconnaissance weight, which automatically locks at the specified depth. Degassing of water samples is performed by an ultrasonic vacuum deaerator. The deaerator consists of an ultrasonic generator, flask, thermocouple vacuum gauge, forevacuum pump, and bellows with mercury and glass ampoule to extract gas.

The degassing of seawater samples is carried out in the following way. First, the extracted (at the expense of pressure differences) gas from the probe is delivered to the emptied deaerator, and then the partly degassed water is passed into the flask. Then the ultrasonic generator is switched on and the extracted gas is pumped from the deaerator system to the ampoule by a mercury pump. A mercury pump represents a glass flask with its top outlet connected via an intermediary cock with the ampoule for gas and with its bottom outlet connected with a stainless-steel sylphon filled with mercury. Its volume is changed with the aid of an adjoining lever. The degree of vacuum in the deaerator, the amount of gas extracted, and the amount of gas delivered to the ampoule are controlled by a thermocouple vacuum gauge. After the gas is delivered, the ampoule is soldered by a torch. The volume of the ampoule is designed so that after degassing 1 L of seawater, some vacuum would remain. After degassing, the gas can be either analyzed on the spot with the aid of a portable chromatograph or delivered to the laboratory for analysis on a stationary chromatograph and mass spectrometer. It can be concluded from the first measurements, given in Table 2.2.4.2, that testing only seawater near the bottom is insufficient.

TABLE 2.2.4.2
Content of Dissolved Gases in the Near-Bottom Layer of Seawater

Station Number	Sea Depth (m)	Oxygen (%)	Nitrogen (%)	Methane (%)
9/79	810	13.6	86.436	
8/79	200		29.1	70.5
1/79	54	15.5	84.547	
10/79	610	17.8	82.2	
7/79	60	17.3	82.0	
2/79	128	17.4	81.2	

Due to the influence of the currents and the variations of temperature near the sea bottom, water masses are mixed, partly or completely smoothing gas concentrations in anomalous areas. Therefore, for more reliable detection of anomalous areas it is reasonable to determine dissolved gases not only in the near-bottom water layer, but also in pore waters of bottom sediments.

The complex profiling of the experiment site of Gudauta, as well as hydrochemical and isotopic analyses of pore solutions and seawaters, and geothermal, gaseous and filtration investigations on the sea bottom, enabled detection of six areas with SGD (see Figure 2.2.4.4). All these areas are located chiefly in the coastal zone at a short distance from the shore. The groundwater discharge is confirmed by the anomalies in the distribution of all the parameters under consideration. For example, the largest groundwater discharge area, detected during thermo- and resistivimetric profiling by an anomaly of salt content, is associated with the place where Paleogene bedrocks are exposed to the bottom surface, which was revealed by seismo-acoustic profiling. Furthermore, the seismo-acoustic profile shows a series of faults in this area. Stations 11/81 and 12/81, located in the zone of anomaly, recorded high-filtration coefficients of bottom sediments ($6 \cdot 10^{-2}$ and $4 \cdot 10^{-2}$ m/day). The same stations registered a decrease of chlorine-ion to 8.4 and 8.8 g/l, respectively. Isotopic testing of pore solutions and seawater has established that at station 12/81, tritium concentration undergoes a slight change and amounts in the near-bottom layer at 4 m from the bottom surface to (48 ± 4) TE, on the bottom (51 ± 6) TE, and in the pore solution within an interval of 20–75 cm (59 ± 5) TE. Tritium concentration in the pore water at this station is close to its concentration in rivers and precipitation. This means that the discharge of fresh groundwater is of atmospheric origin with a short period of water exchange.

At station 11/81, isotopic analyses were carried out only for seawater. High tritium content in a sample from the near-bottom layer (104 ± 5) TE compared with the surface sample (45 ± 5) TE confirms the presence of groundwater discharge on the sea bottom. Station 4/81 registered a low tritium content of (9 ± 2) TE in the pore water of anomalous area VI. This indicates a discharge of deep groundwater of Cretaceous age with considerable periods of water exchange.

Of interest are the data of stations 3T11 and 8/79 located at the southwestern margin of the experiment site near an isobath of 200 m. These stations registered high values of heat flux (82 mBt/m) and abnormally high methane content (to 70.5%) in a sample of the near-bottom seawater. Unfortunately, complex hydrogeological profiling was not carried out in this area. Therefore, it can be only surmised that these anomalies are related to SGD, possibly, with gas and oil deposits.

The desalinizing influence of SGD on pore water can be exemplified by the chlorine–ion distribution along the ground cross-section at station 8/81. In the near-bottom seawater, the chlorion content is 9.42 g/l, in pore water at a depth of 90 m from the bottom surface it is 4.27 g/l, and at a depth of 160 cm it is 2.09 g/l. Station 8/81 is located 500 m from the Lidzava River mouth and surely subjected to influence of the subchannel runoff.

Thus, the hydrogeological investigations, carried out in the Black Sea in the area of the Gudauta bank, made it possible to assess the feasibility of different methods for studying SGD and to determine optimal ones among them for different stages of sea hydrogeological surveying.

2.2.5 REMOTE SENSING METHODS

The use of remote sensing methods to study, assess, and protect groundwater provides the possibility to investigate extensive territories, obtain operatively different aspect information, and examine repeatedly objects under study. In recent years, remote sensing has an ever-growing application in hydrogeology. This is connected with the advantages of images of the earth obtained by this method. First, such photographic images can cover vast territories, enabling investigation into local and regional hydrogeological conditions of the Earth's surface. Another advantage is the ability of "translucence" of geological formations, when loose sediments can easily be seen on the rocks of other compositions. The technical means of remote sensing provide the possibility of taking photos of underlying layers in different narrow ranges of the visible spectrum (from blue to near infrared) and this, in turn, reveals various hydrogeological phenomena and processes that are distinctly seen within one or another interval of the multizonal photograph. Along with black-and-white or color photos of the Earth's surface, information on groundwater can be also obtained in television surveying, photographing infrared rays, and microwave and radiolocation ranges. These methods make it possible to reveal the interrelations of different hydrogeological objects and processes among the elements of the landscape, which are reflected on cosmic images. The study of existing interrelations between hydrogeological features and different components of the landscape is the first and foremost task for the development of scientific principles of hydrogeological decoding of cosmic images (Dzhamalov et al. 1977, 1978).

Basic tasks of hydrogeological decoding of aerocosmic photos of the Earth's surface are to: (1) reveal areas of distribution and determine the depth and mineralization of groundwater; (2) delineate recharge and discharge areas of groundwater; (3) clarify the kind and degree of interrelations between groundwater and surface-water; (4) assess the influence of artificial and natural factors on groundwater dynamics; (5) assess the hydraulic capacity of water-bearing rocks and their relative permeability, based on distinguishing lithologic-stratigraphical complexes and features of tectonic structure.

At any given stage of investigation, the most complicated among the above-listed tasks is to determine the depth and mineralization of groundwater, as well as obtaining quantitative characteristics of water-bearing rock permeability in order to have preliminary regional values of SGD. A solution for this task is possible, based on the integrated use of remote sensing results in visible, infrared, and microwave ranges, which in combination give certain knowledge on certain physical characteristics of a surface.

For determination of reliable decoding indicators of hydrogeological phenomena and processes, numerous different-scaled cosmic images taken from different carriers (satellites, manual crafts, and cosmic stations) were analyzed, with different equipment, during different times of the day and year. It was established through comparative analysis of these sources of information that the solution of the tasks of hydrogeological decoding should be based on direct indications that are typical of the hydrogeological objects themselves, and indirect indications manifested in interconnections of hydrogeological processes with different components of the landscape. Also, those complex indications are to be used for decoding, which reflect by pattern and structure of a photographic image, typical external features of concrete territories, as well as evidence of their internal structure and processes. The evidence results from a linkage between the features of a photographic image and the geologic-hydrogeological conditions of the territory. Hydrogeological decoding should include both study of the physionomic features of landscapes directly imaged on cosmic photos and geologic-hydrogeological interpretation of regions under investigation on the basis of the established landscape-indication links.

Usually direct decoding indicators (color, tone, size, and shape) are used to show concentrated groundwater discharges to the surface and on the sea bottom, swamps, solonchaks, takyrs, karst forms, icings, and large submarine springs in the shelf zone. Among indirect indicators are relief, soil moisture, vegetative communities or sets of plants, and geological formations located in the same area of the Earth's surface and closely interacting with each other and with conditions of the physical-geographical medium. For example, groundwater discharges to the surface and swamps on cosmic photos appear as dark spots. The shapes and sizes of these spots indicate the intensity of groundwater discharge associated with different relief forms and certain types of rocks, as well as the depth of groundwater occurrence. Photos distinctly reveal ancient river channels, mort-lakes, and abandoned irrigation constructions and accumulators of groundwater. Tectonic features of territories, decoded on photos, help in revealing and verifying locations of artesian basins and the directions and paths of groundwater flow. Often a specific displacement of rock layers, an elongated chain of groundwater discharge, or linear accumulations of different vegetation species make it possible to determine exact rupture disturbances characterized by an increased jointing and cracking of rocks along which groundwater moves. However, direct decoding indicators are often not enough to recognize hydrogeological features of the underlying surface. Therefore, for interpretation of photos, it is necessary to use indirect decoding indicators that enable understanding of some hydrogeological processes and objects which are not reflected directly on a photo and cannot be determined by direct indicators. Such indirect indicators include relief, vegetation, soil moisture, conditions of origination and melting of snow and ice covers, density of hydrographic network, geologic-structural features of territory, lithologic-facial characteristics of sediments, shoreline and type of seas, and photo-tone features of seawater. Applicability of indirect decoding indicators are greatly restricted by total knowledge of territory, thus, revealing and purposeful interpretation of them should be carried out on the basis of all existing information of the region under study.

Complex decoding indicators are connected with tone, shape, and pattern of a photographic image of a study surface and reflect a type of landscape as a whole. The landscape has outside and inside elements, the first among which (relief, vegetation, hydrographic network, traces of human activity) serve as physionomic components of the landscape and find a direct reflection on photos. Inside elements and, in the first turn, hydrogeological features of territory usually do not find reflection on photos and can be recognized only by indirect indicators, for which it is necessary to reveal an indicating role of outside elements of the landscape.

It is known that photographing of a surface can be performed in different areas of the electromagnetic radiation spectrum. Presently, of special interest is cosmic and television photographing in visible and near-infrared areas of the spectrum from 0.4–1.1 mkm. Using spectral differences with reflection of sunlight by elements of the Earth's surface, it is sufficiently possible to precisely identify different natural formations. However, in this case, one should distinguish original photos of the Earth's surface and its television image. Original photos are characterized by high resolution; whereas television images have the advantage of being obtained quickly, regularly, and repeatedly if necessary in spite of less resolution than the original photo.

The basic goal of specialized decoding in a given step is to determine the informational degree of different photographing ranges that would be enough to recognize those of other objects. Especially effective in this respect are synchronous investigations in the system with a "satellite–airplane–on-ground team of specialists," each providing various and comparable data on studied landscapes, processes, and phenomena with different degrees of detail and generalization.

The visual decoding of positive imprints of multispectral cosmic surveying in the ranges of 0.5–0.6, 0.6–0.7, 0.7–0.8, 0.8–1.1 mkm reveal the basic features of reflected direct and indirect indicators of hydrogeological conditions of the territory. The results of such a decoding with description of types and structure of imaged objects and indications of the most informational ranges and their combinations are presented in detail in Dzhamalov et al., 1979.

The investigations have shown that the structural plan of large territories, where it is tightly connected with their relief features, is decoded most reliably in the range of 0.6–0.7 mkm. The range of 0.8–1.1 mkm distinctly reflects tectonic rupture disturbances, individual blocks and their boundaries, and cracking and ruggedness of structures. Due to this, it is recommended to interpret photos jointly in these two ranges during hydrogeological decoding of large territories. When manifestations of hydrogeological processes are to be recognized, special attention should be paid to such direct and indirect decoding indicators as groundwater discharge to the surface, lithology of rocks, features of vegetation cover, presence of naturally wetted areas, conditions of origination, and melting of snow and ice covers, etc. Water bodies are seen most distinctly in the range of 0.8–1.1 mkm, loose Quaternary sediments are within a range of 0.5–0.6 mkm, and the genetic forms and structural features of bedrocks within 0.6–0.7 mkm. To recognize different vegetation communities and to verify their phonologic state, it is required to compare the photos in the ranges of 0.6–0.7 and 0.8–1.1 mkm. Areas of melting wet snow and ice are recognized in the range of 08–1.1 mkm; moistened territories differ by a relatively dark phototone and keep their optical contrast in all the ranges of a studied area of the spectrum, from 0.4–1.1 mkm.

As mentioned above, one of the most informational types of cosmic surveying is the infrared survey to obtain images of thermal radiation of the Earth's surface. Information obtained from satellites with the aid of this survey gives a picture of temperature on surfaces of the Earth and water, as well as on relative moisture content in soils and rocks. Temperature contrasts help to recognize availability and spreading of shallow aquifers, groundwater discharges to the surface, groundwater discharges on the bottoms of rivers, lakes, and in the shelf zone of seas and oceans, and to study and map volcanically active areas. As the temperature of groundwater usually differs from the temperature of riverwater or seawater, so the groundwater discharges in seas are seen on infrared photos as feathery shapes or as horse tails in a dark phototone. Such photos give the possibility to delineate where on land areas groundwater can be used for domestic purposes. Physionomic reflection of different components of the landscape on the photos depends usually on the season of photographing. Study of phonological evolution of vegetation, changes in natural moisture content of rocks, conditions of melting and origination of snow and ice covers, hydrologic dynamics of constant and temporary water streams, all require a season for taking photos and with repeated frequency. Thus, decoding of vegetation covers, seasonal changes in moisture content, detection of temporary water streams, salted soils, solonchaks, and takyrs is most effective for photos taken in the spring and summer. Structural, tectoni, and lithological features of an underlying surface in forests, forest-steppes, and steppes can be reasonably decoded from photos taken in spring; whereas in deserts, the summer is when the hidden role of vegetation cover is least significant. Relief morphology of plains is reflected distinctly on photos taken during warm times of the year, whereas dissectedness of mountainous countries is best seen on winter photos when snow cover outlines basic elements of the relief.

As an object of study, groundwater and sources of its discharge in the coastal zone of seas usually do not find a direct representation on photos, so the following approximate scheme of hydrogeological decoding of cosmic images is suggested. At first, direct, indirect, and mixed indicators are distinguished, which are not usually met, but form regular combinations in different coastal landscapes and geologic-hydrogeological structures. These indicators are then correlated and verified with available published data on the territory under study, as well as with different general purpose and specialized maps. For more detailed treatment of decoding results, the revealed indicators and their indicating role should be verified on the spot by means of on-ground reconnaissance routes and special complex works along with several profiles and in key areas of the coast.

In the second phase of decoding, based on the identified decoding indicators and determination of their indicating role for conditions of SGD formation, a series of analytical maps and schemes are compiled, which reflect certain indications of geologic-hydrogeological features of coastal areas. In the first turn, these are the maps of submarine groundwater discharge; density of river network and erosion-induced dissecting of relief; vegetation cover; and geomorphological, tectonic, and lithologic-facial elements of the landscape. This entire set of analytical maps gives the possibility with less or more detail to assess objectively an influence of different factors on formation, transport, and discharge of groundwater and, based on the already-known natural links, to obtain a preliminary picture of submarine discharge conditions.

The third phase includes compilation of synthetic landscape-typological schemes based on analysis and generalization of the analytical maps. The schemes are constructed on the geological-geomorphologic principle. They incorporate all the obtained information and actually represent a geologic-hydrogeological zone of coastal territories according to conditions of SGD on the basis of distinguishing large geologic-geomorphologic elements with typical sediments. The landscape-typological schemes are added with landscape-indicating ones showing different decoding indicators typical of each geological-geomorphologic element and showing evidence of conditions of groundwater discharge within it. Such elements include mean-sized and small elements of the relief, availability of a typical vegetation cover and its density, lithologic-facial characteristics of covering sediments, exposures of bedrocks, water phenomena and their location, direction of tectonic disturbances and degree of rock jointing, and manifestation of different natural processes and phenomena on the surface (karst, landslides, debris flows, icings, taliks, swamps, etc.).

The described schemes of landscape zoning, which reflect a linkage of physionomic components of the landscape with geologic-hydrogeological conditions, enable the closing phase — the compilation of general purpose and specialized hydrogeological maps of groundwater discharge. Informational density of the hydrogeological maps depends on the degree of detailed study of natural complexes with the use of remote sensing and can include contours of hydrogeological zoning, sizes, and characteristics of aquifers and aquifer complexes, some quantitative values of hydraulic capacity and transmissibility of water-bearing rocks, and relative estimates of groundwater discharge. Thus, the landscape schemes collect together all the decoding indicators typical of identified hydrogeological taxonomic units.

Hydrogeological maps, compiled with the aid of cosmic images, contain, as a rule, information which is 1.3–1.5 times larger than maps compiled in a traditional way. Information on the maps increases with distinguishing new elements not reflected on earlier compiled maps, as well as the possibility to assess the hydrogeological role of these structures with more exact laying out contours of objects and correspondence of the contours to a specified scale of surveying.

The basis of hydrogeological zoning by different-scaled cosmic photos must be the geologic-geomorphologic or structural-hydrogeological principle, which makes it possible to distinguish hydrogeological structures of different ranks. During interpretation of cosmic photos to be used for hydrogeological zoning, it is necessary to take into account that when passing to more small-sized images, a natural generalization of an image occurs because of optical, geometric, and thematic integration of structural features of the territory under study. Optical or spectral integration of elements of the Earth's surface, occurring with an increase in height of the carrier, is caused by an influence of the transferring function of the atmosphere and by averaging of the reflecting and radiating abilities of on-ground objects different by their spectral characteristics.

Thematic integration occurring with an increase in surveying scale results in consolidation of distinguished taxonomic units, which, on the one hand, is accompanied by a decrease in the number of direct and indirect decoding indicators due to a reduction of detail on images, and, on the other hand, makes the number of details not enough for reliable recognition and distinguishing of natural objects of a larger rank. However, it should be noted that thematic generalization during small-scale surveying, occurring at the expense of generalized and reduced decoding indicators, leads to a loss of informational degree of images necessary for hydro-geological interpretation.

Geometric integration on cosmic images means a decrease of their details and a change in structure and texture of a photographic image and, hence, a general decrease of photographing with an appropriate increase in its viewing coverage and in resolution. The basic criterion of geometric generalization is spatial resolution of photographing. Practice of decoding shows that during studying of elements of the landscape one should distinguish two kinds of resolution: (a) demarcative spatial for dividing different natural objects and (b) recognizing spatial resolution, enabling reliable identification of concrete objects and investigation of their inside structure. Demarcative resolution is, as a rule, higher than recognizing one approximately by one order of magnitude.

Geometric generalization for the purposes of hydrogeological zoning can be composed of several levels, depending on spatial resolution: global, with a resolution of 3–10 km, at which hydrogeological provinces and areas are identified; regional, with a resolution of 1–100 m, at which large artesian basins and hydrogeological massifs are recognized; landscape, with a resolution of 100 m to several hundred meters, with the possibility of identifying relatively small artesian structures and particular aquifers or aquifer systems with subsequent more detailed investigation of their hydrogeological features. Thus, of special interest in hydrogeological zoning for regional estimation of SGD are regional and landscape levels of generalization and, hence, scales of aerocosmic surveying corresponding to them. Here, on small-scale images the basic role belongs to combined and landscape-like indicators, but with an increase in details of an image, of growing importance are direct and indirect decoding indicators.

During hydrogeological decoding of aerocosmic images, the zoning should be carried out beginning from larger units to smaller ones using various combinations of decoding indicators at each phase. The locations of artesian basins and hydro-geological structures with subsequent identification of areas of groundwater recharge, transient, and discharge within them is determined by structural-tectonic features of territory. Therefore, in identifying these relatively large taxonomic units, relief and geologic-tectonic indicators serve the basic role. A special note should be given here to clarify the hydrogeological significance of lineaments — rupture disturbances, which, to a great degree, determine hydrogeological features of coastal territories, are predominantly in areas of wide surface spreading of crystalline bed-rocks. A stretch of rupture disturbances is often controlled by groundwater discharges to the surface, chemical, gaseous and thermal regimes, which enable understanding of depth and current activity of identified faults.

For identification of smaller taxonomic hydrogeological units (small artesian basins, particular aquifers and aquifer systems, etc.), special attention should be paid to such direct and indirect decoding indicators as groundwater discharges to the surface, rock lithology, features of vegetation cover, availability of naturally wetted areas, conditions of origination and melting of snow and ice covers, etc. The indicators given are reliably decoded on images at the landscape level of gen-eralization. Detailed analysis of aerocosmic images makes it possible to verify location and boundaries of artesian basins and hydrogeological structures of various ranks, to give relative characteristics of hydraulic capacity and transmissibility of water-bearing rocks, and to compile some pictures of groundwater discharge.

Use of remote sensing methods for studying groundwater is most useful and informative in poorly studied regions where principal data on geologic-hydrogeological conditions of large territories are in need. The data on groundwater, obtained in this way, then serve as the basis for definition of more detailed prospecting-exploratory works aimed at estimation of groundwater reserves and prospects for their practical use.

3 Groundwater Contribution to Global Water and Salt Balance

CONTENTS

3.1 GROUNDWATER AND HYDROGEOLOGICAL STRUCTURES OF SEA AND OCEAN FLOORS*

3.1.1 GROUNDWATER OF THE SILT SEDIMENTS OF SEAS AND OCEANS

Two types of groundwater can be determined on the floors of seas and oceans: water of silt sediments and water of consolidated rocks. Groundwater of the first type is closely connected with the history of the given sea basin, and especially with paleogeography of the Quaternary period. Solid compounds, being transferred to oceans from the land, deposit mainly in the shelf zone. The rate of sedimentation is great and equal to 9 sm/yr (Korotkov et al. 1980). In the deep-sea part of the ocean this value is 1–1.5 times lower. Silt water has the most variable composition in the shelf zone. At the first stage of clay deposit formation on the shelf, its wetness amounts to 80–120%. Sediment dehydration begins even at the first stage of diagenesis. Within the upper zone of 10–15 m, wetness decreases 2–3 times and density increases from 1.4–1.9 g/m^3. Then wetness decreases slowly with depth. Squeezing of pore water back to the sea or forming sandy and gravel sediments begins at a depth of several meters and continues during the long-term geological history of sedimentary layer formation.

Complicated hydrochemical and gas regimes of shelf seas influence groundwater chemical composition in silt deposits. A relatively high rate of sedimentation and rapid lithofacies decreases diffusive permeability. This leads to diffusive leveling off in chemical composition of pore water in silts and, hence, it can significantly retard changes in seawater chemical composition, especially for interior seas. The diffusive coefficient for sodium chloride (common salt) in silts is equal to n·10^{-6}sm^2/sec, and can be estimated as 2·10^{-6}sm^2/sec on average (Zatenatskaya 1965; Smirnov 1974). Under such values of diffusive coefficient duration of diffusive leveling off, chemical composition of groundwater in the upper layer of the shelf sediments can be estimated as tens but mostly hundreds of thousands of years (Korotkov et al. 1980). This allows one to carry out paleoreconstruction of events of the Quaternary Period, Neogene. An example is the Black Sea, which is an interior sea connected to the Mediterranean Sea by the straits of the Bosporus and Dardanelles.

During the maximum stages of the Valday glaciation, the level of the World Ocean was 150–200 m lower than in modern times (Vigdorchek 1980). As a result, the Novoeksinsky basin represents a drainage lake with intensive glacial water alimentation separated from the ocean. The water level in this basin was significantly lower than today. At the beginning of the Holocene (~10,000 years ago) seawater again started to penetrate the area of the modern Bosporus strait and approximately 4000 years ago sea level reached close to modern levels. These events were detected in the columns of pore water composition, one of which is given in Table 3.1.1.

* Section 3.1 was written by Vladimor A. Kiryukhin and Alexey I. Korotkov (Saint-Petersburg State University, Russia, spmi.hgig@mail.ru).

TABLE 3.1.1
Chemical Composition of the Silt Sediments Pore Water of the Central Part of the Black Sea (g/kg), Depth of 1700 m

Depth from the Sea Bottom (sm)	Wetness (%)	Eh (mV)	Cl^-	Br^-	SO_4^{2-}	HCO_3^-	Na^+	K^+	NH_4^+	Ca^{2+}	Mg^{2+}	Total	Cl/Br
The sea, depth of 0.270 m			12.3	0.032	1.46	0.24	6.73	0.27	—	0.30	0.82	22.1	384
10–20	64.8	+176	12.2	0.041	1.53	0.32	6.69	0.23	0.005	0.27	0.83	22.1	298
198–220	50.0	—	11.0	0.037	0.16	0.51	5.86	0.20	0.019	0.21	0.58	19.6	297
360–380	47.0	−151	9.59	0.030	0.043	0.68	4.87	0.10	0.025	0.41	0.53	16.3	320
620–635	43.0	−166	7.76	0.031	0.062	0.48	3.59	0.02	0.036	0.76	0.38	13.1	250
704–720	—	−101	7.13	0.030	0.082	0.19	3.13	0.03	0.049	0.84	0.28	11.8	238
1013–1030	37.5	+14	5.31	—	0.062	0.21	—	—	—	0.88	0.25	—	—
1171–1192	41.2	+74	4.54	—	0.082	0.12	1.54	0.01	0.052	0.79	0.24	7.3	—

From Shishkina (1972); Babinets et al. (1973).

The available data can clearly determine the following regularities. Pore water freshening with depth is connected with paleogeographical reasons. Processes of silt water metamorphisation in the "straight direction" are first of all marked by a constant increase of calcium concentration with depth, even with a total mineralization decrease. Processes of sulfate reduction are accompanied by noticeable increases in concentration of hydrocarbonate ions ($SO_4^{2-} + 2H_2O + 2C_{org} \leftrightarrow 2HCO_3^- + H_2S$) at a depth of 2–6 m lower than sea bottom which causes pollution in the deep layers of the Black Sea by hydrogen sulfide. Reducing processes are more precisely reflected in the migration of nitrogen forms, mainly in increasing of ammonium concentration with depth that reaches the highest values equal to 50 mg/l in the lower part of the discussed column. A relatively constant chlorine-bromine ratio, which is similar to ocean water and independent from the above mentioned processes, indicates the marine genesis of the pore water. In the northwest shallow part of the Black Sea, it was possible to trace another interesting effect of the deltas of the Danube and Dnieper Rivers moving in the Holocene on the base of pore water chemical composition. In Korotkov et al. (1980), examples of such detailed reconstruction for the Baltic Sea, Arctic Sea, and seas of the Far East are given.

Processes of certain microcompound accumulation in silt layers are also of great interest. Bromine is the most conservative component. The main mechanism of its accumulation is evaporative concentration. Bromine concentrations are stable and remain the same in the seawater. The chlorine–bromine ratio increases in the areas of salt cupola development. Iodine intensively accumulates and is enriched by organic matter shelf sediments. If in ocean water its average concentration is about 0.06 mg/kg, then in silt water it can reach higher. Its average value in silt pore water of the ocean is estimated at 0.7 mg/kg (Kudel'skiy 1976). Average concentration of fluorine in the seawater is 1.3 mg/kg. In silt water it can increase somewhat. But it decreases due to increase of calcium concentration and because of decrease in solubility of precipitated fluorite (CaF_2). In the deep-sea part of the Black Sea, where silt water is enriched by calcium, fluorine disappears entirely (Shishkina 1972). It is interesting to discuss bromine behavior in the pore water of sea sediments. Its average concentration in the seawater equals 4.6 mg/kg, but in some areas (e.g., in areas with mud volcanism) it can increase to 40–50 mg/kg.

Moving away from the continent in the deep-sea part of the ocean chemical composition of the pore water becomes more uniform and closer to the ocean water. However, it is possible to determine areas (e.g., Kuril-Kamchatsky trench) with intensive sulfate-reduction processes. Processes of cation exchange occur intensively in some areas. Detailed material with such data can be found in the reports of apparatus "Glomar Challenger."

3.1.1.1   Hydrogeological Structures of the Land–Ocean Fringe Zone

The land–ocean fringe zone consists of four parts: the shelf, margin sea, volcanic curve, and interior sea. Each indicated part has its characteristics and it is advisable to discuss each one separately.

The Shelf. The shelf adjoins the continent and its inner boundary separates shelf and continental slope. The shelf depth usually does not exceed 200 m. Hydrogeological structures, specific to the land (e.g., hydrogeological massifs, artesian and volcanic basins) are developed in this area. In comparison with land these hydrogeologic structures usually are semimarine, that is, partly moving on land and partly situated under sea level. The following characteristics of the semimarine hydrogeological structures can be noted:

1. One part of the structure situated on the land is characterized by subaerial conditions and recharges due to atmospheric precipitation. Another part of the basin is situated below sea level and experiences abrasion and erosion. Thus, in comparison with the land, it can be significantly washed out and changed.

2. The area of the aquifer recharge is on the land. Hence, groundwater flow is directed toward the sea. It means that groundwater discharges as subaquatic or concealed sources on the shelf bottom. The most significant sources are observed in the karst areas. Such sources can be determined at a distance of 1000 meters and even some kilometers. Methods for assessing debits of such sources are developed (Korotkov et al. 1980). On the Caucasus coast of the Black Sea, the yield of some submarine sources situated at some kilometers from the coast reaches 1.6 m^3/sec (Yurovskiy 1993).

3. Semimarine structures can be also called coastal-shelf structures because the sea can regress from the modern shoreline tens of kilometers, because of water levels dropping to tens and even hundreds of meters. This means that the shelf can become the land. Such situations often occurred during Quaternary glaciations.

4. Groundwater of semimarine basins is characterized by a high variety in chemical composition and mineralization. This is due to possible freshwater burial on the one hand, and lagoon forming at some stages of development where evaporates, precipitation, saltwater, and brine formation occur.

5. It is possible for intrusions of saltwater of sea genesis to the land. This can occur due to tide effects in river valleys and also in areas of interconnection between freshwater and saltwater. During groundwater withdrawal, depression cones are formed and tightening of saltwater from deep aquifers can occur. Such problems are noticed in the Netherlands, Israel, Russia, and other coastal countries.

Semimarine structures are mostly continuations (extensions) of artesian basins. For example, toward the Barents Sea, the Severo-Dvinsky and Pechorsky basins are continued; toward the Karskoe Sea is the West Siberian basin; and toward the East Siberian Sea is the Yakutsk basin. Continuation of continental geological structures is observed in all ocean coasts. They are of great interest for exploring oil and gas fields and technical mineralization.

The Marginal Seas. The Barents, Karskoe, and East Siberian Seas on the north of Russia; the Chukotskoe, Beringovo, Okhotsk, and Japan (also East China and Yellow) Seas along the boundary of the Pacific Ocean; and seas along the north boundary of Canada and Alaska can be mentioned as examples of such seas. Complicated combinations of different types of hydrogeological structures are observed at the bottom of these seas. First, artesian basins of the semimarine type continuing on the land and shelf occupies a significant place here. Second, artesian basins, hydrogeologic massifs, and volcanic basins of submarine type, which are almostly covered by sea, are developed on these territories. Third, the boundary parts of ocean structures (e.g., subocean basin, marginal arches, etc.) can be noticed here. These structures can occur in the core of the ocean or as a transient type. The bottom of marginal seas is under study and most of them have high perspectives for exporting and development of oil and gas fields.

Volcanic Arches. The main characteristics of volcanic arches (Figure 3.1.1.1) include the following: They are situated on the fringe of ocean structures–deep-sea trenches. Modern volcanism and active magmatic and gas hydrothermal activity is widely developed. Different structural and hydrogeological conditions (occurrence of hydrogeological massif, small artesian and volcanic basins) are noted. They are distinguished by favorable conditions for forming groundwater of different chemical composition, temperature, and characteristics. The following conditions are particularly developed: freshwater of infiltration origin, saltwater of sedimentation origin (widely distributed in areas of intrusion into sea coasts), and mineral water of fumarole and solfatara types of magmatic and infiltration origin in areas of volcanic sources.

Interior Seas. Hydrogeological structures of the interior sea bottoms (e.g., the Mediterranean, Black, Caspian, Baltic, North, Caribbean, Mexican Seas, etc.) occupy the specific place where submarine and semimarine artesian basins can be determined. Oil and gas fields were discovered and exploited in some of them. Wings of some basins appear on land and are on the subaerial stage of development characterized by the infiltration regime of the upper aquifers. The interior parts of these structures lay under sea level and are characterized by a submarine stage of development and an elision regime for sedimentary water-bearing layer of the cover.

Groundwaters of exterior (subaerial) and interior (submarine) parts of artesian basins are closely interconnected. This connection is reflected in the chemical composition of submarine springs situated near the coastal zone of interior seas. Karstwater discharge is noted in many parts of the Mediterranean Sea at depths of 120–700 m. Mineralization and the composition of water of submarine structures of interior seas are different. Sometimes fresh infiltration water of atmospheric alimentation is found at significant distances from the shoreline. Also, saltwater of different mineralization and oil and gas fields are widely distributed. Abundance of hydrogen sulfide is determined in some submarine basins (e.g., the Black Sea's complicated artesian basin). In other basins (e.g., the Alboran, Algier-Provence, Tyrrhenian, and other artesian structures of the Mediterranean Sea) strong brines and evaporites are noticed. Diapiric structures (salt stocks [bodies]) are found at the bottom of the Mexican Bay. Here, strong brines are widely spread and oil fields are developed and

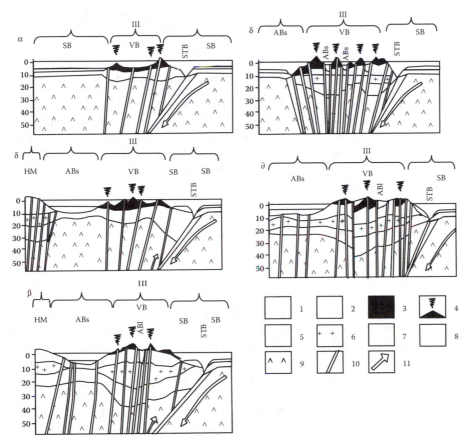

FIGURE 3.1.1.1 Hydrogeological structures of volcanic arches. Types of island arches: (a) Marianskiy; (б) Kurilsk; (в) Japanese-Yavansk; (г) Australaziiskiy; (д) Kamchatsk-Sumatrian; (ABl) artesian basins of land; (ABs) artesian basins of sea bottom; (VB) volcanic basins; (SOB) subocean basins; (STB) subocean trench basins; (HM) hydrogeological massif; (I, II, III, and IV) boundaries of continents, marginal seas, island arches, and oceans, respectively; (1) sea and ocean water; (2) sedimentary cover of artesian basins; (3) modern volcanic sediments; (4) active volcanoes; (5) folded basement; (6) granite layer; (7) basalt layer; (8) serpentinous rocks; (9) subcrustal mantle; (10) fractures; (11) zone of subduction (lithosphere plates and blocks underthrust).

exploited. Also oil and gas fields were discovered and exploitation was carried out at the bottom of the North Sea.

Variability in chemical composition of groundwater can be explained by the sharp variety in the level and chemical regime of seas. During glaciations, seas could transgress far from the modern seashore line. Moreover, as they regressed back, they could bring water of another chemical composition. At present, mineralization and composition of water of interior seas differ significantly from those of oceans. Semimarine structures are widely distributed in interior seas and have an exterior area of alimentation. At the same time, hydrodynamic pressure can contribute to freshening water movement along water-bearing layers at a distance of several kilometers from the shoreline. As a result interstratification (interbedding) of water with different mineralization and composition are observed in the hydrochemical profile.

3.1.1.2   Hydrogeological Structures of the Ocean Bottom

Continents and oceans need to be discussed as hydrogeological structures of the highest range. They differ from each other by Earth crust structure, thickness, composition, groundwater regime, and other factors (Kiryukhin and Tolstikhin 1988). The continental crust has a three-layer profile, but in the ocean crust, absence of the "granite" layer is observed. Different types of sedimentary (terrigenous, carbonate, saliferous [salt-bearing]), volcanogenic, intrusive, and metamorphic rocks with bedding, fractured-vein, and lava water are developed on the continents. The ocean crust profile is more homogeneous. It consists mainly of terrigenous and volcanic sediments; organogenic (biogenic) and chemical sediments are of secondary importance. An aeration zone is almost absent in seas and oceans. Silt water of the ocean bottom contacts directly with ocean water, expressing its influence. The regime of groundwater at the ocean bottom is closely connected with ocean water and differs by relative stability, having deep zonality and latitude zoning of the ocean water. Flow of the ocean groundwater follows gravitational consolidation of bottom sediments. Groundwater can move toward more deep parts of the ocean bottom, as well as the positive relief forms from depressions and deflections, where sediments deposition and compression occurs. Chemical composition of ocean groundwater is homogeneous and mainly depends on the composition of water-bearing rocks. Usually this is saltwater of the ocean type and rarely brines. The temperature of the ocean groundwater occurring near surface sediments changes insignificantly: from 2–3°C at the Equator to –0.7°C at the Polar seas. The solid phase of groundwater is absent and is noticed only on shallow shelves of the Arctic seas. The lithogetic regime determines the dynamic of groundwater at the World Ocean bottom. Tectonic and volcanic processes significantly influence it in some areas.

Within the continents, the following above-range structures are determined: active zone (hydrogeological folded zone) and tectonic stable platforms, where shields and plates correspond to hydrogeological folded zones and artesian basins. Also, above-range structures are determined at the ocean bottom. Active (general) hydrogeological zones of middle-ocean ridges and uplifts on the one hand, and stable hydrogeological areas of the ocean platforms on the other ought to be placed in that category. According to its hydrogeological characteristics, areas of subocean ridges and uplifts differ precisely from zones of ocean platform. On the continents hydrogeological massifs and artesian and volcanic basins are the structures of the first type. At the bottom of the World Ocean, the structures of the first range are the following: subocean massifs of the fractured water, subocean basins of the sedimentary cover, and ocean volcanic basins (submarine and insular).

Subocean massifs (SOM) of the fractured water mainly consist of basalt rocks, overlapped by unconsolidated sediments of insignificant thickness. At ocean bottoms, subocean massifs form positive relief forms. They are arches, ridges, and elevations, composed mainly by basalt rocks. Here, waters with mineralization and chemical composition referred to as the ocean type are noticed. They are related to the fractures of different origin in the base rocks. The upper layer consists of unconsolidated sediments with silt water and overlaps the lower layer. Sometimes the middle layer occurs between the basement and the upper layer with silt water. The middle layer consists of lithified sedimentary rocks with porous-fractured water and basalts with layer and fracture water. Subocean massifs of the fractured water are mainly distributed within the mean-ocean ridges where they form systems large in distance and width. These systems encircle the Earth. They extend in submeridional direction in the Arctic, Atlantic, Indian, and Pacific Oceans.

Subocean basins (SOBs) of the water of sedimentary cover are subdivided on large ocean basins, smaller subocean rift basins (SRBs), subocean trench basins (STBs), subocean basins of depressions (SBDs), subocean fracturing basins: transverse (SFBtr) and transformed (SFBt), etc.

Subocean basins are usually situated on both sides of mean-ocean uplifts and ridges. Lower basalt layers are composed of the basement of the basin and likely contain fractured water of different types and genesis. Basement of the basins rises up to the ocean surface in the peripheral part. In some areas it appears on the bottom forming SOB of the fractured water, which separate one SOB from another. Here, terrigenous material is washed away from the continents and large river deltas (Amazon, Ganges, Nile, Hwang Ho, Volga, etc.). This process is especially intensive in areas of humid climate distinguished by intensive atmospheric precipitation.

Critical depth (~4800 m) separates deep-sea areas with slow, red clay sedimentation from less deep-sea areas with comparative rapid carbon silts sedimentation. Presence of the critical depth in SOB causes forming of a thick sedimentary cover in the peripheral part and a thin cover in its central part. For example, in the Central Pacific SOB the thickness of the cover consists of carbon silts <1 km in the uplifts and peripheral areas, but in its central deep-sea parts it is <200 m and consists of red deep-sea clays.

Silt water occurs in the upper layer of the SOB, whereas layered, fracture and layer, and layer and fracture water is developed in the medium part of the basin and fracture-and-vein water in the lower part of the basement. In areas where deep fractures cross the SOB, fracture and vein water, sometimes hot with metals and gases of mantle origin, is distinguished. The composition of this water differs from silt water of the upper layer, which is similar to the water of the ocean type changed due to the diagenesis processes.

Situated far from the continent distal wing, the SOB adjoins to the slope of the mean-ocean ridges and uplifts. Here volcanic sediments such as pyroclastic rocks, lavas, and tuffs, overlapped by silt cover-containing silt water, are developed. The proximal wing of the SOB is located near the continent. Groundwaters of different composition and forming conditions are typical in the hydrogeological structures of the continent (e.g., hydrogeological massifs, artesian and volcanic basins) and can flow there.

Subocean basins are the largest and the most complicated hydrogeologic structures. Within these structures and in their peripheral volcanic subocean, island basins overlap them and deep (up to mantle) fractures cross SOB. The size of other SOB basins is significantly smaller.

Subocean rift basins are mainly connected with the axial parts of the mean-ocean ridges and uplifts. They are stretched along ridges and uplifts and replace each other on the great distance. These structures are the youngest among SOBs. Sediments of turbidity flows, slides, and collapses are typical to SRBs. According to Rozanova (1971), among them it is possible to determine mixtures of coarse and thin material, noncarbonate, slightly carbonate, and carbonate turbides, volcanic and sedimentary deposits with gradational layering (stratification), and traces of hydro-thermal activity.

Rift origin develops inconstantly in time and space. Rift zones are distinguished by the thickness of sedimentary cover, its form of layering on the basement, Earth crust structure, and also the anomalous structure of the magnetic field, heat flow distribution across the ocean bottom, local hydrothermal developments, and anom-alies in the age character of the base sediments and magmatic rocks, etc. Rocky sediments of the second and first low layers, represented by alkaline rocks, are exposed on the slopes of the rift zones.

It can be concluded from the above discussion that on the bottom and slopes covered by sediments, silt water is developed and fractured water is observed in the rocky sediments. Finally, fracture and vein water and, in some cases, thermal water are noticed by the fractures of the rift. Some of it comes from a significant depth and can also be mantle ore-bearing solutions (Kiryukhin and Tolstikhin 1988). Fractures are also channels of magmatic fusion movement to the surface of the ocean bottom. Thus, young volcanic forms, called ocean volcanic basins (OVBs), are closely connected with rifts.

Subocean trench basins, similar to rifts, stretch significantly and differ by a great variety. In the Pacific Ocean they occur widely, but in the North Arctic Ocean they are absent. Hydrogeology of the STB is complicated and varied. At present it can be described according to data of marine geological studies. Base rocks of the ocean crust are exposed on the slopes of trenches. Thus, the basement of the STB is characterized by fracture and fracture and vein salt and thermal waters. On the bottom of the usually flat trench, sediments with thicknesses of 2 km can sometimes be found. For the bottom sediments of STBs, silt water is typical. In some places, trench slopes are covered by the sediments, which are similar to turbidities, clays, and others, including agglomerates similar to aleurolites and sands. Cases were noted where trench slopes join and leave no place for silt sedimentation. STBs are usually developed in the periphery of the oceans and conjugated with island arches. The latter are the source of the terrigenous and volcanic sediments. A specific feature of STBs is the combination of two types of the Earth's crust: ocean crust from the ocean and a continental crust from island arches. Another feature consists of island arches from the side of marginal seas that swell up from the ocean side, and are called the marginal swell.

Marginal swells are specific hydrogeological structures of the ocean bottom. Their length reaches ≥1000 km, with a width of 300–500 km, and heights above deep-sea basin bottom at 2–5 km. Marginal swells are characterized by the arch structure formed by the Earth's crust of the ocean where there are situated layers typical of the ocean crust (i.e., the upper layer of silt water, the middle of layer and fracture and fracture and vein water, and the lower basalt layer of fissured and fracture and vein water).

In contrast to the typical marginal STB interior, marginal deep-sea trenches in land–ocean areas are determined. They include Weber, New Britain, and Bougainville in the Solomon Sea; and San Cristobal, Santa Cruz, and New Hebrides in the Coral Sea. Another type of ocean trench is the trench fracture, for example, Romansh in the Atlantic Ocean; Vim, East Indian, and Diamantia in the Indian Ocean; and Lira, Mussau, and Macquarie (Hior) in the Pacific Ocean.

Subocean depression basins, especially those in front of continents, are stretched along the continents of the Western and Eastern Hemispheres of the Atlantic Ocean and on the east of the Pacific Ocean. Depressions of a basement with fractured water are overlapped by sedimentary cover with silt water. Besides a basement of an ocean type, the basement of the continental type is determined. Between basement and sedimentary cover the layer of lithify deposits of different composition likely with layered water is observed. Besides saltwater of the ocean type brines are observed. There is possible fresh groundwater flow from the continent proximal wings of STBs and SBDs.

STB and SBD are usually situated on the boundaries of the oceans and are closely connected with the transient land–ocean zone. They are characterized by complicated water exchange and unique conditions of groundwater formation. Significant transformation of silt water due to silt sedimentation and submergence is detected here. At the same time submergence of the ocean crust and its interaction with the mantle lead to degradation and forming of a great amount of thermal water. Hydrothermal fluids enriched by silica, alkali, and volatile components and metals migrate to the surface of the ocean bottom and participate in the metamorphozation processes of the overlying sedimentary covers. A significant amount of water and vapor discharges through existing volcanoes in the ocean (Sudarikov 1992).

SFBtr are more strongly determined in the eastern part of the Pacific Ocean. They are trenches up to 3500 km and more in length, widths up to 20 km, and depths of 5–6 km with silt water in the bottom and fracture and vein water in the fractured zones. They have mainly near-latitude extension. Volcanoes and hydrotherms are related to cracks. The fractures Mendocino, Pioneer, Murray, and others can be referred to SFBt.

Ocean volcanic basins are widely developed in the oceans. Within the near-ocean ridges and uplifts they are related to rift basins and axial and transverse fractures. On the bottom of the ocean basin they form uplifts of significant size, stretching along fractures for a great distance: the northwest (Empire) mountains, subaqueous volcanic ridges, the Hawaiian Archipelago, the system of volcanic basins, etc. OVBs are found as a single mountain, sometimes with a cut top (guyots). Often volcanic forms appear above ocean level as volcanic islands that are found in the Pacific, Indian, Atlantic, and other oceans. On some islands, overlapping of modern lava flows and volcanics on sedimentary deposits of the Neogenic, Paleogenic, and older periods are observed.

Quite often OVBs are situated on the boundary between ocean basins, along ocean coasts, and on the islands. Characteristics of the volcanic basins of the ocean islands consist in the fact that lenses of freshwater occur directly on the saltwater of ocean origin. Freshwater of vulcanite recharges are due to atmospheric precipitation. These characteristics define the specificity of fresh groundwater exploitation for water supply on islands. It concludes with the necessity of not allowing saltwater to intrude into freshwater of atmospheric origin. Besides OVBs of volcanous cone and guyot (cut-top cones) types during eruption on significant depth (>2 km) from ropy (wavy) covers of pillow lavas typical of great depths of the World Ocean and seas are formed. They demand specific hydrogeological study. This specific type of hydrogeological structure of the ocean bottom can be determined as ocean volcanic basins of lava covers (OVBlc). Finally, studies ought to determine ocean volcanic basins of subaqueous volcanic ridges (OVBvr). Sills and laccoliths often originate at great depths between overlying unconsolidated sediments and underlying basalts. Due to this the surface of the ocean bottom becomes wavy and hilly.

Coral reefs and islands are found on the bottom of the World Ocean and marginal seas. Numerous coral islands are often found within an OVB. The thickness of coral reefs reaches 1000 m and more, and porosity is 15–40%. They contain saltwater of the ocean type in subaquatic parts and freshwater of atmospheric origin in the islands. This freshwater occurs as a thin flat layer on saltwater. Its exploitation has to take into account the possibility of saltwater intrusion and requires the necessary preventive measures. In many cases coral islands are superstructures over volcanic basins that are typical of warm and clear waters of the World Ocean.

A brief review of the hydrogeological structures of the World Ocean bottom indicates that some of them are similar to the continent structures. Among them, for example, it is possible to note massifs of fractured water and subocean massifs and volcanic basins and subocean volcanic basins. The others are particularly analogous, for example, with subocean basins and artesian basins. Finally, such hydrogeological structures of the ocean bottom as subocean trench basins, marginal arches and water of coral islands, and transform and transient fractures have no similar features with continental structures and cannot be found on continents. This allows detecting ocean structures as independent in mean and denomination. The scheme of hydrogeological zoning of the bottom of the Pacific Ocean is given in Figure 3.1.1.2.

3.1.1.3   Some Hydrogeological Characteristics of the Ocean Bottom

Characteristics of the Regional Hydrodynamics. Collecting properties of the sediments of the ocean bottom have not been studied, but some ideas on this score can be formulated. Reef structures have the best filtrate properties. Coral structures having thickness up to 1.5 km and porosity up to 40% absorb and yield water like sponges. Well capacities in the areas of development are extremely high. Significant flooding is observed in some types of lava formation, for example, effusive of acid and neutral composition with pumice texture, etc. The role of clay sediments grows with the ocean depth. For example, at the critical depth of 4–5 km, foraminiferous carbon silts alternate with red deep-sea clays. Thus, the thickness of clay rocks, being a waterproof layer (confining bed), increases from the ocean margin to its central parts. Good collectors can form in the deep-sea parts of the ocean from a

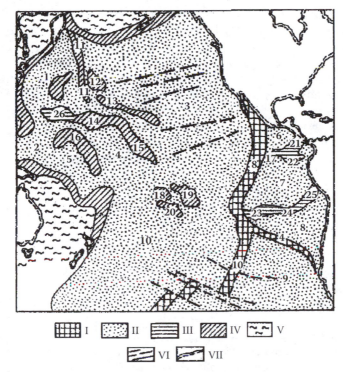

$\boxed{\text{I}}$ $\boxed{\text{II}}$ $\boxed{\text{III}}$ $\boxed{\text{IV}}$ $\boxed{\text{V}}$ $\boxed{\text{VI}}$ $\boxed{\text{VII}}$

FIGURE 3.1.1.2 Hydrogeological zoning of the Pacific Ocean: (I) mobile hydrogeological areas (MHA): East Pacific (E), Southern (S), and Chilean (Ch); (II, III, and IV) stable hydrogeological areas (SHA), correspondingly: subocean basins (SOB), subocean massifs (SOM), and ocean volcanic basins (OVB); (V) hydrogeological structures of the shelf, opened and interior seas (semimarine artesian basins and hydrogeological massifs); (VI) areas of main fractures; (VII) land boundaries; (1–10) subocean basins: Northwest (1), East Marianas (2), Northeast (3), Central (4), Melanesian (5), Guatemalan (6), Peruvian (7), Chilean (8), Bellinsgauzen (9), South (10); (11–20) ocean volcanic basins (OVB): Empire ridge and Obruchev uplift (11), Hess (12), Hawaiian (13), Marcus-Necker (14), Lain Island (15), Marshall Islands (16), Galapagos Islands (17), The Society Islands (18), Tumatou Islands (19), Tubuan Islands (20); (21–26) subocean massifs of fractured water: Cocoas (21), Carnegies (22), Easter (23), Sala and Gomes (24), Naska ridge (25), Shatskiy (26).

depth >2 km. For such a type it is possible to refer turbidities (deposits of suspended flows represented by interstratification of sands, aluerites, and pelites).

Sediment compression and lithogenic water squeezing take place in the upper layer, consisting of unconsolidated deposits together with growth of their thicknesses. Movement likely has a lateral and rising character and is directed from areas with great geo- and hydrostatic pressure to areas with lower pressure. The rate of groundwater movement is very small. It is determined usually by a very slow rate of sedimentation on the ocean bottom. Even in the most favorable conditions, adjoining humid areas of the land, the rate of sedimentation does not exceed 3–10 sm per century.

It is especially necessary to note conditions appearing in the areas of avalanche sedimentation, as singled out by Lisitsyn (1974). In these areas, due to sliding and collapsing at depths of 4–5 km, about 3–5 km^3 (sometimes up to 30 km^3) of sediments are moved. The distance of suspended flow distribution reaches 2–3 ths km. Such sharp hydrodynamic blowing causes local area and impacts relatively short-time stratum pressure skips, and increases elision processes.

In the layer of consolidated sediments, elision processes are likely to be finished. In this case, groundwater moves under geostatic and tectonic pressures, changing fracture and porous space capacity. Other reasons (global, e.g., tides; regional, e.g., bottom deflection and rising; and local, e.g., seismic processes, volcanism, etc.) can influence the hydrodynamic conditions of the layer of consolidated sediments. Their role in groundwater dynamics had not yet been studied.

The rate of stream flow of effusive of basalt layer is in total small, but all lavas of underwater eruption are porous. Content of constitution water in lavas is ~0.01%. Pore diameter and porosity changes from 0.5 up to 0.1 mm and from 33.9 up to 0.1%, consequently with increasing of depth from 490–5000 m. In marine lavas, two systems of fractures (radial and peripheral) and also silt-like cavities are noted. Considering the great amount of volcanic rocks on the ocean bottom, it is possible to present the significance of their roles in the geological circulation on the Earth. According to Ronov (1993), the total amount of volcanic eruption on the ocean bottom beginning from the Jurassic period is $521 \cdot 10^6$ km^3. The volumes of water lost during magma eruption accounts for 3–5% of the mass of vulcatines formed.

On the other side, due to subduction processes of ocean water absorption and its submerging to mantle depths, interconnection with country rocks leading to formation of serpentinous peridotites take place. Pavlov (1977) assessed a rate of water exchange equal to 0.2 km^3/yr, accounting for serpentinous (temporary water conservation) and deserpentinization processes. Some part of this water can get into the ocean as a result of the above described processes — volcanic eruptions.

A special role belongs to the zones of ocean plate spreading. For spreading zones, retention of cold ocean water by fractures on mean-ocean ridge slopes and hot water rising in their axial parts are typical. Intensity of such processes is very great. According to Edmont and fon Dann estimations, every 8 mln years a water mass equal to the volume of the World Ocean passes through the rift zone of the axial parts of mean-ocean ridges. To check such suppositions is difficult, but it is obvious that the rift zone is the most active water exchange system on the ocean bottom.

The influence of surface flows and submarine groundwater discharge (SGD) is detected in shelf areas. Submarine groundwater discharge is possible at a distance of 100 km and more from the shoreline. On the continental slope and its foot, groundwater movement is caused by turbidity flows, the presence of underwater canyons and valleys, and aquifers thinning out. On the abyssal valley, groundwater movement has a rising character and is connected with elision processes. Mean-ocean ridges are not only areas of active water exchange in the zone close to rifts, but also an area of the start of downward convection from rift zones to trenches. In these areas ocean waterholding occurs. Peridotites serpentinization is one of the important components of geological water circulation. Trenches are part of the active zone of ocean margins where ocean crust submerging under continents occurs. This is the lowest global level below which gravitational transferring of sedimentary cover is impossible. This is the lowest draining level of water-bearing rocks that can exist on our planet.

Characteristics of the Regional Hydrochemistry. Homogeneity in chemical composition of groundwater of the ocean crust is illusory. With the standard of saltwater of the Earth, ocean water can be the initial material for its formation. In the ocean bottom, its composition changes, sometimes significantly. Hydrochemical processes occurring in the ocean crust can be subdivided into four groups connected with sediments lithification, peridotites serpentinization, and volcanic and hydro-thermal activity in the zones of modern rift formation.

Changes in porous water composition at sediment diagenesis depends on the components' presence, organic matter enrichment, microbiological processes, and intensity and rate of sedimentation. Thus siliceous, volcanic, and clay sediments of groundwater composition are close to those in the benthic layer. In carbon and biogenic sediments due to diagenesis, calcium and strontium concentrations increase and potassium and magnesium decrease. Usually in the first as in the second case, processes of sulfate reduction are observed. Also some increase in porous water salinity with depth is noted. Thus it is possible to determine the slight brains in the lower part of the sedimentary layer of the ocean crust, and in some areas they are found.

The geochemistry of thermal water in areas of modern volcanism is described in the work of Kononov (1983). He determined thermal water of hydrosulfide-carbonate, carbonate-hydrogen, carbonate, methane, and nitrogen types. The underwater eruptive conditions predominate water of hydrosulfide-carbonate, carbonate-hydrogen, and methane types. They are formed under reduction conditions and are characterized by chloride-sodium (hydrogen, aluminum, ferrous) composition with rich complexes of different microcomponents. Likely, some of the underwater hydrotherms experience an influence of mantle flows. The main role in solutions plays salts of ocean origin and also H_2S, SiO_2, Mn, CO_2, H_2, CH_4. As well water is enriched by K, Li, Rb, Ba, and other metals. Cone-shaped structures as high as 10 m and consisting of sulfides, sulfates, and oxides are formed in areas of hydrotherm exposure.

Fluids of "black smokers" at a temperature ~350°C form magnetite, ferrous chlorite, pyrrhotine, anhydrite, copper, zinc, lead, and nickel sulfides in the contact zone with cold ocean water. Moving away from areas of black smoke vents hydrox-ides of iron and manganese are deposited. Sedimentation of silica, barite, and other minerals occurs during cooling "white smokers" at temperatures between 100–350°C (Sudarikov 1992).

3.2 QUANTITATIVE ASSESSMENT OF SUBMARINE GROUNDWATER DISCHARGE TO THE WORLD OCEAN FROM CONTINENTS

Regional assessment of groundwater discharge to seas and oceans should be based on quantitative analysis of the conditions of groundwater formation and flow in coastal regions. Usually two tasks have to be solved in the first phase: (a) selection of a method of calculation applicable to available data, and (b) the determination of coastal areas drained directly by sea. Most appropriate for these objectives are hydrogeodynamic methods for the calculation of lateral groundwater flow and an integrated hydrologic-hydrogeological approach to regional groundwater discharge assessment in combination with analogic methods. However, insufficient or very poor knowledge of hydrogeological conditions in coastal areas of the continents considerably limits or sometimes precludes the use of hydrogeodynamic methods because of the absence of initial filtration parameters. Therefore, the integrated hydrologic-hydrogeological method was chosen as the basis for quantitative assessment in this case (Section 2.1.3). It enables us to give sufficiently reasonable regional estimation of groundwater discharge to the World Ocean, using already-known specific characteristics of groundwater discharge to rivers from basic aquifer systems with an intensive water-exchange zone.

Of special interest for the given investigations were the works of Kudelin and L'vovich and their followers who through hydrograph separation of the river runoff have defined its subsurface constituent, which represents the subsurface runoff in the intensive water-exchange zone. When using these values, two assumptions should be taken into account. First, based on the analogic method, specific characteristics of groundwater discharge to rivers are also used in areas drained directly by sea. Secondly, if so, then subsurface discharge to the World Ocean is to be estimated only from the upper hydrodynamic zone of intensive water exchange. At the current phase of investigations these assumptions seem sufficiently true. Groundwater discharge to seas and oceans is estimated from areas that are not drained by rivers. These areas occupy certain small sites of river valleys and interfluves. They are associated with a regional zone of discharging transient runoff formed within the main and local groundwater recharge areas in hydrogeological structures that open seaward. In other words, in these coastal zones the conditions of groundwater discharge formation in the intensive water-exchange zone within areas drained both by rivers and seas are very similar. On the basis of this concept, specific characteristics of groundwater discharge into rivers have been used in neighboring geologic-hydrogeologically similar areas drained by sea.

Detailed investigations and assessment of groundwater discharge into particular seas and large lakes show that the main volume of groundwater discharges to the seas from the upper vadose hydrodynamic zone. This is due to the active water exchange in upper aquifers, favorable conditions of their recharge, and, as a rule, high filtration capacity. Contribution of deep aquifers in zones of slow and very slow water exchange to the total groundwater discharge to seas and oceans is not large. The groundwater movement in these aquifers is very slow, and water-exchange periods reach hundreds of thousands or millions of years. Therefore, discharge of deep waters from the lower hydrodynamic zones can be neglected in regional estimation of annual groundwater runoff from land to seas and oceans. But at the same time, in some areas, when studying an influence of groundwater runoff on geochemical processes in bottom sediments, the role of highly mineralized groundwater in deep aquifers on land should be taken into account. These problems require a separate discussion and are not described in detail in this book.

Before starting to calculate groundwater discharge to seas and oceans from particular coastal drainage areas, it is necessary to estimate possible scales of this natural process, which will enable the determination of the accuracy of the values obtained by the hydrologic-hydrogeological method. Such estimation can be carried out in two ways: (1) by estimating total groundwater discharge into oceans using expertly selected average hydrogeological parameters, or (2) by obtaining a quantitative picture about this process from general water-balance constructions. In the literature, there is only a single publication that is devoted to the study of groundwater discharge to the World Ocean, Garrels and Mackenzie (1971), wherein the groundwater discharge to seas is estimated at ~10% of the total surface runoff.

The first attempts to assess this process on a global scale from hydrogeological positions were undertaken by Nace (1967, 1970). In his publication "Are We Running Out of Water?", he states without any evidence that groundwater discharge to seas amounts to 1600 km^3/yr or approximately to 4% of the surface runoff (38,000 km^3/yr). In his next publication "World Hydrology Status and Prospects" (a report from IASH-UNESCO-WHO Symposium on World Water Balance in 1970), Nace (1970) reduces this figure by 7 times (to 230 km^3/yr) and gives some substantiation for this. Namely, he accepts that the groundwater discharge to seas and oceans occurs from water-saturated rocks 4 m thick. Porosity of these rocks is 25%, real groundwater velocity 3 m/day, and specific discharge (per 1 km of shoreline) is 35 l/sec. Having multiplied the specific discharge and length of the World Ocean's shoreline, which, except the shores of the Antarctic and Arctic regions and Greenland composed of permafrost, is accepted to be 200,000 km, Nace obtains the groundwater discharge equal to 230 km^3/yr. We have given critical analysis of the works of Nace elsewhere (Dzhamalov et al. 1977).

Still, it should be outlined once more that the average values of the basic calculated parameters taken by Nace have evinced some serious objections. First of all, the thickness (only 4 m) of the zone where groundwater discharge is formed is extremely underreported. Furthermore, if we use the porosity of 0.25 accepted by Nace and a real groundwater velocity of 3 m/day, the calculated filtration velocity would be 0.75 m/day, which is typical only of karstic and highly fissured or coarse-clastic loose rocks (in order for the filtration velocity to be equal to 0.75 m/day, the hydraulic conductivity, for example, must be 75 m/day and the hydraulic gradient or slope flow of 0.01). The average porosity according to Nace is equal to 0.25, which is typical of sands, is also highly estimated. Thus, the data of Nace cannot be considered as sufficiently substantiated.

We have calculated an approximate groundwater discharge to the World Ocean from the intensive water-exchange zone of land, using the following estimates of hydrogeological parameters: average total thickness of all aquifer systems from which groundwater is discharged to seas (200 m); active porosity of rocks in this zone (0.1); average hydraulic gradient (or slope of flow) (0.005); and coefficient of filtration (10 m/day).

The given figures of hydrogeological parameters have been adopted from concrete values in different seaside regions of the USSR with different hydrogeological conditions (Hydrogeology of the USSR 1966–1972). At these average values, the estimated filtration velocity would be 0.05 m/day, and real groundwater velocity

0.5 m/day. The figures are quite reasonable and generally typical of the upper hydrogeological cross-section of land. Thus, groundwater discharge per 1 km of flow length, i.e., per 1 km of the shoreline of the World Ocean, will amount to about 10,000 m³/day.

It is rather difficult to calculate the length of the shoreline because of its very long extent and sinuation, and also because of transfer from calculations by maps of different scales to the real length of the shoreline. The Laboratory on Ocean Geomorphology at Moscow State University has completed special calculations, which for different geomorphologic regions of the world provide for the possibility of passing from measurements by different-scaled maps to real length of the shoreline. It appears that the length of the shoreline (without Antarctica) is about 650,000 km. Taking into consideration that this figure also includes some territories occupied by deeply frozen rocks which cannot have a significant groundwater discharge, further calculations of the shoreline is taken as equal to 600,000 km. Thus, the total groundwater discharge from the continents to the World Ocean amounts to $6{-}10^9$ m³/day or 2200 km³/yr. These values are indicated in the book *World Water Balance and Water Resources of the Earth* (Hydrometeoizdat 1974).

According to calculations of the water balance in the atmosphere carried out for the territory of the USSR (Kalinin and Kuznetsova 1972), groundwater discharge to the seas sourrounding the USSR amounts to ~300 km³/yr. Extrapolating this value for the entire Earth's globe proportional to the ratio of the total discharge from land, it becomes possible to obtain the total groundwater discharge to the World Ocean in an amount of ~2500 km³/yr, which is close to the earlier given value. However, taking into account the above-mentioned remarks about low accuracy of water balance calculations, it should be noted that the latter calculation should be considered with discretion.

It is interesting to compare characteristics of the groundwater discharge to the World Ocean with those of the discharge to some seas and large lakes that are presented in Chapter 4 and Chapter 5. Although these seas and lakes are located in quite different climatic and geologic-hydrogeological conditions, the groundwater discharge directly into a sea or a lake varies within a small range and amounts, on average, to about 2% of the riverwater inflow (except the Caspian Sea). If the same proportion is used for the World Ocean, then the total river runoff of the planet equals ~40,000 km³/yr, the total groundwater discharge directly to the World Ocean will be ~800 km³/yr. However, this value can be considered only as the lower limit of similar estimates, because formation of groundwater discharge to the World Ocean occurs from a much larger area and actually involves the entire hydrogeological cross-section of land. Thus, by data, the groundwater discharge from land to the World Ocean is within a range of 800–2500 km³/yr. Despite the given values, nevertheless, they reflect the general scales of the phenomenon properly. Subsequent and more detailed calculations will correct values of the SGD, but the order of the sought-for magnitude is already known.

The use of the integrated hydrologic-hydrogeological method enables us to obtain substantiated values of groundwater discharge to seas and oceans for particular drainage areas of the entire coastal zone of continents and large islands. The coastal zones of land were divided into separate drainage areas or calculating areas, from which groundwater discharge is directed to seas without the involvement of the river network. Boundaries of subsurface drainage areas, from where groundwater is discharged directly to seas, are usually determined by maps of the water table. However, there are no such maps for the major part of coastal territories. At the same time, it is known that dividing borders between surface and groundwaters in the intensive water exchange zone coincide with each other in the majority of cases. Due to this, sizes of drainage areas of the groundwater discharge to seas were determined by the World Hypsometric Map of a scale 1:2,500,000. In total, 480 calculation areas were defined.

It should be emphasized that during regional and global assessment of the SGD, determination of aquifers and aquifer systems within specified calculating areas is very difficult due to the poor knowledge of many territories. Therefore, it is reasonable to define in the coastal zone the so-called geofiltration media-genetic types of rocks with uniform conditions of formation and spatial distribution of filtration characteristics (Vsevolozhsky 1983). The type of medium is determined by the genesis of water-bearing rocks or by genetic type of free space (porous, fissured, karst, etc.), subtype, by processes (or conditions of formation) that control character and scales of the filtrative heterogeneity of a medium. Definition of the geofiltration media on the basis of such an approach gives the possibility to apply for different hydrogeological conditions a certain quantitative regularity in the spatial changes of filtration characteristics for geologically complicated regions, taking into account the genesis of water-bearing rocks.

So, geofiltration media are defined on the basis of the genetic approach, i.e., when the regional variability of their basic and, first of all, filtrational properties is determined (predicted) through the analysis of the conditions of sedimentation, processes of tectogenesis, and epigenetic changes of rocks. Thus, on the basis of such analysis, four basic types of geofiltration media are distinguished in the coastal zone of oceans: (1) sedimentogenic-porous, (2) sedimentogenic-fissured, (3) karstic, and (4) magmatogenic-metamorphogenic.

The sedimentogenic-porous type includes incoherent and lowly coherent sediments of sands, gravel, pebbles, and other rocks that were not subjected to compaction and lithification. Permeability of such rocks is determined by the initial sedimentogenic porosity. Among the rocks of this type are marine, coastal-marine, sandy and sandy-clayey sediments, glacial and aqua-glacial formations, and alluvial and lacustrine-alluvial strata forming the marine and continental subtypes of sedimentogenic-porous media. The filtrative properties of this type of media are characterized by the conditions of sedimentation. In regional scale, there are areas with relatively regular filtration properties or with properties altering according to a certain law.

Mountain-folded areas have a widely met sedimentogenic-fissured type of geofiltration media. They are composed of sandstones, argillites, aleurolites, marls, and other coherent sedimentary rocks. Permeability of these rocks is determined by the degree of their jointing that is formed under influence of exogenic and tectonic factors. Subtypes of these media are distinguished according both to lithological composition of water-bearing rocks and conditions for formation of a filtrating space. Due to this, there is a sandy subtype (sandstones, conglomerates, breccia, and others) and a subtype of clayey coherent rocks (argillites, clayey shale, aleurolites, marls, and others) that possess, in general, more lower filtration properties.

Special conditions are for formation of groundwater discharge in karstic rocks including limestone, dolomites, gypsum, anhydrites, etc. They have a high filtrative ability because of dissolving processes of water-bearing rocks and formation of large fissures and karst cavities. The karstic type of a geofiltration medium is subdivided into carbonate and sulfate-carbonate subtypes determining different hydrogeodynamic and geochemical conditions of groundwater formation.

Metamorphic, intrusive, and volcanogenic rocks are incorporated into a single magmatogenic-metamorphogenic type, because they have similar conditions of forming filtration heterogeneity. They are characterized by a fissured permeability. Here, the most intensive regional flows are associated with linear rupture dislocations, whereas local discharge is formed in tectonic disturbances of lower intensity and exogenic jointing.

Thus, the use of the genetic approach to determine the origin of water-bearing rocks and the use of filtration space to define types and subtypes of geofiltration media seems to be a reasonable and effective way for assessment and mapping of groundwater discharge in large regions and of the earth as a whole. The proposed approaches were tested in compilation of the *World Map of Hydrogeological Conditions and Groundwater Flow* on a scale of 1:10,000,000 (Hydroscience Press, 1999). This map reflects the conditions of groundwater discharge to seas and oceans with quantitative characteristics of this process in moduli and absolute values. Occurrence and extent of the geofiltration media are shown within the most studied part of the water area — the continental shelf. Submarine geofiltration media in the continental shelf usually represent the continuation of coastal hydrogeological elements. Hydrogeological structures, that extend into the sea, have unified conditions of groundwater occurrence, which forms on land and discharges directly to the sea. Such specific features of coastal hydrogeological structures to a great degree facilitate regional generalization and assessment of the SGD.

Special attention in this case is given to hydrogeological zoning with continuation of its elements into water bodies, and to distinguishing basic geofiltration media in the shelf. The basis for such seismo-geological cross-sections was obtained as a result of seismological and, first of all, complete acoustic profiling on the shelf, and data from offshore drilling. It is desirable that the seismic profiles are limited to coastal structural or hydrogeological wells and include wells drilled in the sea. Such location of profiles enables reliable determination and observation of submarine geofiltration media, correlation, and jointing of them with coastal hydrogeological elements. For more based spatial spreading of distinguished media in the shelf, their distribution in plane and cross-section should be correlated with specific features of the bottom relief (e.g., its basic geomorphologic and tectonic elements). Thinning out of submarine geofiltration media, reduction of their thickness, change of lithological composition, and possible dislocations are most likely to occur in zones of jointing of geomorphologic elements and at boundaries of tectonic structures. Integrated analysis of seismogeological profiles and geomorphologic and tectonic schemes resulted in the compilation of the map of the distribution of submarine geofiltration media.

Compilation of the given map took into account the principle of hydrogeological zoning of coastal territories with determination of calculating areas (regions) sufficiently homogenous by character of distribution and features of groundwater discharge formation. To achieve this, it was required to take into consideration, alongside the structural-geological principle reflecting specific features of a medium, also the hydraulic principle of zoning that reflects the hydrodynamic nature of the groundwater discharge process. As well, it is important in hydrogeological zoning to reflect relative balance isolation of adjacent areas. Therefore, as already mentioned above, basic boundaries of calculating areas are represented by hydrodynamic (hydraulic) watersheds of different scales. Groundwater basins characterized by their unique hydraulic and water balance features, and a unified basis of draining (sea or ocean) can be used as elements for hydrogeological zoning.

The used pattern of hydrogeological zoning of large territories, namely, according to the conditions of groundwater discharge formation, is based on the principle of the unity of surfacewater and groundwater. Constant interactions between surface and groundwater runoffs enable consideration of groundwater basins in their unity with large systems of the total discharge, which include, first of all, basins of seas' discharge. This creates the basis necessary for complex study of water resources and total water balance. According to this, provinces of groundwater discharge should be subdivided into discharge areas that would correspond to the discharge basins of seas and large lakes. In turn, the discharge areas consist of hydraulically isolated groundwater basins including artesian basins and hydrogeological massifs.

Conditions of groundwater discharge formation are considerably determined by basic features of modern relief, the degree of erosional dissection, and mesoclimatic features. It is known that groundwater discharge in the upper hydrodynamic zone depends, to a great degree, on orographic features of territory, which, with other things being equal, determine the intensity of groundwater flow and control the conditions of its discharge. During regional investigations of groundwater discharge, it is reasonable to distinguish within groundwater basins the so-called zones of groundwater discharge (calculating areas), which correspond to different-order river basins. Boundaries of these zones should be orographic watersheds controlled by internal tectonic structural features of the distinguished structure. In other words, distinguished calculating areas are the basins of submarine discharge of the second order relative to artesian structures and hydrogeological massifs (Shestopalov 1981). Their close linkage with river systems enables consideration of groundwater discharge formation on the basis of the principle of natural waters unity, as well as to monitor by the water balance equation the specific and absolute values of the SGD, the intensity of which is greatly determined by the type and structure of geofiltration media.

Moduli of groundwater discharge for calculating zones (areas) were taken from the *World Map of Hydrogeological Conditions and Groundwater Flow* (Hydroscience Press, 1999); *Maps of Groundwater Flow* for specific continents (L'vovich, 1974); and for the territory of the former Soviet Union, *Maps of Groundwater Flow in the USSR*. For each calculating area, not only the total groundwater inflow into the World Ocean was calculated, but also specific characteristics: areal modulus of groundwater discharge directly to seas from 1 km^2 of a drainage area of land and linear discharge of groundwater per 1 km of shoreline. This enabled the comparison of concrete areas with each other, to correlate the values of groundwater discharge with different natural factors, and allows us to reveal general regularities in the formation of this process. The calculation results for specified large geographical regions are given in Table 3.2.1.

Thus, the total groundwater discharge to the World Ocean, calculated by the hydrologic-hydrogeological method, amounts to 2400 km^3/yr, including the groundwater discharge from the continents (1485 km^3/yr) and from large islands (915 km^3/yr). The high discharge from the islands — over a third of the total submarine discharge — can be explained by a number of reasons. First, the largest oceanic islands (New Guinea, Java, Sumatra, Sakhalin, Madagascar, West Indies, etc.) are located in tropical and humid regions of the earth where abundant precipitation creates favorable conditions for groundwater recharge. Second, mountainous relief, high hydraulic capacity of fissured hard rocks and terrigenic formations, and a poorly developed river network greatly contribute to the formation of considerable SGD there.

Differential assessment of groundwater discharge makes it possible not only to calculate groundwater discharge from different continents, but also to assess submarine discharge to particular oceans. It can be seen from Table 3.2.2 that the total groundwater discharge to the Atlantic Ocean is equal to 830 km^3/yr, to the Pacific Ocean 1300 km^3/yr, and to the Indian Ocean 220 km^3/yr. The groundwater discharge to the Arctic Ocean is estimated only from the European territory and is equal to not more than 50 km^3/yr. On the Asian and American coasts of this ocean, there is almost everywhere a well-developed layer of permafrost, which excludes the formation of SGD from the upper hydrodynamic zone.

The estimated specific characteristics of groundwater discharge to seas and oceans (modulus of groundwater runoff and groundwater flow rate per 1 km of shoreline) give the possibility to analyze and compare the specific features of SGD formation in different physical-geographic and geologic-structural conditions. The highest values of groundwater discharge are typical of mountainous coastal areas of tropical and humid zones where the moduli reach 10–15 l/sec·km^2, and groundwater flow rate in these areas is measured in tens of thousands of cubic meters per day per 1 km of shoreline. The lowest values of groundwater discharge (0.2–0.5 l/sec·km^2) are formed in arid and arctic regions of the Earth, as a result of unfavorable climatic conditions. The general distribution of groundwater discharge values, controlled by latitudinal physical-geographic zonality, is additionally influenced by local hydrogeological conditions, the role of which can be revealed through more detailed analysis of the conditions of groundwater discharge to seas for concrete areas.

Comparison of the values of the total groundwater discharge to the World Ocean calculated by two methods — hydrogeodynamic (2200 km^3/yr) and hydrologic-hydrogeological (2400 km^3/yr) — shows that their difference is slight. It indicates that the obtained values of the groundwater discharge to seas and oceans are sufficiently reliable and should be taken into account in studying the water balance of the World Ocean.

TABLE 3.2.1
Groundwater Discharge to the World Ocean

Region	Subsurface Drainage Area (ths km²)	Total Groundwater Discharge (km³/yr)	Specific Characteristics of Groundwater Discharge		
			Layer[a] (mm)	Areal Modulus (l/sec·km²)	Linear Discharge (ths m³/day·km)
		Australia			
To Pacific Ocean		7.14			
Major Watershed Ridge:	200.25	7.14	$\frac{-}{35}$	1.1	4.6
Southern part	40.9	1.22	$\frac{20-50}{28}$	0.9	2.9
Central part	49.9	2.92	$\frac{30-100}{60}$	1.9	6.8
Northern part	109.45	3.0	$\frac{15-40}{28}$	0.9	4.2
To Indian Ocean		16.36			
Major Watershed Ridge:	6.5	0.23	$\frac{-}{35}$	1.1	2.2
Lower part of Murray River basin	72.6	0.73	$\frac{-}{10}$	0.3	2.4
Hallarbor Plain and Eyre Peninsula	620.9	3.1	$\frac{-}{5}$	0.2	2.8

(Continued)

TABLE 3.2.1 (*Continued*)
Groundwater Discharge to the World Ocean

Region	Subsurface Drainage Area (ths km²)	Total Groundwater Discharge (km³/yr)	Specific Characteristics of Groundwater Discharge		
			Layer[a] (mm)	Areal Modulus (l/sec·km²)	Linear Discharge (ths m³/day·km)
Darling Ridge	63.25	0.63	$\dfrac{--}{10}$	0.3	2.5
West Australia coast	638.0	3.2	$\dfrac{--}{5}$	0.2	3.0
Kimberleys Plateau	126.0	1.89	$\dfrac{--}{15}$	0.5	5.2
Arnhem Land	234.8	2.13	$\dfrac{8-10}{10}$	0.3	3.2
Gulf of Carpentaria coast	588.5	4.45	$\dfrac{5-20}{5}$	0.2	7.8
Continent Total		23.5			
Africa					
To Atlantic Ocean		208.68			
Caspian Mountains	57.8	0.58	$\dfrac{5-20}{10}$	0.3	1.4
Namib Desert	234.5	1.17	$\dfrac{--}{5}$	0.2	2.6
Interfluve of Kuyene and Congo Rivers	212.7	2.06	$\dfrac{5-15}{10}$	0.3	4.1

Interfluve of Congo and Cross Rivers	124.6	27.91	$\dfrac{35\text{–}400}{225}$	7.1	35.2
Interfluve of Cross and Niger Rivers	19.1	7.63	$\dfrac{-}{400}$	12.7	59.7
Interfluve of Niger and Gambia Rivers	453.6	163.72	$\dfrac{50\text{–}600}{360}$	11.5	99.1
Interfluve of Gambia and Senegal Rivers	99.4	0.99	$\dfrac{-}{10}$	0.3	5.2
Western Sahara	374.7	1.87	$\dfrac{-}{5}$	0.2	2.9
Atlas Mountains	117.7	2.75	$\dfrac{7\text{–}35}{22}$	0.7	7.3
To Mediterranean Sea		5.08			
Atlas Mountains	85.4	3.49	$\dfrac{15\text{–}80}{40}$	1.3	5.0
Libyan Desert	318.4	1.59	$\dfrac{-}{5}$	0.2	1.7
To Indian Ocean		22.14			
Drakensberg Mountains	128.1	3.46	$\dfrac{20\text{–}50}{28}$	0.9	5.4
Lower part of Limpopo River	136.3	4.25	$\dfrac{30\text{–}50}{31}$	1.0	10.9
Lower part of Zambezi River basin	152.3	4.57	$\dfrac{-}{31}$	1.0	10.0
East African Plateau	212.7	5.47	$\dfrac{20\text{–}30}{25}$	0.8	7.5

(Continued)

TABLE 3.2.1 (*Continued*)
Groundwater Discharge to the World Ocean

Region	Subsurface Drainage Area (ths km²)	Total Groundwater Discharge (km³/yr)	Specific Characteristics of Groundwater Discharge		
			Layer[a] (mm)	Areal Modulus (l/sec·km²)	Linear Discharge (ths m³/day·km)
Somali Peninsula	488.7	3.65	$\frac{5-20}{5}$	0.2	3.1
Red Sea coast	147.8	0.74	$\frac{—}{5}$	0.2	0.8
Continent Total		235.9			
		Asia			
To Mediterranean Sea		8.32			
Near East	34.1	1.42	$\frac{5-75}{40}$	1.3	5.0
Peninsula of Lesser Asia	74.0	6.9	$\frac{75-100}{95}$	3.0	7.6
To Indian Ocean		65.32			
Lower part of Irrawaddy River basin	40.3	8.06	$\frac{—}{200}$	6.3	46.8
Rakhain Ridge	36.2	7.24	$\frac{—}{200}$	6.3	35.4
Lower part of Ganges River basin	41.5	7.59	$\frac{150-200}{183}$	5.8	27.4

Deccan Plateau	238.5	30.97	$\dfrac{40-200}{130}$	4.1	24.4
Lower part of Indus River basin	130.9	4.48	$\dfrac{15-75}{35}$	1.1	9.1
Coast of Arabian Sea and Persian Gulf	235.4	4.46	$\dfrac{15-40}{20}$	0.6	4.9
Arabian Peninsula					
Interfluve of Shuttle-el-Arab and Masila Rivers	228.1	1.14	$\dfrac{-}{5}$	0.2	0.8
From Masila River to Suez Canal	276.4	1.38	$\dfrac{-}{5}$	0.2	1.0
To Pacific Ocean		254.28			
Malay Peninsula	251.1	40.18	$\dfrac{150-200}{160}$	5.1	29.3
Interfluve of Menam and Mekong Rivers	77.7	15.53	$\dfrac{-}{200}$	6.3	35.8
Annamean Mountains	151.2	30.24	$\dfrac{-}{200}$	6.3	58.5
South Chinese Mountains	238.7	47.72	$\dfrac{-}{200}$	6.3	45.6
Great Chinese Plain	148.1	15.74	$\dfrac{75-200}{107}$	3.4	21.4
Lower part of Liao River basin	34.7	2.6	$\dfrac{-}{75}$	2.4	134.4
Liaosi Mountains	17.0	1.28	$\dfrac{-}{75}$	2.4	13.4

(Continued)

TABLE 3.2.1 (*Continued*)
Groundwater Discharge to the World Ocean

Region	Subsurface Drainage Area (ths km²)	Total Groundwater Discharge (km³/yr)	Specific Characteristics of Groundwater Discharge		
			Layer[a] (mm)	Areal Modulus (l/sec·km²)	Linear Discharge (ths m³/day·km)
Liaodong Peninsula	70.0	5.25	$\frac{}{75}$	2.4	24.5
Korea	66.5	6.65	$\frac{}{100}$	3.2	8.8
Sikhote Alin Ridge	82.8	8.30	$\frac{}{100}$	3.2	9.3
Coast of Okhotsk Sea	262.6	16.54	$\frac{}{63}$	2.0	15.3
Kamchatka	156.3	52.96	$\frac{}{340}$	10.8	46.9
Koryakskoye Upland	70.7	9.55	$\frac{}{135}$	4.3	16.1
Anadir Lowland	55.2	1.74	$\frac{}{31}$	1.0	3.8
Continent Total		327.92			
Europe					
		71.22			
To Atlantic Ocean					
Pyrenean Peninsula	68.9	9.86	$\frac{50\text{–}200}{143}$	4.5	14.4

Garon and Loire Lowlands	23.3	$\frac{--}{150}$	3.49	4.8	13.9
North French Lowland	53.4	$\frac{80-150}{120}$	6.51	3.8	12.3
North Sea coast (Netherlands, Denmark) and North Germany Lowland	93.3	$\frac{--}{80}$	7.47	2.5	10.8
Polish Lowland	44.1	$\frac{--}{80}$	3.53	2.5	
Baltic Sea coast	110.6	$\frac{75-100}{95}$	10.33	3.0	9.8
Scandinavian Peninsula	147.5	$\frac{100-400}{205}$	30.03	6.5	27.7
To Arctic Ocean			47.52		
Scandinavian Peninsula	111.5	$\frac{250-400}{363}$	40.42	11.5	41.9
Cola Peninsula	41.2	$\frac{--}{47}$	1.93	1.5	4.3
White Sea coast	75.5	$\frac{--}{55}$	4.03	1.7	5.2
Barents Sea coast	41.2	$\frac{--}{28}$	1.14	0.9	2.4
To Mediterranean Sea			48.65		
Balkan Peninsula	78.5	$\frac{50-400}{175}$	13.84	5.6	11.7
Dinara Mountains	14.5	$\frac{250-500}{403}$	5.84	12.8	17.8

(Continued)

TABLE 3.2.1 (*Continued*)
Groundwater Discharge to the World Ocean

Region	Subsurface Drainage Area (ths km²)	Total Groundwater Discharge (km³/yr)	Specific Characteristics of Groundwater Discharge		
			Layer[a] (mm)	Areal Modulus (l/sec·km²)	Linear Discharge (ths m³/day·km)
Lombardic Lowland	16.0	3.20	$\frac{—}{200}$	6.3	52.8
Apenninsky Peninsula	91.5	18.52	$\frac{100–250}{202}$	6.4	20.0
South of France coast	24.4	4.88	$\frac{—}{200}$	6.3	25.9
Pyrenean Peninsula	44.5	2.37	$\frac{50–70}{55}$	1.7	5.5
Continent Total		167.39			
North America					
To Atlantic Ocean		219.4			
Central America	124.8	43.54	$\frac{300–600}{350}$	11.1	60.4
Yucatan Peninsula	131.7	26.34	$\frac{—}{200}$	6.3	43.9
Spurs of Sierra Madre Ridge	117.3	14.45	$\frac{5–200}{123}$	3.9	29.8
Mississippi Lowland	323.5	16.33	$\frac{10–100}{50}$	1.6	21.9

Florida Peninsula	205.1	20.51	$\dfrac{\overline{}}{100}$	3.2	29.2
Interfluve of Savanna and Hudson Rivers	198.1	20.54	$\dfrac{100-150}{105}$	3.3	17.2
Northern spurs of Appalachians	159	28.59	$\dfrac{150-200}{180}$	5.7	27.1
New Scotland Peninsula	71.4	14.3	$\dfrac{\overline{}}{200}$	6.3	20.0
Interfluve of Sant Laurenty and Hamilton Rivers	174	34.8	$\dfrac{\overline{}}{200}$	6.3	52.6
To Pacific Ocean		124.58			
Central America	93.2	29.14	$\dfrac{200-400}{313}$	9.9	31.8
Mexican Plateau	194.8	6.74	$\dfrac{5-150}{55}$	1.1	4.8
California	176.7	0.88	$\dfrac{\overline{}}{5}$	0.2	0.7
Cordillera (U.S.)	88.9	15.82	$\dfrac{15-400}{175}$	5.6	20.3
Cordillera (Canada)	128.4	51.36	$\dfrac{\overline{}}{400}$	12.7	78.0
Gulf of Alaska coast	51.6	20.64	$\dfrac{\overline{}}{400}$	12.7	33.0
Continent Total		343.98			

(Continued)

TABLE 3.2.1 (*Continued*)
Groundwater Discharge to the World Ocean

Region	Subsurface Drainage Area (ths km²)	Total Groundwater Discharge (km³/yr)	Specific Characteristics of Groundwater Discharge		
			Layer[a] (mm)	Areal Modulus (l/sec·km²)	Linear Discharge (ths m³/day·km)
South America		185.29			
To Atlantic Ocean					
Patagonia	353.4	2.83	$\frac{7-15}{10}$	0.3	3.0
La Plata Lowland	603.4	20.60	$\frac{10-80}{35}$	1.1	26.1
Brazilian Plateau	456.6	34.34	$\frac{25-100}{75}$	2.4	19.2
Amazonian Lowland	118.2	36.78	$\frac{150-400}{313}$	9.9	39.5
Spurs of Guianian Plateau	192.3	76.96	$\frac{—}{400}$	12.7	117.7
Orinolco Lowland	54.9	1.2	$\frac{5-25}{22}$	0.7	2.5
Northern spurs of Andes	152.1	12.58	$\frac{10-200}{83}$	2.6	12.7
To Pacific Ocean		199.59			
Northern part of Andes	96.0	44.66	$\frac{250-600}{465}$	14.8	50.7

Southern part of Andes	457.0	154.93	$\frac{5-800}{340}$	10.8	61.6
Continent Total		384.88			
Large Islands		913.67			
Ireland	124.1	24.81	$\frac{-}{200}$	6.3	39.6
Great Britain	125.6	24.62	$\frac{75-300}{196}$	6.2	18.5
Puerto Rico	9.0	0.45	$\frac{-}{50}$	1.6	2.8
Haiti	73.3	3.7	$\frac{-}{50}$	0.6	14.9
Cuba	107.2	8.17	$\frac{50-100}{75}$	2.4	9.3
Jamaica	11.0	0.55	$\frac{-}{50}$	1.6	2.7
Newfoundland	99.0	14.85	$\frac{-}{150}$	4.8	13.1
Tierra del Fuego	29.7	1.64	$\frac{35-75}{57}$	1.8	3.4
Sicily	24.9	1.25	$\frac{-}{50}$	1.6	4.4
Sardinia	31.8	3.18	$\frac{-}{100}$	3.2	12.4
Corsica	8.4	1.27	$\frac{-}{150}$	4.8	7.8

(Continued)

TABLE 3.2.1 (*Continued*)
Groundwater Discharge to the World Ocean

Region	Subsurface Drainage Area (ths km²)	Total Groundwater Discharge (km³/yr)	Specific Characteristics of Groundwater Discharge		
			Layer[a] (mm)	Areal Modulus (l/sec·km²)	Linear Discharge (ths m³/day·km)
Vancouver	33.8	13.52	$\frac{—}{400}$	12.7	33.3
Kyushu	36.4	18.22	$\frac{—}{500}$	15.9	33.6
Shikoku	11.5	5.74	$\frac{—}{500}$	15.9	21.9
Honshu	141.6	70.79	$\frac{—}{500}$	15.9	44.5
Hokkaido	50.6	25.29	$\frac{—}{500}$	15.9	36.2
Mindanao	63.0	65.84	$\frac{1000-1500}{1045}$	33.2	81.1
Luzon	73.3	34.34	$\frac{400-600}{470}$	14.9	36.6
Kalimantan	501.3	60.75	$\frac{50-300}{120}$	3.8	40.4
Sulawesi	162.9	27.48	$\frac{100-400}{168}$	5.3	18.9
Sakhalin	56.6	10.11	$\frac{—}{180}$	5.7	11.8

New Zealand	202.2	116.08	$\frac{200-1500}{574}$	18.2	63.5
New Guinea	375.0	264.5	$\frac{600-800}{705}$	22.4	104.7
Madagascar	293.6	28.55	$\frac{5-200}{98}$	3.1	17.7
Java	127.5	44.63	$\frac{-}{350}$	11.1	54.8
Sumatra	226.5	25.4	$\frac{100-130}{112}$	3.6	23.1
Sri Lanka	51.9	15.42	$\frac{200-400}{295}$	9.4	39.3
Tasmania	33.5	2.52	$\frac{20-200}{75}$	2.4	4.8
Continents and Islands Total of Earth		2397.24			

[a]In numerator, limits of variation; in denominator, weighted-average for drainage area.

TABLE 3.2.2
Groundwater Discharge to Oceans from Continents and Large Islands

Ocean/Sea	Continent/Islands	Water Discharge			Ionic Discharge		
		Areal Modulus[a] l/(sec·km²)	Linear Discharge[b] (ths m³/day·km)	Total (km³/yr)	Areal Modulus[a] (t/yr·km²)	Linear Discharge[b] (ths t/yr·km)	Total (mln t/yr)
Pacific Ocean	Australia	0.9–1.9 1.1	2.9–6.8 4.6	7.14	16.4–46.9 24.9	0.7–2.0 1.2	4.99
	Asia	1.0–10.8 4.8	3.8–134.4 27.2	254.28	18.9–179.7 98.2	1.4–98.1 6.5	165.21
	North America	0.2–12.7 5.4	0.7–78.0 21.9	124.58	2.0–125.1 50.1	0.1–8.5 2.4	36.73
	South America	10.8–14.8 11.5	50.7–61.6 58.7	199.59	46.6–67.8 64.1	1.8–4.5 3.8	35.47
	Large Islands	3.8–33.2 13.0	11.8–104.7 51.0	714.72	36.3–360.9 159.8	3.0–11.5 7.3	278.15
	Total			1300.31			520.55
Atlantic Ocean	Africa	0.2–12.7 3.9	1.4–99.1 40.4	208.68	12.1–447.8 99.9	0.6–25.7 12.0	169.19
	Europe	2.5–6.5 4.2	9.8–27.7 15.4	71.22	24.0–71.6 47.8	1.1–3.0 2.0	25.84
	North America	1.6–11.1 4.6	17.2–60.4 31.9	219.4	10.1–200.0 74.6	0.6–16.0 6.0	112.22
	South America	0.3–12.7 3.0	2.5–117.7 28.2	185.29	2.4–240.1 40.2	0.3–25.8 4.3	77.66
	Large islands	1.6–6.3 4.4	2.7–39.6 12.0	77.67	20.0–137.2 76.0	0.4–7.2 2.4	42.88

Mediterranean Sea	Africa	0.2–1.3 / 0.4	1.7–5.0 / 3.1	5.08	20.0–40.9 / 24.4	1.8–2.4 / 2.2	9.85
	Asia	1.3–3.0 / 2.4	5.0–7.6 / 7.0	8.32	46.0–139.9 / 110.3	2.0–4.2 / 3.6	11.92
	Europe	1.7–12.8 / 5.7	5.5–52.8 / 15.6	48.65	53.3–282.1 / 101.8	2.0–5.8 / 3.2	27.42
	Large Islands	31.6–4.8 / 2.8	4.4–12.4 / 8.1	5.7	29.8–39.9 / 34.9	0.6–1.8 / 1.2	2.27
	Total			830.01			479.25
Indian Ocean	Australia	0.2–1.1 / 0.2	2.2–7.8 / 3.7	16.36	4.6–74.9 / 28.4	0.4–15.4 / 5.5	66.72
	Africa	0.2–1.0 / 0.6	0.8–10.9 / 5.1	22.14	14.9–150.2 / 38.7	1.4–8.8 / 4.1	48.95
	Asia	0.2–6.3 / 1.7	0.8–46.8 / 10.7	65.32	6.0–400.0 / 97.2	0.3–34.2 / 7.1	119.25
	Large islands	3.1–11.1 / 5.1	17.7–54.8 / 27.7	115.58	29.2–356.4 / 84.7	1.9–17.2 / 5.3	60.63
	Total			219.4			295.55
Arctic Ocean	Europe	0.9–11.5 / 5.6	2.4–41.9 / 17.8	47.52	4.6–36.2 / 26.6	0.2–1.5 / 1.0	7.16
	Total			47.52			7.16
World Total				2397.24			1302.51

[a] In numerator, limits of variation; in denominator, weighted-average for area.
[b] In numerator, limits of variation; in denominator, weighted-average along shore.

3.3   THE ROLE OF GROUNDWATER FLOW IN THE FORMATION OF A SALT BALANCE AND THE CHEMICAL COMPOSITION OF SEAS AND OCEANS

The World Ocean and peripheral and inland seas serve as the main draining basin for surface and groundwater runoff. In this connection, the salt balance is formed due to salt losses from rivers and groundwater.

River ion runoff is sufficiently well studied and easily determined between these two sources of incoming dissolved substances into seas (Gordeyev 1983). Assessment of the influence of subsurface ion runoff was associated with earlier difficulties caused by the basin lack (till the recent time) of data on regional groundwater runoff directly to seas. After groundwater runoff to the seas and the World Ocean had been evaluated, it became possible to determine quantitatively the subsurface ion runoff and to reveal the basic regularities of its formation and its role in the salt balance of the seas (Zektser et al. 1984).

Under subsurface ion runoff one should understand both total salt loading and in particular dissolved chemical elements transferred by groundwater to draining areas. The transfer of dissolved substances with subsurface flow is one of the most important processes of chemical element migration in the Earth's crust. According to existing estimates, within the former USSR, about 280 mln t of dissolved substances are redistributed annually by underground water flow, whereas surface runoff transfers, on average, amount to 161 mln t (Makarenko and Zverev 1970a). Subsurface ion runoff is defined usually as the product of total mineralization or concentration of a particular component and the total groundwater discharge from specified aquifers or aquifer systems. Here, the most reliable data on denudation activity of groundwater can be obtained from direct measurements of its mineralization and the computation of its runoff.

It has been established that groundwater runoff directly to the seas (leaving aside the river network) is small relative to total river runoff and does not usually exceed a few percent. At the same time, the role of underground ion runoff in the salt balance of inland seas is significant and reaches tens of percentages when compared with the salt income from rivers. Thus, salt transfer from groundwater to the Caspian Sea amounts to about 30% of the salt contribution by rivers, whereas underground runoff is equal to only a few percent of the total salt income from river waters.

Distribution of the subsurface water and ion runoff to the seas is characterized by a common vertical hydrodynamic and hydrochemical zonality of groundwater, which with depth causes an increase in total salt extraction, despite a general reduction of the underground flow modulus. This is explained by a considerably higher mineralization of groundwater in deep aquifers when compared with that found in upper aquifers. Sometimes this general regularity is disturbed due to the influence of local hydrogeologic conditions connected with widely developed karst rocks, continental salinization processes, and availability of salt-bearing sediments. Thus, the low subsurface ion runoff moduli in the Baltic Sea (48.5 t/yr km^2) are typical of the seaside part of the Silurian–Ordovician Plateau, where the basic subsurface runoff to the sea is formed from water-bearing karst limestones and dolomites. The ion runoff from this region is equal to half of the total ionic runoff to the Baltic Sea from the intensive water-exchange zone of the Baltic artesian basin.

It should be noted that the role of subsurface ion runoff in the formation of a sea salt regime can grow significantly with a decrease in total river runoff caused by natural factors and man-made activity. In this case, the salinization of deep parts of the sea, where the time of water exchange is long, will proceed more intensively. The salinization of some deep depressions of inland seas, that takes place at the present time, can be caused, along with other reasons, by an increasing influence of subsurface ion runoff from deep aquifers. This influence affects not only the total salt balance of the seas, but also it can be the basic reason for formation of large geochemical anomalies in the near-bottom layer of water and sea sediments. In turn, as mentioned above, the anomalies in the geochemical fields on the sea bottom serve as indicators of SGD.

3.3.1 COMPUTATION OF TOTAL SALT TRANSFER
WITH GROUNDWATER FLOW TO SEAS AND OCEANS

Regional and global evaluation of total salt transfer was completed for the same specific water-catchment areas that were used to determine water runoff (Table 3.3.1). The groundwater mineralization in the intensive water-exchange zone was taken from domestic and international publications, with a specified value accepted as an average mineralization level chosen from all the available data of groundwater chemical analyses for each area. In addition to total ion runoff, specific values were determined for each area, namely areal modulus and linear transfer of underground ion runoff, i.e., amount of dissolved substances transferred from 1 km^2 of water-catchment area or per 1 km of the shoreline. This made possible to compare the role of different areas in the salt transfer and to reveal the basic regularities of this process. The results are given in Table 3.3.1.

The total salt transfer from groundwater to the World Ocean reaches 1300 mln t/yr, which amounts to 52% of the salt contribution of the river runoff of 2480 mln t/yr. The ratio of salt transfer with the ground and riverwater for the continents is shown in Table 3.3.2 (the total ion river runoff does not include the salt transfer with the riverwater from large islands).

Subsurface ion runoff to the oceans is determined, first of all, by submarine runoff, because groundwater mineralization in the zone of intensive water exchange is higher than 1 g/l. The diagram of the distribution of groundwater mineralization for all the studied water-catchment areas shows that in major cases it varies within a range of ~1 g/l, reaching in certain cases to 4–6 g/l. High mineralization (15–40 g/l) is typical only in particular regions of Africa and Australia (Figure 3.3.1; Table 3.3.1). Based on the above values of total subsurface water and ion runoff, the average mineralization of the groundwater discharging to the seas from the upper hydrodynamic zone of intensive water exchange does not exceed 0.6 g/l (Table 3.3.2), which also confirms the reliability of the computations made and the reality of the values obtained.

The high ion runoff from the continent of Africa is explained, first of all, by the availability of evaporates in the coastal sediments (artesian basins of the Red Sea,

TABLE 3.3.1
Total Salt Transfer from Groundwater Flow to the Oceans from the Continents

Region	Subsurface Water-Catchment Area (ths km²)	Subsurface Runoff (km³/yr)	Ion Runoff (Salt Transfer) (mln t/yr)	Areal Modulus of Ion Runoff (t/yr·km²)	Linear Loss of Ion Runoff (ths t/yr)
To Pacific Ocean					
		Australia			
		7.14	4.99		
Major Watershed Ridge	200.25	7.14	4.99	24.9	1.2
Southern part	40.9	1.22	0.85	20.8	0.7
Central part	49.9	2.92	2.34	46.9	2.0
Northern part	10.45	3.0	1.8	16.4	0.9
To Indian Ocean		16.36	66.72		
Major Watershed Ridge	6.5	0.23	0.12	18.5	0.4
Lower part of the Murrey River basin	72.6	0.73	1.1	15.2	1.4
Nallarbor Valley and Eyre Peninsula	620.9	3.10	46.5	74.9	15.4
Darling Ridge	63.25	0.63	0.63	10.0	0.9
West Australia coast	638.0	3.2	6.4	10.0	2.2
Kimberley Plateau	126.0	1.89	3.78	30.0	3.8
Arnhem Land	234.8	2.13	1.07	4.6	0.6
Gulf of Carpentaria coast	588.5	4.45	7.12	12.1	4.6
Continent Total		23.5	71.71		
		Africa			
To Atlantic Ocean		208.68	169.19		
Cape Mountains	57.8	0.58	0.70	12.1	0.6
Namib Desert	234.5	1.17	3.51	15.0	2.8
Interfluve of Kunene and Congo Rivers	212.7	2.06	10.3	48.4	7.4

Interfluve of Congo and Cross Rivers	124.6	27.91	55.8	447.8	25.7
Interfluve of Cross and Niger Rivers	19.1	7.63	3.05	159.7	8.7
Interfluve of Niger and Gambia Rivers	453.6	163.72	81.86	180.5	18.1
Interfluve of Gambia and Senegal Rivers	99.4	0.99	2.97	29.9	5.7
Western Sahara	374.7	1.87	9.35	25.0	5.3
Atlas Mountains	117.7	2.75	1.65	14.0	1.6
To Mediterranean Sea		5.08	9.85		
Atlas Mountains	85.4	3.49	3.49	40.9	1.8
Libyan Desert	318.4	1.59	6.36	20.0	2.4
To Indian Ocean		22.14	48.95		
Drakensberg Mountains	128.1	3.46	2.42	18.9	1.4
Lower part of Limpopo River basin	136.3	4.25	4.25	31.2	4.0
Lower part of the Zambezi River basin	152.3	4.57	4.57	30.0	3.7
East African Plateau	212.7	5.47	8.21	38.6	4.1
Somali Peninsula	488.7	3.65	7.30	14.9	2.3
Red Sea coast	147.8	0.74	22.2	150.2	8.8
Continent Total		235.9	227.99		
Asia					
To Mediterranean Sea		8.32	11.92		
Near East	34.1	1.42	1.57	46.0	2.0
Lesser Asia Peninsula	74.0	6.9	10.35	139.9	4.2
To Indian Ocean		65.32	119.25		
Lower part of the Iravadi River basin	40.3	8.06	16.12	400	34.2
Rakhain Ridge	36.2	7.24	3.62	100	6.5
Lower part of the Ganges River basin	41.5	7.59	3.8	91.6	5.0

(Continued)

TABLE 3.3.1 (Continued)
Total Salt Transfer from Groundwater Flow to the Oceans from the Continents

Region	Subsurface Water Catchment Area (ths km²)	Subsurface Runoff (km³/yr)	Ion Runoff (Salt Transfer) (mln.t/yr)	Areal Modulus of Ion Runoff (t/yr·km²)	Linear Loss of Ion Runoff (ths t/yr)
Deccan Plateau	238.5	30.97	18.58	77.9	5.3
Lower part of the Indus River basin	130.9	4.48	10.30	78.7	7.6
Arabian Sea and Persian Gulf coasts	235.4	4.46	10.26	43.6	4.1
Arabian Peninsula					
Interfluve of Shatt al Arab and Masila Rivers	228.1	1.14	1.37	6.0	0.3
From Masila River to Suez Canal	276.4	1.38	55.2	199.7	15.1
To Pacific Ocean		254.28	165.21		
Malay Peninsula	251.1	40.18	32.14	128.0	8.5
Interfluve of Menam and Mekong Rivers	77.7	15.53	10.87	139.9	9.1
Annam Mountains	151.2	30.24	12.1	80.0	8.5
South Chinese Mountains	238.7	47.72	42.9	179.7	14.9
Great Chinese Valley	148.1	15.74	23.61	159.4	11.7
Lower part of Leibhe River basin	34.7	2.6	5.2	149.8	98.1
Liaosi Mountains	17.0	1.28	0.77	45.3	2.9
Liaodong Peninsula	70.0	5.25	3.15	45.0	5.4
Korea	66.5	6.65	3.99	60.0	1.9
Sikhote Alin Ridge	82.8	8.30	3.32	40.1	1.4
Coastine of Okhotsk Sea	262.6	16.54	4.96	18.9	1.7
Kamchatka	156.3	52.96	10.6	67.8	3.4
Koryak Ridge	70.7	9.55	2.9	41.0	1.8
Anadyr Lowland	55.2	1.74	8.7	157.6	6.9
Continent Total		327.92	296.38		

Europe

		Europe			
To Atlantic Ocean		71.22	25.84		
Pyrenean Peninsula	68.9	9.86	4.93	71.6	2.6
Garonna and Luara Lowlands	23.3	3.49	1.40	60.1	2.0
North French Lowland	53.4	6.51	2.60	48.7	1.8
North Sea coast (Netherlands, Denmark and North German Lowland)	93.3	7.47	3.74	40.1	1.7
Polish Lowland	44.1	3.53	1.06	24.0	1.7
Baltic Sea coast	110.6	10.33	3.10	28.0	1.1
Scandinavian Peninsula	147.5	30.03	9.01	61.1	3.0
To Arctic Ocean		47.52	7.16		
Scandinavia Peninsula	111.5	40.42	4.04	36.2	1.5
Cola Peninsula	41.2	1.93	0.19	4.6	0.2
White Sea coast	75.5	4.03	2.02	26.8	1.0
Barentsovo Sea coast	41.2	1.14	0.91	22.1	0.7
To Mediterranean Sea		48.65	27.42		
Balkan Peninsula	78.5	13.84	8.30	105.7	2.6
Dinara Mountains	14.5	5.84	4.09	282.1	4.5
Lombardic Lowland	16.0	3.20	0.96	60.0	5.8
Apennines Peninsula	91.5	18.52	9.26	101.2	3.6
South of France coastline	24.4	4.88	2.44	100.0	4.7
Pyrenean Peninsula	44.5	2.37	2.37	53.3	2.0
Continent Total		167.39	60.42		

(Continued)

TABLE 3.3.1 (*Continued*)
Total Salt Transfer from Groundwater Flow to the Oceans from the Continents

Region	Subsurface Water Catchment Area (ths km²)	Subsurface Runoff (km³/yr)	Ion Runoff (Salt Transfer) (mln.t/yr)	Areal Modulus of Ion Runoff (t/yr·km²)	Linear Loss of Ion Runoff (ths t/yr)
To Atlantic Ocean					
		North America			
		219.4	112.22		
Central America	124.8	43.54	21.77	174.4	11.0
Yucatan Peninsula	131.7	26.34	26.34	200.0	16.0
Sierra Madre Ridge branches	117.3	14.45	11.56	98.6	8.7
Mississippi Lowland	323.5	16.33	3.27	10.1	1.6
Florida Peninsula	205.1	20.51	8.2	40.0	4.38
Interfluve of Savannah and Hudson Rivers	198.1	20.54	2.05	10.3	0.6
Northern branches of Appalachians	159.0	28.59	25.73	161.8	9
New Scotland Peninsula	71.4	14.3	2.86	40.1	1.5
Interfluve of Saint Lawrence and Hamilton Rivers	174.0	34.8	10.44	60.0	5.8
To Pacific Ocean		124.58	36.73		
Central America	93.2	29.14	11.66	125.1	4.6
Mexican Upland	194.8	6.74	2.02	10.4	0.5
California Peninsula	176.7	0.88	0.35	2.0	10.1
Cordillera (U.S.)	88.9	15.82	3.16	35.5	1.5
Cordillera (Canada), Gulf of Alaska shoreline	128.4 51.6	51.36 20.64	15.41 4.13	120.0 80.0	8.5 2.4
Continent Total		343.98	148.95		

South America

To Atlantic Ocean		185.29	77.66		
Patagonia	353.4	2.83	0.85	2.4	0.3
La Plata Lowland	603.4	20.60	8.24	13.6	3.8
Brazil Plateau	456.6	34.34	3.43	7.5	0.7
Amazon Lowland	118.2	36.78	14.71	124.5	5.8
Branches of Guiana Plateau	192.3	76.96	46.18	240.1	25.8
Orinoco Lowland	54.9	1.2	0.48	8.7	0.4
Northern branches of Andes	152.1	12.58	3.77	24.8	1.4
To Pacific Ocean		199.59	35.47		
Northern part of Andes	96.0	44.66	4.47	46.60	1.8 4.5
Southern part of Andes	457.0	154.93	31.0	67.8	
Continent Total		384.88	113.13		

Large Islands

Ireland	124.1	24.81	12.41	100.0	7.2
Great Britain	125.6	24.62	17.23	137.2	4.7
Puerto Rico	9.0	0.45	0.18	20.0	0.4
Haiti	73.3	3.7	1.48	20.2	0.7
Cuba	107.2	8.17	6.54	61.0	1.1
Jamaica	11.0	0.55	0.22	20.0	0.4
Newfoundland	99.0	14.85	4.45	44.9	1.4
Tierra del Fuego	29.7	1.64	1.15	38.7	0.9
Sicily	24.9	1.25	0.75	30.1	1.0
Sardinia	31.8	3.18	1.27	39.9	1.8

(Continued)

TABLE 3.3.1 (*Continued*)
Total Salt Transfer from Groundwater Flow to the Oceans from the Continents

Region	Subsurface Water Catchment Area (ths km²)	Subsurface Runoff (km³/yr)	Ion Runoff (Salt Transfer) (mln.t/yr)	Areal Modulus of Ion Runoff (t/yr·km²)	Linear Loss of Ion Runoff (ths t/yr)
Corsica	8.4	1.27	0.25	29.8	0.6
Vancouver	33.8	13.52	12.2	360.9	11.0
Kyushu	36.4	18.22	9.11	250.3	6.1
Shikoku	11.5	5.74	2.87	249.6	4.0
Honshu	141.6	70.79	35.40	250.0	8.11 '
Hokkaido	50.6	25.29	12.64	249.8	6.6 1
Mindanao	63.0	65.84	19.75	313.5	8.9 1
Luzon	73.3	34.34	17.17	234.2	6.7 1
Kalimantan	501.3	60.75	18.22	36.3	4.4
Sulawesi	162.9	27.48	16.49	101.2	4.1
Sakhalin	56.6	10.11	7.08	125.1	3.0
New Zealand	202.2	116.08	46.43	229.6	9.3
New Guinea	375.0	264.5	79.35	211.6	11.5 1
Madagascar	293.6	28.55	8.57	29.2	1.9
Java	127.5	44.63	22.3	174.9	10.0 ^
Sumatra	226.5	25.4	10.16	44.9	3.4 -
Sri Lanka	51.9	15.42	18.5	356.4	17.2 1
Tasmania	33.5	2.52	1.76	52.5	1.2 1
Islands Total			383.93		
World Total		2397.24	1302.51		

TABLE 3.3.2
Ratio of Total Ion River Runoff and Total Ion Underground Runoff for the Continents

Continent	River Ion Runoff (mln t)	Average Mineralization of Riverwaters (g/l)	Underground Ion Runoff (mln t)	Average Mineralization of Groundwater (g/l)
Europe	240	0.077	60	0.4
Asia	850	0.065	296	0.9
Africa	310	0.072	228	1.0
North America	410	0.069	149	0.4
South America	550	0.053	113	0.3
Australia (including the islands of Tasmania, New Zealand, and New Guinea)	120	0.060	199	0.5
World Total	2480	0.063	1045	0.6
River ion runoff (%)			42	

Data on the river ion runoff and average mineralization of riverwaters are given after L'vovich (1974).

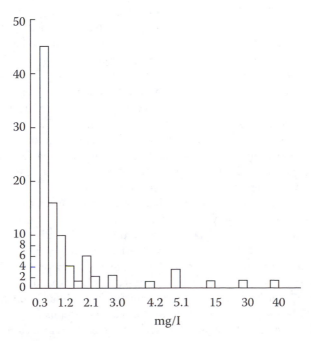

FIGURE 3.3.1 Histogram of the distribution of groundwater mineralization in coastal zones of the World Ocean.

Gulf of Aden, Cameroon-Gabon, and others), as well as by the prevailing areas with arid and semiarid climates that causes the salinization of the upper aquifers due to unfavorable recharge conditions and processes of continental salinization.

The sizable dissolved substance transfer from groundwater from the large islands, which is caused by favorable conditions for the formation of submarine water flow, should be noted. Since no data on mineralization of river water are available for the major islands, it is impossible to compare salt transfer by river and groundwater. However, river networks on islands are poorly developed; therefore, there are grounds to suppose that ion runoff exceeds the salt transfer by the rivers on islands.

The salt transfer from subsurface runoff to the Atlantic Ocean amounts to 479 mln t/yr, the Pacific Ocean to 521 mln t/yr, the Indian Ocean to 296 mln t/yr, and the Arctic Ocean (from the drainage areas evaluated) to 7 mln t/yr (see Table 3.3.1). The figures provided evidence that SGD can exert a significant influence on the salt and hydrobiological regimes of seas and oceans, and on the processes of biogenic sedimentation and formation of mineral deposits.

Thus, the newly recognized high role of salt transfer from groundwater to the World Ocean (52% of salt income by rivers) alters the understanding that the primary biological products of the oceans and the scales of biogenic sedimentation are limited only by the salt incomes from the river runoff.

Analysis of Specific Characteristic Distribution of Total Salt Transfer. Global and regional distribution of the areal modulus and linear discharge show that the formation of subsurface ion runoff is influenced, first of all, by climatic factors, structural-hydrogeological conditions of water-bearing rocks, their lithological composition and hydraulic capacity, and modern and paleohydrogeological conditions for the formation of chemical groundwater composition. The analysis of such a distribution of the specific values and influences of different factors and conditions was used to assess the salt transfer with the groundwater along the coastal zones of Africa, Australia, South America, and other continents. The results obtained make it possible to compare individual coastal areas, reveal specific features in the formation of the subsurface ion runoff in different natural conditions, assess the role of the subsurface flow in the water and salt regime of particular regions, as well as to determine the cause-and-effect links of subsurface flow with natural and anthropogenically disturbed processes on land and in the sea.

Thus, for the Anadir Lowland (Russian coastal zone of the Pacific Ocean) the average modulus of ion runoff was 158 t/yr·km^2, despite the fact that the subsurface water runoff was low (1.7 km^3/yr). The groundwater mineralization in the region reaches 10 g/l and higher as the frozen rocks that are widely developed there and paleohydrogeological conditions are unfavorable for formation of fresh groundwater. At the same time on the peninsula of Kamchatka the submarine runoff is equal only to 53 km^3/yr, and the average ion runoff modulus is only 68 t/yr·km^2. The geological history of structure formation, wide development of a mountainous and highly dissected relief, intensive development of rupture tectonics, high precipitation, and absence of continuous permafrost are factors that determine low groundwater mineralization (0.2–0.3 g/l) in the zone of intensive water exchange in Kamchatka.

The submarine groundwater from an infiltration origin gets its chemical composition chiefly within the land and is distinguished by a different degree of mineralization. In the upper hydrodynamic zone the water is usually low mineralized, whereas in arid and subarctic areas, as well as in places with developing evaporates the mineralization can be high. With movement downward, the infiltration waters gradually replace the sedimentation water. This process takes a long time.

Constant interaction of the infiltration water with sedimentation water and directly with seawater leads through convective-diffusive processes and physico-chemical reactions to a gradual equalization of the chemical composition and mineralization of the submarine waters with earlier different geneses. In this connection, it is possible that at certain depths there is a transient zone between waters of different geneses, the spatial characteristics of which are greatly determined by hydrodynamic and physico-chemical gradients of the counter submarine water fluxes. Depending on geofiltration properties of water-bearing rocks in the cross-section, the submarine waters of different geneses may have a bi-layer mode of occurrence.

In general, infiltration waters, inflowing from the land, are distributed mainly in the shelf zone. Their current infiltration recharge per unit of time is higher than the amount of elision recharge (Dyunin 2000). Therefore, when conditions are favorable, the tongues of confined infiltration waters can penetrate far into the sea, reaching the continental slope and replacing on their way sedimentation waters. The influence of infiltration water from the coastal artesian structures is noticed in the sea area for many tens of kilometers. Thus, an edge of fresh-brackish waters extends under the Atlantic Ocean in the area of the Florida Peninsula for 120 km from the coast and for a depth of over 600 m. Freshwaters under the Atlantic Ocean floor in the marginal part of the United States continental shelf have been detected a relatively long time ago and described in numerous publications. Distribution of the groundwater mineralization in this region vividly shows that the infiltration water gradually replaces the sedimentation water from the aquifers. The influence of the infiltration water is weakening with depth and with distance from the coast line where the water mineralization becomes higher and reaches over 35°C in some parts of the cross-section. The high mineralization levels (50–200°C) that are found there are explained by the availability of the salt-bearing rocks in the cross-section. Evaporite brines are formed due to an interaction of submarine waters of different geneses with salt-bearing rocks. These brines are not only found together with the mother rocks, but also in the upper and underlying aquifers, where they penetrate through convective-diffusive processes. Thus, in the influence zone of coastal artesian structures, submarine groundwater usually undergoes desalinization. On the other hand, the occurrence of submarine waters with a mineralization of over 35‰ mostly indicates the availability of salt-bearing rocks at certain depths.

In the mostly submerged parts of the continental slope and within the sea floor, there are predominantly developed submarine sedimentation waters. In these areas there is not any influence from infiltration water exchange, and the vertical filtration of groundwater prevails to depths of a few hundred meters. Because of specific press-out of pore solutions with consolidation of sea-bottom clayey sediments at these depths, it is likely to meet there strong lateral sedimentation water fluxes. The vertical filtration of the sedimentation water provides a constant connection with the seawater. The availability of appropriate gradients leads to the appearance of notable convective-diffusive currents within the sea sediments, which, in turn, not having constant perturbative sources or runoff, can lead during geological time to the equalization of concentrations of dissolved components in sea and sedimentation waters. Exceptions are found in areas that lie close to the bottom evaporite surface, and those with incoming hot brines or waters, enriched with some components, in rift or other weakened zones of the oceanic or continental crust.

The estimated average annual values of the buried and pressed-out sedimentation waters (Dzhamalov et al. 1999) indicate that over 100 mln t of salts are buried annually with sedimentation waters, about 60 mln t of which return to the seas and oceans from the upper lithogenic zone of the sea sediments (to depths of 250–300 m). It is obvious that presently the processes of formation and press-out of sedimentation waters in the upper part of the cross-section of the sea sediments actually do not have an influence on the modern salt balance of the World Ocean.

Transfer of Dissolved Chemical Elements with Groundwater Flow to Seas and Oceans. Aside from the total salt amount transferred with the underground runoff, apparent interest is to determine the income of concrete chemical elements to seas and oceans, namely: macrocomponents, biogenic elements (BE), and particular microcomponents. The main difficulty in solving this task was connected with the limited data on the groundwater chemical composition in the coastal zone. These data were obtained through:

- Analysis and generalization of experimental results on the groundwater chemical composition in particular testing areas of the coastal zone
- Evaluation of possible chemical element content in the groundwater, proceeding from its mineralization and in the coastal zone and composition of the water-bearing rocks
- Use of the data on riverwater composition during a stable mean-water season when groundwater is the basic source of river recharge

With such an approach, the underground runoff of particular chemical elements was estimated by direct measurements and indirect data of chemical elements in the groundwater and generalization of the measured results in different scales. The indirect data can introduce a certain error into the computations made. However, such a wide involvement of existing information and adequate calculating methods allow the possibility of obtaining the most reliable data on the regional and global underground ion runoff and reveal general regularities in this poorly studied process.

Subsurface water exchange between land and sea is one of the elements of the general water circulation in nature. As mentioned above, in a genetic respect submarine waters are subdivided into infiltration waters that are formed on land at the expense of precipitation and surface runoff, and sedimentation waters formed directly in the sea due to accumulating processes of sediments and their subsequent diagenesis. At submarine discharge, the composition and mineralization of the groundwater of the infiltration type are transformed due to mixing and physico-chemical reactions of these waters with seawater. Currently groundwater composition is subjected to significant changes due to agricultural and industrial development of many coastal territories. Thus, nitrate concentrations in the shallow groundwater at its submarine discharge in some coastal areas of the United States, Australia, Jamaica, and Guam vary from 20 mg/l to 380 mg/l. Such high BE concentrations in submarine water cause an anomalous growth of biomass in the seawater (origination of specific species of microorganisms and algae), which, in turn, can serve as an indicator for submarine discharge of groundwater of a certain composition. In other words, intensive man-made activity in coastal water-intake areas, through directed changes in shallow groundwater composition, influence significantly the seawater environmental state due to the transfer of BE, pesticides, heavy metals, and other toxicants. Thus, submarine waters serve as one of the basic agents and mediums of chemical element migration in the Earth's crust.

The data on groundwater mineralization, concentrations of microcomponents, BE, and microelements are reported in a number of monographs and articles (Krainov et al. 2004; Shvartsev 1999; Krainov and Shvets 1980; Gordeyev 1983; Korzh 1991). Additionally, chemical element content in the groundwater was determined by its mineralization and pH values taking into consideration the composition of water-containing rocks, using for this purpose the known (in hydrogeochemistry) programs for computing an equilibrium chemical composition in the "water-rock" system under certain thermodynamic conditions.

The factually determined data on the groundwater chemical composition in the coastal zone of the continents show that in major cases the groundwater belongs to the hydrocarbonate class. Only in dry areas of Africa, Australia, and Asia, or in areas under intensive anthropogenic impact anions of sulfates and chlorides predomintae in groundwater chemical composition. However, taking into account that submarine runoff from these areas is relatively low, therefore, in general, their influence on the total ion runoff from these continents is slight.

Taking into consideration the regionally and globally generalized *in situ* determinations (Krainov et al. 2004; Krainov and Shvets 1980; Shvartsev 1999), as well as in accordance with the distinctive composition of the river mean-water runoff (Gordeyev 1983; Korzh 1991) and on the basis of computations made with the use of equilibrium thermodynamic models, Table 3.3.3 presents the varying ranges of

TABLE 3.3.3
Total Mineralization and Concentration of Macrocomponents in Submarine Waters

Continent	M (g/l)	Concentrations of Macrocomponents (g/l)					
		Ca	Mg	Na + K	HCO$_3$	SO$_4$	Cl
Africa	0.2–10	40–420	10–120	30–4200	100–900	20–1200	20–4000
Australia	0.4–12	40–360	20–120	80–4200	200–750	30–1500	100–5000
South America	0.2–1.2	40–120	10–40	20–350	160–450	30–150	30–400
North America	0.2–0.9	45–170	12–60	10–45	140–250	35–250	15–170
Europe	0.1–1.2	20–120	10–40	20–350	90–450	20–150	10–400
Asia	0.4–10	40–260	20–120	60–3200	200–750	80–1500	60–4000

TABLE 3.3.4

Calculated Average Values of Total Ion Runoff and Transfer of Particular Macrocomponents

Continent	Total Ion Runoff (mln t/yr)	Transfer of Particular Macrocomponents (mln.t/yr)					
		Ca	Mg	Na+K	HCO$_3$	SO$_4$	Cl
Africa	142.2	27.2	7.4	26.3	44.1	18.1	19.1
Australia	73.8	3.9	2.2	20.7	10.4	7.9	28.7
South America	113.1	16.2	4.1	6.3	61.1	21.5	3.9
North America	148.9	21.9	6.8	8.9	74.6	28.5	8.2
Europe	60.4	11.5	3.1	2.2	23.9	7.5	12.2
Asia	296.4	55.8	19.6	13.8	82.0	69.4	55.8

total mineralization and macrocomponent concentrations in submarine waters for the continental areas under study.

Based on the average mineralization and macrocomponent contents in the groundwater, the total transfer of salts and particular macrocomponents were calculated with the submarine groundwater flow from the continents of the earth in mln t/yr (Table 3.3.4). The resulting average values of salt transfer with the submarine runoff agree well with the above-given (Table 3.3.1) detailed computations of the total ion runoff for individual continents and truly reflect the scale of this natural process.

The calculated average groundwater mineralization in the upper hydrodynamic zone is equal to 0.6 g/l, and the following concentrations of Ca, Mg, Na+K, HCO$_3$, SO$_4$, CL, and SiO$_2$ are typical (mg/l): 90, 20, 45, 320, 65, 50, and 25, obtained with the aid of thermodynamic models in accordance with the above-mentioned mineralization, accepted pH value, and known composition of the water-containing rocks. These average contents of the macrocomponents agree well with the generalized concrete values of the groundwater composition in accordance with its genetic nature (Shvartsev 1999; Perelman et al. 1999).

At the given total mineralization and concentrations of the macrocomponents, the total underground ion runoff and transfer of Ca, Mg, Na+K, HCO$_3$, SO$_4$, Cl, and SiO$_2$ with the groundwater from land to the seas and oceans were estimated in a global scale as 1400, 210, 45, 105, 735, 150, 115, and 60 mln t/yr, respectively. It can be seen from the comparison of the calculated values of total ion runoff and transfer of the macrocomponents (Table 3.3.4) that they have a satisfactory similarity, which confirms once more the reality of the calculations and reliable estimation of the scale of this natural phenomenon.

It should be noted that the intensity of chemical element migration in water is suggested to be characterized with the use of the water migration coefficient (K_x) equal to a ratio of the content of a particular chemical element in water (in dry residue) to its content in rocks of the given water-catchment area (Perelman et al. 1999).The higher the coefficient of water migration, the stronger the element is leached from rocks and the more intensive is its migration in water. According to the existing estimates of water migration coefficients, the highest migration ability in groundwater in the zone of intensive water exchange are Cl, Br, and J; with an increased migration ability are Na, Ca, Mg, F, Mo, Sr, and Zn; and the elements Cu, Ni, K, P, and Mn have a mean ability. The indicated migration abilities of the specific chemical elements were used in calculations of their transfer with the submarine flow and especially of their precipitation in the zone of mixed groundwater and seawater.

Transfer of the dissolved organic substances with the underground runoff to seas was determined based on the data of organic carbon (C_{org}). content in the groundwater of different climatic zones. It should be noted that the water-soluble organic compounds include the representatives of actually all the chemical groups and classes (carbohydrates, proteins, lipides, humus substances, carboniferous acids, hydrocarbons, etc.). The sources of organic substances in groundwater are soils, silts, sedimentary rocks, and oil and gas deposits.

An integral value of the total amount of all dissolved organic substances in a liter of water is taken usually as the content of their basic element — C_{org}. Here, the total C_{org} value amounts in most cases only to about 50% of all the organic substances, enclosed in water, as recalculated to their full form, but can sometimes reach to 80%.

C_{org} content in the groundwater varies from a few units to 3000 mg/l and depends on its origin and conditions of occurrence. The highest contents are connected with oil- and gas-condensate deposits (400–3000 mg/l). Outside the oil and gas deposits the deep groundwater contains, on average, 40 mg/l of C_{org}, whereas in groundwater of the upper hydrodynamic zone its content increases from 10 to 25 mg/l in the arid zone to 20–40 mg/l in humid areas (Krainov and Shvets 1980). At an average C_{org} concentration of 20 mg/l, the transferred dissolved organic matter with the direct underground runoff to seas can reach 45–50 mln t/yr, which corresponds to ~25% of their transfer with river runoff (Gordeyev 1983).

Transfer of BE with groundwater to seas is difficult to calculate because of rare and not always reliable data. Due to this, the average concentrations of, first of all, nitrates and phosphates in the groundwater were determined from the models of equilibrium chemical compositions of solutions in the water-rock system under thermodynamic conditions in aquifers of the upper hydrodynamic zone. According to modeled computations, the phosphate concentration in the shallow groundwater does not usually exceed 0.1 mg/l and nitrate concentration 1–2 mg/l with a groundwater mineralization of 0.5–2 g/l. In this case, it should be taken into consideration that the P-compounds are met only in unconfined aquifers tightly connected with surfacewaters and subjected to a considerable anthropogenic impact. Nitrates, due to their conservative nature, can be found at depths of a few tens and even hundreds of meters, but their concentration is increased in the agricultural territories. Recently it was reported that considerable pollution of groundwater in agricultural areas by nitrates with concentrations in the coastal zone, exceeding the maximum allowable content (50 mg/l). But for global assessment of the above-mentioned BEs transferred with the underground runoff to seas, the above-given minimal concentrations of phosphates (0.1 mg/l) and nitrates (1.5 mg/l) were used to make more reliable computations. In this case, the total transfer of phosphates with the underground runoff is not higher than 0.25 mln t/yr, and nitrates 3.5 mln t/yr. We note for comparison that according to Gordeyev (1983), the amount of inorganic phosphorus compounds, transferred with the riverwater runoff to the seas, is 0.4 mln t/yr, and the total inorganic nitrogen is 11.3 mln t/yr. In the other words, the BE transfer with the groundwater to seas amounts to ~50% of the transfer by the rivers. It should be taken into account that in the coastal areas with intensive agricultural activities, the transfer of BE with the submarine runoff considerably exceeds their income with poorly developed riverwater runoff.

The transfer of microelements with groundwater to seas is the part of the submarine ion runoff that is most difficult to determine. There are few data on microcomponent concentrations in the coastal groundwater, which cannot be directly used to calculate global amounts of microelements transferred to seas and oceans. For this reason, the data on the chemical elements that are transferred with the river runoff to seas (Gordeyev 1983; Korzh 1991) and the average contents of the microelements in the groundwater of the upper hydrodynamic zone (Shvartsev 1999) were used as the initial concentrations of the most widely spread microcomponents in natural waters. Proceeding from the understanding of the formation of the natural-water chemical composition in the upper part of the Earth's crust, one can suppose that the major concentration of microcomponents in riverwater is formed, first of all, at the expense of river recharge by groundwater. Just groundwater, leached during filtration in the water-containing rocks, is the basic source of microcomponents (especially in relatively high concentrations) penetrating to surfacewaters. However, it should be taken into consideration that in accordance with the regional and global estimates of groundwater runoff to rivers, underground river recharge amounts, on average for the continents, is 24%–36% of the total river runoff (L'vovich 1974; Hydrometeoizdat 1974), i.e., ~30% for the earth as a whole.

Thus, the average concentration of microcomponents in groundwater of the upper hydrodynamic zone is not less than 3 times higher than its average concentration in the riverwaters. Based on data reported in Korzh (1991), only 15% of the river-runoff salts were transferred from the ocean to the rivers due to atmospheric moisture transfer. Therefore, it should be outlined once more that actually within the entire land the composition of surface and groundwater is being formed at the expense of leaching chemical elements from rocks. The commonly observed high mineralization of the groundwater as compared with surfacewater mineralization is explained by a much longer interaction of the former with water-containing rocks and, hence, by a much longer physico-chemical process of leaching chemical elements from rocks. The investigations (Shvartsev 1999) carried out during recent years on the groundwater hydrochemistry of the hydrodynamic zone confirm the above-stated concept and the possibility to accept for the majority of microcomponents their average concentrations in the groundwater of the upper hydrodynamic zone.

The proposed concept for substantiation of average concentrations of microcomponents in groundwater and, respectively, for assessment of their direct transfer to seas is tentative and requires a further specification. However, as numerous reports show, regional pollution of the natural waters and, particularly, groundwater, caused by intensive man-made activity at water-intake areas during the recent years, has

TABLE 3.3.5
Transfer of Microcomponents with Subsurface Runoff to Oceans

Element	Concentration (mg/l)	Total Transfer (mln t/yr)
Fe	0.55	1.26
F	0.45	1.04
Al	0.28	0.64
Sr	0.19	0.44
Br	0.18	0.41
Ba	0.12	0.28
B	0.06	0.14
Mn	0.05	0.12
I	0.02	0.05
Zn	0.03	0.07
Cu	0.006	0.014
Pb	0.003	0.007
Ni	0.003	0.007
Cr	0.003	0.007
As	0.002	0.004
Co	0.001	0.002

led to an increase in concentrations of Pb, Zn, Hg, Cu, Cd, As, Sb, Mo, and other heavy metals by an order of magnitude and higher. All this indicates that the calculated and accepted concentrations of microcomponents are sufficiently substantiated and reliable.

According to the data reported in Gordeyev (1983), among the wide spectrum of the microcomponents in river waters 11 elements (Al, B, Ba, Br, Cu, F, Fe, I, Mn, Sr, and Zn) amount to almost 99% of the total runoff to the seas. Just the concentrations of these 11 elements with a slight addition of other ones were used for the global-scaled assessment of their transfer with the submarine runoff to the seas (Table 3.3.5). Thus, the basic microelements transferred with the groundwater to the seas amount to ~4.5 mln t/yr or ~30% come with the river runoff. No reliable data on other microelements are available, but, as mentioned above, their share in the total submarine ion runoff is very slight.

3.3.2 BEHAVIOR OF CHEMICAL ELEMENTS IN THE ZONE OF MIXING GROUNDWATER AND SEAWATER

Mixing of salt seawater with fresh and low-mineralized submarine waters occurs, mainly, in bottom sediments and near-bottom water layers in shelf zones. Due to this, near-bottom shelf areas of seas and oceans serve as an important geochemical and biological barrier for dissolved chemical elements migration with submarine flow. Short distances from the zone of mixed seawater and groundwater one can observe changes in the salt content, pH, Eh, and other physico-chemical characteristics that exert a high influence on the conditions of the existence and preservation of chemical elements in the mixed solutions. It has been established by field and experimental investigations that during mixing of salt and freshwaters in estuaries (which is very similar to mixing of submarine groundwater with seawater), the dissolved substances are subjected to flocculation, deposition, chelation and complexion, and adsorption and desorption (Gordeyev 1983).

Many dissolved microelements are removed from the solution due to the mechanism of flocculation, i.e., transformation of dissolved organic and inorganic substances into suspended amorphous particles with an increase of salt content and pH of water. This results in the formation of, mainly, ferrohumate floccules, which are deposited, due to adsorption, together with Al, Mn, Zn, and other microelements. Along with adsorption, a change in the conditions can cause a desorption and even an equilibrium state between these two processes. It is established that removal of different elements from the solution due to formation of ferro-organic floccules with subsequent adsorption and co-settlement of elements occurs in accordance with the following sequence (Gordeyev 1983): Fe>Cu>Al>Ni, Mn, Si>Co>Cd>Ca, Mg.

Note that the active flocculation of organic colloids shows that the dissolved organic substance, coming with groundwater, actually does not reach the upper layers of seawater due to its precipitation. The average C concentration in the oceanic water (1.6 mg/dm^3) is much lower than in river and groundwater. However, the total C_{org} content in the World Ocean reaches $2.13 \cdot 10^{12}$ t. The total organic substances in the groundwater of the Earth is $2.5 \cdot 10^{12}$ t, which is 10 times higher than the oil reserves and only twice lower than coal reserves (Krainov et al. 2004).

Removal of biogenic elements (N, P, Si) from the solution with mixing of submarine groundwater and seawater occurs mainly due to processes of their utilization by microorganisms, i.e., due to biological processes that are controlled by the amount of plankton and its development activity. Active assimilation of biogenic elements by microorganisms in mixed solutions can reduce the BE concentration to minimal levels. Moreover, the BE concentration often depends significantly on salt contents in the mixed solutions, which is most typical of nitrogen and phosphorus. Active removal of dissolved phosphates and nitrates takes place at a salt content of 8–10‰ and continues until 15‰ when the nitrate concentration drops almost to zero.

The basic elements of submarine water (Na, K, Ca, Mg, SO4, HCO3, and Cl) behave themselves conservatively in the zone of mixing. They can be used (especially Na and Cl) as indicators of the degree to which fresh and saltwaters are mixed. However, at active formation of primary products of the phytoplankton (chiefly, diatoms), the concentrations of calcium and magnesium can decrease considerably. The latter is difficult to notice, as the submarine water discharge is slight, and calcium and magnesium are determined usually by their content in seawater. Therefore, it can be supposed that the basic ions of submarine groundwater overpass the geochemical barrier at the boundary between seawater and groundwater without significant losses. At the same time, it is established that in the arid zone of the earth chemical precipitation of hemogenic calcium carbonate is observed because of a disturbance in the equilibrium and photosynthesis process.

Losses of chemical elements in the zone of mixed submarine groundwater and seawater are presented in absolute and relative values in Table 3.3.6. The elements lost in the mixing zone are caused mainly by hydrogeochemical processes, while the poorly studied biological processes connected to the formation of primary products are difficult for quantitative assessment. At the same time, origination of microorganisms and their life considerably alter the physico-chemical characteristics within short distances in the mixing zone and exert a sizable influence on preservation or transformation of almost all incoming elements.

Numerous wells, drilled during recent decades into the floors of seas and oceans, show the similarity of the chemical composition of submarine and seawaters in the areas adjacent to median oceanic ridges and in areas with rather slow sedimentation. In the rest of the deep parts of the oceans, the concentrations of some elements in the chemical composition of the submarine (pore) waters undergo a notable change. Namely, the concentration of calcium usually increases with depth, and magnesium and potassium decrease (Figure 3.3.2). Scientists suppose that just the reaction of the change (destruction) in basalts in the deep parts of oceanic depressions and, to a lesser extent, destruction of diffused volcanic materials in sedimentary rocks serve as a reason for an increase of dissolved calcium and decrease of magnesium in the submarine waters (McDuff and Gieskes 1976). The gradient of the calcium content change, with depth of testing, varies within the World Ocean floor from 25 to 1 mmol/100 m and is equal, on average, to 4 mmol/100 m or 160 mg/l per each 100 m.

The investigations of the oxygen isotope composition in the pore waters of ocean floors have established a good correlation between the changes in contents of calcium and ^{18}O (Figure 3.3.3). On the basis of this correlation, there was calculated the average gradient of ^{18}O-decrease with depth in the pore waters of the World Ocean's

TABLE 3.3.6
Losses of Chemical Elements in the Zone of Mixed Submarine Groundwater and Seawaters

Element	Groundwater Transfer (mln t/yr)	Losses in Mixing Zone (%)	Transfer, Taking into Account Losses (mln t/yr)
Ca	210	0	210
Mg	45	0	45
Na+K	105	0	105
Si	60	20	48
$C_{org.}$	50	5	48
Al	0.64	30	0.45
Cu	0.014	40	0.008
Fe	1.26	80	0.25
Mn	0.12	20	0.1
Ni	0.02	20	0.016
Zn	0.07	10	0.06

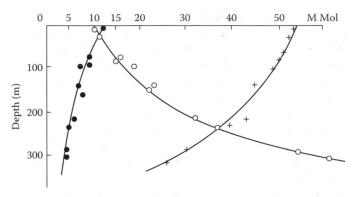

FIGURE 3.3.2 Change with depth of the concentration in some components of porous water in sea bottom sediments (Gieskes and Lawrence 1981).

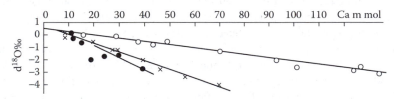

FIGURE 3.3.3 Relation between calcium concentration and oxygen isotope in the porous water of sea bottom sediments in separate areas of deep-sea drilling (Gieskes and Lawrence 1981).

sediments as equal to 0.32‰ per each 100 m (Gieskes and Lawrence 1981). The average gradient of the ^{18}O-decrease with depth enables the tentative evaluation of scales to which this isotope of oxygen penetrates into sedimentary and basalt layers of the oceanic crust. If one accepts diffusion as the basic process of pore solution migration, then at its average coefficient of $6 \cdot 10^{-6}$ cm^2/sec, the ^{18}O-transfer together with the H$_2$^{18}O-flux from the oceanic water mass to the oceanic crust will amount to ~2–105 mol per a million years. However, in order to keep the ^{18}O-balance in the World Ocean, a considerable return flow is needed, which may occur in the zones of spreading and subduction (zones of distribution and sinking of the Earth crust's blocks relative to each other). Gieskes and Lawrence (1981) have determined that if the O^{18} return flow occurs due to hydrothermal processes in the central parts of the median oceanic ridges, then over 40% of the 6-km thick basalt oceanic crust is required to form a compensation flux. The balance ratios given are approximate and tentative to a great degree. Nevertheless, they indicate how scaled the hydrogeochemical processes occur in the ocean sediments.

Of greatest interest in a hydrogeochemical respect are the regions of modern submarine volcanism in the zones of the median oceanic ridges. The magma flows out there to the surface or approaches the ocean floor, causing a powerful heat flux. The young oceanic crust is highly fissured there due to the processes of cooling, compaction, and extension. The seawater saturates the fissured zone, cools it, and destructs the magma body. This process proceeds approximately by the following scheme: cracking $\rightarrow$ water penetration $\rightarrow$ convection $\rightarrow$ cooling. The similarly forming hydrothermal solutions of the rift zones are usually enriched with CO_2, He_3, H_2, metals, and other components. The temperature of these solutions can be close to the point of magma hardening (980°C) but at the floor surface, due to mixing processes, it amounts usually to 10°–30°C and only in particular cases reaches 350°C (Corliss et al. 1981).

The thermal gas- and water-containing fluids, enriched with different chemical elements, serve as the basic source of polymetallic and other mineral deposits. In some cases, the isotope composition of carbon and helium in hydrothermal solutions indicate the juvenile nature of these components. However, this does not give complete grounds to make a conclusion on the juvenile nature of the hydrothermal solutions themselves. Fluids in the oceanic rift zones represent seawater which through contact with magma acquires an increased temperature and specific chemical composition.

Thus, transfer of dissolved substances with the groundwater flow is one of the most important natural processes of chemical elements migration in the Earth's crust. The computations performed give evidence of the significant role of submarine groundwater flow in the formation of peculiar hydrogeochemical conditions in the zone of groundwater discharge. Groundwater transfers to the seas over 25–50% of particular chemical elements out of the elements transferred by the rivers. The total salt transfer with groundwater from land to sea is >50% of the salt transferred by the rivers, which, to a considerable degree, influences the hydrochemical and hydrobiological regimes in the sea coastal zones. Carried-out estimates have regional and global character. The results obtained on the chemical elements transferred with direct groundwater flow to the seas and oceans reflect the scale of this poorly studied phenomenon with the expectation that further investigations will make the calculation more exact.

This is the first time for the submarine water flow, a proposed and quantitatively characterized concept on the existence of a geochemical and biological barrier at the boundary between submarine groundwater and seawater was done. This causes a significant transformation and loss of many chemical elements due to mutually connected physico-chemical and biological processes.

3.4 INFLUENCE OF SUBMARINE SEDIMENTATION WATER ON THE WATER AND SALT BALANCE OF OCEANS

Natural water circulation is an enormous process that involves the entire hydrosphere: atmospheric, surface and groundwater, as well as waters of the biosphere, mantle, and cosmos. The circulation provides and constantly keeps the unity of natural waters. Basic elements of water circulation have existed actually in all the phases of the geological evolution of the Earth. However, their role in groundwater formation has changed depending on the scale and a development degree of different geological processes. In particular, hydrologic or climatic circulation in its present-day manifestation has been formed since the appearance of land amid the ocean.

The geological circulation started developing alongside with the appearance of the first sedimentary rocks, i.e., approximately since Proterozoic age. The beginning in plate tectonic activity has given an impulse to the beginning of the mantle-oceanic circulation. Each type of water circulation (hydrological, geological, mantle-oceanic) can include one or more cycles providing continuous water exchange between the land and seas. The above water circulation types alone provide an interconnection and unity of all the types of water. At the same time, water exchange between land and sea is not limited only to the hydrologic cycle, but embraces also another more enormous process — geological circulation. This type of circulation depends on geological processes such as sedimentation, formation of sedimentation waters, their subsequent diagenesis and metamorphism, tectonic motion of rock layers, blocks and volcanism, rock serpentinization and granitization, etc. Thus, all the circulation elements provide for the formation and distribution of basic genetic types of groundwater.

Based on a groundwater genetic type, one should understand waters which are united by uniform formation conditions, mechanisms of transport, and common conditions of circulation. In accordance with these genetic features, one can distinguish, first of all, juvenile, infiltration, and sedimentation groundwater types which, in turn, are subdivided into a number of subtypes. If one follows the juvenile hypothesis of the origin of the hydrosphere, all groundwater has been formed due to extraction of vapor from magma (mantle) through deep-seated faults with its subsequent condensation in the upper Earth crust. In this way, juvenile water was generated and, for many years Zyuss (1888–1909) was given a basic role in the geological history of the hydrosphere. However, according to data from present-day investigations, the juvenile component in solutions of volcanic areas and middle-oceanic ridges amounts only to not more than 5%, which does not allow this genetic groundwater type to be basic in the modern geological history of the Earth. At the same time, within the composition of juvenile water, the majority of researchers distinguish magmatogenic groundwater which has reached to the upper Earth crust with magma, and matamorphogenic waters formed due to metamorphism at different depths and coming to the Earth's surface through faults and rift zones.

As present-day experimental and theoretical investigations show, the majority of the groundwater in the upper zone of the Earth's crust (to a depth of 2–3 km and deeper) is being formed by precipitation (meteoric or vadose waters) which has infiltrated into the subsurface. The infiltration type of groundwater is closely connected with climatic circulation. This mechanism of how groundwater penetrates into the upper Earth's crust is subdivided into truly infiltrated water and water of condensation which enters the unsaturated zone in the form of aqueous vapor.

Another important genetic type of groundwater is sedimentation water which is formed during sedimentation and lithification of sedimentary rocks. Such water is often called formation, connate, or fossil water. Sedimentation waters are connected with geological circulation, and one can distinguish between truly sedimentation (or connate) and elision waters. Elision waters are those that are formed in marine clayey sediments and which are pressed out during the sinking of the latter into more permeable layers due to the geostatic pressure of the overlying rocks.

It should be specially noted that groundwater at different depths of specific geologic-hydrogeologic structures represents, in major cases, mixed waters of different genesis, which are connected with different branches of the water circulation on the Earth. Submarine water in the rocks of the World's Ocean floor are also subdivided genetically into juvenile, infiltration, and sedimentation waters. Juvenile waters are found mainly in tectonically active zones and play no significant role in the present-day water balance. Infiltration waters are formed on land and discharged fully as they move along the continental slope. Their role in the water and salt balances of seas and oceans is mentioned above (Dzhamalov 1996; Zektser et al. 1984). Sedimentation waters are formed throughout the sea bottom, but their influence on the water and salt balances in oceans has been poorly studied. Usually only very general expert estimations of the formation and pressure driven submarine sedimentation waters are reported in the literature, without any spatial and temporal analysis of data obtained (Timofeyev et al. 1988).

Computation Method and Results. Geological water circulation is caused by a wide spectrum of geological and tectonic processes in the Earth's crust. Its initial phase is connected with sedimentation in water basins in the sea and takes place constantly during the entire geological history of the Earth, becoming especially active during structural transformations in the Earth's crust. Sedimentary rock massif and of itself in source sea basins co-bury vast amounts of sedimentation water, a certain part of which with compaction of the sediments is pressed back out into the basin (Dzhamalov 1996).

Silt sediments are composed of fine-disperse, mainly clayey and sandy fractions having different mineralogical compositions. They contain not only water, but also a certain amount of salts, microorganisms, and gases. The initial porosity and moisture of these sedimentary rocks reach 80% and higher. Thus, the average natural moisture of the modern bottom sediments of the North Caspian Sea basin exceeds 70%, the Greenland Sea 55%, and the Arctic Ocean 70% (Shvartsev 1996).

With sinking of sea sediments, the initial silts get compacted, lithified, and undergo diagenesis due to a rise in geostatic pressure and temperature. These processes lead to a decrease in the rock porosity and extrusion of the enclosed free pore water upward back to the main sea basin. With deeper submerged sea sediments, the compaction velocity becomes gradually lower, and, hence, as a result, isles of the pore water are pressed out. Thus, the porosity of clays at a depth of 400–500 m is equal to 0.35–0.4, at a depth of 2000 m is ~0.2, and at 3000 m still lower at 0.1 (Vassovich, 1986). The generalized porosity curves plotted against a decrease in sea sediments due to their compaction with depth are given in Figure 3.4.1. Drilling data from deep oil boreholes and the super-deep Cola borehole show that even at depths of up to 10 km there are zones which have a relatively high jointing (porosity) and vadose groundwater. As a rule, publications report only general expert estimates on formation and press-out of submarine sedimentation waters, without spatial and temporal analysis of the data obtained.

The method for calculating an amount of pressed-out pore water is based on the analysis of a change in clay (and other rocks) porosity depending on an occurrence depth. Drilling data carried out both on and off shore (deep-sea drilling) show that

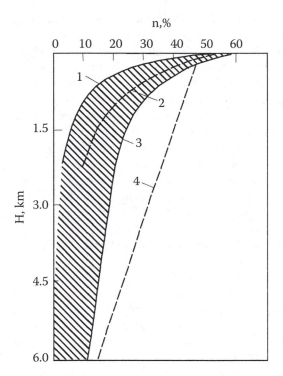

FIGURE 3.4.1 Change in porosity (n) with depth (H) in clay (1–3) and sandstone (4). (Shvartsev, 1996.)

the sediment porosity dependence on depth and lithification degree can be approximated by the following exponential function:

$$n = n_0 e^{-aH} \tag{3.4.1}$$

where n is rock porosity at a depth H; n_0 is initial porosity; a is a numerical coefficient obtained through processing of factual data.

In other words, the basic amount of submarine pore waters is pressed out with clay compaction within the first hundreds of meters. If at depths >250 m, no significant diagenesis occurs in clays and the latter are compacted only due to losses of pore waters, then one will have

$$V(1-n) = V_0(1-n_0) \tag{3.4.2}$$

where V_0 and V are, respectively, clay amounts before and after compaction, m³; n_0 and n are clay porosity before and after compaction correspondingly.

In this case, the amount of squeezed-out submarine pore water (W) will be equal to:

$$W = V_0 - V = V \cdot (n_0 - n/1 - n_0) \tag{3.4.3}$$

Thus, in order to calculate how much sedimentation pore water penetrates to the main sea basin with compaction of clayey sediments within the first hundreds of meters, it is necessary to have data on the amount and porosity of clays at a specified depth, and on their initial porosity. This method was tested while analyzing the initial reports of the Deep-Sea Drilling Project. Over 300 boreholes, drilled in the Atlantic, Pacific, and Indian Oceans and in adjoining seas, were analyzed. Besides the porosity, the lithological composition and thickness of sedimentary rocks were considered (Scripps Institution of Oceanography 1969 and 1972).

The deep-sea drilling data show that the porosity of the youngest sedimentary silts in the upper part of the lithological cross-section is usually 0.8–0.7, and sometimes 0.6. As the sediments submerge approximately to 200 m, their porosity decreases gradually to 0.4. But beginning from depths of 250–300 m, the porosity starts undergoing sharper and sometimes abrupt changes to 0.30–0.25 (Figure 3.4.2). This indicates a change in the physical properties of sedimentary rocks, which at these depths are predominantly argillites and limestone. The porosity of these lithified rocks depends mainly on their jointing. Stratigraphically, the upper part of the cross-section is composed chiefly of clayey sediments, deriving from the Pleistocene, Pliocene, and Miocene periods, i.e., the age of these sediments is 20–25 mln years.

Based on the analysis of deep-sea drilling data, a regional evaluation of the pressed-out submarine sedimentation water was made for the upper 250 m of the cross-section of bottom sediments. This is because only the upper part of the World Ocean's cross-section has been drilled by deep-sea drilling boreholes and, hence has been most studied by seismic investigations. A well-grounded possibility therefore exists to interpolate and extrapolate the borehole data across the ocean taking into account the geomorphologic and geologo-tectonic structure of its bottom. This regional evaluation helped to get a picture showing the distribution of the values obtained within the World Ocean, and to determine the scales and to reveal the basic regularities of this global process (Dzhamalov & Safronova 1999).

Through analysis of available material and computations specific values were obtained of the submarine sedimentation waters pressed out during compaction of each cubic meter of clayey sediments for 20 mln years, with a known porosity distribution at depths of up to 200–250 m. These values vary from $0.2-0.3 \cdot 10^6$ to $2 \cdot 10^6$ m^3, but sometimes can range up to $3 \cdot 10^6$ m^3 from 1 km^2 of the ocean's area, i.e., annually the amount of pressed out submarine sedimentation water from an area of 1 km^2 changes approximately from 0.01–0.1 m^3/yr. This value can be considered as an analogue of the modulus of pressed-out sedimentation water from compacting sea deposits. On the basis of these computations, the amount of the pressed-out sedimentation water was evaluated for the Pacific, Indian, and Atlantic oceans, and the World Ocean as a whole (Table 3.4.1). The distribution of the specific values within the World Ocean is reflected on the schematic map of pressed out submarine sedimentation pore waters on a scale of 1:50,000,000 showing the areal changes in the amounts of pressed out pore solutions (Figure 3.4.3).

The total amount of pressed out submarine sedimentation water within the World Ocean's area of 283 mln km^2 and an average thickness of compacted sediments of 250 m is equal to $36 \cdot 10^6$ km^3 or $0.37 \cdot 10^{23}$ g. Thus, a question arises: How much of the value obtained corresponds to the scales of the process under study? To answer

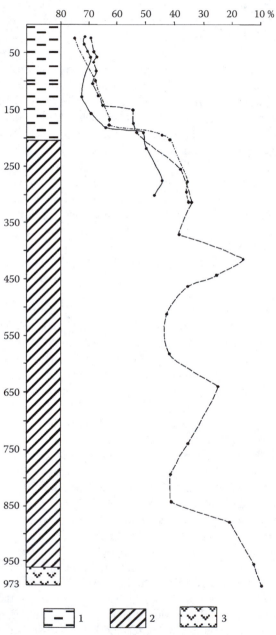

FIGURE 3.4.2 Change in the porosity of sea sediments according to data of deep-sea drilling in (1) unconsolidated sediments; (2) consolidated sediments; (3) basalts.

TABLE 3.4.1
Press-Out of Submarine Sedimentation Water in the World Ocean Bottom

Ocean	Specified Area (mln km^2)	Specific Press-Out of Submarine Sedimentation Water (10^6 m^3 from 1 km^2)	Specified Thickness of Sediments (m)	Total Press-Out of Submarine Sedimentation Water (10^3 km^3)
Pacific Total	130.7			20,300
Zone (area) 1	24.9	0.5	110	800
Zone (area) 2	52.4	0.5–1	170	6700
Zone (area) 3	53.4	1–2	160	12,800
Indian Total	66.7			7500
Zone (area) 1	25.2	0.5	140	1100
Zone (area) 2	23.2	0.5–1	170	2900
Zone (area) 3	18.3	1–2	130	3500
Atlantic Total	85.6			8200
Zone (area) 1	57.3	0.5	150	2600
Zone (area) 2	12.7	0.5–1	200	1900
Zone (area) 3	15.6	1–2	160	3700

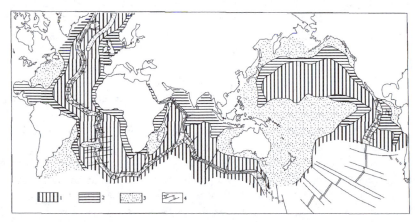

FIGURE 3.4.3 Schematic map of released submarine sedimentary water. Specific values of released submarine sedimentary water (n·10^6 m^3/km^2): (1) <0.5; (2) 0.5–1; (3) 1–2; (4) mid-ocean ridges and faults.

this question, we can compare the value obtained from known (from the literature) total water masses extruded from the sedimentary rocks of the oceans during their lithogenesis. The most reasonable, in this respect, seems to be the value of 0.41·10^{23} g, which has been reported in Zverev (1993) and Timofeyev et al. (1988) and agrees well with the above-mentioned conclusion.

The scale of the process under study can also be determined from expert estimates of the total water mass extruded from sedimentary rocks of the oceans, derived from the average characteristics taken from guidebooks and the initial data from deep-sea drilling in the World Ocean. Thus, if the area of the World Ocean is $360 \cdot 10^6$ km^2, the average velocity of the sedimentation is 0.01 mm/yr, and the density of young clayey sediments is 1.2 g/cm^3 (Lisitsyn 1978), then the total mass of sediments precipitated during a year will amount to $4.3 \cdot 10^{15}$ g. If we assume that the initial porosity of the sediments is 0.7 and in 20 million years it will be twice as low (Figure 3.4.1 and Figure 3.4.2), then the total submarine sedimentation water amount extruded due to sediment compaction will reach $0.38 \cdot 10^{23}$ g. This agrees with the values obtained earlier.

The relationship between the velocity of burying submarine sedimentation water in precipitated sediments and their subsequent press-out with compaction of the sediments is of interest. One can see from the above total amount of precipitating sediments and their porosity that the burial velocity of sedimentation water is equal, on average, to $3 \cdot 10^{15}$, whereas the velocity of pressing out is $1.8 \cdot 10^{15}$ g/yr. In other words, over 60% of pore waters that are being co-formed with sediments return to the source sea basin from the upper zone of the cross-section (200–250 m). It should be noted that all the given values are tentative and can vary in different parts of the World Ocean depending on sedimentation velocity and the intensity of sea sediments compaction. At the same time, they are quite real, reflect properly the scales of a phenomenon under study, and can be used in studying the water balance and geological history of the oceans.

The average annual data obtained on burying and press-out of the sedimentation waters provide the possibility to assess a degree to which these processes influence the salt balance of the World Ocean. Analyses of deep-sea drilling data show that sea and sedimentation waters usually have a close mineralization and similar salt composition (Gordeyev 1983; Zektser et al. 1984). It can be assumed in this connection that mineralization of the sedimentation waters in the upper zone of the cross-section of sea sediments is on average 35 g/kg. In this case, the sedimentation waters co-bury annually over 100 mln t of salts, about 60 mln t/yr of which return to the mother basin as a result of compaction and lithogenesis of sea deposits. If we compare the annual addition of salts to oceans with river and groundwater runoffs from continents (>3500 mln t/yr) (Zektser et al. 1984), it becomes evident that presently the processes of formation and press-out of sedimentation waters actually do not exert any influence on the modern salt balance of the World Ocean.

Regularities in Press-Out of Submarine Sedimentation Waters. The specific values obtained for press-out are submarine sedimentation waters derived from the compaction of clayey sediments and the character of how they are distributed over the floor of the World Ocean makes it possible to reveal the basic regularities of this natural process. When compared with the regularities in the distribution of the subsurface runoff from continents to seas and oceans (Zektser et al. 1984), the process is not studied well and the knowledge of it will be greatly improved with further collection of factual material. At the same time, the current knowledge of the investigations of global regularities of pressed out submarine sedimentation waters is seen as sufficiently clear.

As noted from the schematic map (Figure 3.4.3), the máximum specific values of the pressed-out pore waters are associated, as a rule, with the peripheral parts of oceans adjacent to continents, as well as to middle oceanic ridges and other tectonic uplands on the World Ocean's floor. Thus, the dynamics of sediment compaction and, hence, the press-out of pore waters are controlled by two basic factors: velocity of sedimentation and tectonic structural features of the oceans' bottoms. It has been established by marine geologists and lithologists that sedimentation in oceans depends on physico-geographical, circumcontinental, bathymetric, and tectonic zonalities (Lisitsyn 1978), the most pronounced among which are physico-geographical and circumcontinental. Interaction with the latter creates a faster sedimentation and increased thickness of sediments as one approaches the continents, especially in humid zones of the moderate and equatorial latitudes. In cases when considerable amounts of terrigenic rock materials come from the continents and large islands toward these humid zones, the sedimentation velocity reaches 0.1 mm/yr and more, and the total thickness of the Cenozoic sediments is a few hundred meters. In this connection, the maximum specific values of pressed out submarine sedimentation waters are equal to $2-3 \cdot 10^6$ m³ from 1 km², and generally these zones are characterized by rather high volumes of pressed-out pore water.

Another regional zone which has increased specific values of press-out of pore water ($1-2 \cdot 10^6$ m³ from 1 km²) is related to the middle-oceanic ridges and other tectonic uplands on the bottom of oceans. The middle-oceanic ridges have a thin (to 100 m) sedimentary cover. However, these are the most active zones of the World Ocean and are characterized by high seismicity, volcanic phenomena, large tectonic motions, increased heat fluxes, hydrothermal activity, etc. All these features cause a much faster lithogenesis of sedimentary rocks and, hence, more intensive press-out of sedimentation pore waters when compared with the tectonically calm adjacent parts of the abyssal valleys. The latter are distinguished by the weakest specific press-out of sedimentation water (to $0.5 \cdot 10^6$ m³ from 1 km²) and divide the two regional zones with a more intensive pore-water press-out.

Thus, the values obtained of the specific and total amounts of pressed out submarine sedimentation water and the regional regularities revealed in this poorly studied process agree well with the known features of sedimentation in oceans, tectonic structure, and geological history. This initial phase of geological water circulation, connected with deposition of hydrated sediments and their subsequent diagenesis and lithification, is called *sedimentogenic* by many researchers.

Metamorphogenic, Magmagenic, and Mantle-Oceanic Branches of Groundwater Circulation. The sedimentogenic phase prevails in geological water circulation within the first kilometers of depth. This is explained, first of all, by the lithological composition of the rocks, velocities of their submergence, and the geothermal conditions in the depth interval. However, within the first hundreds of meters of the depth, the pressed-out sedimentation waters return to the main sea basin. At deeper (1–3 km) submergence of the clayey sea sediments, the pore water is pressed out and collected into sandy and chemogenic beds lying among the clays. The water-collecting rocks (sand, sandstone, limestone) keep their relatively high porosity (to 0.2) at rather large depths, and the syngenetic sedimentation waters enclosed in them are under predominant hydrostatic pressure. On the other hand, the compacting plastic clayey sediments have a predominant geostatic pressure due to the weight of the overlying rocks, which is approximately twice as high as the hydrostatic pressure. That is why the sedimentation waters from the clayey sediments enter the fissured water-collecting rocks under an increased geostatic pressure (depending on the depth of submergence of compacting clays) and press out syngenetic gravity waters from them according to the hydrodynamics laws, i.e., from the places of the deepest submergence (sagging) and high pressure to the places of the lowest submergence and pressure lowering up-dip the aquifers (water-collecting rocks). In this way the so-called *elision mode* manifest itself, which has, in recent years, been a determinant role in the formation of groundwater flows and gaseous and aqueous fluids in the submerged parts of young platform-type artesian basins and piedmont troughs. Due to the elision processes, the deep sedimentation waters move gradually toward the surface and into the climatic circulation.

However, the role of the elision processes in the formation of regional lateral streams of deep groundwater requires more detailed investigation. Here, special attention should be given to the analysis of a relationship between incremental rates of geostatic loads and relaxation of stratum pressures depending on the degree of insularity of a specified element of the hydrogeological structure, as well as to the relationship between elision and infiltration recharge per a unit of time and a unit of area (Dyunin 2000).

Elision processes develop especially intensively at depths of 2–3 km. With depth they dampen gradually. At a depth of 6–8 km, the porosity of clayey sediments decreases to 0.1 and lower. The volume is reduced and they start undergoing deep lithological transformations up to the formation of a lithified clayey shale. The rock lithification occurs due to the growth of geostatic and geodynamic pressures, as well as with the gradual rise of temperature. Complete compaction and lithification of primary sedimentary rocks is accompanied by heating in the zones of low- and high-temperature (progressive) metamorphism. Due to recrystallization in the metamorphism zone, the sedimentary rocks extract not only the remaining 5–6% of pore water, but also the physically bound water, and the crystallized and constitutional waters. In this case, both free water in the form of H_2O molecules and OH hydroxyl group and ions of hydrogen and oxygen are extracted, which, under certain thermodynamic conditions, synthesize new molecules of water. Thus, in deep layers of the Earth's crust (15–30 km) the meteoric waters that have penetrated to these depths as a result of sedimentation, lithogenesis, and recrystallization of sedimentary rocks undergo regeneration or revival. This second phase of geological circulation, with complete extraction of meteoric waters and their partial revival, is called metamorphogenic.

The zone of metamorphism is subjected to active tectonic processes, such as the formation of regional faults and crustal block, their sinking, uplifting, and drawing apart, volcanic manifestations with intrusion of magmatic bodies, streams of lava, and overheated solutions. Regional faults serve as large drains for abyssal gas–water fluids which through tectonic dislocations and volcanic vents migrate to the surface. Thus, the metamorphogenic phase is tightly connected with the *magmatogenic* phase; they interact with each other, supplement each other, and virtually close the geological water circulation, advancing the completion and migration of abyssal fluids from the crust and crustal and mantle interiors of the Earth to its surface.

The invasion of magmatic intrusive bodies to a depth of several kilometers leads to the gradual cooling of magma, its crystallization, consolidation, and extraction of gas. Further formation of a gas-aqueous fluid takes place at the final stages of rock crystallization. The pressure of an invading magmatic body, which can exceed the geostatic (lithostatic) pressure of the overlying rocks, is completely or partly transferred to a gas-aqueous phase. Under action of this high pressure, the regenerating waters (gas-aqueous fluids) from the cooling magmatic (metamorphic) melt move through a system of communicating canals (faults) upward to a zone of low pressures. It should be outlined that movement of the abyssal waters (their press-out) occurs also due to a gradual reduction of the porous space in consolidating magmatic melts and rocks subjected to metamorphism. The computations fulfilled with the use of real initial parameters have shown that the consolidation (crystallization) of magmatic and metamorphic bodies and their consolidation under pressure of overlying strata cause formation of crystalline rocks with a porosity of 0.1–1.0% and a permeability of about 10^{-7} mkm^2, which is quite typical for hard rocks at these depths. Then, the pressed-out revival waters migrate to zones of abyssal faults and adjacent fissured rocks. Such zones of tectonic dislocations are characterized by higher permeability and exert, hence, a draining effect on regenerating abyssal gas-aqueous fluids and therefore advance their vertical transport upwards (Peck 1968).

Junction and interaction of the above phases of the geological water circulation (sedimentogenic, metamorphogenic, magmatogenic) advance formation of various hydrothermal, oil- and gas-bearing, mineral, and other types of groundwater. These waters of deep circulation migrate gradually to the Earth's surface (sometimes very slowly) and move into the hydrological circulation. Thus, the geological circulation cannot be considered independently. Because of natural processes with different intensity, this type of circulation interacts not only with the hydrological (climatic), but also with mantle circulation of matter and water in its different phase states.

The *mantle–oceanic water circulation* is connected with global plate tectonics. However, all the moving forces and co-occurring processes of the plate tectonics are studied and known to be open ended. The same refers also to mantle–oceanic water circulation. Therefore, the basic points of view on the mechanisms of this circulation are given and will be further specified with details later.

In accordance with theoretical statements of plate tectonics, the oceanic floor represents, from the geologo-tectonic point of view, a lithospheric plate formed by the oceanic crust and mantle with a total thickness of 400–600 km.

The very complicated and large lithospheric plate moves as a uniform structure from the middle oceanic ridges toward the continents. In the zone of junction with the continental crust (zone of subduction), the oceanic plate is sinking under it, forming large dislocations, faults, and islands. Under the lithospheric plate there is a counter-flux of mantle matter which in the zone of middle ridges is moving upward to the surface and effusing in the form of magma into the rift zone of the central part of oceanic ridges, thereby forming a young oceanic crust. Because of fast cooling, the oceanic crust in this zone is highly fissured. Seawater saturates the fissured zone, cools, and destroys the magma body. This process follows the scheme: cooling, cracking, penetration of water, convection, and cooling. The hydrothermal solutions formed (by this scheme) in rift zones (possibly with a juvenile component) have a temperature close to the point of magma consolidation (980°C). However, because of the temperature mixing processes at the bottom surface, it is usually equal to 10–30°C and in some cases may reach 350°C. At high temperatures the water has a low density and viscosity. This creates an active convection in fissured basalts and an increased ability to dissolve and leach water-bearing rocks. Thermal gas-aqueous fluids, enriched with different chemical elements, serve as a source of polymetal and other mineral deposits in ocean ridges. The isotope composition of hydrocarbon and helium in the hydrothermal solutions sometimes indicates the juvenile nature of these components. However, this does not give enough of a basis to conclusively define the juvenile nature of the hydrothermal solutions. Fluids of the oceanic rift zones represent a mixture of juvenile and mostly seawaters, which through contacting with magma acquire an increased temperature and peculiar chemical composition.

The mantle matter and the co-related gas-aqueous fluids while rising to the bottom surface interact with the oceanic water and form a serpentinous peridotite containing a large amount of chemically bound water. This serpentinous layer moves, together with the litospheric plate, and submerges beneath the continents. This leads to its deserpentinization with the extraction of water. According to existing estimates, the oceanic crust of the World Ocean contains approximately $1.8 \cdot 10^{23}$ g of free and physically bound water. Under the action of alternating cycles of serpentinization and deserpentinization of perodotites, this water contributes to the formation of the granite crust of the continents and is involved by subcrustal backflows into the ocean where it discharges in the rift zones of oceanic ridges and once again moves into the mantle–oceanic circulation (Pavlov 1977).

However, some researchers (Sorokhtin, 1974) feel that the water-enriched serpentinite layer of the oceanic crust is formed as a result of the direct influence on the peridotites and dunites of the juvenile fluids in the central part of the middle oceanic ridges. Besides, according to the recent seismic tomography data, at the boundary of the upper and lower mantles (670 km), descending and ascending whirls or plumes are formed that transfer the mantle matter to the Earth's core or lifts it to the boundaries of the mantle and crust. This approach is being developed by Japanese scientists (Magara, 1982), whose opinions are supported by Russian geologists (Dobretsov et al. 1994).

One should once again outline the genetic interconnection and interdependence of geological and climatic water circulations which have a single source — seas and oceans. Often, the final step of one can be the initial stage of the other, which provides mutual transition and involvement of natural water from one branch of the circulation to another. But the mechanisms, leading processes, and basic manifestations create circulations which are significantly different. First of all, water within moves under the action of different sources of energy, namely: climatic circulation occurs at the expense of energy of the sun, whereas the leading source for geological circulation is the energy of the Earth's interior. Climatic circulation provides for the transport of vaporous moisture from oceans to continents, and then gravitation water, moving in the free space of rocks under the action of hydrostatic pressure, is again drained by oceans. The geologic circulation begins with sedimentation and, under the action of geostatic pressure and tectonic forces, transfers the water upward and downward along the rock cross-section. Investigations, results, and revealed regularities of the press-out of submarine sedimentation waters make it possible to quantitatively express the zonal distribution of submarine flow on a global scale.

3.5 NATURAL REGIONAL AND GLOBAL REGULARITIES OF SUBMARINE GROUNDWATER FORMATION AND DISTRIBUTION

Comprehensive investigations of SGD to the seas and the World Ocean show that this global process closely linked with generation, distribution, and transformation of the subsurface part of the total water circulation has certain regularities. The World Ocean serves to be the global draining basis of groundwater discharge. Submarine waters of infiltration origin are chiefly met in the shelf zone and adjacent part of the continental slope. In this zone of influence of coastal artesian systems, submarine groundwater has both lateral and vertical movement. With distance from the coastline, movement of submarine groundwater becomes predominantly vertical due to a gradual decrease of horizontal flow gradients and deterioration of hydraulic capacity of water-bearing rocks. Only vertical movement in major cases determines the total SGD. Submarine infiltration waters are characterized by a variety of chemical composition and mineralization and often exert a considerable influence on salt regime and balance of inland and coastal seas.

Subsurface runoff is an integral factor of groundwater recharge, which is determined by filtration properties of the unsaturated zone and water-bearing rocks, and by a proportion of heat and moisture within the territory under study. Moisture degree of the drainage area depends on gains and losses of its water balance. Therefore, abundant precipitation, high permeability of rocks, as well as poorly developed river network and low evaporation are factors that cause in the coastal zones an intensive groundwater discharge directly to the sea. Such conditions of groundwater discharge formation are mostly really met in mountainous areas. Coastal mountainous structures greatly screen the atmospheric moisture transfer between ocean and continent and favor an increase of precipitation in seaside areas. With height, air temperature, and evaporation decrease, condensation of atmospheric moisture becomes speedier, and precipitation increases to a certain level — all of which in combination leads to an improvement of groundwater recharge conditions. Besides, this process is often advanced by favorable geologic-structural and hydrogeological conditions, such as large slopes of aquifers, high absorbing ability of deluvial-proluvial sediments and fissured bedrocks, karst development, and availability of water-conducting tectonic disturbances.

The specific characteristics obtained of the groundwater discharge to particular seas and oceans (areal modulus and linear flow rate) make it possible to analyze and compare the specific features of its generation in different physical-geographical and structural-hydrogeological conditions (see Table 3.2.1). Linkage of the SGD with its basic natural forming factors is most distinctly seen when comparing specific values of groundwater discharge from concrete coastal areas of particular continents. Analysis of conditions of SGD formation is carried out by continents. This gives the possibility to consider each continent as a single megaregion of groundwater discharge formation where the features of distribution of total and specific submarine discharge reveal the regularities of its formation in regional and global scales.

Asia. Groundwater discharge from Asia to the Arctic Ocean is actually absent because of widely spread permafrost in the coastal zone. At the same time the SGD from the same continent to the Pacific Ocean reaches 254 km^3/yr. Here, the lowest groundwater discharge (to 1 l/sec·km^2) is observed in the very northeast with severe subarctic climate. Southwards, on the coast of the Okhotsk Sea the modulus of SGD increases to 2 l/sec·km^2, which occurs due to an increase in the annual average temperature and a rise of territory moisture. In general, background low values of groundwater discharge to the seas in the north of Asia, the Kamchatka Peninsula is in sharp contrast with its submarine discharge moduli reaching 10–11 l/sec·km^2 and a groundwater flow rate equal to tens of thousands of cubic meters per day per 1 km of the shoreline. This is explained by a high total precipitation, especially in the warm season of the year, and by the mountainous relief of the peninsula, and a high permeability of covering effusive and terrigenic formations.

The coastal areas of the southern Far East and the Peninsula of Korea are characterized by a warmer and more humid climate due to the influence of monsoons, which, alongside the screening effect of the mountainous structures, causes an increase of the submarine discharge modulus to 3.2 l/sec·km^2. Relatively high submarine discharge moduli (5–6 l/sec·km^2) are typical of Sakhalin Island where the water-containing terrigenic formations possess high filtration and accumulating properties.

The total chemical discharge from the entire eastern coast of Russia amounts to 37.6 mln t/yr. The moduli of chemical discharge gradually increase from the south towards the north, and in the area of Anadyr Lowland reach 158 t/yr·km^2 (Table 3.3.1). The sharp increase of salt transfer with the groundwater in the very northeast of Asia is explained, in the first turn, by the availability of permafrost, which hampers the water exchange conditions and leads to an increase of groundwater mineralization to 5–10 g/l. In the remaining areas of the Far East the groundwater at a depth of 500 m is chiefly fresh, but with mineralization that often changes sharply within short distances. The groundwater hydrochemical zonality has been formed there under the influence of paleo-hydrogeological conditions; presently its outlook is determined by climatic-specific features, and by conditions of groundwater recharge. The salt transfer with groundwater to the Pacific Ocean from the eastern areas of Russia amounts to 35% of the salt amount transferred by local rivers.

Exclusively favorable conditions for the formation of groundwater discharge are observed on the Japanese Islands. The influence of monsoons, the combination of latitudinal and altitudinal zonalities, and an excess of total annual precipitation (to 2000 mm) above evaporation (to 1000 mm/yr) — all these provide abundant wetting of the mountainous coastal areas. Wide development of well-permeable Quaternary alluvial and marine formations (pebble, sand, sandstone) leads to formation of intensive groundwater discharge, the moduli of which may reach 16 l/sec·km^2. The groundwater in the well-washed water-rich Quaternary rocks is fresh, as a rule. The underlying Neogene rocks usually have a hindered water exchange, because of which the groundwater mineralization sometimes reaches ≥20 g/l (Hydrogeology of Asia 1974). However, in general the water content in these rocks is not high, and it can be proposed that the submarine chemical discharge is formed chiefly at the expense of freshwaters from the Quaternary formations, the thickness of which is 250–300 m. The modulus of chemical discharge for all the Japanese Islands amounts on average to 250 t/yr·km², and the specific salt transfer per 1 km of the shoreline varies from 4–8 ths t/yr.

In the northeast and east of the People's Republic of China, in the coastal areas of the north Chinese and Liao-He artesian basins, the groundwater discharge moduli gradually increase southward from 2.4–3.4 l/sec·km². Though the territory has a plain-type relief and good draining by large rivers, groundwater discharge directly to the sea is not high. At the same time, as gradual precipitation increases southward to 1500 mm/yr and some improvement of the conditions for groundwater recharge in the same direction, lead to a growth of the SGD modulus. It should be also kept in mind that within the plain territories that are usually transient areas of the artesian basins, only a small part of the regional confined water is discharged directly to the sea. This is connected with natural and artificial groundwater discharge within the land. Waters in the upper aquifers are associated with the Quaternary sandy-clayey, alluvial-lacustrine sediments of thickness reaching in some places 1000 m. Due to processes of the continental salinization, the coastal areas show a distinct reverse hydrochemical zonality. The shallow saltwaters with a mineralization of several grams per 1 liter downward from a depth of 50–100 m are changed into brackish and low-brackish confined waters with a mineralization to 1.5–2 g/l (Hydrogeology of Asia 1974; UNESCO 2004). Because of the increased mineralization, the total submarine chemical discharge reaches 29 mln t/yr, which determines its high moduli, from 150–160 t/yr·km².

In the mountainous coastal zones where the groundwater recharge areas are very near to the regional draining basins (i.e., sea or ocean) the SGD sharply increases. Significant influence of the structural-hydrogeological and hydrogeodynamic features on the conditions of groundwater discharge to seas is distinctly seen when comparing specific characteristics of the discharge from the plain and mountainous drainage areas located within the same climatic conditions (see Table 3.2.1). A bright example of how mountainous structures influence the conditions of SGD formation is the mountainous drainage areas of Southeast Asia. Several hydrogeological massifs and relatively small artesian structures are distinguished there. The modulus of SGD reaches 6.3 l/sec·km²; the groundwater flow rate per 1 km of the shoreline varies from 30–58 ths m³/day. Screening action of mountainous ridges on the atmospheric moisture transfer, abundant precipitation (>2000–2500 mm/yr), and relatively low evaporation (700–1200 mm/yr) are the factors that create in these subtropical and tropical mountainous areas very favorable conditions for groundwater recharge. The groundwater is chiefly associated with well-permeable effusive-sedimentary rocks of the Neogene–Quaternary and more seldom of the Mesozoic. Along the valleys of large rivers the aquifers are associated with the Quaternary water-rich alluvial formations of up to 300 m thick. The favorable conditions for groundwater recharge and well-washed rocks make the groundwater low-mineralized with a salt content of not more than 1 g/l almost everywhere within this large region (Hydrogeology of Asia 1974; UNESCO 2004). The moduli of submarine chemical discharge vary from 80–180 t/yr·km², depending on water contents in the water-bearing rocks and intensity of SGD; the specific salt transfer per 1 km of the shoreline varies from 8.5–15 ths t/yr.

Favorable conditions for formation of groundwater discharge are also observed on the islands of Southeast Asia. Constant monsoon influence of the Pacific and Indian Oceans in this tropical zone leads to almost year-round wetting of the islands. The rain amount there is on average 2000–3000 mm/yr, and on the windward mountainous slopes of the Malay Archipelago islands it sometimes reaches 4000–5000 mm/yr. Low evaporation (to 1000 mm/yr) at such a level of rains leads to intensive surface and subsurface runoff. The latter varies 4–15 l/sec·km^2, reaching 33 l/sec·km^2 on Mindanao Island. Intensive groundwater discharge to the Pacific and Indian Oceans from the islands of Luzon, Mindanao, and Java is provided not only by high wetting of their mountainous territories, but also widely spread, easily permeable karst carbonate rocks and fissured volcanogenic formations, and by low draining influence of small river valleys. At the same time, predominance of plain territories with a well-developed erosional network within large islands, such as Kalimantan and Sumatra, and widely spread lower-permeable terrigenic sandy-clayey formations lead, at the same water-balance structure, to a decrease of the submarine discharge modulus to ≤4 l/sec·km^2. Salt transfer from the islands of Southeast Asia (117 mln t/yr) is caused, first of all, by intensive submarine discharge of low-mineralized groundwater from the upper part of well-permeable, Neogene–Quaternary sediments. The small drainage area of the islands and intensive groundwater discharge provide high moduli of the chemical discharge, reaching on some of the islands 200–300 t/yr·km^2.

Westward from the overwetted areas of Southeast Asia, the groundwater discharge to the Indian Ocean gradually decreases, amounting from the Hindustan Peninsula already to >4 l/sec·km^2. This is caused by a decrease of precipitation (1000–1500 mm/yr) and an increase of evaporation (1300–1400 mm/yr). The conditions of groundwater recharge are significantly influenced by a seasonal regime of precipitation connected with an active action of monsoons on the coastal areas. In dry periods the groundwater is greatly depleted due to evaporation and active drainage by the erosional network. Besides, crystalline rocks (traps), developed in some areas of the coast, do not contribute to formation of deep groundwater discharge. The major groundwater in the upper fissured zone of the crystalline strata is tapped by river valleys and participates little in SGD. In the structural-hydrogeological respect, several independent artesian basins are distinguished within the peninsula (Godavarsky, Polksky, West Gatsky, and others) the coastal areas of which are mainly composed of sands, sandstone, conglomerates, and shale of different ages. Water content of these rocks entirely depends on recharge conditions. As a rule, they are well washed and contain freshwaters to a depth of 400 m and deeper (Hydrogeology of Asia 1974; UNESCO 2004). Because of this, the chemical discharge there is generally small with moduli varying within 70–90 t/yr·km^2.

Further to the west the climate becomes drier; groundwater discharge from the semidesert and desert coasts of the Arabian Sea and Persian Gulf does not exceed 0.6 l/sec·km^2. Rainfalls amount to 200–300 mm/yr, decreasing sometimes to 100–150 mm/yr, whereas evaporation sharply increases and is usually equal to >2000 mm/yr. The driest region in the west of the Asian continent is the Arabian Peninsula where precipitation very seldom exceeds 100 mm/yr. In connection with this, the ground-water discharge to the seas decreases to 0.2 l/sec·km². Scarce rains and high evapo-ration make the groundwater recharge poor. The shallow groundwater in these areas is spent chiefly for evaporation and does not form submarine discharge; only sub-channel fluxes of a few constant and temporary streams are sometimes discharged directly to the sea. Therefore, the major SGD is represented there by regional runoff of confined groundwater. It is associated with dolomites, limestone, sandstone, and conglomerates, as well as sedimentary-volcanogenic formations mainly of the Neogene–Quaternary and Mesozoic. These sediments compose several independent artesian basins of different sizes, the largest among which are Ind, Oman, Aden, and the Red Sea. The specific feature of these artesian structures is availability of salt-bearing rocks. The average groundwater mineralization gradually increases from east to west from 2 g/l (Indus artesian basin) to 40 g/l (artesian basin of the Red Sea). In connection with this, the chemical discharge increases in the same direction from 10–55 mln t/yr, and its modulus reaches on the shore of the Red Sea 200 t/yr·km² at a salt transfer with groundwater per 1 km of the shoreline at 15 ths t/yr.

The groundwater amounts discharged to the Pacific and Indian Oceans from the territories of Asia and adjacent islands show that they gradually increase from subtropic regions to moderate zones and then sharply rise in humid subtropics and tropics, decreasing after that in semiarid and arid regions. Thus, the climatic factor exerts the basic influence on conditions of SGD and determines its dependence, on a global scale, on total latitudinal physical-geographic zonality (Figure 3.5.1). This general background of distribution of the groundwater discharge to seas is over-lapped by the influence of local relief, geologic-structural, and hydrogeological features, which cause variations (sometimes very significant) of specific values of submarine water and chemical discharges within a single climatic zone. A high groundwater discharge to oceans is formed on territories adjacent to large islands located in humid and tropical zones. This is connected both with favorable climatic conditions, mountainous coastal relief of many islands, high hydraulic capacity of near-surface rocks, and terrigenic formations and low draining ability of the local erosional network.

Submarine chemical discharge is determined by intensity of submarine ground-water runoff, a degree to which water-bearing rocks are washed, paleo-hydrogeo-logical and modern conditions for formation of groundwater mineralization and composition, availability of evaporates, and processes of evaporating concentration in arid regions. It can be seen from Table 3.3.2 that from the Asian coast the submarine fresh groundwater dischage is formed with an average mineralization of 0.9 g/l. However, in arid and semiarid regions, alongside a decrease of groundwater discharge, the mineralization becomes higher at the expense of continental saliniza-tion. This results in increasing total and specific values of the submarine chemical discharge. The anomalously high salt transfer with groundwater is observed in the areas where evaporates are developed in water-bearing rocks (e.g., Red Sea, Persian Gulf), which serves as the determining azonal factor of formation of groundwater mineralization and composition. The above-described features of SGD formation are typical not only on the Asian continent, but they have a general global character, as will be shown below.

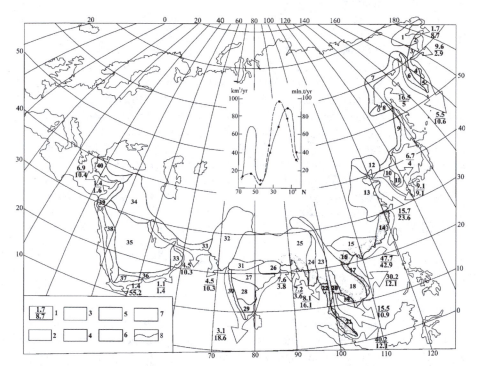

FIGURE 3.5.1 (Please see color insert following page 236.) Schematic map of groundwater discharge to the ocean from Asia: (1) in numerator groundwater discharge (km^3/yr), in denominator–ion groundwater discharge (mln t/yr); areal module of groundwater discharge (l/sec·km^2): (2) 0.1–0.5, (3) 0.5–1.0, (4) 1.0–3.0, (5) 3.0–5.0, (6) 5.0–7.0, (7) 10.0–20.0, (8) boundaries of artesian basins and hydrogeological massifs: hydrogeological areas: (7) Okhotsk, (9) Sikhote-Alinsk, (10) North-Korean, (11) South-Korean, (14) South-Chinese, (15) Sitszyansk, (17) East-Vietnam, (21) East-Malacca, (22) West-Malacca, (25) Ganges, (30) West-Ganges, (31) Narbad, (32) Indsky, (33) Omansky; artesian basins: (1) Anadyr, (2) Opukhsko-Pikulneisky, (3) Olyutorsky, (4) Central-Kamchatka, (5) East-Kamchatka, (6) West-Kamchatka, (8) Udsky, Toromsky, Tugursky, Ulbansky, Usalginsky, (12) Liao-He, (13) North-Chinese, (16) Hanoi, (18) Indosiniysk, (19) Pursatsk, (20) Menemsky, (23) Khticharsky, (24) Burma, (26) Ust-Mahanadi, (27) Godavari, (28) Cuddapah, (29) Polksky, (34) Mesopotamia, (35) Arabian, (36) Hadramsutsky, (37) Adensky, (38) Red Sea, (39) Lebanon-Sinai, (40) Adana, Antalya, Menderes. Distribution of water (I) and ion (II) groundwater discharge into the ocean from the continent by latitudinal zones is shown on the graph.

Africa. Groundwater discharge to oceans and seas from the major territory of Africa is rather small, which is connected with the hot and dry climate of many coastal areas of this continent. Minimum moduli of submarine discharge (0.2–0.3 l/sec·km^2) are typical of the coasts of the Red Sea, Somali Peninsula, and West Sahara and Namib Deserts (see Table 3.2.1). These areas have unfavorable climatic conditions for groundwater recharge. Shallow water runoff is actually absent. As a result, the SGD is formed entirely at the expense of confined deep aquifers. Regional recharge areas of artesian basins are usually located far apart from the coasts. On the way to the ocean, the major part of artesian waters are spent for processes of natural and artificial discharge. Because of this, deep artesian waters in only rare cases form a considerable submarine discharge.

A special accent should be concentrated on formation of the submarine chemical discharge to the Red Sea. In the tectonic respect, this region is located in a rift zone, the banks of which represent a system of grabens of different origin and age. The grabens are mainly composed of sedimentary rocks (limestone, sandstone, clay, and marl) with widely spread evaporites. The basic tectonic elements go on actively developing at the present time, which is confirmed by an increased heat flux, volcanic phenomena, and high seismicity. Because of this, surface springs of thermal waters with an increased mineralization are observed here, which are mainly discharged in the depressions of the Red Sea. Such natural conditions make this region unique and interesting in the hydrogeological respect. Numerous publications are devoted to the problems of origin of the thermal brines in the Red Sea rift zone. Our investigations confirm the most widely accepted opinion that high-mineralized thermal waters have an infiltration origin. In particular, this is indicated by the areal and vertical hydrochemical zonalities formed within the land: i.e., with nearing the shoreline of the Red Sea and with depth, the groundwater mineralization increases from 4–50 g/l and from 4–380 g/l, respectively (Hydrogeology of Africa 1978; UNESCO 2004). The unfavorable recharge conditions result in low moduli of the SGD at 0.2 l/sec·km^2. At the same time, the high mineralization causes a significant chemical discharge of 22.2 mln t/yr at a modulus of 150 t/yr·km^2.

The southward-located Somali artesian basin is characterized by widely developed (in the upper part of cross-section) fissured effusives, karstic limestone, dolomites, and sandstones, among which are gypsum interbeds. The water-bearing rocks possess a high hydraulic capacity. However, due to the unfavorable recharge conditions, the groundwater discharge is small, the moduli of which seldom exceed 0.2 l/sec·km^2. The groundwater circulates weakly in relatively washed fissured-karst hollows and has a low leaching ability. In this connection, despite availability of gypsum-bearing rocks the groundwater mineralization does not exceed 5 g/l and amounts on average along the cross-section to 2 g/l. The submarine chemical discharge is generally slight (~7 mln t/yr) and characterized by low areal moduli and linear losses (see Table 3.3.1).

Southward from the Somali Peninsula, the Dar-es-Salaam artesian basin is located where the groundwater discharge to the Indian Ocean gradually increases to 1 l/sec·km^2. Precipitation increases here to ≥1000 mm/yr, but intensive evaporation hampers formation of large groundwater resources. Wide spreading of laterites with low filtration properties provides weak filtration of rainwater. The seasonal character of territory wetting leads to high groundwater depletion during dry periods. Availability of evaporites in the water-bearing rocks provide an increased groundwater mineralization to 1.5–2 g/l. Due to this, the average modulus of chemical discharge is 39 t/yr·km^2.

More favorable conditions for formation of groundwater discharge are observed on Madagascar Island. The screening influence on atmospheric circulation in the eastern coast of the island causes an increase in precipitation to 3000 mm/yr. It leads to intensive groundwater recharge. The major groundwater is discharged directly to the ocean. However, to the west and southwest where the rain amount sharply decreases, the relief becomes more plain-like; more surface streams appear, which have well-developed valleys. These factors in combination cause a decrease of SGD more from the western coast of the island than from the east. Due to this, the average modulus of the groundwater discharge to the ocean from all of Madagascar slightly exceeds 3 l/sec·km². The groundwater in all the aquifers, from Quaternary to Cambrian inclusively, is fresh with a mineralization of 0.2–0.9 g/l. The submarine chemical discharge is 8.6 mln t/yr, its areal modulus reaches 29 t/yr·km², and linear flow rate does not exceed 2 ths t/yr·km².

Unfavorable modern conditions for groundwater recharge exist in the areas of the West Sahara and Namib Deserts. In the structural-hydrogeological respect, the artesian basins of Dra, Rio-de-Oro, and Senegal are distinguished in the West Sahara. The aquifers in the zone of intensive water exchange in the first two artesian basins are connected with Neogene–Quaternary limestone and sandy-clayey formations. A low submarine discharge of 0.2 l/sec·km² is formed by brackish groundwater with an average mineralization of 5 g/l. Hence, the modulus of chemical discharge reaches 25–30 t/yr·km². The specific feature of the Senegal artesian basin is the wide development of water-rich Cretaceous (Maastrichtian) sands and sandstones. They are located at a depth of 200–500 m, reliably isolated from above by Paleogene clays and chiefly contain freshwaters. But with nearing the Atlantic Ocean these waters are replaced by brackish waters with a mineralization of 2–3 g/l. The overlying continental Neogene–Quaternary strata have low water content and contain waters of an increased mineralization to 15 g/l (Hydrogeology of Africa 1978; UNESCO 2004). The SGD modulus slightly increases from the north to the south (0.3 l/sec·km²). The total salt transfer with groundwater to the Atlantic Ocean from the coast of West Sahara is slightly over 12 mln t/yr.

In the Namibian artesian basin (Namib Desert) the groundwater is mainly associated with sandstones of the Pre-Cambrian basement located at a shallow depth. The water mineralization is 3–5 g/l. The submarine groundwater and chemical discharge is slight. Northward from the Namibian artesian basin, the poorly studied basins of Angola-Congolese (interfluve of Guyana and Congo Rivers) and Cameroon-Gabonese (interfluve of the Congo and Cross Rivers). Their specific feature is the wide development of evaporites. This gives the grounds to consider that Paleogene and Cretaceous sediments may contain saltwaters which under certain hydrogeological conditions can exert an influence on water mineralization in the overlying aquifers. This influence can be especially considerable within the Angola-Congolese artesian basin where the unfavorable climatic conditions cause a low groundwater discharge (to 0.3 l/sec·km²) in the zone of intensive water exchange. The conditions of groundwater recharge in the Cameroon-Gabonese artesian basin become gradually better from the south to the north at the expense of increasing rainfalls to 1500–2000 mm/yr. According to data by Lvovich (1974), groundwater discharge to the rivers in the northern areas of this basin reach 200–300 mm/yr. Such an intensive groundwater discharge is typical, probably, only of the upper Neogene–Quaternary aquifers where the mineralization is highly varied, but does not exceed on average 2 g/l. Thus, the influence of evaporites on formation of submarine chemical discharge in the zone of intensive water exchange of the Cameroon-Gabonese artesian basin is greatly leveled by the favorable conditions of groundwater recharge.

The greatest SGD in the African territory is observed on the coast of the Gulf of Guinea (Nigerian artesian basin) where the modili reach 13 l/sec·km^2. In these tropical highly wetted areas the precipitation reaches 3000–4000 mm/yr, whereas the evaporation usually does not exceed 1100–1200 mm/yr. This leads to formation of a rather considerable groundwater discharge, which is additionally favored by widely developed, well-permeable sandy rocks, and by the regulating influence of evergreen tropical forests. The aquifers in the Neogene–Quaternary, Paleogene, and Cretaceous sands, sandstone, and limestone have inclusively high water content (water well yields are 8 l/sec). The mineralization does not exceed 0.6 g/l (Hydrogeology of Africa 1978; UNESCO 2004).

In contrast to the low groundwater discharge from the African territory to the Indian Ocean and on the major coast of the Atlantic Ocean, the Gulf of Guinea is sharply distinguished by its intensive SGD. Favorable combination of climatic, physical-geographic, and geologic-hydrogeological conditions leads to active groundwater recharge, the considerable part of which is discharged directly to the ocean. Whereas the total SGD from the African continent reaches 236 km^3/yr, the discharge from the coast of the gulf is equal to 171 km^3/yr, and the flow rate per 1 km of the shoreline varies from 60–100 ths km^3/day. The total chemical discharge amounts here to 85 mln t/yr; its moduli vary from 160–180 t/yr·km^2. This area is located in a humid tropical zone having over its entire territory a high groundwater discharge to the World Ocean. However, so considerable groundwater discharge allows this area to be considered as unique on the Earth.

The major discharge to the Mediterranean Sea from the African continent is formed in the Atlas Mountains; its basic regularities are described in Section 4.5.

Analysis of distribution of groundwater discharge to seas and oceans from the territory of Africa shows that it is also governed by the latitudinal physical-geographic zonality (Figure 3.5.2). Predominance of hot and dry climate on the major part of the continent somehow smoothes the difference between the specific discharge values from the drainage areas of the basic latitudinal zones. But the more detailed analysis of groundwater discharge formation within concrete coastal areas makes it possible to observe a gradual, though slight increase of the moduli of groundwater discharge to the Indian Ocean southward from the Sahara Desert to more wetted tropics and subtropics of East Africa. On the Atlantic coast the submarine discharge also gradually increases from the desert areas in the far north and south of the continent equatorwards, especially increasing in the zone of wet tropics of the Gulf of Guinea.

Europe. Formation conditions of the groundwater discharge to the Mediterranean Sea are described in Section 4.5. It should be discussed here that SGD to the sea is formed within the coastal zones of three continents, and the inland position of the sea makes it possible to compare SGD with other sources of the water balance. At the same time it is known that in the Mediterranean Sea the processes of SGD are developed very intensively. Numerous concentrated bursts of groundwater are observed on the sea bottom, which form submarine springs with considerable yields. By a number of such springs the Mediterranean Sea represents a unique sea basin (Zektser et al. 1972).

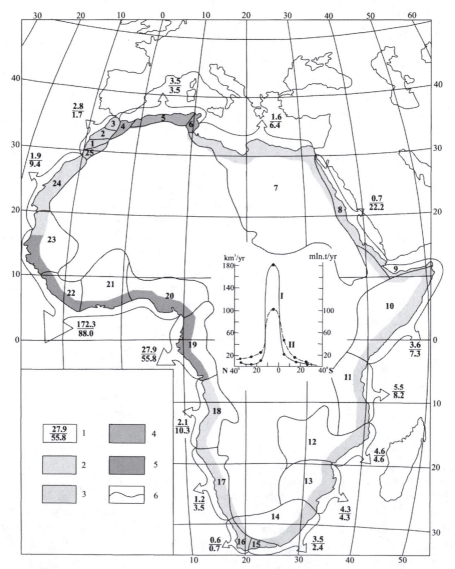

FIGURE 3.5.2 (Please see color insert.) Schematic map of groundwater discharge to the ocean from Africa: (1) in numerator: groundwater discharge (km³/yr), in denominator: ion groundwater discharge (mln t/yr); areal module of groundwater discharge (l/sec·km²): (2) 0.1–0.5, (3) 0.5–1.0, (4) 1.0–3.0, (5) 7.0–10.0, (6) boundaries of artesian basins and hydrogeological massifs: artesian basins: (1) Susah, (2) Casablanca, (3) Er-Reef, (4) Muluinsk, (5) Tell-Atlas, (6) Tunisian, (7) Libya-Egypt, (8) Red Sea, (9) Aden, (10) Somali, (11) Dar-es-Salaam, (12) Zambezia, (13) Limpopo, (14) Karroo, (15) Cape, (16) Cape Town, (17) Namib, (18) Angola-Congolese, (19) Cameroon-Gabonese, (20) Nigerian, (21) Volta, (22) Guinea-Liberian, (23) Senegal, (24) Rio-de-Oro, (25) Dra. Distribution of water (I) and ion (II) groundwater discharge into the ocean from the continent by latitudinal zones is shown on the graph.

The total groundwater discharge to the Mediterranean Sea reaches almost 68 km³/yr, which amounts to about 24% of the river runoff. This volume includes the groundwater from the European territory equal approximately to 49 km³/yr; the territory of Asia, 8 km³/yr; Africa, 5 km³/yr; and the territory of the largest islands, 6 km³/yr (see Table 3.2.1). Estimation of groundwater discharge to some lakes and seas of the former USSR shows that it amounts usually to the first percents of the river runoff (Dzhamalov et al. 1977). At the same time, the total groundwater discharge to the Mediterranean Sea amounts to 24% of the riverwater inflow. This again outlines the uniqueness of the given sea basin and points to the necessity of taking into account the groundwater component in calculating water balance.

The groundwater discharge to the Atlantic Ocean from the European territory gradually decreases from the south to the north from 4.5 l/sec·km² in the coastal areas of the Pyrenean Peninsula to 2.5 l/sec·km² in the Baltic Sea coast. A significant influence on the conditions of SGD formation and distribution is exerted here by relief features of the drainage areas. The plain territories on the northwest of the continent are well drained by large river valleys, and only a slight part of the groundwater is discharged directly to the sea. Within the mountainous coasts the groundwater discharge is large. Thus, in the coastal part of the Pyrenean Penin-sula three large hydrogeological structures are distinguished: the Pyrenean mountain-folded area, the Meseta hydrogeological massif, and the West Portugal artesian basin. Within these hydrogeological massifs the groundwater is associated chiefly with Paleozoic sedimentary rocks. Of the highest water content are the interbeds of karstic limestone where the groundwater discharge moduli reach 5.0–6.5 l/sec·km². The waters are mainly fresh, but in some small areas with developed salt-bearing rocks their mineralization varies from 1–30 g/l. However, the contribution of these small-sized areas to formation of the submarine chemical discharge is not large. The West Portugal artesian basin consists of differently composed rocks of the Mesozoic to Quaternary. Depending on the filtration properties of water-enclosing rocks, the groundwater discharge moduli in the coastal part vary from 1.6–4.8 l/sec·km². The waters are fresh with an average mineralization of 0.5 g/l. The submarine chemical discharge from the Pyrenean Peninsula amounts to 4.9 mln t/yr. Its major portion (3.6 mln t/yr) is formed within the hydrogeological massifs with an increased submarine discharge (7.1 km³/yr).

On the Atlantic coast of France there are distinguished the Akvitansky (Garonne and Loire Lowlands) and Parisian (North French Lowland) artesian basins divided by the Armorikansky hydrogeological massif. The basic water-bearing system of artesian basins is the karstic strata of Jurassic–Cretaceous carbonate rocks with a thickness of 1500–2000 m. Fissured-karst waters of the Mesozoic age form a single water-bearing complex with waters of the overlying Quaternary terrigenic formations and have often a free surface. With nearing the shoreline the waters acquire a considerable pressure head and are discharged within the entire wide shelf zone of the Biscayne Bay and La Manche Strait. The high filtration properties of carbonate rocks and favorable conditions for groundwater recharge lead to formation of considerable natural groundwater resources in these basins, leading, in turn, to significant submarine discharge, the moduli of which vary from 3.8–4.8 l/sec·km². In the Armorikansky hydrogeological basin, the groundwater is associated with the upper fissured zone of Pre-Cambrian and Paleozoic metamorphic rocks broken through by intrusives. The most watered are the tectonic disturbances and contact zones with the intrusives. The hydrogeological structure of the basin is such that it excludes development of extensive aquifers and aquifer systems. But the SGD is comparable with the discharge of confined groundwater from the artesian basins and amounts in total to 4.8 km³/yr. The groundwater within the entire French coast is fresh with an average mineralization of 0.4 g/l. The submarine chemical discharge amounts to 4 mln t/yr, with the chemical discharge modulus in the areas of karst-forming carbonate rocks ≥60 t/yr·km².

The groundwater of the artesian structures associated with the North Germany and Polish Lowlands is chiefly related to the sandy and pebble interbeds in alluvial, alluvial-marine, and glacial sediments of the Neogene–Quaternary with a thickness of 200 m and thicker. Frequently met clayey interbeds of moraine formations lead to formation of confined aquifers beginning from the depth of 20–40 m. Of significant hydrogeological importance are the ancient buried valleys composed of well-graded fluvioglacial sands and characterized by a groundwater discharge modulus equal to 5 l/sec·km². The plain-type territory and dense network of developed river valleys provide a good draining ability of this territory, making, thus, the SGD modulus not exceeding 2.5 l/sec·km². The waters in the upper hydrodynamic zone are, as a rule, fresh with an average mineralization of 0.3–0.5 g/l. However, the salt-bearing sediments in the Mesozoic rocks sometimes form salt domes and stocks breaking through the Paleogene–Neogene layers. This leads to an increase of groundwater mineralization in some areas to 3–5 g/l. The total submarine chemical discharge amounts to 4.8 mln t/yr; its modulus varies from 20–40 t/yr·km².

The Scandinavian coastal regions belong, due to an active influence of air masses from the Atlantic, to the most wetted territories of Europe. Low evaporation and wide development of fissured bedrocks cause an intensive infiltration of rainwater. The combination of so favorable natural factors leads to an abnormally high SGD (6.5–11.5 l/sec·km²), not typical of this zone transient from the moderate to subarctic belt. Of a regional distribution in this region is the aquifer associated with the fissured zone of metamorphic Archean–Proterozoic rock. Exogenic jointing of these rocks is considerably intensified in the zones of numerous tectonic disturbances located within a wide range of depth. Different-shaped tectonic and eroded sinks composed of fluvioglacial and glacial sediments of upto 100 m thick have also a significant hydrogeological importance, being usually a source of large groundwater accumulations. One should especially mention the availability of synclinal structures in the south of Sweden (Malmo, Christianstad), composed of Cretaceous carbonate rock, sandstone, and sand with thicknesses to 200 m. The groundwater discharge there sharply increases, reaching in some areas 500 l/sec. In crystalline rocks the waters are ultra-fresh, in Cretaceous sediments, brackish. The total chemical discharge amounts to 9 mln t/yr; its modulus reaches 60 t/yr·km² at the expense of intensive SGD.

The European coast of the Arctic Ocean differs by its relatively low groundwater discharge, the moduli of which gradually decrease from the west to the east from 1.5–0.9 l/sec·km². The decrease of SGD to the east is connected, in the first turn, with more severe climate. Sporadically distributed permafrost in the near-surface strata hampers or completely excludes infiltration of precipitation, because of which the groundwater discharge in the subarctic areas sharply decreases. The groundwater is ultra-fresh and fresh with a mineralization of 0.1–0.5 g/l. The total chemical discharge amounts to 7.2 mln t/yr; its modulus usually does not exceed 30–40 t/yr·km².

A special accent should be given to the conditions for formation of SGD from the islands of Great Britain and Ireland. Submarine springs are known here from ancient times. As a rule, they are associated with water-bearing systems of karstic limestone. The groundwater within the islands is associated with the entire strata of the Quaternary to Pre-Cambrian. But just the limestone of Jurassic, Cretaceous, and Carboniferous ages are everywhere characterized by high water content and form a considerable submarine discharge, the moduli of which exceed 6 l/sec·km². Thus, according to estimates of English specialists, in southeast Kent in the area from the town of Duvra to Falkstone, the SGD reaches 23 ths m³/day, which is equal to 42% of the natural groundwater resources of southeast Kent (Reynolds 1970). The groundwater mineralization usually does not exceed 0.5–0.7 g/l, but in some areas is composed of salt-bearing rocks and at large depths it reaches 14 g/l and greater. The total submarine chemical discharge from both islands amounts to almost 30 mln t/yr; its modulus varies from 100–140 t/yr·km², linear discharge rate reaches 7 ths t/yr·km.

Thus, the groundwater discharge to the seas from the European continent is also governed by the latitudinal physical-geographic zonality (Figure 3.5.3). The local geologic-hydrogeological and relief features of the drainage areas make more complicated the general situation of discharge distribution and can sometimes cause their sharp deviations from typical average values. The determining influence of the local factors on groundwater discharge formation conditions can be exemplified by the coastal areas of Scandinavia and the Mediterranean Sea where the screening effect of mountainous structures and wide development of karst and fissured rocks lead to an azonally high submarine discharge.

The Americas. Groundwater discharge to the seas and oceans from the territory of America is formed under the influence of the same factors like within the above-described territories. However, the American continent in total differs from Eurasia by a high wetting at the expense of extensive penetrating influence of air oceanic masses. In connection with this, the American territory has the highest SGD to the oceans (780 km³/yr).

Minimum groundwater discharge moduli characterize small areas on the coast of Hudson Bay. On this background of the low discharge values in the north of the continent the coastal areas of South Alaska and the Labrador Peninsula are distinguished by their high moduli (5–7 l/sec·km²) that are caused, like in Scandinavia, by the screening influence of mountainous structures on the atmospheric moisture transfer. The near location of the recharge areas to the oceanic coasts and incomplete draining of water-bearing rocks by the eroded network provide an intensive SGD. Favorable combination of the above-mentioned factors greatly intensifies the groundwater discharge to the Pacific Ocean from the drainage areas of the Canadian Cordillera, where due to a much warmer and more humid climate the average modulus of SGD reaches 13 l/sec·km²; the groundwater flow rate per 1 km of the

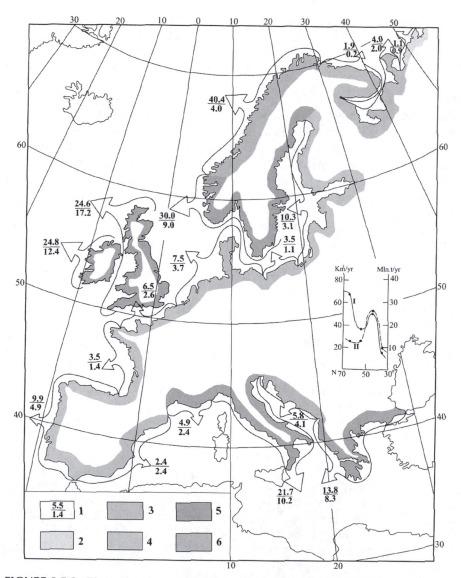

FIGURE 3.5.3 (Please see color insert.) Schematic map of groundwater discharge to the ocean from Europe: (1) in numerator: groundwater discharge (km³/yr), in denominator: ion groundwater discharge (mln t/yr); areal module of groundwater discharge (l/sec·km²): (2) 0.5–1.0, (3) 1.0–3.0, (4) 3.0–5.0, (5) 5.0–7.0, (6) 10.0–20.0. Distribution of water (I) and ion (II) groundwater discharge into the ocean from the continent by latitudinal zones is shown on the graph.

shoreline exceeds 70 ths m³/day. The groundwaters in these regions are associated, as a rule, with the upper fissured zone of different-aged crystalline rocks, as well as with highly dislocated sedimentary sediments, among which sandstone and conglomerates have the highest water content. Of significant importance are the eroded and tectonic valleys composed of washed pebble and graded sands of alluvial-glacial genesis with a thickness reaching sometimes several hundreds of meters (Hydrological Atlas of Canada 1975).

Groundwater is fresh with an average mineralization of 0.2–0.4 g/l. But in the northern part of the Appalachian Mountains, mineralization reaches 2.5–10 g/l at the expense of salt-bearing rocks developed in the lower part of the cross-section. The area of sites with increased mineralization is generally small, and the average mineralization for the entire Appalachian hydrogeological region amounts to 0.9 g/l. The SGD modulus varies from 10–60 t/yr·km², reaching in the area of Appalachia 160 t/yr·km² (Water Atlas of the United States 1973).

Mean groundwater discharge values are typical of the moderate zone of the Atlantic coast of North America (see Table 3.2.1). These values can be considered as most characteristic for moderate latitudes of the Earth with typical (for these regions) conditions of SGD formation. At the same time, prevailing of any natural factors of discharge formation can lead to abnormally low or high values of SGD. An example of azonally low moduli of groundwater discharge to the ocean (1.6 l/sec·km²) can be the coast of the Mississippi Lowland. There, the groundwater is chiefly discharged in the Mississippi River valley and only a part of the discharge reaches the Gulf of Mexico. At the same time, the wide development of karstic carbonate rocks and fissured sandstones on the Florida Peninsula lead to an abnormally high groundwater discharge from the peninsula to the ocean, the modulus of which in some areas reaches 6.3 l/sec·km². A wedge of fresh and brackish groundwater, being formed on the peninsula in Paleogene limestone and partly of the Cretaceous, stretches under the bottom of the Atlantic Ocean at a distance of over 120 km and to a depth of over 600 m (Manheim and Paull 1981). This region, with Long Island, New York, has become a classic example of SGD. Wide development of fresh groundwater over the entire coast of the Atlantic Ocean of the United States (0.2–0.4 g/l) creates a slight chemical discharge, the moduli of which gradually increase toward the Florida Peninsula from 10–40 t/yr·km².

The hot and dry climate of the Mexican Plateau does not favor formation of large groundwater resources; therefore, the SGD moduli amount to only 1–2 l/sec·km². This zone of minimum discharge is sharply changed to the south into the humid tropical belt of Central America where the groundwater discharge moduli increase to 10–11 l/sec·km². The mountainous coasts, abundant precipitation (2000–3000 mm/yr), and wide development of karst and fissured effusives lead to active recharge of groundwater, the considerable part of which is discharged directly to the Pacific and Atlantic Oceans. The groundwater mineralization varies from 0.3–1 g/l (Illades 1976). The moduli of submarine chemical discharge reach 200 t/yr·km² in the area of the Yucatan Peninsula, which is explained by a wide development of karstic carbonate rocks with thin interbeds of evaporites.

The most wetted regions of South America include the Amazon River basin and territory of Guyana. The water-balance structure is favorable for abundant groundwater recharge, as the precipitation (>2000 mm/yr) is approximately twice higher than evaporation (900–1000 mm/yr). Therefore, in this region a rather intensive groundwater discharge is formed, a considerable part of which is discharged, despite the draining influence of the river valleys, directly to the Atlantic Ocean. The moduli of SGD vary within 10–13 l/sec·km², and the groundwater flow rate per 1 km of the shoreline sometimes exceeds 100 ths m³/day. In this part of the Atlantic coast (within Venezuela and Guyana) a series of so-called coastal artesian basins are distinguished. The most water-abundant aquifers in these basins are associated with the cavernous limestone of Paleogene and graded sands of Neogene (Gascoyne 1977). Large thickness of water-bearing rocks (from 50 m to several hundreds of meters), their high filtration properties (water conductivity of over 2000 m²/day), and favorable recharge conditions provide formation of significant groundwater resources and intensive discharge (Worts 1963). Groundwater in the well-washed sediments has an average mineralization of 0.4–0.6 g/l. However, owing to prevailing carbonate rocks and exclusively high values of the groundwater discharge, the moduli of submarine chemical discharge amount to 120–240 t/yr·km².

The Brazilian Plateau is characterized by irregular wetting of the territory; short rainfalls and widely developed laterites do not provide active groundwater recharge. Submarine groundwater discharge gradually decreases to the south and usually does not exceed 25 l/sec·km². Further to the south a sharp decrease of the groundwater discharge is observed, and in the area of Patagonia groundwater discharge moduli decrease to 0.3 l/sec·km². This is connected, in the first turn, with an extremely dry climate of the southeastern marginal part of the continent where precipitation does not exceed 100 mm/yr, and evaporation sharply increases. In this part of South America a whole number of artesian structures are distinguished, which are composed of sedimentary rocks of different genesis. These structures usually inherit tectonic and erosional sinks in crystalline metamorphic and volcanogenic rocks of Pre-Cambrian age (Szikszay et al. 1981). About 20 such artesian basins are located only within the Brazilian Plateau. Among the coastal artesian structures opened toward the Atlantic Ocean, one should mention the basins of Potiguar, Almada, Santos, San-Paulo, and others (Szikszay and Teissedre 1978). The water-bearing rocks in the sedimentary cover of these basins are sandstones, argillites, aleurolites, volcanogenic formations, and partly sands of the Paleozoic–Mesozoic and Quaternary. The lithological composition of these rocks is unfavorable for formation of large groundwater resources. The thickness of the freshwater zone often reaches 1000 m. Therefore, the submarine chemical discharge modulus varies from 2.5–13.5 t/yr·km².

Groundwater discharge to the Pacific Ocean from the large Andes Mountains system varies within a wide range. Besides the zonal climatic factors, a great influence on conditions of groundwater formation is exerted by the altitudinal zonality. Large slopes of the territory, bedrock jointing, and high permeability of deluvial-proluvial sediments contribute to active groundwater recharge. The groundwater discharge modulus in the very south and north of the Andes reaches 24–30 l/sec·km², decreasing in the driest central regions actually to zero. In this connection, the average values of SGD in the northern part of the Andes reach 15 l/sec·km², decreasing sharply in the area of the Atacama Desert and then again increase to 11 l/sec in the Patagonian Cordillera. The groundwater is chiefly associated with the zone of exogenic and tectonic jointing of crystalline rocks. The most water-abundant areas are connected with tectonic disturbances, and with small artesian structures composed of Quaternary sediments. The area of such artesian structures within the Peruvian coast of the Pacific Ocean varies from 10–500 km². The thickness of alluvial aquifers reaches 150 m. Actually all the groundwater of these structures is discharged directly to the ocean (Gilboa 1971). The waters are fresh and ultra-fresh with a mineralization of 0.2–0.3 g/l. The total submarine chemical discharge amounts to 35.5 mln t/yr; its moduli vary from 45–70 t/yr·km².

Owing to the elongated shape of the South American continent from the north to the south, one can distinctly see here the latitudinal zonality in distribution of the specific values of the groundwater discharge to the seas (Figure 3.5.4 and Figure 3.5.5). This indicates to the active influence of climatic and general physical-geographic factors, determining a potential possibility of groundwater recharge, on the submarine runoff. The obtained values of SGD depend not only on the above-mentioned factors, but to a considerable degree they are determined by concrete structural-hydrogeological and hydrodynamic conditions of drainage basins, filtration properties, and hydraulic capacity of water-bearing rocks. An active influence of purely geologic-hydrogeological factors can cause azonal values of groundwater discharge to the seas, which are observed in the areas of the Mississippi Lowland, Florida Peninsula, and Guyana Plateau. Submarine chemical discharge depends on the intensity of submarine groundwater runoff, the leaching ability of groundwater, the solubility of water-containing rocks, the processes of continental salinization, and other conditions for formation of groundwater chemical composition. On average, the modulus of submarine chemical discharge within the continent varies from 40–60 t/yr·km², decreasing in low-water dry areas to ≤10 t/yr·km². The highest submarine chemical discharge (100–200 t/yr·km²) is typical of humid subtropic and tropic areas, where aquifers are composed of easily-soluble carbonate rocks.

Australia. The dry climate of Australia and the predominance of plain- and desert-type territories are factors that do not favor formation of considerable groundwater discharge to oceans, the total amount of which from the continent as a whole does not exceed 25 km³/yr. The specific discharge values to the Indian Ocean seldom reach 0.5 l/sec·km² (Kimberly Plateau) and usually amount to 0.2–0.3 l/sec·km² (see Table 3.2.1). At the same time, the submarine discharge to the Pacific Ocean from the drainage areas of the Great Dividing Range increases to 1 l/sec·km², and in some areas of the Australian Alps reaches 3 l/sec·km² (Zektser and Dzhamalov 1981a). Such a distribution of the submarine discharge specific values is caused, in the first turn, by the climatic and orographic factors within the continent (Figure 3.5.6).

Screening influence of the Great Dividing Range causes an increase of precipitation on its slopes to 2000 mm/yr, which favors more active groundwater recharge. In the structural-hydrogeological respect, there are distinguished large hydrogeological massifs composed of fissured volcanogenic, metamorphic, and sedimentary rocks, as well as several relatively small artesian basins opened toward the Pacific Ocean. The largest artesian structures on the eastern coast are the basins of Sydney, Moreton Clarence, and Laura (AWRC 1976). The water-bearing rocks are fissured sandstones, argillites, and shale of the Permian and Mesozoic ages. In general, they have a low hydraulic capacity and low water content. The groundwater has various mineralization usually growing downward in the cross-section, becoming at a depth of ~200–800 m unsuitable for domestic water supply. In the upper part of the cross-section where the major SGD is formed, the mineralization usually

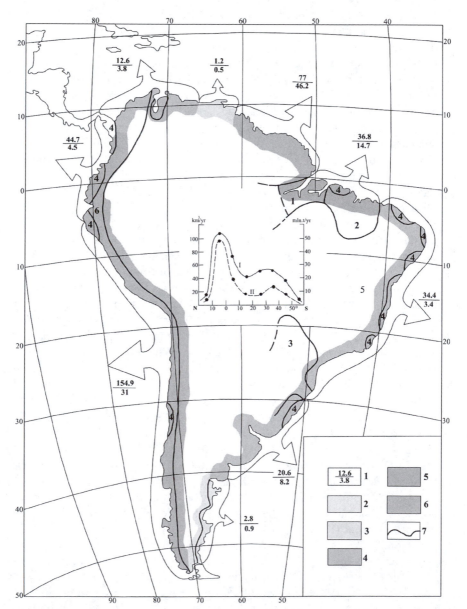

FIGURE 3.5.4 (Please see color insert.) Schematic map of groundwater discharge to the ocean from South America: (1) in numerator: groundwater discharge (km^3/yr), in denominator: ion groundwater discharge (mln t/yr); areal module of groundwater discharge (l/sec·km^2): (2) 0.1–0.5, (3) 0.5–1.0, (4) 1.0–3.0, (5) 7.0–10.0, (6) 10.0–20.0, (7) boundaries of artesian basins and hydrogeological massifs: (1) Amazon River basin, (2) Maranhão, (3) Paraná, (4) coastal; hydrogeological massifs: (5) Brazilian Plateau, (6) Andes. Distribution of water (I) and ion (II) groundwater discharge into the ocean from the continent by latitudinal zones is shown on the graph.

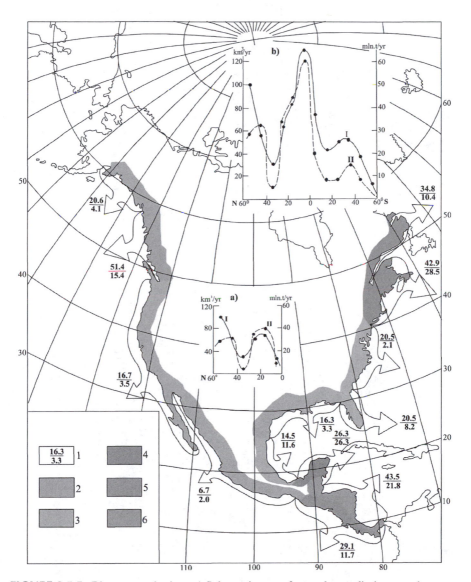

FIGURE 3.5.5 (Please see color insert.) Schematic map of groundwater discharge to the ocean from North America: (1) in numerator: groundwater discharge (km³/yr), in denominator: ion groundwater discharge (mln t/yr); areal module of groundwater discharge (l/sec·km²): (2) 1.0–3.0, (3) 3.0–5.0, (4) 5.0–7.0, (5) 7.0–10.0, (6) 10.0–20.0. On the graphs: (a) distribution of water (I) and ion (II) groundwater discharge into the ocean from the continent by latitudinal zones, (b) the same in total from the continents of North and South America.

does not exceed 1 g/l. Due to the variable groundwater mineralization, the calculated total submarine chemical discharge (5 mln t/yr) can be considered as the lower limit of salt transfer with groundwater to the Pacific Ocean. The weak solubility of water-bearing rocks and low discharge determine low moduli of the submarine chemical discharge, i.e., they vary from 20–40 t/yr·km².

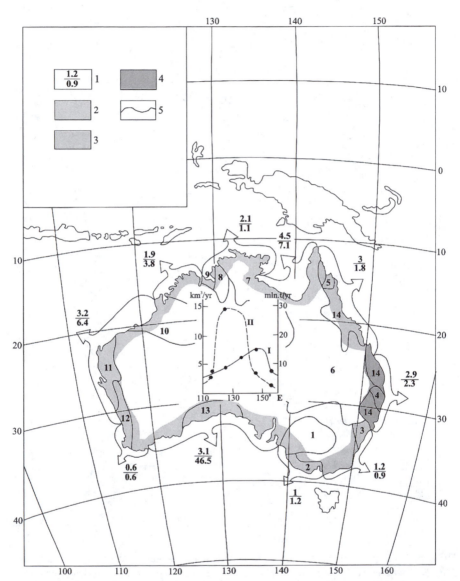

FIGURE 3.5.6 (Please see color insert.) Schematic map of groundwater discharge to the ocean from Australia: (1) in numerator: groundwater discharge (km³/yr), in denominator: ion groundwater discharge (mln t/yr); areal module of groundwater discharge (l/ sec·km²): (2) 0.1–0.5, (3) 0.5–1.0, (4) 1.0–3.0; (5) boundaries of artesian basins and massifs: (1) Murray, (2) Otuey, (3) Sydney, (4) Moreton-Clarence, (5) Laura, (6) Great Artesian Basin, (7) Georgina, (8) Victoria, (9) Bonaparte Gulf, (10) Canning, (11) Carnarvon, (12) Perth, (13) Yukla; (14) hydrogeological massif of the Great Dividing Range. Distribution of water (I) and ion (II) groundwater discharge into the ocean by longitude is shown on the graph.

Among the artesian structures on the Australian coast of the Indian Ocean there are distinguished the basins of Murray, Yukla, Perth, Carnarvon, Canning, Georgina, and Carpentaria (AWRC 1976). The Murray basin contains in its sedimentary cover a series of aquifers of Pleistocene to Eocene–Paleocene ages, composed of sands, sandstones, limestones, and marl. The total thickness of these sediments is 350–450 m. The basic drain of the artesian basin is the Murray River with its numerous tributaries, and only a small part of the groundwater discharge is forwarded directly to the Great Australian Bight. A considerable part of this SGD is tapped by coastal swamped spaces and spent for evapotranspiration. Because of this, the average SGD modulus for the entire coastal part of the artesian basin does not exceed 0.3 l/sec·km^2. Mineralization of the groundwater varies within wide ranges from 1–35 g/l and amounts on average to 1.5 g/l (AGPS 1975). The modulus of submarine chemical discharge, on average for the basin, does not exceed 15 t/yr·km^2.

The sedimentary cover of the Yukla basin, occupying the Nullarbor Plain and Eyre Peninsula, begins with Permian and Cretaceous sediments. However, the most water-abundant aquifer is associated with the Eocene–Miocene cavernous loams with a total thickness of 150 m (AGPS 1975). This basin is a typical example of artesian structures in arid desert areas of Australia where potentially water-rich and well-permeable aquifers actually do not get recharged, because total precipitation is not higher than 180 mm/yr. In this connection, SGD amounts to only 0.2 l/sec·km^2. The groundwater mineralization rapidly increases with depth and amounts on average to 15 g/l. The modulus of chemical discharge reaches, at the expense of high groundwater mineralization, 75 t/yr·km^2.

On the coast of West Australia two artesian basins are distinguished, Perth and Carnarvon. They are studied irregularly. Only in the large populated areas, a series of aquifers from Quaternary to Cretaceous and Permian ages are drilled through by wells. The aquifers are composed of sands, sandstones, and limestones located at depths of 60–750 m (AGPS 1975). Precipitation is relatively low (200–800 mm/yr), which does not contribute much to formation of groundwater resources. Zones with the highest water contents are associated with the modern and buried valleys of constant and temporary water streams, the discharge of which is forwarded directly to the Indian Ocean. Low infiltration recharge, draining influence of the local erosional network, and arid climate on the major territory lead to a slight SGD, which from the entire western coast does not exceed 0.2 l/sec·km^2. Freshwaters are met sporadically, mainly within the Perth basin. On the rest of the territory the mineralization reaches 4–10 g/l. Salt sources are, alongside the water-containing rocks, also precipitation with a mineralization of 15–80 mg/l. As a result of a high evaporation, salts accumulate in the soils and unsaturated zone and salinize the aquifers. The total chemical discharge amounts to 6.4 mln t/yr, its modulus 10 t/yr·km^2.

The artesian basins of Georgina and Carpentaria, located in the north of the continent, are opened toward the Gulf of Carpentaria. The Georgina basin is one of the largest basins in Australia, covering 325 ths km^2. Its specific feature is the wide development of karst-generating carbonate rocks of Early Paleozoic and mainly Cambrian age. Eastward these rocks are gradually overlain by carbonate terrigenic sediments of the Lower Cretaceous. The total thickness of aquifers is not known.

The deepest wells strip them at a depth of 750 m. The basin does not have pronounced regional recharge areas. In general, the carbonate rocks have low water content. The well yields usually do not exceed 2–3 l/sec (AGPS 1975). The mostly watered upper part of the carbonate rocks is intensively drained by constant and temporary water streams. Because of this, the total SGD amounts on average to 0.2 l/sec·km². Here, the SGD is usually manifested in the form of concentrated karst submarine springs with a low yield. The groundwater mineralization varies within wide ranges from 0.3–11 g/l.

The artesian basin of Carpentaria is a part of the Great Artesian Basin, within which the elevated zones of the Paleozoic basement form the inside boundaries of 2nd- order basins. Confined aquifers are associated with sedimentary rocks of the Cretaceous, Jurassic, and Triassic ages with a total thickness of 2000–2500 m. The water-containing rocks consist usually of sandstone. The Carpentaria basin is actually not studied. Fragmentary information evidences that the aquifers are water-poor with a water mineralization varying from 1–6 g/l. The total chemical discharge from the entire coast of the Carpentaria basin is slightly higher than 7 mln t/yr.

World Ocean. Analysis of formation of the groundwater recharge to the World Ocean within the basic continents shows that this global process depends on a complicated combination of different natural factors, the main role among which belongs to climate, relief, and structural-hydrogeological features of the coastal territories (Zektser and Dzhamalov 1981a). A significant influence on SGD is exerted also by the hydrodynamics of groundwater discharge, filtration properties, and hydraulic capacity of the unsaturated zone and water-containing rocks. All these factors are closely linked with each other and determine conditions of recharge, transport, and discharge of groundwater in different natural zones.

Groundwater discharge depends on a structure of recharging and discharging sources of the water balance of drainage areas, which, in turn, are determined by a proportion of heat and moisture as a basic index of natural physical-geographic zonality. In connection with this, distribution of specific values of the groundwater discharge to the World Ocean on the global scale is governed by the latitudinal physical-geographic zonality. They gradually increase from subarctic areas to moderate zones, sharply increasing in humid subtropics and tropics and decreasing in semiarid and arid regions (Figure 3.5.7). The local orographic, geologic-structural, hydrogeological, and hydrogeodynamic features of the coastal drainage areas make the discharge distribution more complicated and can sometimes cause considerable deviations of it from the average values typical of the given latitudinal zone. However, azonally high or low values of SGD, connected with a screening influence on atmospheric circulation in mountainous structures, wide development of karst, draining action of river valleys, and other local factors are associated with local coastal areas and, in general, do not disturb the general dependence of groundwater discharge to the World Ocean on the latitudinal physical-geographic zonality. It can also be seen that distribution of total submarine chemical discharge depends on the latitudinal physical-geographic zonality, because salt transfer with groundwater is determined, first of all, by SGD. As it is seen from Figure 3.3.1, the average submarine groundwater mineralization

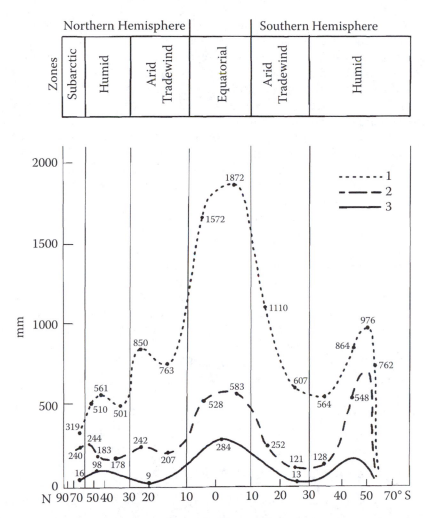

FIGURE 3.5.7 Distribution of atmospheric precipitations and surface and groundwater discharge into the World Ocean by latitudinal zones of the land: (1) atmospheric precipitations; (2) surface discharge; (3) groundwater discharge.

varies within 0.3–2.5 g/l and only seldom reaches 15–40 g/l. In other words, predominintly fresh and low-brackish groundwater from the upper hydrodynamic zone is transported to the World Ocean; the chemical discharge of this groundwater is entirely determined by the latitudinal physical-geographic zonality of its distribution on a global scale.

The increased groundwater mineralization depends, in the first place, on availability of salt-bearing rocks, processes of continental salinization, stagnant filtration regime, or the low degree to which aquifers are washed. Such conditions of groundwater chemical composition formation are most often met within the coasts of the African and Australian continents, which will lead in some areas to azonally high values of submarine chemical discharge. The complicated character of the distrubution of the submarine chemical discharge indicates that this natural process is, to a great degree, determined by geologic-hydrogeological Paleocene and modern conditions of groundwater formation. In other words, the regional influence of latitudinal physical-geographic zonality on distribution of basic discharge-forming factors is smoothed or complicated in some coastal areas by concrete conditions of groundwater chemical composition formation.

Principles of the theory on zonal distribution of basic components of the geographic environment, including groundwater, were lain by Dokuchaev in the 1880s. This theory has a basic methodological importance for the understanding of general regular laws in natural processes and phenomena. Investigations of groundwater zonality in the zone of intensive water exchange have been successfully developed in subsequent decades by Il'in, Lichkov, Kamensky, Lange, Kudelin, Lvovich, and other Russian hydrogeologists. At the present time, the idea of zonal groundwater distribution in the Earth's crust has come to stay in the study of regional hydrogeology. Investigations carried out on groundwater discharge to the World Ocean obtained results and revealed regularities in formation of this process and contributed to the development of hydrogeological science. It is all the more important that the zonal distribution of submarine groundwater and chemical discharge has received, for the first time, a quantitative expression on a global scale. Thus, the obtained values of groundwater and chemical discharge to the World Ocean and basic regularities of this complicated natural process give grounds to consider the revealed latitudinal zonality of submarine dischage as the basis for scientific prediction and study of conditions for formation of groundwater discharge of the Earth as a whole.

3.6 SUBMARINE GROUND DISCHARGE AND MODERN-AGE CLIMATIC CHANGE*

This subchapter's objectives are twofold: (1) to review central issues concerning modern-age (that is, in the Holocene, or last 10,000 years) climate change and its implications on SGD (this borrows from Loaiciga 2003); and (2) propose estimates of probable climate-change impacts on SGD and its total dissolved solids output. Earlier work on the matter of modern-age climatic change and SGD and TDS output was reported in Zektser and Loaiciga (1993).

3.6.1 THE CLIMATE-CHANGE PUZZLE

The term "climate change" has become synonymous with post-Industrial Revolution changes in global mean surface-air temperature that are hypothesized to have been caused primarily by increased atmospheric concentrations of carbon dioxide (CO_2) — an active greenhouse gas — during the 18th, 19th, and 20th centuries. The average CO_2 atmospheric concentration was about 280 ppmv (parts per million

---

* Section 3.6 was written by Hugo Loaiciga, Ph.D. (University of California, Santa Barbara) with the participation of Igor S. Zekster, D. Sc. (Water Problems Institute, Russia).

by volume) in 1765*; and in the year 2000 it averaged 364 ppmv. Human-induced changes in land use, burning of biomass (primarily in the form of dead plant matter) and fossil fuels (coal, gas, petroleum-based products), and associated gaseous output to the atmosphere took global significance since the invention of agriculture, some 10,000 years ago. Human impacts on the global environment accelerated dramatically since the second half of the 18th century, from the combined effect of rapid population growth and greater use of fossil fuels. Cumulative landscape changes throughout the Holocene have altered the relationship between precipitation and runoff, between precipitation and aquifer recharge, as well as the evapotranspiration at regional scales. These changes may, in turn, affect regional precipitation, thus establishing feedbacks among hydrologic processes and regional climates that vary greatly due to geographical, topographic, vegetative, geologic, and other biotic factors (fauna and soil microbial activity, for example). To complicate matters, while all these human-induced changes have taken place, it is well established that the Earth's mean surface temperature has increased several degrees since the end of the last glacial age about 20,000 years ago, a trend that is of natural origin.

The superposition of natural warming with that that may be caused by increased greenhouse-gas concentrations due to human activity triggered a frenzy of research during the last quarter of the 20th century, which continues to this day, aimed at sorting out the contributions of natural and human causes to the slight increase in the Earth's mean surface temperature manifested in instrumental records during the last century and a half (on the order of 0.3°C and 0.6°C) (Intergovernmental Panel on Climate Change [IPCC] 2001).**

Another dimension of climate change-related research has been the numerical simulation of the future climate, and related hydrological, ecological, and economic phenomena, under the assumption that CO_2's atmospheric concentration may double in the next few centuries relative to its 1990 mean concentration (the reference concentration used by the IPCC in 2001). These simulations, and the reliability of their results, are bedeviled by the complexity of the Earth's biogeochemical processes, not to mention uncertainties concerning population and economic growth, human adaptation, and technological and scientific changes and innovation. In the realm of hydrologic consequences of climate change, see Loaiciga et al. (1996) and Loaiciga (2003) for a review and regional-scale impact analysis on groundwater, respectively. In the socio-economic realm, see Yohe et al. (2004), for a study of CO_2-management strategies to cope with climate change. These scientific studies occur while the international community strives to reach consensus about curtailing future greenhouse emissions according to specific rules laid down in the Kyoto Protocol, a proposed treaty aimed at curtailing the levels of greenhouse emissions by humans*** — a goal that, if achieved, could exert substantial economic, technological, and social impacts worldwide. The ultimate consequences are difficult to predict.

---

* This is the year when the steam engine was patented, launching the Industrial Revolution.

** Other researchers put the temperature increase in the 0.5°C to 1.5°C range. The debate continues over this matter.

*** Future CO_2 emissions are to be kept at or near the 1990 reference mean atmospheric concentration.

3.6.2   THE HYDROLOGIC PERSPECTIVE IN RELATION
TO CLIMATE CHANGE

To the hydrologist, the question of whether or not global mean surface temperature has increased or will continue to increase, say, at an average rate of 0.5°C every 100 years, is of secondary importance. The scope of work of the hydrologist is delimited by his capacity to measure hydrologic fluxes (water, substances, energy), to analyze them, and to make meaningful and useful inferences and predictions about them at relevant spatial scales. In the practical realm, where most hydrologic work lies, the intersection of hydrologically relevant spatial scales and administrative/political boundaries defines a clear context to the study of hydrologic processes, with or without climate change. Specifically, hydrologic studies are largely circumscribed to the watershed and the regional aquifer system. Typically, this entails working with regions of less than 10^6 km², and in the great majority of cases, the typical perimeter of the watershed or groundwater basin encloses areas well under 10^5 km² (Loaiciga 1997). These spatial scales are referred to in this work as "regional scales." Therefore, to the hydrologist, climate change must be resolved in terms of precipitation, surface-air temperature, evapotranspiration, sediment transport, groundwater levels, groundwater recharge, water quality, and runoff and SGD changes at the relevant spatial scales.

As for the temporal resolution of climate change, hydrologic processes encompass a wide spectrum of meaningful time scales. In the case of flood studies, the relevant temporal scales of precipitation changes range from minutes to days; while for drought-impacts studies, the precipitation and temperature temporal scales of interest can vary from days to years, depending on the inter- and intraseasonal disposition of water in the natural or human-occupied environment (Loaiciga et al. 1993).

The following sections provide a review of the state of the art in the analysis of climate change and its hydrologic consequences. The review is focused on the groundwater component of the hydrologic cycle. An example is presented using estimates of SGD and its TDS output under a climate-change scenario.

3.6.3   CLIMATE-CHANGE SCENARIOS AND SIMULATION MODELS

Early studies (many were produced between 1975 and the mid 1990s) of regional-scale hydrologic consequences of climate change were mostly based on simple scenarios for precipitation and temperature under a warmer climate (see Gleick 1989 for a review of representative articles). Precipitation was increased or decreased a certain percentage relative to historical values (a ±10% range was commonly used). Historical temperature was increased a few degrees (1 – 5°C, typically). With these two forcing variables, hydrologic models were then implemented to carry out simulations in the region of interest. It was assumed that a calibrated hydrologic model under the historical climate remained valid under a modified climate. The results so obtained for important fluxes such as sediment output, groundwater recharge, stream flow, or other variables of interest were then compared with those that corresponded to the historical-climate simulations. The differences between the two sets of results

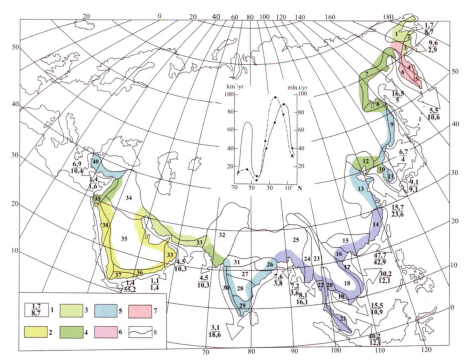

FIGURE 3.5.1 Schematic map of groundwater discharge to the ocean from Asia. Please see text for complete figure caption.

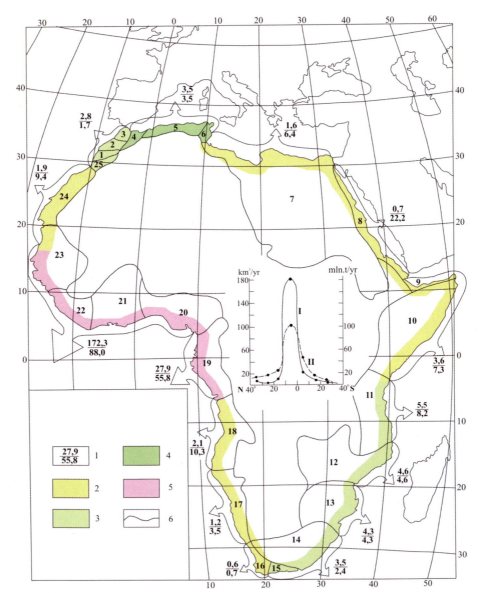

FIGURE 3.5.2 Schematic map of groundwater discharge to the ocean from Africa. Please see text for complete figure caption.

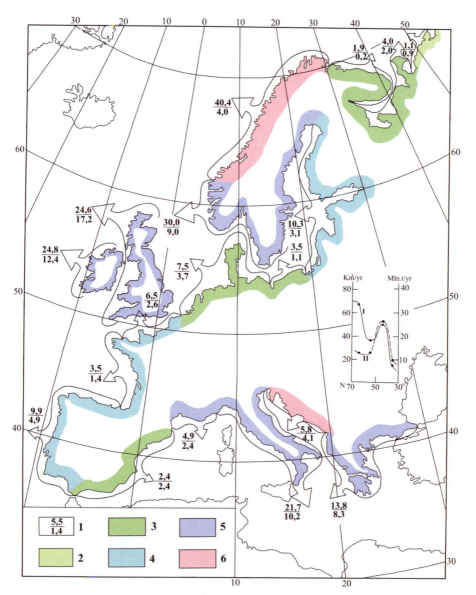

FIGURE 3.5.3 Schematic map of groundwater discharge to the ocean from Europe. Please see text for complete figure caption.

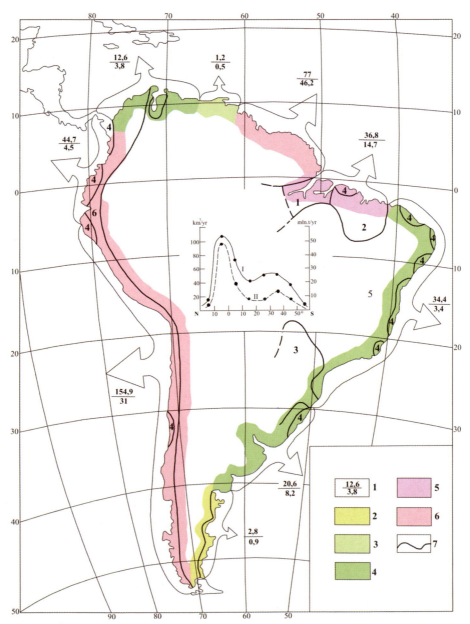

FIGURE 3.5.4 Schematic map of groundwater discharge to the ocean from South America. Please see text for complete figure caption.

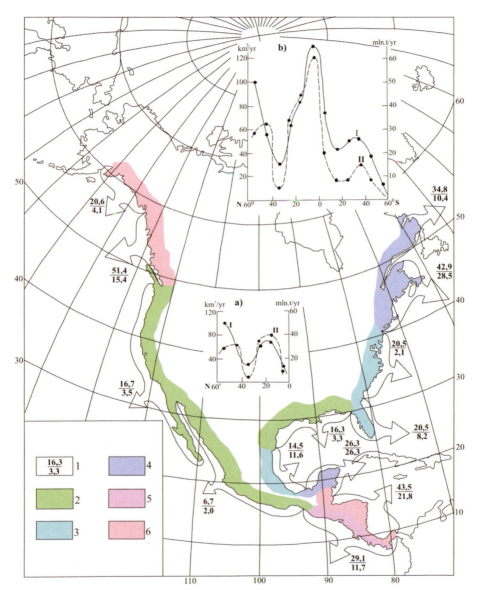

FIGURE 3.5.5 Schematic map of groundwater discharge to the ocean from North America. Please see text for complete figure caption.

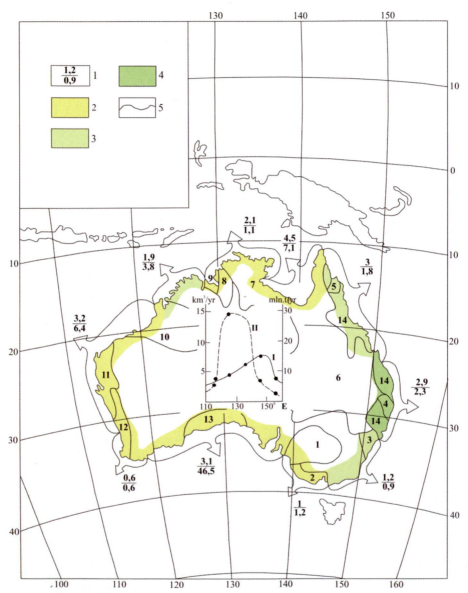

FIGURE 3.5.6 Schematic map of groundwater discharge to the ocean from Australia. Please see text for complete figure caption.

FIGURE 4.3.1.1 Aerospace image of the Aral Sea in September 2004 (according to data from the Institute of Space Explorations, Kazakhstan). Please see text for complete figure caption.

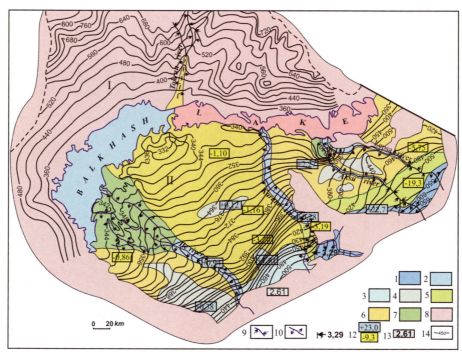

FIGURE 5.2.2.2 Schematic map of the formation of submarine groundwater discharge (SGD) to Lake Balkhash. Please see text for complete figure caption.

were then attributed to climate change, other things being equal (e.g., population, water use, cropping patterns, and water technology).

A second wave of studies emerged in the refereed literature in the mid 1980s. They were based on the linkage between climate predictions from general circulation models (GCMs) and regional climate models (RCMs) (see Henderson-Sellers and Pitman 1992; Giorgi et al. 1993a,b, 1994; Vörösmarty et al. 2000). GCMs, which emerged in the global-scale climatic modeling community in the late 1960s, have been steadily improved in their physically based structure and numerical solution algorithms. They also have evolved by incorporating refined spatial resolution of their numerical grids. RCMs have the same physical basis as GCMs but with a much greater spatial resolution and are confined to synoptic-scale and mesoscale simulation regions rather than planet-wide simulations. At present, a GCM, and there are several leading ones (see Loaiciga et al. 2000 for a list of GCMs in hydrologic studies), may have a spatial grid with cells on the order of 200×200 km, while the RCMs have achieved resolutions on the order of 20×20 km.* The RCMs rely on the coarser output from GCMs, which they use as initial and boundary conditions to drive their spatially refined simulations of climate change.

Many GCM and RCM climate-change simulations are based on the so-called steady-state $2 \cdot CO_2$ scenario, whereby the 1990 CO_2 atmospheric concentration (about 355 ppmv, a base level adopted by the climate-change community; see, e.g., Houghton et al. 1996) is doubled and that value is used in the GCMs and RCMs to simulate the $2 \cdot CO_2$ warmer climate. The climate models simulate various relevant climate-forcing variables of hydrologic interest at the land–atmosphere interface: precipitation, air temperature, radiant-energy fluxes, wind speed, atmospheric pressure, absolute humidity, latent-heat flux, and runoff averaged over the models' surface grid cells. The RCM key output variables, such as precipitation, surface air temperature, ground-level radiant-energy fluxes, water-vapor pressure, and wind speed, become the forcing input variables to hydrologic models, which then calculate in a classical fashion the dependent hydrologic variables of greatest interest. Stream flow and groundwater recharge are examples of the former (see, e.g., Loaiciga et al. 2000).

In some instances, GCMs and RCMs have undergone "subgrid" parameterizations which introduce approximate numerical representations of hydrologic processes at the land–surface interface, which allows them to make calculations of hydrologic fluxes at fine spatial resolution or at selected locations (e.g., stream gages, or zones of influence of a water well). However, watershed-scale hydrologic models are better suited to carry out fine-resolution hydrologic simulations (e.g., stream flows at selected gages, spring flows, groundwater recharge, groundwater levels) due to their more realistic, physically based structure and internal parameterization (Vaccaro 1992; Panagoulia 1992). This is particularly true when attempting to simulate groundwater response to climate change, because groundwater flow, transport, and geochemical processes are poorly represented, if at all, by the subgrid parameterizations thus far proposed.

---

* See the Pennsylvania State University/National Center for Atmospheric Research numerical model home page at http://www.mmm.ucar.edu/mm5/mm5-home.html for a description of the RCM MM5.

Substantial contemporary efforts on climate-change studies is devoted to the transient, numerical simulation of the future Earth's climate, in which CO_2 atmospheric concentrations are allowed to rise according to subjective scenarios. Concurrently, studies of the Earth's carbon cycle have taken central stage as scientists struggle to understand the fate of natural- and human-produced atmospheric CO_2. Long-term, transient simulation of the Earth's future climate face daunting challenges. One is the chaotic nature of climatic dynamics. This is typically referred to as the "butterfly effect." This effect is allegorical concerning the chaotic nature of weather: if a butterfly were to flap its wings in some part of the world, it would disturb the atmosphere around it, thus, triggering a divergent weather pattern (Lorenz 1963, 1967). This is an extreme view of weather and climate sensitivity. It has, however, significance concerning the sensitivity of climate and hydrologic change to initial conditions. Slight differences in the initial conditions (say, a 1% difference) of variables simulated by GCM and RCM lead to large divergence and inaccuracy in their predicted values as time goes on, say, after ≥ 10 years of simulation. Minor variations in model structure, boundary conditions, and key model parameters usually produce disproportionate change in the models' output as well. The hurdles facing future climate simulations (that is, several decades or longer ahead) pose serious questions about model-simulated scenarios of an eventual 2·CO_2 climate, given that the likelihood of reaching such an equilibrium level of atmospheric concentration may not happen within the next 100 years or even in the next few centuries.

3.6.4 ON CLIMATE-HYDROLOGIC FEEDBACKS AND 2·CO_2 UNCERTAINTY

Climate feedbacks constitute a second complication in the accurate simulation of 2·CO_2 or transient climate scenarios. Several authors (see, e.g., Ramanathan and Collins 1992; Loaiciga et al. 1996) have identified several important feedbacks in the climate system that are not well captured, and in some cases not represented at all, in GCMs and RCMs. By feedback it is meant the interaction established between climate forcings, such as increased CO_2 atmospheric concentration or surface-air temperature increase, and the response of the terrestrial and/or marine environment, which, in turn, acts to either accentuate or dampen those forcings. A positive feedback on surface temperature, for example, would act to increase it, while a negative would reduce it. The following are key feedbacks on climate-change forcing.

The Water-Vapor Feedback. The water-vapor feedback on climate is positive. Increased CO_2 concentrations produce larger infrared radiation to the Earth's surface, raising its temperature. That rise in temperature evaporates more water and the atmosphere's absolute humidity increases. The evaporated water transfers latent heat to the atmosphere, which increases tropospheric temperature after the water vapor condenses. The greater tropospheric temperature enhances the water-holding capacity of the atmosphere and hence its humidity. Water vapor, an effective greenhouse gas, traps Earth-emitted infrared radiation and, along with the warmer troposphere as a whole, increases infrared emissions back to the Earth's surface and heats it up higher. This heat input contributes to evaporation, completing the water feedback loop of water vapor. Black-body cooling of the warmer surface-atmosphere system (i.e., by long-wave radioactive emission to outer space according to the Stefan–Boltzmann law) impedes runaway warming of the Earth's surface. In addition, the water-vapor (positive) feedback forcing has its own natural "brakes," such as the (negative) lapse-rate feedback (see Lindzen 1990).

The Cloud Feedback. In the present climate, in which the global mean cloudiness is 50%, clouds have both a positive forcing as well as a negative one. Positive climate forcing arises from the trapping of Earth-emitted infrared radiation, while the negative forcing is caused by the reflection of incoming solar radiation back to space, thereby reducing the flux of radiant energy reaching the surface compared to clear-sky areas. Earth radiation studies seem to indicate that clouds have a net negative forcing which is responsible for a surface temperature between 10–15°C cooler than it would otherwise be (Ramanathan and Collins 1992). Most GCMs predict a decrease in global mean cloudiness in a warmer climate (despite the increased atmospheric humidity expected in that case). As a result, because of the strong albedo effect of clouds (i.e., they reflect incoming solar radiation), the overall predicted drop in cloudiness has a net positive (warming) feedback on the climate under the $2 \cdot CO_2$ scenario according to most GCMs.

GCMs are simplistic in their treatment of cloud physics. Issues that are not properly modeled by GCMs in relation to the cloud feedback include diurnal and seasonal cloud cover shifts, changes in latitudinal cloud cover (low latitudes to high latitudes), cloud cover shifts related to variations in surface albedo, and changes in cloud optical thickness. The water vapor and cloud feedbacks interact among them, and with the lapse rate, the albedo and other feedbacks (see below). Variations in atmospheric humidity, which are the result of changes in the radiation balance under $2 \cdot CO_2$ warming, play a central role in the ultimate equilibrium state of a potential warmer climate.

Surface Albedo, Soil Moisture, and Vegetation Feedbacks. The surface albedo feedback refers mainly to ice-mass modifications in a warmer climate. A warmer climate, with surface temperatures magnified at high latitudes according to GCMs, melts ice and snow. These two surfaces are more reflective than water or land surfaces generally. Therefore, a warmer climate may reduce the surface planetary albedo (the ratio of reflected to incoming shortwave radiation), thus increasing the absorption of incoming solar radiation, which, in turn, warms the surface further, giving rise to the ice-albedo feedback.

The soil-moisture feedback is bound to occur in regions where precipitation is predicted to decline, at least seasonally, such as during the summer in central North America. With lower precipitation and a warmer surface, the soil surface becomes drier. Evapotranspiration is reduced as the soil moisture drops. This produces lower cloud formation. As evapotranspiration declines, so does the evaporative cooling associated with latent heat removal from the surface. Lower cloud formation enhances surface warming as the atmosphere becomes less reflective to solar radiation. The reduced evaporative cooling and the enhanced atmospheric transparency start the soil-moisture feedback on surface warming, which is positive according to this reasoning.

Vegetation feedbacks on climate are poorly understood. The vegetation feedback which results from surface warming may be triggered by changes in vegetated areas and in the type of vegetative cover. These changes can alter the surface-atmosphere temperature through modifications in the surface albedo and in the CO_2 exchange between the atmosphere and the Earth's surface. Plant growth and respiration depend on atmospheric CO_2, surface temperature, and soil moisture. The interactions among these variables introduce several degrees of freedom and uncertainties in the biosphere–climate system. Under these circumstances, even the prediction of the sign of the vegetation feedback remains elusive.

Ocean–Atmosphere Interactions. The ocean–atmosphere coupling is second only to the Earth's rotation in its importance to affect global atmospheric circulation patterns. Yet, the ocean–atmosphere coupling remains poorly understood and poses substantial challenges to accurate numerical modeling, whether in the present climate or under the $2 \cdot CO_2$ scenario (Willebrand 1993; Forest et al. 2002). Interestingly, hydrologic fluxes could play an important role in large-scale oceanic circulation and, then, in atmospheric phenomena affected by such circulation. Changes in precipitation, evaporation, and oceanic surface temperature could modify the ocean salinity and its density in the high latitudes (for example, in the North Atlantic), thus causing major shifts in the rate of deep-ocean water replenishment by surfacewater. The latter phenomena could modify present ocean–atmosphere circulation patterns and bring about unpredictable changes in regional hydrologic regimes. Such changes could take place over tens, hundreds, or even thousands of years, given the relevant time scales involved in global oceanic circulation.

The Aerosol Feedback. Aerosols (fine solid particles found in gaseous suspension in the atmosphere) introduce other complexities in the analysis of climate feedbacks. Aerosols may be either anthropogenic or natural — those produced by volcanic eruptions being ubiquitous. The former are of particular interest in the quest for sorting out the human contribution to global warming during the last two centuries. On one hand aerosols reflect incoming solar radiation, which would cause a negative climate feedback. On the other hand, they serve as cloud-formation nuclei; and clouds, depending on their particular characteristics, may induce either positive or negative climate feedbacks. A still-controversial clue about the net aerosol feedback may be found in climate simulations. GCM predictions based on increased concentrations of greenhouse gases overestimate the global mean temperature in the post-Industrial Revolution era. Improved resemblance between the GCM-predicted and observed temperatures was achieved with the introduction of a negative radiate forcing attributed to sulfate aerosols in climate simulations (Mitchell et al. 1995). The technique used by those authors relied on adjusting the parameterization of surface albedo to account for the backscattering of solar radiation by sulfate aerosols. The simplicity and the ex post facto nature of the aerosol fix to a fundamental GCM shortcoming has raised questions concerning circular logic in the restructuring and recalibration of climate models (Demeritt 2001). The role of aerosols in modern climate change continues to be actively assessed in the climate research community (Forest et al. 2002).

3.6.5 A STEADY-STATE APPROACH TO SGD AND TDS RESPONSE TO CLIMATE CHANGE

This section contains an analytical treatment of groundwater hydraulics and recharge interaction, which helps in understanding one possible mechanism of climate-change effects on groundwater flow in general, and on SGD and its TDS in particular. To this end, consider steady-state groundwater flow driven by spatially-variable recharge ($N(\bar{x}, \bar{y})$) in a 2-dimensional, rectangular, domain with homogeneous and isotropic hydraulic conductivity (K). The (deterministic) equation of groundwater flow in this instance is (b denotes the confined-aquifer thickness; ϕ is the hydraulic head):

$$\frac{\partial^2 \phi(\bar{x}, \bar{y})}{\partial \bar{x}^2} + \frac{\partial^2 \phi(\bar{x}, \bar{y})}{\partial \bar{y}^2} = -\frac{N(\bar{x}, \bar{y})}{Kb} \qquad (3.6.1)$$

Equation (3.6.1) applies in the aquifer domain defined by the spatial coordinates: $0 < \bar{x} < A, 0 < \bar{y} < B$ and the boundary condition is:

$$\phi = 0 \quad \text{along} \quad \bar{x} = 0, \bar{x} = A, \bar{y} = 0, \bar{y} = B \qquad (3.6.2)$$

that is, the hydraulic head is zero along the edges of the rectangular domain. Problem (3.6.1)–(3.6.2) is referred to herein as the "island" aquifer, wherein the aquifer is surrounded by a constant water level arbitrarily set to a reference elevation equal to zero. The solution of problem (3.6.1)–(3.6.2) using Fourier series is expedited by scaling the spatial coordinates. Letting $x = \bar{x} \cdot (\pi / A)$ and $y = \bar{y} \cdot (\pi / A)$ scales the rectangular domain to $0 < x < \pi, 0 < y < L$, where $L = B\pi/A$, and the redefined boundary-value problem (BVP) becomes:

$$\frac{\partial^2 \phi(x, y)}{\partial x^2} + \frac{\partial^2 \phi(x, y)}{\partial y^2} = -\frac{1}{K} \frac{N(x, y)}{b} \left(\frac{A}{\pi}\right)^2 \quad 0 < x < \pi, 0 < y < L \qquad (3.6.3.)$$

with boundary condition:

$$\phi = 0 \text{ along } x = 0, x = \pi, y = 0, y = L \qquad (3.6.4)$$

The solution of the BVP (3.6.3)-(3.6.4) solution is (see, for example, Weinberger 1965):

$$\phi(x, y) = \frac{\Omega}{K} \int_0^\pi \int_0^\pi G(x, y; v, \mu) N(v, \mu) \, dv \, d\mu \qquad (3.6.5)$$

in which:

$$\Omega = \frac{1}{b}\left(\frac{A}{\pi}\right)^2 \qquad (3.6.6)$$

and Green's function is given by (letting $\sinh[z]=(\exp(z)-\exp(-z))/2$ represent the hyperbolic sine, (see, for example, Gradshteyn and Ryzhik 1994):

$$G(x,y;v,\mu) = \begin{cases} \dfrac{2}{\pi}\displaystyle\sum_{n=1}^{\infty} \dfrac{\sinh[n\cdot(L-y)]\cdot\sinh[n\mu]\cdot\sin[nx]\cdot\sin[n\cdot v]}{n\sinh[nL]} & \text{for } \mu \le y \\[2ex] \dfrac{2}{\pi}\displaystyle\sum_{n=1}^{\infty} \dfrac{\sinh[ny]\cdot\sinh[n\cdot(L-\mu)]\cdot\sin[nx]\cdot\sin[n\cdot v]}{n\sinh[nL]} & \text{for } \mu \ge y \end{cases} \qquad (3.6.7)$$

The solution (3.6.5) shows that the hydraulic head is proportional to the recharge and inversely proportional to the hydraulic conductivity. If climate change modifies recharge by a factor $\lambda > 0$ throughout the domain $0 < y < A$, $0 < y < B$, it follows that the hydraulic head would change by the same factor λ. Moreover, the SGD in the island aquifer would increase by the same factor λ, easily established from the fact that groundwater discharge is proportional to the hydraulic-head gradient. The response of SGD to a recharge modified by climatic change can be established directly once a steady-state condition is reached. Under this condition all recharge must discharge to the surrounding sea floor:

$$SGD_{2\times CO_2} = \int_0^B\int_0^A N(x,y)_{2\times CO_2}\,dxdy = \int_0^B\int_0^A \lambda\cdot N(x,y)\,dxdy = \lambda\cdot SGD \qquad (3.6.8)$$

in which the subscript $2\cdot CO_2$ denotes climatic conditions pertinent to a future, steady-state, climate. Equation (3.6.8) is valid regardless of the value of the hydraulic conductivity.

The mass output of dissolved solutes in SGD equals:

$$M = TDS\cdot SGD \qquad (3.6.9)$$

Climate change could alter the mass output, which, in its most general form can be expressed as follows:

$$\Delta M = \Delta(TDS)\cdot SGD + TDS\cdot\Delta(SGD) \qquad (3.6.10)$$

That is, climate change could affect mass output through modifications in TDS (a concentration) and in SGD (a flow rate). SGD are TDS represent baseline values.

TABLE 3.6.1
Submarine Groundwater Discharge (SGD) and Its Total-Dissolved-Solids (TDS) Output

Oceans, Continents, Major Islands	SGD (km³/yr)	TDS (10⁶ t/yr)
Pacific Ocean		
Australia	7.1	5.0
Asia	254.3	165.2
North America	124.6	36.7
South America	199.6	35.5
Major Islands	714.7	278.1
Total	1300.3	520.5
Atlantic Ocean		
Africa	208.7	169.2
Europe	71.2	25.8
North America	219.4	112.2
South America	185.3	77.7
Major Islands	77.7	42.9
Total	762.3	427.8
Mediterranean Sea		
Africa	5.1	9.9
Asia	8.3	11.9
Europe	33.9	18.4
Major Islands	5.7	2.3
Total	53.00	42.5
Indian Ocean		
Australia	16.4	66.7
Africa	22.1	49.0
Asia	65.3	119.2
Major Islands	115.6	60.6
Total	219.4	295.5
Artic Ocean		
Europe	47.5	7.2
World Total	2382.5	1293.5

From Zektser and Loaiciga (1993).

Assuming that the latter are dominant (as done in Zektser and Loaiciga 1993), the mass output in a $2 \cdot CO_2$ climate would be given in the approximate manner by:

$$M_{2 \times CO_2} \approx TDS \cdot \int_0^B \int_0^A \lambda \cdot N(x,y)dxdy = \lambda \cdot TDS \cdot SGD \qquad (3.6.11)$$

It is seen in Equations (3.6.8) and (3.6.11) that the recharge-modification factor λ plays a central role in the fate of SGD and TDS output in a changing climate. The factor λ is meaningful at the regional or aquifer scale, because of the substantial geographic variability in recharge and TDS. Further progress in the analysis of climate-impacted SGD and TDS output is possible by introducing the recharge ratio $0 < \gamma = N/P < 1$, in which N and P are the annual average recharge and precipitation, respectively. A climate-induced change in average (regional) precipitation, ζ P, in which $\zeta > 0$, produces a modified recharge $\gamma \zeta P$ if the ratio γ remains constant. Using the last assumption, the following holds $\lambda N = \gamma \zeta P$, and, therefore, $\lambda = \zeta$. Equations (3.6.8) and (3.6.11) can then be written in terms of changes in average precipitation:

$$SGD_{2 \times CO_2} = \int_0^B \int_0^A \gamma \zeta P(x,y)dxdy = \zeta \cdot SGD \qquad (3.6.12)$$

$$M_{2 \times CO_2} \approx TDS \cdot \int_0^B \int_0^A \gamma \zeta P(x,y)dxdy = \zeta \cdot TDS \cdot SGD \qquad (3.6.13)$$

The precipitation-change factor ζ can be estimated by climatic simulations or set as a presumed climate scenario. If, for example, $\zeta = 1.1$, tantamount to a 10% increase in average precipitation, the SGD and TDS output estimates in Table 3.6.1 would be increased by 1.1, also. It is probable that in some parts of world ζ might be larger than 1, whereas it could be less than 1 in others.

The results (3.6.12)–(3.6.13) might seem overly simplistic at first. Yet, there is a logical thought process behind their derivation, and the constancy of the recharge ratio γ is a robust assumption. Far more important is to have accurate estimates of the baseline values SGD and TDS, and the setting of the precipitation-change factor ζ.

4 Experience of Study in Groundwater Discharge into Some Seas

CONTENTS

4.1   SUBMARINE GROUNDWATER DISCHARGE TO THE BALTIC SEA

4.1.1   QUANTITATIVE ASSESSMENT OF SUBMARINE GROUNDWATER DISCHARGE

In recent decades many European researchers have carried out investigations on the water balance of the Baltic Sea. Of particular importance is the study of the interaction between groundwater and seawater, the assessment of submarine groundwater discharge (SGD) directly to the sea, and its role in the formation of the current water and salt balance of the Baltic Sea. The results of the quantitative assessment of groundwater discharge to the Baltic Sea from European territory are discussed below, with special attention on detailed data of groundwater inflow to the sea from the territory of the former Soviet Union. It should be noted that the first works referred to here were publicized some 50 years ago.

Gatalsky (1954) in his monograph "Groundwater and Gases of Paleozoic Age in the Northern Part of Russian Platform" brought forth a proposition that although rock layers are dipping east- and southwards — i.e., toward the center of the Middle-Russian Syneclise — groundwater moves to the west and north toward the Baltic Sea, under an influence of a hydrostatic pressure head. He suggested that the Baltic Sea is a drainage area of deep and chiefly chlorine-calcium waters.

Great attention to groundwater discharge to the Baltic Sea was given in the publication by Silin-Bekchurin (1958), "Hydrodynamic and Hydrochemical Regularities on the Baltic Sea Territory." Analyzing the hydrogeological features of Cambrian rocks, Silin-Bekchurin proposed that groundwater from the Gdovsky aquifer drains toward the Gulf of Finland, and considers that the Gulf of Finland can be countered around by a cone of depression opened toward the Baltic Sea, just where the Cambrian groundwater is totally being discharged to. Karst waters of Silurian sediments move chiefly from elevated recharge areas of groundwater to eroded sinks, i.e., to river valleys and lakes. Deep groundwater of the Silurian period moves in the northern part of the Baltic region in two directions: in the west to the Gulf of Riga, in the east to Lake Chudskoe. This has been confirmed by the distribution of piezometric levels. In the opinion of Silin-Bekchurin, the underground watershed is an elongated strip almost in meridional direction through the central part of Estonia, to the west and east from which the hydroisolines of groundwater are lowering. The author presumes that in the lower water-saturated system of the Middle Devonian period groundwater moves east, southeast, and south toward the Gulf of Riga. After having analyzed the hydrodynamic conditions of groundwater of the Pärnu River, Silin-Bekchurin came to the conclusion that on the Pärnu Peninsula the unconfined shallow and artesian waters are drained by the Pärnu River on one hand, and the Gulf of Pärnu on the other. However, Silin-Bekchurin noted that despite the distinct groundwater flow in the lower zone toward the Baltic Sea, there are no natural groundwater springs along the gulfs' shores and, hence, the groundwater discharge takes place beyond the shore zone of the gulfs. The author concluded that a ridge of elevated rocks stretched in the northwestern direction through the central deepwater part of the Gulf of Riga and is a possible zone of groundwater discharge.

Verte (1958) also noted that the Lower Cambrian aquifer discharges into the Gulf of Finland and the Baltic Sea, whereas the other aquifers discharge their waters into rivers and lakes.

Based on 35-year observations, Pastors (1961) determined the discharge of the main rivers of the Gulf of Riga basin and evaluated the heat transfer of rivers into the gulf. His investigations showed that the average annual inflow of surfacewater to the Gulf of Riga is equal to 31.23 km^3/yr or 989 m^3/sec, with 1.73 km^3, i.e., 5.5% of which is discharge of small rivers. The average annual modulus of the total river discharge for the entire Gulf of Riga basin is equal to 7.53 l/sec·km^2, and the annual thickness of discharge is 38 mm. Annually each 1 m^3 of water in the Gulf of Riga gets freshwater in the form of surface runoff, 4 times greater than 1 m^3 of water from the Baltic Sea.

As it can be seen from these studies, the problem of groundwater discharge to the Baltic Sea has, in the literature, only a qualitative reflection, i.e., groundwater inflow to the sea is estimated.

In more recent times, researchers have been investigating the northwestern part of the Russian Platform, an extensive Baltic artesian basin with an area of about 214 ths km^2. The boundary of the Baltic artesian basin in the west and north passes along the contact line of Lower Paleozoic rocks with granites and gneiss of the Baltic Shield. Its eastern boundary passes across the Lakes Chudskoe and Pskovskoe through the Loknovskoe uplift of the basement and to the south toward the Byelorussian–Lithuanian uplift in the area of Minsk, and to the west approximately across the towns of Lida and Grodno. In the geostructural context, the basin opens toward the Baltic Sea. The present-day relief of the Baltic artesian basin territory was formed mainly under the influence of glacial and river activity. In general, the basin surface represents a hillocky plain elevating from the coasts of the Gulfs of Finland and Riga toward inside the continent to absolute elevations of 250–300 m. The landscape is typical of alternating extensive hillocky uplifts adjacent to outwash lacustrine-glacial moraine plains and gently convex sinks. In the northern basin, the marginal scarp of the Silurian–Ordovician plateau-clint composed of limestone can be clearly seen; the scarp is steep (and in some places abrupt), falling north toward the Gulf of Finland. The height of the scarp above the Baltic Sea level reaches 60–80 m.

High amounts of precipitation vary with an annual rainfall of >500–700 mm, as well as almost year-round excess wetting, which creates favorable conditions for the formation of surface and groundwater discharge. The average modulus of the river discharge for the Baltic Sea basin for this territory is 8.4 l/sec·km^2. A general increase of river discharge moduli to the west is 12–14 l/sec·km², which is typical within the South-Vidzemskaya, Zhdudskaya, and Baltic uplifts.

There are numerous lakes in the Baltic artesian basin territory. The area is dominated by the lakes of glacial origin characterized by low, almost flat, sometimes swamped shores, with a depth of 1–3 m to 10–15 m. Also, large and widely known lakes such as Chudskoe, Pskovskoe, and others are located in extensive depressions. Along the coast of the Gulf of Finland, lakes of marine origin can be found.

Two stages of water-bearing systems can be distinctly distinguished in the territory of the Baltic artesian basin, including the coastal zone of the Gulf of Finland (within Russia): the lower one including the Gdovsky aquifer, and the upper one involving the Cambrian–Ordovician and the overlying aquifers and aquifer systems. The lower stage is separated from the upper one by a thick and extensive stratum of laminarian and blue Cambrian clays. Groundwaters of the upper stage are linked hydraulically with each other. The presence of freshwater in the upper stage is associated with the zone of intensive water exchange, which is influenced by the draining action of rivers. The recharge areas of the water-bearing systems of the upper stage are represented by the elevated northern and southeastern wings of the artesian basin, and by local water-dividing highlands.

The groundwater discharge from the zone of intensive water exchange was assessed with the use of an integrated hydrologic-hydrogeological method. Analyzing the available hydrogeological information, using maps of hydrocontour and hydroisopiestic lines and hydrogeological profiles plotted along boreholes drilled into the main aquifers and the aquifer systems of the upper stage, the total drainage area was determined; the groundwater discharge from which was forwarded directly to the sea without involving the rivers.

This area, estimated using planimetry and various map scales, consists of separate subareas located along the seashore and is characterized by different hydrogeological conditions. The area covers the western coast of the Karelia Isthmus adjacent to the Gulf of Finland, the northern part of the fore-clint lowland, the western part of the Silurian–Ordovician plateau, the northwestern and western peripheries of the coastal plains of Latvia, and the coastal zones of the Saaremaa and Khuiumaa islands, as well as the western parts of the basins of the Borta, Minin, Neman, and Pregoli Rivers. The total drainage area of the groundwater which discharges directly to the sea is ~16,000 km².

In the late 1960s, based on the analysis of river-runoff hydrographs, the average multiyear moduli of groundwater discharge were determined for all the main aquifer systems in the zone of intensive water exchange for this territory (Zektser 1968). The values of these moduli were used over hydrogeologically similar subareas of the drainage area, discharge from which is forwarded directly to the sea. Through the multiplication of the average multiyear moduli of groundwater discharge from the basic aquifers and the relevant sizes of such subareas, and through the summation of the values obtained, total groundwater discharge to the sea was estimated. In total, the fresh groundwater inflow to the Baltic Sea from the territories of the former USSR is ~1.2 km³/yr. The calculations for groundwater discharge to the sea from freshwater-bearing complexes, for separate areas, and the sea basin as a whole are presented in Table 4.1.1.1.

The calculations show that the largest discharge to the sea from the upper water-bearing systems occurs in the coastal zone of the Silurian–Ordovician Plateau and the islands of Khuiumaa and Saaremaa, which are composed of highly fissured and karstic carbonate rocks and thus have high moduli of groundwater discharge (average annual modulus of groundwater discharge for a long-term period equal to 3.2 l/sec· km^2. Groundwater inflow to the sea from the zone of intensive water exchange is highest in the Quaternary and Ordovician water-bearing systems, 425.2 and 264.6 mln m^3/yr, respectively. The least amount of groundwater flow into the sea is observed in the Upper Permian and Cambrian–Ordovician systems, 6.3 and 22.0 mln m^3/yr, respectively (Babinets et al 1973).

Analysis of the general geostructural and hydrogeological conditions in the northwestern part of the Russian Platform shows that the deep groundwater discharge to the Baltic Sea is formed from the freshwaters of the Cambrian–Ordovician and Gdovsky aquifers, which are desalinated on the Estonian territory during the infiltration of precipitation through "hydrogeological windows," as well as from highly pressured and mineralized waters from Paleozoic and Mesozoic sediments. Discharge of deep confined groundwater occurs because of its seepage through low-permeable layers and the overlying aquifers of the coastal zone, and through the bottom sediments within the shelf zone of the sea. To delineate a coastal zone where discharge of the Cambrian–Ordovician aquifer is possible through seepage, piezometric profiles were constructed in several directions. These profiles correspond to a logarithmic curve, the equation of which has the following form: $y = a + b \ln [1/(x + c)]$. Coefficients of the a, b, c equation for each piezometric curve were calculated with a computer standard program.

The piezometric curves, plotted according to the obtained approximating equations, were extended further over the sea area to an absolute elevation of the seawater line, which revealed areas of the most intensive water discharge from the aquifers within the sea. The calculations show that discharge of confined groundwater of the Cambrian–Ordovician aquifer occurs over an area ~5.5 ths km^2 within a 20- to 25-km-wide strip, narrowing in some places to 10–15 km. In general, it can be assumed that the discharge of deep groundwater to the sea is small compared with groundwater discharge from the zone of intensive water exchange. However, quantitative assessment of deep groundwater discharge to the Baltic Sea can be made only after special investigations.

Besides the quantitative assessment of groundwater discharge to the Baltic Sea from the territory of the former USSR, the authors approximated an evaluation of the total groundwater discharge to the sea from the European territory was approximately assessed. The evaluation was carried out by the method described above. The moduli of groundwater discharge from the main water-bearing systems drained by the sea were taken from the literature. The drainage areas were determined by means of planimetry, based on the available hydrocontour lines and hypsometric maps.

The total discharge of fresh and low-brackish groundwater from the zone of intensive water exchange to the Baltic Sea from the European territory is on average ~10.3 km^3/yr, with about 3.6 mln m^3/yr of water discharge per 1 km of the shoreline. The area of groundwater drainage on the Baltic Sea coast, where the groundwater

TABLE 4.1.2.1
Groundwater Discharge to the Baltic Sea[a] (from the Territory of the USSR)

Geographic Region	Area of Groundwater Drainage (km²)	Water-Bearing System (Aquifer)	Area of Water-Bearing System (km²)	Groundwater Discharge to the Sea (mln m³/yr)	Total Mineralization (g/l)	Ion Discharge Total (t/yr)	Modulus, (t/yr·km²)	Linear Discharge (t/yr from 1 km of Shoreline)
Shoreline between the southern boundary of Neman River basin and state boundary of Russia and Poland	700	Quaternary	700	22.1	0.370	8200	11.6	130
		Paleogene	700	12.6	0.350	4700	6.7	
Western part of Neman River basin	2000	Quaternary	2000	63.1	0.340	21,800	10.9	265
		Cretaceous	2000	75.7	0.480	36,400	18.2	
Shoreline between Venta River and northern boundary of Neman River basin	2100	Quaternary	2100	66.2	0.350	23,200	11.0	220
		Upper Permian	250	6.3	0.400	2500	10.0	
		Upper Devonian	900	31.5	0.440	13,900	15.4	
		Upper and Middle Devonian	500	18.9	0.450	8600	17.2	
Shoreline between West Dvina and Venta Rivers	1700	Quaternary	1700	53.6	0.350	18,800	11.0	245
		Upper Devonian	1000	34.7	0.440	15,400	15.4	
		Upper and Middle Devonian	1300	50.4	0.490	24,700	19.0	

Shoreline between West Dvina and Pyarnu Rivers	1200	Quaternary	1200	37.8	0.350	13,200	11.0	210
		Upper Devonian	200	6.3	0.480	3000	15.0	
		Upper and Middle Devonian	1000	37.8	0.480	18,300	18.3	
Shoreline between Pyarnu and Keila Rivers, incl. Saaremaa and Khiiumaa Islands	4900	Quaternary	2400	75.7	0.360	27,200	11.3	2800
		Silurian	3900	182.9	0.470	86,000	22.0	
		Ordovician	4900	265.3	0.470	124,500	26.4	
Shoreline between Keila and Luga Rivers	1400	Quaternary	1400	44.2	0.370	16,300	11.6	85
		Cambrian–Ordovician	1400	22.1	0.450	10,000	7.2	
Shoreline from Luga to Neva Rivers	800	Quaternary	800	25.2	0.340	8500	10.6	47
Shoreline from Neva River to state boundary of Russia and Finland	1200	Crystalline rocks of basement	1200	37.8	0.350	13,200	11.0	98
			800	44.2	0.200	8800	11.0	
Total	16,000			1214.4		507,200		

[a]From the territory of the former USSR.

discharge to the sea is formed, is about 100 ths km². The average modulus of the groundwater discharge in this area is equal to 3.0 l/km²).

4.1.2 CHEMICAL GROUNDWATER DISCHARGE

Ion discharge to the Baltic Sea from the basic aquifers and water-bearing systems of the zone of intensive water exchange is determined by groundwater discharge and mineralization. The groundwater discharge is determined, using the means of the above-described method, by the average mineralization (from the chemical analyses of the water from wells and springs).

Ion discharge was calculated for the entire coastal zone of the Baltic Sea (within the shoreline of the former USSR). This zone was subdivided into nine areas for calculating. The subdivision was made taking into consideration similar geologic-hydrogeological conditions, close parameters of groundwater discharge, and similar values of average groundwater mineralization. Eight areas were located within the Baltic artesian basin, the fresh groundwater of which is enclosed in sedimentary sediments. Only the shoreline of the Gulf of Finland stretching from the Neva River to the boundary with Finland belongs to the Baltic crystalline shield, where fresh groundwater is enclosed chiefly in Pre-Cambrian crystalline rocks and loose Quaternary sediments. Calculation results are given in Table 4.1.2.1 and Table 4.1.2.2.

TABLE 4.1.2.2
Ion Groundwater Discharge to the Baltic Sea for Basic Aquifers and Water- Bearing Systems

Water- Bearing System (Aquifer)	Ground- water Drainage Area, km²	Ground- water Discharge (mln m³/yr)	Average Mineralization of Water (g/l)	Ion Groundwater Discharge (t/yr)	Average Modulus (t/yr·km²)	Total Ion Groundwater Discharge
Quaternary	13,500	425.2	0.355	150,400	11.1	29.6
Paleogene	700	12.6	0.350	4700	6.7	0.9
Cretaceous	2000	75.9	0.480	36,400	18.2	7.2
Upper Permian	250	6.3	0.400	2500	10.0	0.5
Upper Devonian	2100	72.8	0.455	32,300	15.4	6.4
Upper and Middle Devonian	2800	107.9	0.460	51,600	18.4	10.2
Silurian	3900	182.7	0.470	36,000	22.0	17.0
Ordovician	4900	264.6	0.470	1 24,500	25.4	24.5
Cambrian–Ordovician	1400	22.0	0.455	10,000	7.1	2.0
Crystalline rocks of basement	800	44.4	0.200	8800	11.0	1.7
Total	16,000	1214.4	—	507,200		100

The total groundwater ion discharge directly into the sea from the coastal area is equal to 16,000 km² which is ~507 ths t/yr. The average modulus of the ion discharge is approximately 31.7 t/yr. The length of the shoreline of the Baltic Sea within this territory is ~2500 km. The average salt transfer with groundwater discharge from the zone of intensive water exchange is estimated at 200 t/yr per 1 km of the coast.

The data reveal the main regularities in formation of the ion groundwater discharge to the sea. Sufficiently distinct here is the influence of different natural factors and, in the first turn, recharge and permeability of aquifers characterizing quantitatively the groundwater discharge, as well as lithological composition and solubility of water-bearing rocks determining groundwater mineralization. The highest moduli of the ion groundwater discharge to the sea (48.5 t/yr·km²) are typical of the coastal part of the Silurian–Ordovician plateau, where groundwater discharge to the sea is formed from water-bearing systems composed of Silurian and Ordovician karstic limestone and dolomites and loose Quaternary rocks. The ion discharge to the sea from this extensive territory is nearly half (46.8%) the total ion discharge to the Baltic Sea from the zone of intensive water exchange.

Aquifers with the highest moduli of ion groundwater discharge are composed of soluble carbonate rocks. Thus, the maximum moduli of the groundwater discharge, from the Ordovician and Silurian water-bearing systems composed of limestone and dolomites, amount to 25 and 22 t/yr·km², respectively. Unexpectedly, lower moduli of ion discharge to the sea (15–19 t/yr·km²) are typical of the Devonian water-bearing systems containing layers and interbeds of limestone. The water-bearing systems of marl-chalky and sandy sediments of the Cretaceous age have approximately the same moduli of ion discharge (18 t/yr·km²) of the Upper and Middle Devonian water-bearing systems. The lowest moduli of ion discharge are associated with the Paleogene (6.7 t/yr·km²) and Cambrian–Ordovician (7.1 t/yr·km²) aquifers, both of which have a small groundwater discharge directly to the sea.

The total ion discharge to the sea from individual aquifers depends also on the distribution area of these aquifers within the coastal zone where discharge is directed to the sea. Thus, mostly spread on the Baltic Sea coast is the Quaternary water-bearing system. The ion discharge from that system amounts to almost a third (29.6%) of the total ion discharge to the sea from all the water-bearing systems of the zone of intensive water exchange, despite its relatively low moduli of the groundwater ion discharge (10–11 t/(yr·km²)). The lowest values of ion discharge to the sea are formed in the Upper Permian aquifer to a limited extent.

Of interest is to evaluate the role of precipitation in the formation of groundwater ion discharge to the Baltic Sea. According to the data of Zverev (1968), the precipitation mineralization in the zone of sufficient wetting, within which the Baltic Sea coast is located, is equal to 0.02 g/l. The total groundwater discharge directly to the Baltic Sea equals 1.2 km³/yr, the total amount of salts brought with precipitation and transferred with groundwater discharge is 24.5 ths t/yr, i.e., ~5% of the total ionic groundwater discharge. Thus, the major part of dissolved salts transferred to the sea (~482.5 ths t/yr) is formed due to the weathering of soils and rocks, including chemical and biological processes.

Results of these investigations have given first-hand knowledge on the sizes and conditions of the groundwater discharge to the Baltic Sea from the upper hydrodynamic zone and the role of this discharge in the formation of the salt balance of the sea. The total ion discharge to the Baltic Sea from the territory covered by the former USSR is equal to 12.4 mln t/yr [Alekin and Brazhnikova, 1964]. The groundwater ion discharge directly to the sea is ~507 ths t/yr, i.e., ~3% of the total salts transferring with total discharge to the Baltic Sea. It should be kept in mind that the only ion discharge from the upper aquifers of the zone of intensive water exchange containing chiefly fresh groundwater was evaluated. Therefore, it can be assumed that the total groundwater ion discharge to the sea including groundwater discharge from deep aquifers containing water with a higher mineralization will be considerably greater.

4.1.3   Ecological Aspects of Estimation of Groundwater Discharge into the Gulf of Finland*

Groundwater discharge is an important part of seawater balance and plays a crucial role in the formation of ecological conditions in coastal areas. Unfortunately the ecological aspect of groundwater discharge is neglected in coastal environment research. We carried out an ecological assessment of the groundwater discharge into the Gulf of Finland, which is the shallowest and most vulnerable part of the Baltic Sea.

The Gulf of Finland lies in the eastern part of the Baltic Sea. The surface area is 29,700 km² and the average depth of the basin is 38 m. The countries of Finland, Russia, and Estonia border the coastal area of the Gulf of Finland. The main rivers of the basins are the Neva, Luga, Narva, and Plussa; and many springs enter these coastal waters.

The Gulf of Finland is on the junction of two major geological basement structures: the Baltic Shield and the Russian Platform. Overlying these basement structures are sedimentary rocks ranging in age from the Vendian to Quaternary. The near-surface crystalline basement lies only in the northern part of the gulf and dips steadily southward to reach a depth of hundreds of meters below the surface. The upper part of the basement rocks contain fissured water, which is sufficient only for local water supplies. The discharge zone of this groundwater is located in the northen part of the gulf. Vendian sandstone rests on the irregular surface of the crystalline basement. The Gdov aquifer is present in this formation. This layer is very well protected against surface contamination by overlying thick Kotlin clay. This horizon is widely spread in the Gulf of Finland. The Gdov aquifer is used for water supplies to the northen part of Petersburg. The aquifer depth varies from 100–150 m before sinking southward to deeper depths.

Along the southern boundary of the gulf Cambrian and Ordovician sandstone and clay lie beneath the Vendian rocks forming the Cambrian–Ordovician aquifer. The Ordovician limestone occurs in the south of the gulf and is known as the Izhora Aquifer, covering 3000 km². Ordovician limestone is greatly fractured and has only

---

* Section 4.1.3 was written by A. Voronov, E. Viventsova, and M. Shabalina (Saint-Petersburg State University, Russia, arkad@Avzoll.spb.edu).

a thin cover of glacial deposits, thus making the aquifer vulnerable to surface pollutants. A thin layer of uranium-rich Dyctionema shale underlies the aquifer. This aquifer is discharged by numerous karstic springs along the southern coastal zone of the gulf.

Quaternary glacial sediments cover the Paleozoic and Pre-Cambrian systems. These sediments are, in general, glacial and marine. They form confined and unconfined aquifers. All the Quaternary period aquifers discharge directly to the Gulf of Finland. The amount of groundwater discharge to the seas can be estimated by several geological and hydrogeological methods (Zektser et al. 1973). The methods are divided into two groups: methods based on investigation of the coastal drainage area and methods based on hydrogeological investigation of the estuary. The first group includes the hydrodynamic, combined hydrogeological, average long-term water balance, and modeling of groundwater discharge methods. The second group of investigatory methods includes approaches using recognition and determination of different anomalies. The most detailed results are obtained from combined or complex methods. Thus, the authors investigating the disharge area in the Gulf of Finland and Lake Ladoga have used several of these methods.

The coastal area of the Gulf of Finland can be divided into four zones with different geological, hydrogeological, and discharge properties (Table 4.1.3.1). The zone I stretches 180 km from the Finnish border to the Russian town of Primorsk. Geologically, the zone consists of basement rocks overlain by Quaternary sediments. The discharging groundwater has a low mineralization with an average value ~0.2 g/l. Groundwater in the urban areas of Primorsk and nearby Viborg is degraded by industrial contaminants such as oil, heavy metals, and sulfates.

Zone II extends from Primorsk to Sestroretsk (~110 km). The groundwater discharge in this zone is mostly from Quaternary aquifers. Some of them discharge along the coastal zone directly into the sea.

Zone III lies in the Greater Petersburg area and includes urban areas from Sestroretsk in the north to Lomonosov in the south. The coastal line extends for 75 km. Borders of the zone approximately juxtapose the line of the protection system against flooding. This is the reason that the eastern part of the Gulf of Finland is the most water-interaction isolated part of the sea and also the reason why ecologists are so interested is this region. One of the specifics of the zone is the presence of the largest river, the Neva, and the large number of islands. Groundwater discharge goes mainly into the Neva River in this area. The upper groundwater contains heavy metals and organic, radioactive, and petroleum compounds.

Zone IV extends for 180 km along the southern shore of the Gulf of Finland from Lomonosov to the Estonian border. Groundwater discharge is mainly derived from Ordovician karsts of the Izhora Plateau, 20–30 km from the coast. Approximately half of the discharging groundwater goes to the central water supply. The groundwater contains a high concentration of nitrogen.

The groundwater discharge is mostly dispersed, aside from a few zones of concentrated discharge. These zones have been discovered in the coastal zone near the Sestroretsk area and in Zelenogorsk, where there is probably concentrated discharge from buried valleys.

TABLE 4.1.3.1
Groundwater Discharge in Different Zones of the Gulf of Finland

Zone No.	Zone Length (km)	Zone Spread (km)	Zone Square (km²)	Module of Groundwater Discharge (l/sec·km²)	Value of Groundwater Discharge[a] (km³/yr)	Value of Groundwater Discharge[a] (%%)	Average Mineralization (g/l)	Value of CSD (t/yr)
1	180	2.5	450	0.8	0.0114	1	0.15	1703
2	110	20.0	2200	2.4	0.1665	12	0.20	33,302
3	75	20.0	1500	2.2	0.1041	7	0.80	83,255
4	180	50.0	9000	4.0	1.1353	80	0.50	567,648
Total	555		13,150		1.4172			685,908

[a]Into the Gulf of Finland
CSD, chemical substances discharge.

TABLE 4.1.3.2
Ion Discharge in Different Zones of the Gulf of Finland

Zone No.	Ion Discharge (t/yr)	Main Contaminants
1	1703	Radon
2	33,302	Iron, barium
3	83,255	Heavy metals, organic compounds, oil products
4	567,648	Nitrate
Total	685,908	

Groundwater discharge to the Gulf of Finland from the Russian part of the gulf equals ~1.4 km³/yr. Compare to the results of the project by Kimmo Peltonen (2002), where the groundwater discharge to the Baltic Sea equals 4.4 km³/yr. Groundwater discharge is uneven in the different zones. The greatest input of groundwater discharge to the Gulf of Finland is from the southern zone IV and the smallest from the northern zone I. Greater than 80% of the discharging groundwater value involves water from zone IV. Zone II provides 12% of the total discharging groundwater to the sea.

The groundwater in zone I is mainly pure (1% of total). The contaminated groundwater that discharges from zone III is ~7% of the total groundwater disharging to the gulf. Considering the presence of the Neva River with an input of 80 km³/yr to the gulf, the amount of other sources of water discharge may be low.

The assessment of ion discharge and main contaminants flowing with the groundwater allows the comparison of the amount of contaminants inflowing with surface and groundwaters. The ecological situation in the Gulf of Finland is determined by a combination of natural and anthropogenic factors. The transfer of different chemical components with industrial (or communal) wastes from Petersburg and effects of intensive navigation can be classified as anthropogenic factors. Wastes and sewage enter the Gulf of Finland with the water of the Neva River. Although they do not exceed 2% of the total water volume, the sewage significantly determines the quality of the water in the gulf. The waste enriches the gulf not only with biogenic matter, but also with suspended precipitate and different pollutants. The reason for this is that about a quarter of the waste of Petersburg flows into the gulf without any treatment. A great amount of suspended materials and pollutants from sewage precipitate in Nevskaya Bay. A considerable amount of oil (petroleum) products and heavy metals have accumulated on its bottom.

The present-day water quality of the Gulf of Finland does not conform to sanitary norms by different hydrochemical, hydrobiological, and bacteriological indicators. The range of contaminants varies significantly. The annual influx of contaminants in Nevskaya Bay, including atmospheric precipitation, is as follows: P_{total} 6328 t, N_{total} 68,625 t, petroleum products 6395 t, phenols 104 t, Pb 81 t, Cr 159 t, Cu 496 t, Mn 7845 t, Zn 826 t, Fe 176,333 t, Al 6426 t. Petroleum products, organic matter, elements of the Al-Fe-Mn group, nitrates, and sulfates are the dominant contaminants. For example, in the Lakhta region, an increased content of heavy metals in the bottom sediments and bottom water is observed. Also, the Cd content in this area reaches 12 µg/l, which exceeds by 12 times the MPC (maximum permissible concentration). This is probably connected with the presence of dumps and buried pits.

The role of groundwater flow in contamination of the Gulf of Finland has not yet been determined. The completed investigations show that the groundwater contribution to gulf contamination is not great. For example, the annual iron content inflow by groundwater in zone I is 34 t. This is 1000 times less than the iron contamination from river flow. The nitrate pollutant from groundwater flow in zone IV is ~15,000 t, which is 500 times less than the total amount of nitrogen compounds coming annually with contaminated surface flow. On the coast of the Gulf of Finland, waste storage, ash dumps, and unapproved sewages are a common sight. The coastal zone is also used for storing metal and site constructions. The sanitary zones are cluttered with communal and industrial wastes. The construction of unapproved ways and other buildings lead to the disturbance of the hydrogeological regime. The rate of groundwater contamination in each case depends on many factors (e.g., the relationship between surface and groundwater flow, the flow velocity, the filtration properties of media, the thickness of the aeration zone, the water composition, etc.).

The amount of annual ferrous pollutants from groundwater in zone I is ~8 t, which is 1000 times less than that from river flow. The amount of nitrogen compounds from groundwater is ~45,400 t, which is 100 times less than the total amount of nitrogen compounds from surface flow.

However, on the local scale groundwater contribution can be significant. This can cause a sharp decrease in the ecological fitness of the coastal zone where biota is actively developed. For example, discharging shallow groundwater in zone III (the Petersburg zone) often represents an anthropogenic solution enriched by heavy metals, organic matter, and radioactive and petroleum compounds that does not correspond to sanitary standards. It causes typological changes in flora and fauna of the coastal zone and irreparable damage of the coasts. The drastic state of water basins causes the emergence of alien species in plankton and benthos that leads to grave alterations in the biodiversity of water organisms. Such an example has been observed in the eastern part of the Gulf of Finland. Water basin contamination also causes the active disintegration of aquatic flora in the coastal zone from the water line to depths of 1.5 m. The geochemical barrier on the boundary of sea and groundwater along the coastal line causes the precipitation of some pollutants coming from urbanized territory and thereby enriches the bottom sediments.

The highly corrosive properties of groundwater and its level fluctuation contribute to the deterioration of coastal constructions. Furthermore, the difference in the dynamic of contaminants coming with surface flow and groundwater discharge should be considered. In practice it can be difficult to divide the way of water-area contamination (e.g., with surface flow or groundwater discharge), because of the high flow-type (velocity) of the Gulf of Finland and water masses dynamicity. Often the negative influence of contaminated groundwater flow is manifested after many years.

The anthropogenic influence in local and in regional scales is varied. The deterioration of coastal conditions as a whole, the change of flora and fauna, the increasing rate of morbidity among coastal populations, and groundwater contamination are a reflection of anthropogenic pollution. The close interconnection of all elements of the geosystem ensures its integrity. Vertical (inner) connections within the geosystem contribute to anthropogenic distribution from one element to another (e.g., from vegetation to soil and through soil to groundwater; from surface and groundwater to vegetation) and as a result causes changes in the complex as a whole. Contaminant accumulation occurs not only in the silt sediments of the reservoirs and in soil layers but also in the biota (vegetation, animals, and microorganisms) and in the water layer. Groundwater plays a significant role in the transfer of compounds, hence, in contaminant distribution.

Anthropogenic influence can cause long-term changes in the groundwater flow, especially if the groundwater regime is connected with overexploitation of aquifers. As a result of changes in groundwater regime, the groundwater inflow to the Gulf of Finland has been steadily decreasing. The Ordovician, Gdovsk, and intermoraine aquifers are the most intensively exploited aquifers within the study area. For example, in the Izhorskoe Plateau approximately half the total spring discharge (Ordovician aquifer) is used for potable water supply. The formation of depression cones in the Gdovsky and intermoraine aquifers and decrease in groundwater piezometric levels cause changes in the interaction between groundwater and surface-water.

The contamination of groundwater discharging into the Gulf of Finland takes place in the total catchment area. The sources of contamination can be subdivided into areal and point contamination. Atmospheric emissions and soil contamination due to agricultural activity are the areal sources of contamination. Point sources of contamination include industrial, atomic, mining, agricultural and cattle-breading complexes, approved and unapproved dumps, sewage discharge, docks, etc.

Atmospheric emission of sulfur dioxide, hydrocarbon oxide, nitrogen dioxide, oxides of hydrogen sulfide, organic compounds, ammonia, and heavy metals have been observed in the study area. Regional soil contamination is typical in areas of industrial and civil development. For example, the total area of civic development in the Kingiseppsk, Lomonosovsky, and Volosovsky regions are 9274, 9187, and 1114 ha, respectively. Increased concentrations of nickel, copper, zinc, arsenic, and cadmium are observed in the soils occupied by industries and settlements. In the industrial areas chromium can be added to these contaminants. Intensive soil contamination is observed in the territories close to the cities of Kingisepp, Slantsy, and Sosnovyi Bor. Emission from auto transport is also an example of regional contamination. In this case soil contamination has a linear character.

Groundwater and soil contamination can occur as a result of the use of mineral and organic fertilizers and pesticides. For example, in the Kingiseppsk region, total area of tillage is 16,296 ha. Annually 1660 t of mineral and 134 t of organic fertilizers (86 kg/ha) are introduced into this region; 3520 hectares of crops are treated by pesticides, including 2075 ha by herbicides against weeds, 536 ha against pests, and 1260 ha against illnesses. Point sources of contamination are mainly connected with the presence of large settlements. The Greater Petersburg area is a specific anthropogenic zone of pollution.

Influence of groundwater discharge in the Gulf of Finland has several levels. Regional influence of groundwater consists in inflowing of solved compounds into surfacewater. The salt balance of surfacewater is mainly connected with the chemical composition of groundwater. Usually the average mineralization of groundwater discharging into the gulf differs slightly from the mineralization of water in the gulf. However, on the local level, groundwater considered as anthropogenic solutions significantly influences surfacewater composition. This influence is chiefly observed in areas with the discharge of highly mineralized artesian water.

The module of ion discharge varies by latitude zoning. The latitude zoning consists in the increase of the module of ion discharge from the north to the south that is connected with increase of water mineralization in the upper hydrodynamic zone. The average module of ion discharge is 20–40 t/yr·km^2. An increase in the module of ion discharge occurs in the areas of easily soluble carbon rock.

According to completed investigations, an average module of ion discharge in the Gulf of Finland is 52.3 t/yr·km^2. Chlorine, sulfates, calcium, magnesium, and sodium are the main components of ion discharge. Total salt volume discharging with groundwater is insignificant in comparison with those in surfacewater. However, as mentioned above, the chemical composition of discharging groundwater plays a significant role on the local level. In this case groundwater discharge contaminates surfacewater.

The situation of Vasilievsky Island is an example of contamination of the Gulf of Finland by groundwater discharge. In the 1950s radioactive waste was buried in the zone of overmorain aquifer discharge. The burial of radioactive waste caused groundwater contamination by radionuclide that caused surfacewater and coastal zone contamination. The discharged groundwater of zone III (Petersburg zone) contains much anthropogenic solutions enriched by heavy metals, organic compounds, and radioactive and oil components.

Along the coastline of the Gulf of Finland, point sources of groundwater contamination have been determined. Among them are burials of different wastes, ash dumps, untreated sewage descent, unapproved dumps of domestic and industrial wastes, etc. Groundwater discharging into the surface reservoirs carries a great number of pollutants. The rate of groundwater contamination in each case depends on different factors (e.g., flow rate and discharge, filtration properties of the rocks, thickness of the aeration zone, etc.). A hydrochemical survey is necessary for determining aureole of contamination caused by groundwater discharge.

In practice it is difficult to subdivide the influence of surface and groundwater discharge on the coastal zone and defined areas of water due to the mixing of the water mass. The total contamination of the Gulf of Finland negatively influences the coastal zone. In most it causes irreplaceable damage for the revitalization abilities of the shoreline. Often the water quality does not satisfy the sanitary demands due mostly to its high microorganisms content. High aggressive groundwater causes destruction of the coastal constructions.

When a source of contamination is situated directly in the coastal zone, the purification properties of the rocks are negligible. Groundwater discharge, which starts in the foreshore zones or before that, leads to coastal zone contamination. The distribution of contamination flowing into the gulf depends on seawater dynamics that determine the amount of suspended matter entering the gulf and its redeposition. In areas of intensive water dynamics (e.g., zones of strong roughness, quasistational drift currents in shoaling water), material transferring and sorting occurs constantly. This material inflows jointly with river runoff, groundwater discharge, and because of the vital activity of organisms. Material transfer usually occurs from shoreline to several hundred meters off the coast.

The main hydrochemical indexes determining the situation in the defined water areas are oxygen content, biochemical oxygen demand (BOD), mineral phosphor, nitrate nitrogen, and ammonium nitrogen concentrations. Natural annual changes in concentration of these indexes are usually disturbed. Almost half the samples taken near the north and the south coastal zones of the Gulf of Finland showed values of BOD-5 significantly higher then normal (2.0 mg/l). A regime of groundwater discharge characterized by two maximum values in spring and autumn can be the reason of such disturbances.

The high rate of nitrogen and phosphorus contamination has led to the intensification of eutrophication processes in the ecosystem of the Gulf of Finland. These processes are especially active in the coastal areas where fila-mentous phage seaweeds and water flora are developed in the summer time. This makes coastal areas practically useless for recreational use. As a result of the intensive anthropogenic load, and the exceptionally large water withdrawal from the Gdovsky aquifer, a broad depression cone with a radius of 100 km was formed. As a consequence of piezometric level decreases, the character of the interaction between seawater and groundwater occurred. This statement can be evinced by the fact that the piezometric level of the Gdovsky aquifer in the coastal zone and aquatorial part is lower than the water level in the gulf. Seawater intrusion into the aquifer can occur by disjunctive dislocations and "lithologic openings." Such changes in interaction between ground and surfacewater lead to the rearrangement of the geosystem elements and mechanisms of intercon-nection of these elements.

4.2 GROUNDWATER DISCHARGE TO THE CASPIAN SEA

The recent water level variations in the Caspian Sea have brought considerable damage to different sections of the economy and of the living conditions in the cities, towns, and settlements of the surrounding coastal zone. This problem has been the subject of numerous publications, since changes in the water level have been observed for nearly two centuries. Since 1830, the water level has varied from −25.3 m to −29 m (Figure 4.2.1). Only in the 20th century did the Caspian Sea thrice experience a sharp change in its water level. In 1929 it was −26.0 m. Considerable decrease of sea level began in 1930, decreasing by 2 m by 1941 and equaling −29.02 m by 1977 (the greatest decrease observed in the past 200 years). Since 1978 the sea level began a gradual rise, reaching its maximum (−26.5 m) in 1991–1992. But in 1996–1997 the Caspian Sea level dropped by almost 34 cm, and by 2004 decreased again to −27.0 m (Khublaryan 2000; Panin et al. 2005).

During the steady decrease of sea level (1930–1977), the dried areas were used for the development of industries and municipalities. This turned out to be an error, because during the last rise of the sea level almost all the new sites were destroyed or faced the threat of being flooded. In Dagestan (Russian Federation), 270,000 people settled in the dried coastal areas. To evacuate them from these areas and to compensate them for material losses incurred has become a serious problem due to the difficult economic situation in the country and the reluctance of many of these people to leave their homes. Similar situations have also been observed in other Caspian Sea coastal states. Therefore, gaining a basic understanding of what causes sea level variations is necessary for the prediction of the level dynamics and assessment of its environmental, social, and economic consequences. However, existing research methods and available prediction data on sea level regime show rather contradictory results. According to some data, the sea level will rise by several meters in the next 10–20 years; whereas according to other data, it will decrease by 2–2.5 m (Khublaryan et al. 1995).

A number of hypotheses explain such sharp variations in the Caspian's sea level (neotectonic, climatologic, anthropogenic, etc.). Some geologists and hydrogeologists have linked the fundamental reasons of sea level variations with the geodynamic features of the sea basin, fluid breath the bottom sedimentary strata, and with tectonic processes of compression/extension that cause a considerable increase in groundwater inflow. Other specialists have hypothesized that the increase of groundwater inflow to the sea (up to 40–60 km^3/yr) due to oil and gas exploitation in the basin, underground nuclear explosions, and other tectonic reasons have caused a disturbance in the continuity of regional aquicludes and in intensity of SGD.

It should be noted that activization of geodynamic processes, manifested in an increase of horizontal tectonic stresses (compression), especially in the central and southern parts of the Caspian depression, can lead to a change in the sea bottom configuration. It is therefore necessary to carry out regular morphometric surveying of the Caspian depression bottom. The total value of the earth-crust movements cannot be characterized as a sum of annual values, because the annual values themselves have variable signs and are determined tentatively. A more accurate

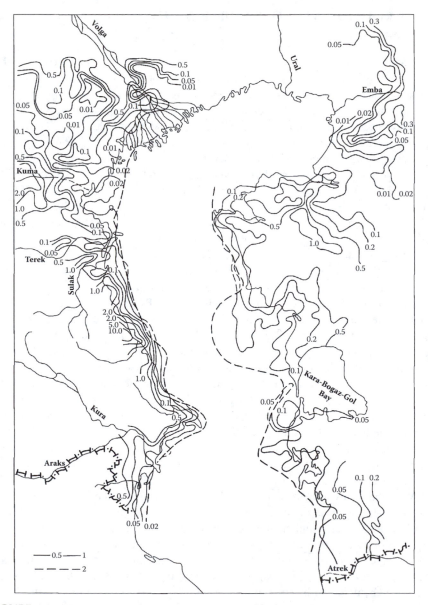

FIGURE 4.2.1 The map of vertical discharge of confined groundwater: (1) isolines of the module of vertical confined groundwater discharge(l/sec·km²); (2) conventional boundary of the areas of submarine discharge of confined groundwater into the sea.

assessment of the influence of geodynamic processes on the sea level over a long period requires carrying out sufficiently exact and regular (at least once a year) instrumental observations. Simultaneously, the application of aerocosmic methods could prove to be useful.

However, hydrologists and climatologists state, based on the analysis of the dynamics of water-balance components, that during the recent two decades a gradual decrease of average annual evaporation with the simultaneous increase of cyclone frequency has been observed in the Caspian Sea basin and water area. Thus, they conclude that during recent decades it is the hydrometeorological factors that have caused an increase in water content and cloud cover in the sea basin, a decrease in the seawater temperature in its northern shallow part, and, hence, a decrease in evaporation, which together are the main determinants of current sea level variations (Khublaryan 2000). At present, one of the basic problems of the Caspian Sea is the development of measures for keeping an optimal water level regime while taking into account social and economic requirements. To predict the water level of the Caspian Sea is possible only on the basis of a combined analysis of water-balance dynamics, neotectonic processes, and human activity within its basin.

It should be noted that the study of groundwater discharge to the Caspian Sea as an element of water balance will enable the resolution of a number of important theoretical and practical problems, including:

- Quantitative characterization of groundwater discharge to the sea and assessment of its role in the formation of water and salt balance of the sea
- Evaluation of the amount of dissolved chemical compounds transferred with groundwater
- Prediction of multiyear changes in groundwater discharge under the influence of changes to other water-balance elements and human activity
- Evaluation of groundwater discharge in particular areas of the sea coastal zone
- Zoning of the most active groundwater discharge to the sea

Analysis of the works on the evaluation of groundwater discharge as a component of the Caspian Seawater balance (Dzhamalov et al. 1977) shows that the data reported on the groundwater discharge to the Caspian Sea and its role in the water balance is extremely different, rather subjective, and even contradicting. The data of different authors differ from each other by over 150 times (from 0.3 to 49.3 km^3/yr). The main reason for this discrepancy is the lack of reliable hydrogeological information and the frequent use of unreliable calculation methods.

4.2.1 NATURAL CONDITIONS OF THE COASTAL ZONE

The Caspian Sea depression is elongated in the meridional direction and has a length of about 1200 km, with a maximum width of about 430 km, and a minimum width of 196 km. The current level of the Caspian Sea is 27 m lower than the level of the World Ocean. The length of its shoreline is approximately 6000 km and the area is about 400 ths km^2 (Panin et al. 2005). The Caspian Sea depression shall be understood hereinafter in the text as an area of the sea and its coastal territory is confined by watersheds, boundaries of water-bearing systems, and geological structures.

Three depressions are distinguished in the Caspian Sea — northern, central, and southern divided, respectively, by the thresholds of the Mangyshlak and Apsheron

Peninsulas. The northern depression (North Caspian water area) is rather shallow with a depth ~5 m and an area ~80 ths km^2. The central depression (Central Caspian water area) is an asymmetric kettle with a steep slope in the west and gentle one in the east. Its maximum depth (788 m) is in the Derbent depression near the western coast. The Central Caspian water area is equal to 138 ths km^2. The southern depression (South Caspian water area) is also asymmetric with steep western and southern banks. Its maximum depth in the Kurinskaya kettle reaches 1025 m. Owing to its large depth, the South Caspian area contains about two-thirds of the entire water in the sea.

Over 100 rivers flow into the Caspian Sea, but they are distributed irregularly over the coast. The largest rivers — Volga, Ural, and Terek — flow into the North Caspian Sea area, with 83% of the total river discharge contributed by the Volga River and only 5% by the Ural and Terek Rivers. The numerous rivers from the western coast supply about 7% of the annual river discharge. The largest among them are Sulak, Samur, and Kura. The rivers of the Iranian coast discharge to the Caspian Sea the remaining 5% of the total river discharge. The eastern coast has no constant water discharge. The annual total river discharge had been gradually increasing during recent years and in 2000–2002 reached approximately 300 km^3/yr. However, the average multiyear river discharge to the Caspian Sea by the end of the 20th century amounted to 285 km^3/yr or 790 mm (Panin et al. 2005).

The climate of the Caspian Sea coast is rather diverse. On the western coast, it varies from the climate of semideserts and dry steppes in the north to subtropical in the south. On the eastern coast, the desert climate prevails. The maximum precipitation (1700 mm/yr) falls in the southwestern part of the coastal territories. On the western coast, the average annual precipitation reaches 300–400 mm with the most abundant rains in autumn. On the eastern coast, precipitation does not exceed 100 mm/yr. This amount increases to the north and reaches 200 mm/yr. The Iranian coast is characterized by high humidity. On the slopes of the Elborus Mountain, precipitation amounts to ~2000 mm/yr.

The Caspian Sea depression is a complicated formation consisting of platform-like and geosynclinal structures. The North Caspian area is located within the southeastern ending of the Russian Pre-Cambrian Platform and occupies a part of the Epihercynian Platform. The Central Caspian area occupies a part of the Epihercynian Platform and a part of the alpine geosynclinal zone. The South Caspian area is entirely located in the alpine folded zone.

The northern part of the territory under study is located within the Caspian syneclise. In the south, it is confined by the Don-Mangyshlak system. Faults and regional flexures serve as the boundary between them. The crystalline basement in the center of the syneclise lies at a depth of 10–14 km. The Caspian syneclise represents a large artesian basin.

Further southward is another large structure, the Karpinsky Arch, being a part of the Don-Mangyshlak folded system. It stretches in the latitudinal direction, running under the bottom of the Caspian Sea in the form of islands, banks, and the North Kulandinsky Ridge, and is exposed on the eastern coast as the Buzachinsky Dome. The Karpinsky Arch is dissected by numerous tectonic disturbances and has a considerable hydrogeological significance. It taps and drains groundwater flowing from the Caucasian Mountains.

Far to the south on the western coast, there is a large structural element — the Tersko-Caspian marginal trough separated from the Karpinsky Arch by the Manychsky trough and connected with the latter by a sublatitudinal fault. The depth of the basement in the Tersko-Caspian marginal trough reaches 10,000–12,000 m. The trough is composed of a very thick sedimentary stratum chiefly Meso-Cenozoic. The trough is opened toward the Caspian Sea and is traced along its bottom as the Derbent depression. The Tersko-Kumsky and Dagestan artesian basins are distinguished within this area.

The next large structure of the western coast is located between the mega-anticlinorium of the Major Caucasus and the Caspian Sea depression — this is the Kusaro-Divichinsky foredeep composed of low-dislocated Upper-Pliocene and Quaternary sediments having a large thickness.

The eastern coast of the Central Caspian water area lies within the Turanskaya Plate confined in the north by the South Embensko-Primugodjarsky fault. In the south, the area is separated from the alpine folded zone by the Fore-Kapetdag and Krasnovodsk-Balkhansky faults and Apsheron threshold. Within this territory, the Central Caspian and Mangyshlak artesian basins have been delineated.

The border between the Central and South Caspian areas is the Apsheron threshold that is an extension of the Apsheron Peninsula, crossing the Caspian Sea toward the city of Nebit-Dag. A large structure of the western coast of the South Caspian water area is the Kurinskaya depression opened toward the sea. In the north, it is bordered by a zone of faults of the Kobystanskaya tectonic zone, in the southwest by the Talyshsky mega-anticlinorium.

The eastern coast of the South Caspian water area is characterized by submergence of the Kopet-Dag folded zone. The largest element here is the South Caspian depression occupying the West Turkmenian Lowland and the adjacent part of the sea. It is a synclinal structure opened toward the sea and in the hydrogeological respect is the West Turkmenian artesian basin. The southwestern border of the South Caspian water area is the Talyshsky mega-anticlinorium composed of Paleogene volcano-genic-terrigenic strata.

In the south, on the Iranian territory the Elborus mega-anticlinorium is located there.

The territory under study has sediments of all the systems, but systematic hydrogeological data are available only on Mesozoic and Cenozoic sediments. Within the territory of the Caspian Sea coast, the formation, movement, and discharge of groundwater is a rather complicated hydrogeological process dependent on many natural factors. As commonly known, the formation of groundwater discharge directly depends on the geologic structure, relief, and climatic and hydrogeological conditions in a region. The Caspian Sea coast is located within the southeastern ending of the Russian Pre-Cambrian Platform, Epihercynian Platform, and alpine geosynclinal area. Relief forms of the Caspian Sea coast are very diverse and consist of extensive flat lowlands and piedmont plains, plateaus, and mountainous ridges. The Caspian Sea coast covers several climatic zones changing from deserts in the east, semideserts and dry steppes in the north, to subtropics in the

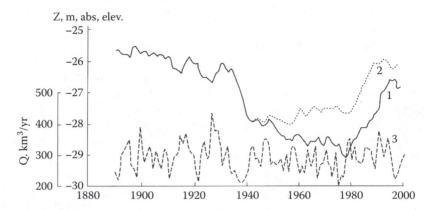

FIGURE 4.2.1.1 Changes in the Caspian Sea level and total runoff of inflowing rivers: (1) real sea level; (2) renewed level (where water consumption in the basin remained at the level of 30 is from the 20th century); (3) total river runoff, Q.

south. The complicated geologic-structural, geomorphologic, and climatic conditions of this region exert a determining influence on formation of groundwater discharge. The zoning of the Caspian Sea coast was carried out to distinguish areas with the most homogenous natural conditions. The basis of hydrogeological zoning was the subdivision of the coast into widely known artesian basins and hydrogeological zones (Figure 4.2.1.1). The hydrogeological conditions of the defined areas are described below.

4.2.2 QUANTITATIVE ASSESSMENT AND PREDICTION OF SUBMARINE GROUNDWATER DISCHARGE TO THE CASPIAN SEA

The quantitative assessment of groundwater discharge to the Caspian Sea was carried out with the use of hydrogeodynamic methods. The use of such methods to solve such a large regional problem like assessment of groundwater discharge to the Caspian Sea along its entire perimeter and for deep aquifers was restricted chiefly by a lack of sufficiently complete initial geological and hydrogeological information. Data, which have been published on the hydrogeological conditions of the region under study, are limited (localized) both in plane and cross-section.

The use of hydrodynamic methods enabled the collection of primary geologic-hydrogeological information for the entire coastal zone of the Caspian Sea, including data for more than 3500 hydrogeological wells. Through analyzing this material, the following basic water-bearing systems were distinguished, which have a regional spreading within the coast and contribute the main part to the groundwater balance of the sea.

- Shallow aquifer in modern and Quaternary sediments
- Water-bearing system of Quaternary sediments
- Water-bearing system of Upper Pliocene sediments
- Water-bearing system of Sarmatian sediments

- Water-bearing system of Middle Miocene sediments
- Water-bearing system of Cretaceous sediments

For the distinguished water-bearing systems, maps of water levels and water conductivity were compiled, which reflect general regularities in formation of the groundwater discharge to the Caspian Sea.

Total groundwater discharge to sea from the systems over the entire shoreline of the Caspian Sea was determined on the basis of Darcy's relationship for the groundwater flow rate. The calculations were fulfilled by flow net. Here, the basic hydrogeological parameters were not averaged within large areas, but were taken directly from the appropriate hydrodynamic maps. This traditional method, though simpler, makes it possible to obtain a sufficiently reliable value of groundwater discharge to the sea from each water-bearing system. Special attention should be paid to the reliability and representativity of the initial hydrogeological parameters and hydrodynamic maps, and in connection with the latter, to the processing and analysis of primary geologic-hydrogeological information of the mentioned maps. The maps show a number of calculating areas with similar hydrogeological conditions within the basic hydrogeological zones. The flow rate within these areas was calculated by flow nets, using hydrogeological parameters that are typical only of a relatively narrow groundwater flow net from each water-bearing system.

The width of the flow nets was longer than 30 km, amounting chiefly to 15–25 km. The hydrogeological parameters within all the considered flow nets undergo expected changes along the flow, i.e., from the piedmonts toward the sea, water productivity decreases, causing the flattening of the piezometric surface. This is explained by a change in facie composition of the terrigenic sediments composing the water-bearing systems under study, i.e., the coarse-clastic rock material is replaced there by fine-grained sandy sediments in direction from the piedmonts to the central areas of the Caspian Sea depression (Klenova and Solovieva 1962). Such a situation is observed both within the coast and the seawater area. Along the strip width, the hydrogeological parameters actually do not change. This was due to the specified selection of the flow nets. Thus, the values of the calculated parameters were not averaged for large areas, but determined more precisely and in more detail for each flow net.

As mentioned above, as we approach the Caspian Sea in almost all the water-bearing systems, the piezometric surface flattens and the hydraulic capacity of rocks decreases. The latter is caused by a decrease in the thickness of water-bearing layers due to an increase of clayey interbeds in the total geological cross-section, facie replacements of different-grained sands by clayey fine-grained sands in terrigenic formations, as well as by the attenuation of rock jointing. All this indicates that closer to the shoreline, the groundwater discharge decreases, which is explained by natural and artificial discharge of the groundwater within the land. However, artesian wells are present in the Caspian Sea coast; therefore, the natural discharge of confined water occurs through seepage upward through the dividing clayey beds. This is confirmed by the increasing pressure heads with increasing depth of the aquifers in the coastal zone. Areal seepage of groundwater is possible both within the land and seawater area, where the piezometric levels of the artesian aquifers are located considerably higher than the seawater line. To delineate this part of the sea, piezometric profiles for several flow lines are plotted on the basis of pressure head distribution maps in a water-bearing system. Then, these profiles are extended into the seawater area. To do this, it was necessary to find the y = f(x) function that describes the profiles with sufficient accuracy. Such approximation of the piezometric curves of different types was fulfilled by the least squares method using typical programs.

Through equations of approximating curves, the piezometric profiles can be extended into the seawater area to any specified point (Figure 4.2.2.1). Such extrapolation of factual data in space on the basis of the obtained relationship is usually prone to certain errors. However, the insufficient hydrogeological knowledge of the

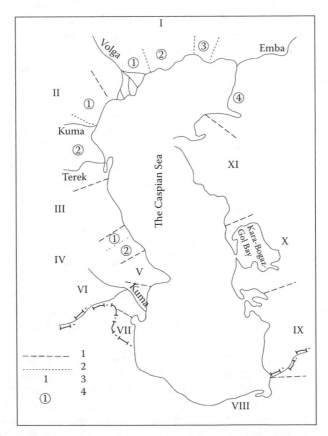

FIGURE 4.2.2.1 The schematic map of the hydrogeological zoning of the Caspian Sea shore: (1) boundary of the hydrogeology areas; (2) boundary of hydrogeology sections; (3) the number of the area; (4) the number of the section. Areas: (I) Pricaspian artesian basin, (II) Tersko-Kumsky artesian basin, (III) Dagestan artesian basin, (IV) Kusaro-Divichinsky piedmont plain, (V) the area of the southeast submergence of the Major Caucasus, (VI) Kurinskaya depression, (VII) Lenkoranskaya lowland, (VIII) Iranian shore, (IX) West Turkmensky artesian basin, (X) eastern part of the Middle Caspian artesian basin, (X) Mangyshlak artesian basin.

submarine sediments under consideration forces the researchers to extrapolate experimental data obtained on land. To check the precision of the approximation of
piezometric profiles and the selection of reference parameters, horizontal sectional
models were constructed for several piezometric profiles, also taking into consideration seepage processes. By changing the filtration resistance of the dividing beds
in the model, it was concluded that the latter coincided with the appropriate piezometric profile of the aquifer under study. When the most complete juxtaposition of
the modeled and real piezometric curves were achieved, the obtained filtration
coefficients of the dividing beds were used for the horizontal filtration models. The
groundwater balance was calculated for the most representative directions of groundwater discharge to the sea in each calculating block of the horizontal sectional
models. At the end of the profile models, the boundaries with constant pressure
heads were specified: a maximum pressure head on land and a minimum one (–27 m)
in the sea. The values obtained for horizontal and vertical groundwater discharges
were also controlled by the areal distribution of pressure heads along the modeled
profile. These approaches enabled the compilation of sufficiently accurate piezometric maps of the aquifers within the sea, which were later used to determine
groundwater discharge to the sea, and to study its distribution both in plane and
cross-section over the entire Caspian Sea coast.

The groundwater discharge was estimated on stationary quasi 3-dimensional
filtration models with the use of the finite difference method for the determination
of hydrogeodynamic conditions of artesian basins (modification of the MODFLOW
Code). This method enabled (at a specified distribution of pressure heads and water
conductivities over an area) us to obtain the horizontal and vertical components of
groundwater discharge at any measuring point, i.e., to give a quantitative assessment
of groundwater seepage. Taking into consideration that in our case the assessment
was carried out for six water-bearing systems, the coefficients of groundwater seepage and vertical discharge were calculated in succession beginning with the lowest
one. The boundary conditions in plane were specified along the measured isopiestic
lines. The water conductivity values that were set into the measuring points of the
grid were taken directly from the wells drilled into the aquifer systems under study
or from the appropriate maps by means of interpolation. The groundwater discharge
values obtained in this way need to be checked and verified. For this purpose, within
a measuring area, the flow nets are constructed along several representative directions, and within these flow nets the horizontal and vertical components of the
groundwater flow are determined. The comparison of these values with the values
obtained in the same points on the plane model determines the reliability of the
latter. As mentioned above, this book includes the calculations by flow net for the
entire coast of the Caspian Sea, except Iran.

The total groundwater discharge to the Caspian Sea, determined by the Darcy
relationship, is equal to 3.2 km^3/yr. Data for particular regions, areas, and water-
bearing systems are presented in Table 4.2.2.1.

While calculating the groundwater discharge to the Caspian Sea, the focus
should be concentrated on its possible change under the influence of sea level
variations. It is acknowledged that the Caspian Sea depression has been occupied
by the sea since the ancient geological epochs, but the sizes and water regimes

TABLE 4.2.2.1
Salt Transfer with Groundwater Flow to the Caspian Sea

Area No.	Water-Bearing System	Flow Rate (km³/yr)	Calculated Mineralization (g/l)	Salt Transfer (ths t/yr)
I. Caspian artesian basin	Shallow aquifer	0.024	2–25	350
	Ancient Quaternary	0.01	20	200
	Upper Pliocene	0.012	30	360
	Cretaceous	0.006	10	60
	Total	0.052		970
II. Tersko-Kumsky artesian basin	Shallow aquifer	0.025	10–25	250
	Ancient Quaternary	0.025	4	100
	Upper Pliocene	0.025	2–20	50
	Sarmatian	0.05	30	1500
	Middle Miocene	0.025	30	750
	Total	0.15		2650
III. Dagestanian artesian basin	Shallow aquifer	0.05	5	250
	Ancient Quaternary	0.1	5	500
	Sarmatian	0.15	5	750
	Middle Miocene	0.2	15	3000
	Cretaceous	0.07	30	2100
	Total	0.57		6600
IV. Kusaro-Divichinskaya piedmont plain	Shallow aquifer	0.65	1	650
	Ancient Quaternary	0.13	1	130
	Upper Pliocene	0.12	6	720
	Total	0.9		1500
V. Southeastern subsidence of Major Caucasus	Shallow aquifer	0.15	10	1400
	Ancient Quaternary	0.05	10	500
	Upper Pliocene	0.02	10	250
	Total	0.22		2200
VI. Kurinskaya depression	Shallow aquifer	0.004	10	40
	Ancient Quaternary	0.01	10	100
	Upper Pliocene	0.006	10	60
	Total	0.02		200
VII. Lenkoranskaya Lowland	Shallow aquifer	0.005	2	10
	Ancient Quaternary	0.03	2	60
	Total	0.035		70
VIII. Iranian coast		1.00	5	5000
IX. West-Turmeninan artesian basin	Shallow aquifer	0.05	30	1500
	Ancient Quaternary	0.02	20	400
	Total	0.07		1900

X. Eastern part of Central Caspian artesian basin	Shallow aquifer	0.02	8	160
	Ancient Quaternary	0.007	10	70
	Upper Pliocene	0.002	10	20
	Sarmatian	0.002	10	20
	Middle Miocene	0.002	10	20
	Cretaceous	0.002	20	40
	Total	0.035		330
XI. Mangyshlaksky artesian basin	Shallow aquifer	0.03	20	600
	Sarmatian	0.135	7	950
	Middle Miocene	0.008	10	80
	Cretaceous	0.026	10	260
	Total	0.2		1900
Sea Coast Total		3.25		23,320

of the sea have since changed many times. Two different periods of the Caspian Sea can be distinguished in its geological history: the first includes the longest early period of the existence of the open sea connected to the Mediterranean basin, and the second one includes the present-day existence of the closed Caspian Sea. It has become a lake not joined to the Black Sea due to the closure of the Manychsky Strait in the Late Khvalynskaya epoch. The above-mentioned periods of the sea history were confirmed once more by the results of the recent drillings into the bottom of the Central and Southern Caspian water areas, carried out by a joint Russian-French project. The investigations of the isotope compositions of oxygen and carbon in the carbonate cores (with a height of 10 m) taken from the near-bottom sediments have helped to verify the scales of the sea level variations and the changes in the salt content of the seawater during the last 25,000 years, caused by geologic-tectonic and climatic processes (Ferronsky et al. 1999).

As mentioned above, during the current period of the Caspian Sea, its water level continues to undergo variations of ~1 m caused by the earth's crust movements, changes in climatic conditions and surface runoff, as well as by human activity at waterwell fields. Thus, the tentative assessment of the groundwater discharge was carried out at seawater levels of −26 and −28 m. With an increase in the sea level by 1 m, the total submarine discharge will slightly decrease at the expense of some flattening of water levels in the upper aquifers. Otherwise, the submarine discharge will increase approximately by the same total value (0.1 km³/yr) with a decrease in the sea level by 1 m. If it is assumed that on the Caspian Sea coast the evaporation from the Quaternary aquifers will increase because of the expansion of the coast area or the rise of shallow groundwater to the surface, and that the slopes of the groundwater flows will not change greatly, then it can be expected that at water level variations by 1 m the groundwater discharge to the Caspian Sea will change slightly.

4.2.3   Regularities in Formation and Distribution
of Groundwater Discharge to the Caspian Sea

This section contains the description of the hydrodynamic scheme of groundwater formation, movement, and the possible ways of discharge in the Caspian Sea depression, based on the completed investigations and analysis of existing opinions on the mechanism of groundwater discharge.

As previously noted, the Caspian Sea depression is located in various climatic, geologic-structural, and tectonic conditions, and incorporates a system of different-order artesian basins. The calculations of the groundwater discharge to the sea show that the major part of the groundwater discharge (2 km^3/yr) is formed within the western seacoast. This is explained by the favorable physical-geographic and geologic-hydrogeological conditions in this part of the coast, which is characterized by relatively high precipitation, considerable differences in elevations, the coarse-grained composition of terrigenic sediments, and by the increased jointing of hard rocks in some areas. The relative proximity of the recharge areas in the Caucasian area of the coast could be another factor.

The hydrogeological regions that are the richest in water on the western coast are the Dagestanian artesian basin and Samur-Kusarchaiskoye interfluve where the groundwater discharge to the sea is equal to 0.6 and 0.9 km^3/yr, respectively. The major part of the discharge within the Dagestanian artesian basin falls in the Sarmatian and Middle Miocene water-bearing systems (0.35 km^3/yr). This is connected, alongside favorable climatic conditions, with the high water conductivity and high pressure-head gradients. In the Samur-Kusarchaiskoye interfluve, the major part of the groundwater discharge is formed in the Quaternary aquifers (0.75 km^3/yr), which is explained by the favorable conditions of recharge and high water abundance of these coarse-grained sediments. A considerable portion of the groundwater discharge to the sea is formed within the Tersko-Kumsky artesian basin and the southeastern subsidence of the Major Caucasus (respectively, 0.15 and 0.25 km^3/yr).

The least amount of groundwater discharge on the western coast is registered in the area of the Kalmykian steppes (0.01 km^3/yr), in the Kura River (Kurinskaya) depression (0.02 km^3/yr), and on the Lenkoranskaya lowland (0.03 km^3/yr). This can be explained by the unfavorable conditions of recharge of the water-bearing systems, the low gradients of the groundwater flow, the sandy-clayey composition of water-bearing rocks, and the notable predominance of clays in some areas. The poor hydrogeological knowledge of deep aquifers, especially in the Lenkoranskaya lowland does not allow assessment of the groundwater discharge across the entire cross-section.

The eastern coast of the sea has heterogeneous conditions of groundwater formation and discharge to the sea, which amounts to 0.35 km^3/yr. Such a small groundwater discharge is caused, first of all, by the hot and dry climate, the low hydraulic capacity of water-bearing rocks over the major part of the coast, and by the slight slopes of piezometric surface, except in the Mangyshlaksky artesian basin where the discharge reaches 0.2 km^3/yr. Karst limestone of the Sarmatian water-bearing system is developed here where the major part of groundwater discharge is formed from this hydrogeological region — 0.15 km³/yr. Within the West Turkmenian artesian basin, the groundwater discharge does not exceed 0.1 km³/yr. This is connected with the presence of sandy-clayey sediments, with the predominance of clays and the slight slopes of the groundwater flow. A small part of the groundwater discharge (0.05 km³/yr) is formed in the east of the Central Caspian artesian basin.

The groundwater inflow to the sea from the upper part of the Caspian artesian basin does not exceed 0.05 km³/yr. This area is characterized by low relief, unfavorable conditions of groundwater recharge because of the dry climate, and the occurrence of loam and loess sediments near the surface. The groundwater in the region is drained to a considerable extent by the Volga and Ural Rivers.

It can be seen from the given condensed analysis of the conditions of the formation of groundwater discharge to the Caspian Sea that its natural factors within the western, eastern, and northern coasts exert a different influence on the groundwater discharge from the main water-bearing systems. The groundwater discharge to the Caspian Sea is formed under the influence of the interaction among the climate, relief, structural-hydrogeological, and hydrodynamic factors. On the western coast, which is characterized by favorable climatic and geomorphologic conditions of groundwater recharge, the main influence on the groundwater discharge directly to the sea comes from geologic-lithological and hydrodynamic factors. The reason for this is that a considerable part of the discharge is formed from relatively deep water-bearing systems, since the Quaternary Period. The influence of physical-geographic factors is lessened by the distance of the recharge areas and the thick overlying strata. The main influence is from hydrogeological factors: high water content, relatively good hydraulic capacity, high pressure heads, and slopes of the piezometric surface which quantitatively determine groundwater discharge. These conditions of the formation of groundwater discharge are probably typical of the Iranian coast as well. However, on the Iranian coast a more significant role is played by the climatic factors because of the large amounts of precipitation and favorable conditions for its infiltration.

On the eastern and northern coasts of the Caspian Sea, the groundwater discharge is formed chiefly under the influence of climatic and geomorphologic factors. The small amount of precipitation, high evaporation, and the extensive occurrence of low-permeable sediments near the surface create unfavorable conditions for groundwater recharge and the formation of regional groundwater flow. A considerable influence on the groundwater discharge is exerted by numerous deep drainless depressions (*sors*: partly submerged valleys) and eroded entrenchments of temporary water streams, which catch the groundwater flow. The Mangyshlaksky artesian basin is an exception to the general trend due to the presence of karst formations. A major part of precipitation accumulates in the karst and fissured limestones of the Sarmatian water-bearing system. This causes a relatively high groundwater discharge to the sea from these sediments.

As mentioned above, groundwater discharge within the sea area and within the sea coast occurs mainly because of water seepage upward through dividing low-permeable beds. This is confirmed, in particular, by the increasing pressure heads of groundwater with depth in the coastal zone, the flattening of the piezometric surface of all the water-bearing systems as they approach the shoreline, and by the deterioration of the hydraulic capacity of the water-bearing rocks in the same direction, which leads to a decrease of flow rate and decrease in groundwater velocity. The areal groundwater discharge through seepage does not preclude its concentrated outflow on the sea bottom in the form of submarine springs in areas of tectonic disturbances and exposed rocks. However, the existing data indicate that the latter kind of discharge is possible only within small places of the sea bottom where tectonic disturbances are near the present-day surface and cover a sufficiently thick hydrogeological cross-section. Other sources of SGD are areas of mud volcanism, the role of which is considered by some researchers as significant in the formation of groundwater contribution to the water balance of the Caspian Sea. However, the groundwater discharge through mud volcanoes on the sea bottom is a local process that is difficult to determine quantitatively and requires further study from the marine hydrogeology perspective. At the same time, the quantitative assessment of water seepage shows that this process determines the dynamics of entire artesian basins or their large parts. The interaction was assessed between the water-bearing systems and groundwater seepage within the coastal zone and sea area on the basis of the above method. The obtained values of vertical discharge were compared with the total groundwater discharge obtained on the basis of the Darcy relationship.

The areas where seepage of infiltrated groundwater was most intensive were distinguished by means of the extrapolation of the piezometric profiles of each water-bearing system to an absolute elevation mark of 27 m. But due to the inaccuracy of the extrapolation, the boundary of the most intensive groundwater discharge in the sea is subjective. The seepage processes probably cover more extensive areas, but reach their maximum values within the defined areas of the coastal zone. This conventional boundary of SGD through seepage on the western coast lies at a distance from the shore from a few kilometers for the Quaternary water-bearing system to 100 km for the Sarmatian system. On the eastern coast this band expands in some places to 120 km for the Sarmatian and underlying water-bearing aquifers.

The areas of SGD through seepage amount to, on the western coast 29 ths km^2 for the Quaternary, 23 ths km^2 for the Upper Pliocene, 31 ths km^2 for the Sarmatian, and 10 ths km^2 for the Middle Miocene water-bearing systems; on the eastern coast 9 ths km^2 for the Quaternary, 20 ths km^2 for the Upper Pliocene, 34 ths km^2 for the Sarmatian, 38 ths km^2 for the Middle Miocene, and 40 ths km^2 for the Cretaceous water-bearing systems.

The quantitative assessment of the interconnection between the water-bearing systems shows that the major discharge of groundwater by seepage occurs within land areas where the moduli of vertical flow varies from units to 0.5 l/sec·km^2, gradually decreasing toward the sea to 0.2 l/sec·km^2. This is because in the zones adjacent to the recharge areas or close to them, the water-bearing systems are mostly water abundant and possess a considerable vertical pressure-head gradient, whereas the low-permeable beds dividing them are thin and filled with sand. The seepage coefficient is on average $5 \cdot 10^{-5}$ l/day; vertical discharge reaches a few tens of thousands of cubic meters per day from the measuring area of 100 km^2.

Therefore, only a relatively thin layer of groundwater flow reaches the sea, the discharge of which is possible (as the carried-out calculations show) at the expense of seepage into the previously defined sea areas. The moduli of vertical flow here vary from 0.2–0.1 l/sec·km², gradually decreasing toward the conventional boundary of the distinguished areas to 0.05 l/sec·km². The coefficients of seepage as a rule do not exceed $5 \cdot 10^6$ 1/day, decreasing in some places to $5 \cdot 10^{-7}$ 1/day. This is an indication of the substantial thickness and the high clay content of the dividing beds. The total discharge of confined groundwater through seepage in the coastal zone under assessment is ~1 km³/yr. This value does not include the coastal zone of the Caspian artesian basin and the Cretaceous aquifer system on the western coast.

The total groundwater discharge, obtained according to the relationship of Darcy, to the sea within the area under calculation, except for the entire shallow aquifer and Cretaceous aquifer system on the western coast, amounts to 1.2 km³/yr. As a result, almost the entire flow of confined groundwater directed from the surrounding land to the sea can be discharged at the expense of seepage directly within the sea area. The analogous comparison of the vertical discharge and the lateral flow rate along the shoreline was carried out for each aquifer system. It showed a good agreement of SGD values determined by the two methods. Thus, for the Cretaceous aquifer system within the eastern coast, the flow rate along the shoreline on the eastern coast amounts to 0.03 km³/yr, i.e., the entire flow is discharged by seepage. The vertical groundwater discharge from the overlying aquifers is greater than the groundwater flow rate along the shoreline from each aquifer system. This is due to the fact that when assessing the interconnection of water-bearing systems, each of them includes not only the seepage formed within the system under measurement, but also the total vertical groundwater discharge from the underlying aquifers (Figure 4.2.3.1).

The total SGD by seepage from the Quaternary aquifer system (being the upper one in the accepted measuring scheme) must be close to the total value of

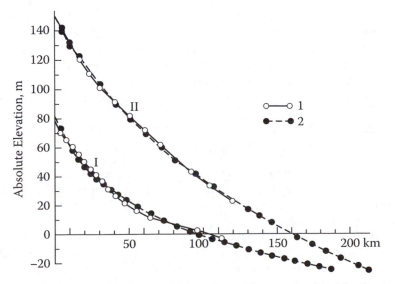

FIGURE 4.2.3.1 Piezometric profiles of the aquifers of the coastal zone of the Caspian Sea: (I) Upper Pliocene aquifer of Tersko-Kumsky artesian basin; (II) Middle Miocene aquifer of the Mangushlak artesian basin; curves of the piezometric profiles: (1) actual, (2) approximate.

groundwater discharge to the sea from all the water-bearing systems within the area under measurement. Thus, on the western coast the flow rate of all the confined water-bearing systems amounts approximately to 1 km³/yr, and groundwater seepage from the Quaternary system directly to the sea reaches 0.85 km³/yr.

The eastern coast can be conventionally subdivided into two measuring areas: a northern area (up to Kara-Bogaz-Gol), including the aquifers from the Cretaceous to Sarmatian systems, and a southern one where the hydrogeological cross-section begins with the Upper Pliocene aquifer system. In the northern measuring area, the flow rate for these water-bearing systems amounts to 0.18 km³/yr, and the SGD by seepage reaches 0.16 km³/yr. Within the southern area these values are almost the same and amount to 0.5 km³/yr.

The slight disagreement in the values of the vertical flow rate on the western coast and within the northern measuring area of the eastern coast may be caused both by possible groundwater discharge along tectonic disturbances and exposed bedrocks and by the inaccuracy of the calculations of the interconnections between the water-bearing systems.

The investigations of the groundwater discharge to the Caspian Sea is closely connected with the studies of the processes and conditions of the interaction between infiltrating groundwater, sedimentation waters, and modern seawaters. The data on the dynamics of the groundwater in large artesian structures adjacent to the inland seas and large lakes, in particular to the Caspian Sea, show that the major groundwater flow from the aquifers is directed seaward. The discharge of infiltrated groundwater seeping through low-permeable covers of aquifers and sea floor sediments occurs in the part of the sea where piezometric levels are located much higher than the seawater line so that sufficient pressure-head gradient for vertical groundwater discharge can be created. As a rule, the coastal zones have water-bearing systems with a direct relationship of pressure heads when below-lying aquifers are discharged into overlying ones and then the total vertical flow is drained by sea. Under these circumstances, the entire infiltrated groundwater from the coast is discharged within the sea area. This process is greatly favored by a hydrostatic pressure head created by sedimentation waters occurring at different depths. This in part depends on the geologic-structural conditions of the water-pressurizing system.

4.2.4   Role of Modern Tectonic and Mud Volcanoes in Water and Salt Balances of the Caspian Sea

On the floor and coasts of the Caspian Sea, there are scores of mud volcanoes which are, in principle, separate or group sources of subaqueous groundwater discharge from deep aquifers. Therefore, it is important to assess the role of mud volcanism in the water and salt balance of the sea.

According to data of morphometric surveys, 142 active mud volcanoes were registered on the Caspian Sea floor (Yakubov et al. 1983). As a rule, they are associated with zones of tectonic disturbances; in particular, they are stretched as a chain along the deep-seated faults of the Apsheron threshold. The mud volcanoes are formed in a sedimentary cover containing thick clayey sediments, in conditions favoring creation of high anomalous pressures (tectonic motions, generation of gases with decay of organic substances, inflow of overheated waters along zones of tectonic disturbances, etc.). These conditions are present in the South Caspian depression. The total thickness of the sedimentary cover of the depression is 20 km. The roots of the mud volcanoes are formed mainly in the upper structural layer of Cenozoic sediments. These sediments include a stratum of intensively dislocated sandy and clayey rocks complicated by numerous diapir folds, with a total thickness of 6–7 km in the peripheral areas to 12–14 km in the central part of the depression. The presence of tectonic rupture dislocations in the sedimentary cover of the South Caspian depression creates favorable conditions for the inflow of overheated mineralized waters and hydrocarbonous gases from beneath, a necessary condition for the formation of mud volcanoes. The volcano cores composed of brecciated rocks reach in plane large sizes (6–8 km) (Kulakova and Lebedev 1983).

The tentative quantitative assessment of water inflow from mud volcanoes to the Caspian Sea was first performed in 1972 (Glazovsky et al. 1976). It was established that at a calm regime of mud eruptions the total inflow to the sea does not exceed 0.001 km^3/yr; at an intensified regime (about two eruptions per year) discharges of submarine waters reach 0.01 km^3/yr. The maximum salt amount transferred to the sea by the waters from the mud volcanoes is $2 \cdot 10^5$ t/yr.

The investigations of the submarine mud volcanoes of the South Caspian depression have been fraught with difficulties due to the significant depths (≥ 800 m) and the absence of appropriate measuring equipment. The intensity of a volcano activity can be seen only by the height of the volcano breccia varying from 0.5–500 m and higher. At the present time a sufficient amount of information has been collected on active continental mud volcanoes, which by analogy (with some assumptions) can be extrapolated over submarine mud volcanoes. This enables us to carry out a more detailed assessment of water inflow and dissolved substances transferred to the sea from submarine mud volcanoes. However, in submarine conditions one should take into consideration the pressure of the height of the seawater column, which exerts a certain influence on eruptive processes (Rakhmanov 1987). The water volume ejected from the submarine volcano crater differs from the total ejected from a continental volcano. The quiescent stage of mud volcanoes activity (0.5 l/min) is characterized by low water content. In such a regime, 172 acting on-ground mud volcanoes annually erupt $8.6 \cdot 10^4$ m^3 of water (Rakhmanov 1987). If this value is extrapolated over 142 submarine mud volcanoes, without taking into account the seawater column, the estimated submarine water discharge will be $7.1 \cdot 10^4$ m^3/yr.

It has been determined that the water content in hard rock material erupted from mud volcanoes is equal to 28% of the total volcano breccia mass. For example, the amount of hard rock eruptions from mud volcanoes in Azerbaijan varies from $1 \cdot 10^4$ to $(17–40) \cdot 10^7$ m^3/yr. If we take the density of volcano breccia as 1.85 t/m^3, then the average mass of erupted hard rock material can reach $2.2 \cdot 10^8$ t. Thus, the maximum erupted water volume can amount to $0.6 \cdot 10^8$ m^3. The analysis of the activity of mud volcanoes shows that during the recent 50 years two to three eruptions are observed per year. It means that the maximum erupted water volume does not exceed 0.18 km^3/yr (Rakhmanov 1987; Zektser et al. 1994).

It can be assumed that during the eruptions a relatively small part of submarine water is discharged directly to the sea basin in order to relieve excess pressure.

A large amount of water enclosed in the brecciated rocks of the volcano core is discharged to the water-bearing systems (aquifers) of the Pliocene–Quaternary structural stage, which may be manifested in the changes in the hydrochemical composition of these systems. If the volcano core is schematized in the form of a tube with a radius of 3–4 km and length of 1–2 km and the minimum active porosity of brecciated rocks is defined as ≈5%, then the water amount in the core will be ≈3 km³. Thus, mud volcanoes, despite the low water contents vented, serve as powerful accumulators of mineralized groundwater.

The presented results have an approximate character and indicate only the scales of this phenomenon, the role of which in the water balance of the Caspian Sea is extremely low. Despite this, the submarine eruptions can exert a notable influence on the chemical composition of seawater and the formation of anomalous hydrochemical zones. From the available analyses, it can be surmised that the concentrations of some chemical components and the total mineralization of the erupted waters vary within wide ranges. The average total mineralization of erupted waters from continental mud volcanoes of the Apsheron group is 13.3 g/l; the Baku Archipelago (Azerbaijan), 44 g/l. Extrapolating these data over submarine waters of the submarine mud volcanoes, with a frequency of eruptions of not more than 3 times per year, the estimated salt transfer to the Caspian Sea will reach 2.4–7.9 mln t/yr, which is approximately 9% of the salts transferred with the river runoff.

4.2.5 Chemical Submarine Discharge and Its Influence on Sea Salt Balance

The chemical submarine discharge was determined along the entire perimeter of the Caspian Sea, which was subdivided into 15 measuring areas. Each area is a part of the artesian basin or shallow groundwater and is characterized by individual hydrogeological conditions, values of calculated parameters, and by approximately the same average mineralization of groundwater. The assessment of the chemical submarine discharge was carried out using average values of mineralization and the calculated amount of salts transferred with groundwater flow (see Table 4.2.2.1). Below is a brief description of the measuring areas and the calculated results of transferred salts with the groundwater flow to the sea.

1. In the area from the Ural River to the Volga River, the groundwater discharge from the shallow aquifer amounts to ~0.002 km³/yr. The groundwater has a mineralization equal, as a rule, to over 20 g/l. At a specified calculated mineralization of 25 g/l, the chemical discharge from the shallow aquifer amounts to 50 ths t/yr.

2. In the Volga River delta the groundwater discharge from the sediments of the modern and Upper Quaternary age amounts to 0.001 km³/yr. The groundwater of these sediments has chiefly low mineralization (3 g/l) and contains in the upper part of the aquifer system lenses of freshwater (Hydrogeology of the USSR 1966–1972). At a mineralization of 2 g/l, the chemical discharge is 2 ths t/yr. The groundwater discharge from the ancient Quaternary aquifers is 0.01 km³/yr. At a mineralization of 20 g/l,

the chemical discharge is 200 ths t/yr. The groundwater discharge from the Upper Pliocene aquifer system reaches 0.012 km^3/yr. The average mineralization is 30 g/l; total chemical discharge, 360 ths t/yr. The total chemical discharge in the area of the Volga River delta is 562 ths t/yr.

3. The coast from the Volga River to the Kuma River is located within the Caspian Lowland. The average groundwater discharge from the Quaternary sediments is 0.008 km^3/yr. The groundwater has a mineralization equal to 15–20 g/l, but in some places it increases to 50–100 g/l. The chemical discharge at a specified mineralization of 25 g/l is 200 ths t/yr. The groundwater discharge from the Upper Pliocene aquifer system in this area is equal to 0.007 km^3/yr, calculated mineralization is 20 g/l (Hydrogeology of the USSR 1966–1972). At these values, the chemical discharge amounts to 140 ths t/yr. The total groundwater discharge in the area of the Caspian Sea coast between the rivers of Volga and Kuma from the Quaternary water-bearing system and sediments of Apsheron amounts to 0.015 km^3/yr with a total chemical discharge of 340 ths t/yr.

4. In the coastal area (~300 km long) from the Kura River to the city of Makhachkala, the shallow groundwater discharge from the modern and Upper Quaternary sediments amounts to 0.025 km^3/yr, the ancient Quaternary sediments to 0.025 km^3/yr, the Upper Pliocene sediments to 0.025 km^3/yr, the Sarmatian sediments to 0.05 km^3/yr, and the Middle Miocene sediments to 0.025 km^3/yr. The total groundwater discharge from all of these aquifer systems in the given area reaches 0.15 km^3/yr. According to literature data, the calculated mineralization of the shallow aquifer can be accepted as equal to 10 g/l, in the ancient Quaternary aquifer system to 4 g/l, the Upper-Pliocene system to 2 g/l, the Sarmatian system to 30 g/l, and the Middle Miocene aquifers to 30 g/l (Hydrogeology of the USSR 1966–1972). At these values of average mineralization, the chemical discharge from the shallow aquifer amounts to 250 ths t/yr, the Quaternary sediments to 100 ths t/yr, the Upper Pliocene system to 50 ths t/yr, the Sarmatian aquifer system to 1500 ths t/yr, and the Middle Miocene water-bearing system to 750 ths t/yr. The total groundwater discharge from the listed aquifer systems amounts to ~2650 ths t/yr.

5. The area from Makhachkala to the Samur River represents a piedmont plain. Data on groundwater and chemical discharge within this area are given in Table 4.2.5.1.

6. The coastal area from the Samur River to the Divichai River represents a piedmont plain dissected by a great number of small rivers. The groundwater discharge from the Quaternary aquifers amounts to 0.75 km^3/yr. At a specified mineralization of 1 g/l, the chemical discharge is equal to 750 ths t/yr. The discharge from the Upper Pliocene system is 0.12 km^3/yr. The chemical discharge, at a mineralization of 6 g/l, is 720 ths t/yr. The total groundwater discharge from the approved aquifer systems within this area is about 0.9 km^3/yr, and total chemical discharge reaches 1500 ths t/yr.

TABLE 4.2.5.1
Ion Groundwater Discharge

Water- Bearing System	Groundwater Discharge (km³/yr)	Calculated Mineralization (g/l)	Ion Discharge (ths t/yr)
Dagestanian Artesian Basin			
Shallow aquifer	0.04	5	200
Ancient Quaternary	0.1	5	500
Sarmatian	0.13	5	650
Middle Miocene	0.2	15	3000
Cretaceous	0.07	30	2100
Total	0.54		6450
Krasnovodsky Peninsula			
Shallow aquifer	0.02	8	160
Ancient Quaternary	0.007	10	70
Upper Pliocene	0.002	10	20
Sarmatian	0.002	6	12
Middle Miocene	0.002	10	20
Cretaceous	0.002	20	40
Total	0.035		322

7. In the area of the Apsheron Peninsula to the Maraza River (the southeastern subsidence of the Major Caucasus), the groundwater discharge from the shallow aquifer amounts to 0.15 km³/yr, the Quaternary aquifer system to 0.05 km³/yr, and the Upper Pliocene system to 0.02 km³/yr. The mineralization values there are very diverse, ranging from 2–20 g/l and greater (Hydrogeology of the USSR 1966–1972). At the specified mineralization of 10 g/l, the total chemical discharge is 2200 ths t/yr.

8. In the area stretching between the rivers of Maraza and Yardymla (Kurinskaya depression) the groundwater discharge from the Quaternary sediments amounts to 0.014 km³/yr and from the Upper Pliocene system to 0.006 km³/yr. The mineralization is equal to 10 g/l. The total chemical discharge from the aquifer systems within this area is 200 ths t/yr.

9. The area from the Yardymla River to the town of Astara (Lenkoranskaya Lowland) has a length of ~60 km. The groundwater discharge from the Quaternary sediments is equal to 0.035 km³/yr. The groundwater mineralization varies from 1–3 g/l (Hydrogeology of the USSR 1966–1972). The chemical discharge to the Caspian Sea, at a mineralization of 2 g/l, is estimated at 70 ths t/yr. Thus, the estimated total groundwater discharge on the western coast of the Caspian Sea (areas 1–9) is ≈1.9 km³/yr, with the total chemical discharge from the aquifers and aquifer systems under study ≈14,000 ths t/yr.

10. The area stretching from the Atrek River to the Krasnovodsky Peninsula is located within the West Turkmenian artesian basin. The groundwater discharge to the Caspian Sea is calculated only for the Quaternary sediments and amounts to 0.07 km³/yr. Over the major territory the groundwater has a mineralization of 20–50 g/l and in some places 100 g/l. At a calculated mineralization of 30 g/l for the shallow aquifer and 20 g/l for the ancient Quaternary aquifer system, the chemical discharge to the Caspian Sea from this area is equal to 1900 ths t/yr.

11. On the Krasnovodsky Peninsula the groundwater discharge is calculated for the Quaternary, Neogene, and Cretaceous aquifer systems. Data on the groundwater discharge, mineralization, and chemical discharge are given in Table 4.2.5.1.

12. The coastline from Kara Bogaz Gol Bay to the Gulf of Kochak (Mangyshlaksky artesian basin) consists of Quaternary water-bearing, sandy-clayey sediments, carbonate rocks of the Neogene, and terrigenic formations of the Cretaceous. In the coastal zone, the shallow groundwater discharge amounts to 0.03 km³/yr. The water in the Quaternary sediments has a mineralization of 1–20 g/l, sometimes to 100 g/l. At a calculated mineralization of 20 g/l, the chemical discharge from these sediments amounts to 600 ths t/yr. Groundwater discharge from the water-bearing system in cavernous karst limestones of the Sarmatian stage amounts to 0.15 km³/yr. The mineralization of the groundwater varies from 1–8 g/l and greater (Hydrogeology of the USSR 1966–1972). At a calculation mineralization value of 6 g/l, the chemical discharge amounts to 900 ths t/yr. The groundwater discharge from the Middle Miocene water-bearing system is not large and amounts to 0.01 km³/yr. At a mineralization of 10 g/l, the chemical discharge is equal to 100 ths t/yr. The groundwater discharge from sands and sandstones of the Senomanian-Albian aquifer reaches 0.03 km³/yr. The mineralization of the groundwater in the coastal zone hovers chiefly ~10 g/l. At this mineralization, the chemical discharge amounts to 300 ths t/yr. The total chemical discharge to the Caspian Sea depression from the Quaternary, Sarmatian, Middle Miocene and Cretaceous water-bearing systems of this area reaches 1900 ths t/yr.

13. The Caspian Sea coastal area from the Gulf of Kochak to the Ural River delta stretches along the sea for about 600 km and is located within the Caspian Lowland. The groundwater is associated with the submarine sediments of the Quaternary Period. The total groundwater discharge in these sediments amounts to 0.02 km³/yr. The predominant water mineralization in this area is estimated at 15–20 g/l. At a mineralization of 15 g/l, the chemical discharge from the Quaternary sediments amounts to 300 ths t/yr. The groundwater discharge from the Cretaceous water-bearing system reaches 0.006 km³/yr. The groundwater mineralization of the Senomanian-Albian aquifer varies from 0.5–15 g/l and greater. At a mineralization of 10 g/l, the chemical submarine discharge is equal to 60 ths t/yr. The total chemical discharge for all the estimated water-bearing systems in the given area is 360 ths t/yr.

14. The delta of the Ural River occupies a part of the seacoast stretching for about 30 km. The groundwater discharge from the modern and ancient alluvial delta sandy and sandy-clayey sediments amounts to 0.001 km^3/yr. The water has a mineralization value reaching 10–15 g/l, and in lower places to 1–3 g/l. At a mineralization of 8 g/l, the chemical submarine discharge from the alluvial sediments of the Ural River is equal to 8 ths t/yr. For the eastern coast of the Caspian Sea, the total groundwater discharge reaches approximately 0.35 km^3/yr, and the total chemical discharge from this area is equal to 4200 ths t/yr.

15. On the Iranian coast of the Caspian Sea, the total groundwater discharge from the aquifers is accepted as equal to 1 km^3/yr. Due to the lack of data, the average mineralization is taken as equal to 5 g/l by analogy with some of the areas on the western coast of the Caspian Sea characterized by similar hydrogeological conditions. In this case, the chemical discharge within the Iranian coast will amount to 5000 ths t/yr. Thus, the total groundwater discharge along the entire perimeter of the Caspian Sea, including the Iranian coast, reaches approximately 3.5 km^3/yr. The total chemical discharge amounts to about 23 mln t/yr, including ~14 mln t/yr from the western coast of the Caspian Sea, 4.2 mln t/yr from the eastern coast, and ~5 mln t/yr from the Iranian coast. The given data show that the major submarine chemical discharge is formed within the western seacoast. This is connected with more intensive groundwater discharge on the western coast (~2 km^3/yr) as compared with the eastern coast (0.35 km^3/yr), which is explained by more favorable physical-geographic and structural-hydrogeological conditions of groundwater discharge formation on the western coast.

The distribution of the salts transferred with groundwater in the vertical cross-section looks as follows: from the upper hydrodynamic zone, including the water-bearing systems of the Quaternary sediments, the groundwater discharge along the sea coast within the territory of the former USSR amounts to 1.5 km^3/yr; its chemical discharge is equal to 7.5 mln t/yr. The groundwater discharge from deep aquifers is 0.8 km^3/yr with a chemical discharge of around 10.5 mln t/yr. Thus, despite the decrease of the groundwater discharge with depth, the salt transfer increases. This is related to the higher mineralization of the deep groundwater (Table 4.2.5.1).

Thus, in the Kusaro-Divichinskaya piedmont plain, groundwater discharge to the sea from the Quaternary sediments is 6 times higher than the discharge from the Upper Pliocene sediments, whereas the salt transfer is approximately the same from both water-bearing systems. This is because the groundwater in the Quaternary sediments is relatively fresh with a mineralization of to 1 g/l, whereas the average mineralization of the Upper Pliocene water-bearing system is 4 g/l. The exception is the Mangyshlaksky artesian basin where the groundwater discharge from the karstic Sarmatian limestones (0.15 km^3/yr) is 5 times higher than the discharge from the shallow aquifer (0.03 km^3/yr). Their average mineralization values are equal to 6 and 20 g/l, respectively.

The current salt balance of the Caspian Sea is formed mainly at the expense of salt transfer from surface and underground runoffs. Despite low mineralization of the riverwaters (about 0.3 g/l), the salt transfer with the surfacewaters reaches 86 mln t/yr. This is connected with the large discharge (average multiyear value of 285 km^3/yr) of the rivers flowing into the Caspian Sea. The submarine chemical discharge amounts approximately to 27% of the salt transfer by the rivers, whereas the calculated groundwater discharge is a little bit more than 1% of the total river runoff.

The presented data show that the role of the groundwater discharge, despite its low value, is significant in the formation of the total salt balance of the Caspian Sea.

4.3 GROUNDWATER DISCHARGE TO THE ARAL SEA*

For the last four decades interest in studying the Aral Sea has grown sharply. This is connected with the catastrophic drying up of the sea. First, assessments of groundwater discharge to the Aral Sea as a part of its water and salt balance were considered to be very important in the hydrogeological study of this region. This work was carried out from the beginning of the 1960s by different specialists using various methods, including numerical modeling. In the last decade, the perspective of the total disappearance of the Aral Sea and growing anthropogenic desert with a draining bottom has created problems of salt deposition and ejection by wind on adjacent territories. These cause significant ecological, agricultural, medical, and social problems in the Priaral area. Thus, territories adjacent directly to the Aral Sea are considered to be a zone of ecological catastrophe because of a sharp decrease in the natural conditions and in groundwater quality and the near-surface atmosphere.

4.3.1 NATURAL CONDITIONS OF THE COASTAL ZONE AND SEA

The Aral Sea is uniquely situated within the arid and semiarid bioclimatic zone of the Earth. The atmospheric precipitation amounts to 100–200 and 300–400 mm/yr. This area belongs to the African–Asia arid region (Strakhov 1962).

4.3.1.1 Orography

The sea is located in the central part of the greatest closed drainage of the Eurasia–Turanskaya Lowland and this determines the characteristics of the natural conditions of groundwater discharge of the Aral Sea basin and Priaral region. The main part of the lowland is occupied by the valleys of different genesis with the total inclination to the Aral Sea. It is circled by mountains of different altitude: the Mugodzhar on the northwest, Ulutau on the northeast, and west parts of the Tan-Shan Mountain on the south and east. They represent the outcrop basement of the Ural–Siberian epipaseozoic platform where the Turanskaya plate and similar lowlands are located Kazakhstan 1969).

The western border part of the Priaral area is represented by an uninhabited desert plateau called Ustyurt which has total flattening and inclination in the

---

* Section 4.3 was written by V.V. Veselov and V.I. Poryadin (Institute of Hydrogeology and Hydrophysics, Kazakhstan Republic, v_panichkin@mail.kz).

southwest direction. Here several vast depressions (e.g., depressions Sam, Acman-tay-Matay, and Karbulak) with sand and salt marsh massifs are located. Along the shoreline of the Aral Sea plateau Ustyurt is bordered by chinks inaccessible for passage.

The North Priaral area is situated to the north of the Aral Sea, in the northwest corner of the Priaral area and borders Mugodzhar (the Urals continuation) to the south. This area is formed by a low plateau separated by table residual mountains with clearly expressed gradation. They alternate with lowered deflation basins with numerous solonchaks and sandy massifs of the large and small Barsuki and Priaral Karakumy. Geomorphologic structures of the North Priaral area are an excellent example of inverted relief forms, where negative relief forms correspond to the axis of the positive forms and vice versa.

The northeast corner of the Priaral area is bordered by the Ulutausky low-mountains and isolated hills with an altitude up to 1133 m. It is a body of a destroyed mountainous system and is characterized by shallow but split separation. River systems, including the Kalmakkyrgan, run off to all sides of Ulatau and dry up in summer. The Kalmakkyrgan River flows into the vast acrid Lake Shelkar-Tengiz that is the lowest point of the North Priaral area (~80 m).

The flat, undifferentiated plain is situated between the North Priaral area and Ulutau (the southeast end of Central Kazakhstan). To the south it smoothly passes into moderately domed, slightly separated Nizhne-Syrdariinskoe uplift with a max-imum altitude up to 266 m.

The broad Syrdariinskaya alluvial-delta plain occupies the central sublatitude part of the East Priaral area. It is cut through by the terraced valley of the Syrdariya River, running off from the West Tan-Shan to the Aral Sea, and its old riverbeds Zhamdary, Kuvandarii, etc., forming ancient and modern deltas of the Syrdariya River.

The main morphogenetic types of modern relief control both lithological com-position of the basement and cover rocks and the direction and intensity of water-salt discharge of the Priaral part of the Turnaskaya plate. They are the following types (Kazakhstan 1969): erosion and denudation low-mountainous areas and pla-nation surfaces that border the Ulutau, Karatau, Bukantau, and Sultan-Uizdag moun-tainous systems; layer-denudation elevations, plateau and plains with solonchak internal-drainage basins; alluvial-delta plains of Amudariya and Syrdariya and their old riverbeds with piedmont plain and proluvial circuits (piedmonts of Karatau); lacustrine plains situated in the interfluvial areas of Syrdariya and Chu-Sarysu; marine, lacustrine, and alluvial rewinnow plains and atmogenic drifted plains Kyzyl-kum, Priaral Karakumy, Aryskumy, Large and Small Barsuki, and sandy massifs.

The depth of relief ruggedness determines not only regional, but the local character of groundwater (confined or unconfined) level fluctuations, their intensity in water exchange, quality, resources, ecological conditions, and possibility of their use for economic purposes. The highest depths of relief ruggedness are typical of the low-mountainous areas and ancient planation surfaces (100–400 m) and areas of low rounded isolated hills of residual elevations (60–m). The smallest depths of relief ruggedness are typical of modern piedmont (1–2 m) and ancient (20–50 m) accumulative plains; modern alluvial plains (1–5 m), lacustrine and marine plains

(1–3 m,) and rewinnow and atmogenic drifted plains (3–15 m). Erosion and denudation plateaus and plains occupy intermediate positions.

The altitude of the Aral basin surface decreases from +53 m up to 16.5 m in the eastwest direction. There the deepest part of the Aral basin, named the Predustyurtsky trench, is noted. Two main zones of uplifts are clearly noted in this area: submeridian and sublatitude. Sublatitude elevation divides the Areal Sea basin in two unequal parts: the Small Sea on the northeast and the Large Sea. These seas are divided by the Strait of Berga. Submeridian elevation also divides the Large Sea into two unequal parts: the west (the small and deep sea) and the east (the large and shallow). Both large islands and numerous small ones are noted in the sea. At present, due to the sea level decreasing, islands extend to the west and south and form a unified drained zone (Figure 4.3.1.1).

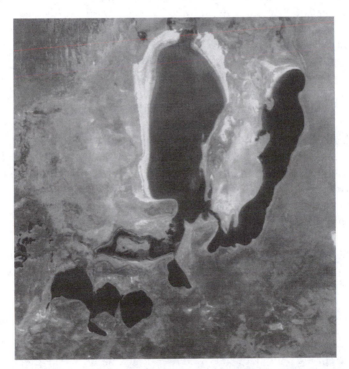

FIGURE 4.3.1.1 (Please see color insert following page 236.) Aerospace image of the Aral Sea in September 2004 (according to data from the Institute of Space Explorations, Kazakhstan). In the right upper corner is the Small Sea; in the center, the Large Sea, divided by the sublatitude Kulandinsko-Nizhneamudariinskoe elevation system on two parts: the west (deep-sea) and the east (shallow); to the right from shallow (east) part, in the elevation zone and also along the fringe of the Small Sea it is noticed the drained bottom of the Aral Sea (1960–2004). The Syrdariya River delta with the lake system is shown in the right boundary of the photograph. In the right lower corner is the Kyzylkum Desert, adjoining the Akpetkinsky Archipelago on the northwest. In the lower boundary is the lower part of the Amudariya River delta which penetrates to the drained bottom. Along the left boundary of the photograph is the Ustyurt plateau located on the west of the deep-sea part of the Large Sea. White color indicates salt deposits.

The sea shorelines are morphologically various. The west coastline is formed by the Ustyurt plateau with an altitude of >200 m and is slightly indented. The north one is also high and abrupt. The coastlines of the east and south parts of the basin are lower, formed by the Syrdariya and Amudariya River deltas and are highly indented by narrow and small gulfs and bays.

4.3.1.2 Climate

The climate of the Priaral area is sharply continental with significant variations of annual and day temperatures. The year range of average monthly temperature reaches 30–40°C (Atlas of Kazakhskaya SSR 1982; Kazakhstan 1969). The main characteristics of this region are large amounts of heat and light.

The high radiation promotes high evaporation from the earth surface (100–200 mm/yr) and leads to the deep groundwater level position, its high mineralization, and soil salinization. The high evaporation from the water surface reaches 900–1200 mm/yr, and within the Aral Sea it equals 900–1100 mm/yr. During vegetation evapotranspiration is more than 2000 mm/yr. This value is significantly greater than the atmospheric precipitation which average annual value varies 80–90 mm in the southwest (in the Priaral Kyzylkumy), to 160–370 mm in the north, and to 400–650 mm in the piedmont and mountainous areas on the southeast (Atlas of Kazakhskaya SSR 1982; Kazakhstan 1969).

The small amount of atmospheric precipitation causes low air humidity, especially in summer (15–35% in the south and west and 40–45% in the north and within the Aral Sea area). In winter its value increases significantly and reaches 60–85%.

The thickness of snow cover, its duration, and changes in regime by seasons and different relief forms are of great importance in the formation of surfacewater and groundwater runoff. According to the average long-term observations, the maximum snow cover thickness within the Aral Sea basin varies from several centimeters in the south, southwest, and west and to 15–20 cm in the north and southwest. In the piedmont and mountainous areas it reaches 30 cm.

4.3.1.3 Hydrogeologic Regime of the Aral Sea

The Aral Sea area equals 67.6 ths km^2 (at sea level, 53.41 m). Until 1961 it occupied fourth place among intercontinental water reservoirs of the world after the Caspian Sea, the Great Lakes (North America), and Lake Victoria (East Africa). The lowest absolute altitude is marked near the west coast of the sea and equals 16.5 m. Before 1961 the Aral Sea was a stable water body with century and seasonal water level fluctuations estimated at ±3 m and ±25 cm, respectively.

It is necessary to note that within geological time the Aral Sea is significantly unstable. Beginning from the Pliocene epoch (>2 mln years ago), the Aral basin was periodically filled by riverwater or drained. Only up to the end of the Upper Quaternary did the basin form within its modern boundaries. The Amudariya and Syrdariya Rivers formed their deltas and discharge into the Aral Sea.

The Syrdariya River springs from the Tan-Shan Mountains. It is 2137 km long (from the river Naryn mouth it is 3019 km) and its catchment area is 462 ths km^2.

The Amudariya River springs from the high mountain system Pamir near the boundary with Afghanistan (Pyandzh River). Its length is 2275 km and its catchment area is 492 ths km². The runoff of the Amudariya and Syrdariya Rivers is regulated by several hydraulic reservoirs.

Concerning water and salt balance of the sea, the Amudariya and Syrdariya Rivers runoff has played the main role during the last 200 years (up to 1961). For the period 1911–1960, the total runoff of these rivers equaled 117 km³/yr (80 km³/yr from Amudariya and 37 km³/yr from Syrdariya) in the formation area (the mountainous regions of Pamiro-Altay and West Tan-Shan). At that time river runoff to the Aral Sea was 56 km³/yr (Table 4.3.1.1), including 42 km³/yr from the Amudariya and 14 km³/yr from the Syrdariya. Average mineralization was 0.47 and 0.55 g/l, consequently, and contributes along with atmospheric precipitation and evaporation processes the Aral Sea water salinity of 9.9‰ (Table 4.3.1.1).

Since 1961 total water consumption in the Amudariya and Syrdariya basins began to grow steadily. This occurred due to intensive irrigation and hydraulic power development of water and land resources of the Aral Sea basin and regulated and irreversible withdrawal of surface runoff. Thus, water discharge to the sea in 1961–1965 was 7.7 km³/yr; 1966–1970, 17 km³/yr; 1971–1975, 30 km³/yr; end of 1980s, 50 km³/yr; and 1990, only 5 km³/yr. Thus, starting from 1960, negative water balance became a norm. For the period 1970–1990, the annual balance deficiency exceeded 30 km³ and the regime of the Aral Sea is considered to be anthropogenic. Such a significant deficiency of the Aral Sea balance is also caused by climatic factors. In the 1970s, water content of the Amudariya and Syrdariya was 20–25% lower than the norm, and total water resources renewing the Aral Sea was 20–25 km³/yr less than earlier (Table 4.3.1.1).

TABLE 4.3.1.1
Water Balance of the Aral Sea

Period	Coming (km³/yr)[a]		Evaporation	Balance
	Discharge	Precipitation		
Natural Regime of the Sea				
1911–1960	56.0	9.1	66.1	−1.0
Anthropogenic Regime of the Sea				
1961–1970	43.3	8.0	65.4	−14.1
1971–1980	16.7	6.3	55.2	−32.2
1981–1990	3.9	6.2	43.7	−33.6
1991–1994	21.0	4.6	33.6	−8.0
1995–2002	4.8[b]	3.5	28.6	−20.3

[a] From data of the regional Scientific-Research Hydrometeorological Institute of Kazakhstan and etc.
[b] Inflow to the Small Sea

Since 1961, when the Aral Sea began drying up, its area and volume decreased continuously with different rates: to 1989 it decreased almost 29 ths km^2 and 600 km^3, respectively. According to data of the water station "Aralskoe more" in 1961–1987, intensity of sea level dropping was 0.3 m/yr on average; but for the period 1987–2004, this value reached 0.5 m/yr (Large Sea). Mainly the sea bottom drained in shallow parts where the sea regressed more than 100 km, i.e., it regressed with the average rate up to 2.7 km/yr.

Despite the fact that the Aral Sea basin (areas of Kazakhstan and Uzbekistan) experienced numerous anthropogenic reservoirs formed constantly due to the discharge of collector-drainage water from the irrigated areas and flood water. This significantly restricted water inflow into the Aral Sea and, hence, lead to rash sea drying up. In 1987 the Aral Sea at level equals 40.19 m divided in two parts: the Large Aral and the Small Aral. After this the rate of level decreasing for the Large Aral Sea was 0.55 m/yr for the period 1989–2004.

After subdividing the Aral Sea inflow and discharge of the water balance, the Small Aral became equal. The value of water flow from the Small Aral to the Large Aral through the Kokaralskaya strait to Berg Bay in years with water abundance (1993, 1994, 1995) reached, respectively, 30, 64, and 40% of the total Syrdariya River runoff, directed to the Small Aral recharge. In 1999 the strongest storm in the Small Aral destroyed the strait between it and the Large Aral that lead to the catastrophic water runoff into the Large Aral and sharp sea level decreased from 42.3 m to 40 m in the Small Aral.

The water balance of the Large Aral is still negative because evaporation from its surface predominates the surface and underground runoff and atmospheric precipitations. After subdividing the sea into two hydrologically independent reservoirs and the almost total discontinuation of the Amudariya River runoff into the Large Aral, its level forms an evaporative regime partly provided by atmospheric precipitations.

As a result, the sea level of the Large Aral continues to decrease abruptly, and since the beginning of 2005 it has dropped to 31.25 m. If the present anthropogenic load on water resources of the Amudariya River exists into the future, then by 2010–2012 the Large Sea will finally subdivide into two parts — the west deep-sea and the east shallow — due to sea level decrease. The latter will probably dry up by 2010. Likely, total disappearance of the west deep-sea part of the Large Aral will occur by 2025–2030 because of the absence of the Amudariya River runoff and abrupt reservoir salinization accompanied by salt deposits.

The hydrochemical regime driven by the drying up of the Aral Sea has changed sharply and become instable. The main components of salt balance determine sea salinity. Salt addition to the reservoir from the runoff of the Amudariya and Syrdariya, atmospheric precipitation, and underground salt runoff form the receipt part of the salt balance. Evaporation, wind, underground water discharge, ion exchange, biota absorption, etc., all determine the demand parts of salt balance. Because of significant salinization of the residual part of the sea after 2005 it is necessary to account for the processes of halogenesis that lead to salt mass precipitation.

4.3.1.4 Hydrogeologic Conditions of the Aral Sea and Priaral Region

Aquifers of the artesian basins of the central part of the Turanskaya Plate discharge into the Aral Sea. Within the sea basin three main hydrogeological floors are noticed in the platform cover: the Middle Jurassic–Upper Cretaceous, the Paleogene–Lower Pliocene, and the Upper Pliocene–Quaternary (Groundwater and Salt Discharge into the Aral Sea 1983).

The Middle Jurassic–Upper Cretaceous floor, especially the Cretaceous layer, is distributed everywhere and has the greatest thickness. Water-bearing deposits of the Cretaceous period consist of three main water-bearing complexes: Senon-Turon, Alb-Senoman, and Apt-Neokom sediments. Total thickness is 0.5–1 km in the Kulandinsko-Nizhneamudarinskaya raising system, 1.2–1.8 km in its western part, and 1.6–2 km in the eastern. The total thickness of water-bearing sands and sandstones of the confined complex is 400–500 m on the west, 100–300 m on the north and east, and 200–300 m on the south. The filtration coefficient is 1–6 m/day. High-pressure water of the upper horizons of the Cretaceous deposits are determined at depths of 200–400 m on the east and in the zone of Kulandinsko-Nizhneamudarinskaya raising system, 400–700 m on the southeast and central parts of the basin, and 600–1000 m on the west. Groundwater mineralization of the Cretaceous deposits varies 2–5 g/l on the east and north of the basin to 25–30 g/l on the south and 100–150 g/l and greater on the west.

Overlying Paleogene–Neogene rocks are mainly marl-clay with thicknesses up to 1100 m in the west (deep-sea) and to 500 m in the east (shallow) parts of the Large Sea. An aquifer of the Upper Quaternary alluvial deposits is distributed in delta areas of the Amudariya and Syrdariya Rivers and is represented by flood-plain and riverbed siltstones and sands with a thickness of 50–70 m. Groundwater mineralization is variegated and varies from 1–50 g/l and greater.

As a result of sea regression, an aquifer of shallow groundwater of newest marine deposits of the Holocene is formed on the drained sea bottom in an area >48 ths km^2 (according to data in 2005). The Holocene horizon of the basin is represented by several facies consisting mainly of rocks of light and hard fractions. Mineralization of the shallow groundwater of the modern marine deposits changes from 20–150 g/l and greater.

4.3.2 QUANTITATIVE ASSESSMENT AND PREDICTION OF SUBMARINE GROUNDWATER DISCHARGE

Groundwater discharge from confined aquifers is directed toward the Aral Sea basin and is formed outside the basin in the Priaral region in the areas of the Cretaceous deposits outlet on the surface or under the water-bearing sediments of the Neogene–Quaternary in the piedmont areas Mugodzhary (on the north of the Priaral region), Ulytau and Karatau (on the east), and Sultan-Yizdaga-Bukan-Nuratau (on the south). Hence, three main regional groundwater flows from the Priaral region toward the Aral Sea basin are determined. The dynamic of groundwater within the basin is determined by the outlet of deposits of the Cretaceous period within the

Arkhangelsky swell, where underground flow has a submeredional regional direc-tion. This occurs due to extremely high tectonic breaking of the rocks within the Arkhangelsky swell. This leads to the forming of numerous high-yield springs (vents) of subareal and subaqual groundwater (Groundwater and Salt Discharge into the Aral Sea 1983; Poryadin 1994; Sadov and Krasnikov 1987; Sydykov et al. 1993). Intensive confined groundwater discharge within the Arkhangelsky swell causes layer pressure decreasing in the confining complex and forming discharge not only from Cretaceous, but also from the underlying Triassic–Jurassic rocks.

Shallow groundwater discharge is mainly determined by morphology and lithol-ogy of the basin. But confined groundwater discharge is mainly defined by sublat-itude Kulandinsko-Nuzhneamudariinskaya system of lifts that in the Aral Sea basin is the end base level of confined groundwater of the Cretaceous deposits of the Priaral area. Here are intensive opened (by tectonic disturbances) and concealed or diffuse (due to confined leakage through Neogene–Paleogene aquiclude) discharges of confined groundwater of Cretaceous deposits in the Aral basin. Subaqueous sources of confined water along the Arkhangelsky embankment form vast thermal anomalies that occur nonfreezing in this part of the open sea area (Groundwater and Salt Discharge into the Aral Sea 1983; Poryadin et al. 1997; Sadov and Krasnikov 1987; Sydykov et al. 1993).

Assessment of groundwater flow into the Aral Sea basin was carried out in several stages (Groundwater and Salt Discharge into the Aral Sea 1983). In the first stage a hydrogeological map of the Aral Sea basin was compiled. It contains data on thickness and transmissivity of the three hydrogeological floors: water-bearing rocks of the Upper Pliocene–Quaternary (the upper floor), the Middle Juras-sic–Upper Cretaceous (the low floor), and the relatively confining stratum of the Paleogene–Lower Pliocene deposits (the second, intermediate floor).

In the second and third stages, groundwater flow (Q) to the sea basin or water area was determined. This was done on the basis of classical filtration of Darcy's equation, separately for shallow and confined water:

$$Q = TLI \qquad (4.3.2.1)$$

where L is the perimeter of the basin or water area for the different periods; T is the average-weighted value of rocks transmissivity along the determined perimeter; I is the average-weighted value of confined gradient, and for shallow groundwater this is the flow slope for different calculated periods.

In the forth stage, areal distribution of vertical (upward) confined flow or module of confined groundwater discharge were calculated separately for the sea and drained sea bottom. For determining this value the Equation of plane stationary filtration was used:

$$\frac{\partial^2 H^*}{\partial x_i^{\,2}} = a^2 H^*, i = 1,2 \qquad (4.3.2.2)$$

where $a = \sqrt{k_0 / Tm_0}$ is the value inverse to leakage coefficient of confined ground-water from the third floor with transmissivity T through relative aquiclude of the second floor with the thickness m_0 and filtration coefficient k_0; H^* is the difference of piezometric levels of confined groundwater and the sea (or the basin surface); x_i is the coordinates of the plane filtration flow. For the basin or water area confined groundwater flow for the calculated period (Q_0) can be determined by the following equation:

$$Q_0 = H_0^* L \sqrt{\frac{Tk_0}{m_0}} \qquad (4.3.2.3)$$

where H_0^* is the difference of piezometric levels of confined groundwater and the sea (or the basin surface) for the calculated counter, where $x_i = 0$.

To simplify the calculations, the values of the filtration coefficient of the relative aquiclude of the intermediate hydrogeological floor ($k_0 = 5 \cdot 10^{-4}$ m/day) and also H_0^* were determined as constant within the basin area.

According to the equation for Q_0, confined groundwater flow for each calculated area of the sea basin (water area or drained bottom) is directly proportional to the transmissivity (T) of the rocks of the third (water-pressure) floor and inversely to the thickness (m_0) of the overlying aquiclude of the second floor. These values vary in a wide range: $T = 50 = 1500$ m²/day, $m_0 = 50 = 1000$ m.

Taking into account areal distribution of hydrogeological parameters T and m_0 elementary calculated blocks with these different values were determined within the sea basin (water area and drained bottom). Particular values of $\left(\sqrt{T/m_0} \right)_i$ were calculated for each block. Then, the average-weighted values of the last parameters were calculated for each block within the basin (water area and drained bottom) using the following equation:

$$\left(\sqrt{\frac{T}{m_0}} \right)_{average} = \frac{\sum \left(\sqrt{\frac{T}{m_0}} \right)_i F_i}{\sum F_i} \qquad (4.3.2.4)$$

where F_i is the block area.

Further average-weighted meanings of the module of groundwater discharge of the confined aquifer into the sea basin (water area and drained bottom) were determined by division of the total influx (Q_0) for the calculated contour fixed by Darcy's equation, on area of the sea:

$$q_{average} = \frac{Q_0}{F} = \frac{\sum Q_i}{\sum F_i} \qquad (4.3.2.5)$$

Finally, the dimensionless coefficient of water inflow was determined by comparison of concrete (in block) and average-weighted values of the left part of Equation (4.3.2.4). The values of module of groundwater inflow (in l/sec·km^2) for each determined block were received by multiplying of this dimensionless coefficient on the average value of module of groundwater inflow. On the basis of these data maps of module of groundwater flow in isoline for the concrete calculated period were compiled. The similar methodology was used for compilation of the maps of salt flow with groundwater discharge. The above methodical approach allows one to determine not only the total shallow and confined groundwater discharge, but also divide it within the total sea basin (water area and drained bottom), accounting for the main criteria which is the leakage coefficient.

Groundwater flow into the Aral Sea basin was carried out for determined periods by using mathematical modeling. This procedure includes compilation of a hydrodynamic grid on the base map of hydroisopieses, using the method of strip fragments. Then the elementary flow rate of confined flow was determined for each cell of the hydrodynamic grid using the following equation:

$$q_f = T_i \frac{\Delta l_i}{\Delta b_i} \Delta h_i \qquad (4.3.2.6)$$

where q_f is the filtration discharge in each cell of the calculated flow band; T_i is the water conductivity of the confined layer; Δl_i, Δb_i are the average stream and piezometric head lines in the cell, respectively; and Δh_i is the piezometric head difference in the cell. Total productivity of groundwater flow is equal to the sum of yield in elementary cells. Flow discharge was determined to the band between isopieses 60–70 m, located within the Aral Sea basin (Groundwater and Salt Discharge into the Aral Sea 1983). Maps of modules of water and salt discharge are absolutely the same as those compiled using such a simplified approach.

Ion discharge was calculated by multiplication of the module of water discharge on groundwater average mineralization in each cell and then summarized by all determined cells within the sea basin (water area and drained bottom).

4.3.2.1   Quantitative Assessment of Groundwater Discharge and Flows of Its Formation

As was determined above, groundwater discharge from the Priaral area into the Aral Sea basin occurs both from unconfined and confined aquifers. Shallow groundwater flow is connected with the upper hydrogeological floor and occurs by lateral filtration. The flow from the confined aquifer occurs with different intensity mainly due to upward vertical dispersed filtration through the relative aquiclude of Paleogene–Lower Pliocene deposits in the water area and drained sea bottom. Moreover, confined groundwater discharge occurs also through the tectonic failures and appears as vents on the drained bottom within the Kulandinsko-Nizhneamudarinskaya swell-like uplift.

Groundwater discharge into the Priaral basin is along the total perimeter of the water area, but the most intensive is noted in areas where Upper Pliocene–Quaternary deposits of alluvial (deltas of Amudariya and Syrdariya Rivers) and aeolian (sandy massifs of the Large and Small Barsuki in the north Priaral region) genesis. In these areas, rock transmissivity is the highest and equals 100–120 m^2/day (Veselov and Panichkin 2004; Groundwater and Salt Discharge into the Aral Sea 1983).

According to 1960 data, groundwater discharge into the Aral basin equaled 137.2 mln m^3/yr (4.35 m^3/sec) and ion groundwater discharge equaled 1.05 mln t/yr with mineralization of 8.4 g/l (Table 4.3.2.1) (Groundwater and Salt Discharge into the Aral Sea 1983). The total mass of shallow groundwater and ion discharge occur in the east and south parts of the seacoast. In the north the main flows are connected with sandy massifs of the Large and Small Barsuki and Priaral Karakumy. The same situation is observed at present. But the difference consists in the fact that due to sea regression and the decreasing groundwater level of 3 m on the border with water area, inclination of groundwater flow increases and water conductivity decreases. It can be explained by the high rate of sea draining that leads to the disturbance of hydrodynamic connection of groundwater with the sea and, as a result, groundwater flow is enclosed within the narrow interior zone of the basin.

Confined groundwater flow is determined both by water conductivity of the rocks of the lower floor and thickness of the overlying aquiclude. Water conductivity of the confined system within the basin varies from 500–1200 m^2/day on the west and east up to 1000–1700 m^2/day and greater in the zone of Kulandinsko-Nizhnea-mudariinskoe uplift. The thickness of relatively impermeable rocks is minimal in the zone of uplift and maximal in the west and central parts of the sea basin. Thus, the maximal value of the module of groundwater flow is 0.15–0.45 l/sec·km^2. They are related directly to the zone of the Kulandinsko-Nizhneamudariinskoe uplift where numerous vents related to the tectonic fractures are noted. They are also noted on the continuation of the uplift zone (Kulandy Peninsula) where they are related to the exposure of the Cretaceous deposits. The module of groundwater discharge decreases ≤0.03 l/sec·km^2 toward the west and east from this zone and in the Small Sea (Figure 4.3.2.1; see Figure 4.3.2.4). Due to increased draining on the Arkhan-gelsky Arch, through 2005, areas with maximal module occur in the draining zone, and total confined groundwater flow becomes significantly predominant under the flow to the sea area.

Under undisturbed conditions the total shallow and confined groundwater flow to the Aral Sea basin was 340.4 mln m^3/yr, among them 137.2 mln m^3/yr and 203.2 mln m^3/yr, respectively, shallow and confined groundwater flow with changes from

TABLE 4.3.2.1

Shallow Groundwater and Ion Discharge into the Aral Sea

Year	Transmissivity (m^2/day)	Perimeter of the Water Area (km)	Flow Inclination	Water Inflow (mln m^3/yr)	Ion Discharge (mln t/yr)
1961	10–120	1850	0.01–0.001	137.2	6.58

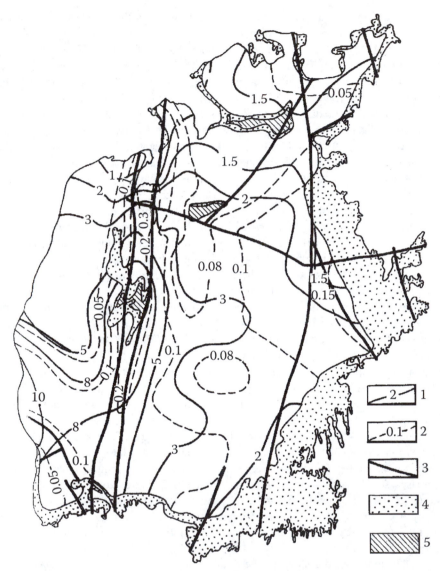

FIGURE 4.3.2.1 Schematic map of groundwater and salt discharge from confined aquifer to the Aral Sea (in 1980). Module of groundwater discharge: (1) salt, g/sec·km²; (2) water, l/sec·km²; (3) tectonic dislocations; (4) drained sea bottom; (5) islands.

330.9 mln m³/yr (1965) to 383.3 mln m³/yr (1990). At the same time if groundwater flow is considered to be practically changeless, then the flow from the unconfined aquifer into the water area and drained sea bottom is inverse: i.e., increasing sea regression causes flow decrease into the water area from 203.2 mln m³/yr in 1960 to 72.5 mln m³/yr in 2000, and flow into the drained bottom increases from 0–139.7 mln m³/yr in 2000 (Table 4.3.2.2).

TABLE 4.3.2.2
Groundwater and Ion Discharge into the Aral Sea

Inflow Elements	Calculated Year						
	1960	1965	1980	1985	1990	2000	2020
Sea level (m)	53.41	52.3	45.76	41.95	38.26	33.45	23.41
Sea salinity (‰)	9.9	10.70	17.01	22.45	31.82	47.20	103.08
Area (ths km²):							
Seawater area	68.5	63.3	51.7	44.4	36.2	22.5	7.9 (2.4)
Drying part	–	5.2	16.8	24.1	32.3	46.0	60.6 (60)
Water discharge (mln m³/yr) into the seawater area:	137.2	–	–	–	–	–	–
Unconfined groundwater	203.2	193.7	163.8	158.1	129.1	72.5	5.7
Confined groundwater on the drying part:	–	137.2	137.2	137.2	137.2	137.2	137.2
Unconfined groundwater	–	11.0	34.7	76.2	117, 2	139.7	197.5
Confined groundwater	340.4	193.7	163.8	158.1	129.1	72.5	5.7
Total							
into the seawater area on the drying part:	–	147.2	171.9	213.4	254.4	276.9	334.7
Total water discharge	340.4	330.9	335.7	371.5	383.3	349.4	340.4
Ion discharge (mln t/yr) on the drying part:							
Unconfined groundwater	–	1.05	1.05	1.05	1.05	1.05	1.05
Confined groundwater into the seawater area	–	0.20	0.66	2.43	4.10	4.87	7.06
Unconfined groundwater	1.05	–	–	–	–	–	–
Confined groundwater	6.44	6.71	6.12	6.61	5.88	4.94	0.65
Total	7.49	6.71	6.12	6.61	5.88	4.94	0.65
into the seawater area on the drying part	–	1.25	1.71	3.48	5.15	5.93	8.11
Total ion discharge	7.49	7.96	7.83	10.09	11.03	10.87	8.76

In parentheses, the Large Sea.

To 2020 almost all flow from the unconfined aquifer in volume equal that in 1960 will be noted at the drained bottom area, which probably increased to 60 ths km² (Table 4.3.2.2). As before it will be used on evaporation in the aeration zone of confined groundwater of the bottom sediments. Prediction modeling, carried out considering the overflow by zones of tectonic fracturing, the most contrasting of which underscores an importance of the Kulandinsko-Nizhneamudariinskoe uplift as a main zone of groundwater discharge into the Aral Sea basin (see Figure 4.3.2.4).

Behavior of the module of ion discharge in many respects is similar to that of water discharge: on the north and northeast its values varies 1–1.5 g/sec·km²; on the east, southeast, and northwest 1.5–3 g/sec·km²; on the south 2–6 g/sec·km²; on the west, within Arkhangelsky Arch and in the central part of the basin, 4–10 g/sec·km²

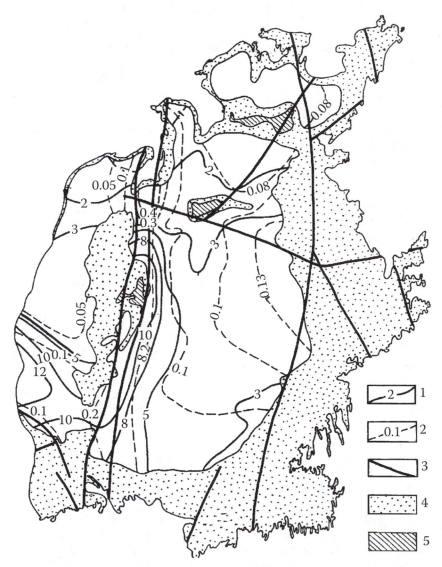

FIGURE 4.3.2.2 Schematic map of groundwater and salt discharge from confined aquifer to the Aral Sea (in 1990). Module of groundwater discharge: (1) salt, g/sec·km²; (2) water, l/sec·km²; (3) tectonic dislocations; (4) drained sea bottom; (5) islands.

and greater. Despite the comparative temporal uniformity of ion discharge of groundwater into the Aral basin (1.05 mln t/yr), discharge into defined areas of water decreases from 6.44 mln t/yr (1960) to 4.94 mln t/yr (2000), and on the drained sea bottom increases from 0 (1960) to 4.87 mln t/yr (2000) (Table 4.3.2.2). The quantity of salts from confined groundwater on the drained bottom was 5.2 mln t/yr in 2005, and probably will increase to 8.11 mln t/yr in 2020. Average year salt inflow with

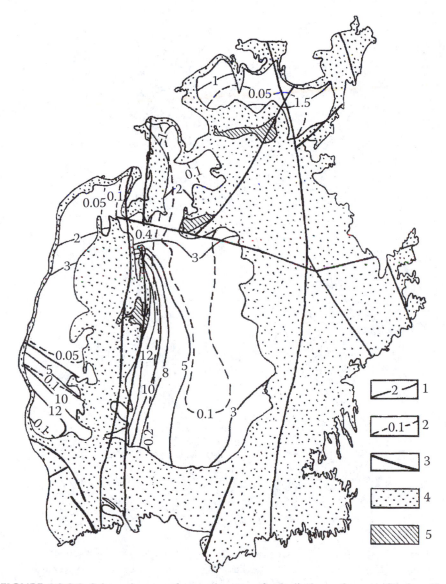

FIGURE 4.3.2.3 Schematic map of groundwater and salt discharge from confined aquifer to the Aral Sea (in 2000). Module of groundwater discharge: (1) salt, g/sec·km^2; (2) water, l/sec·km^2; (3) tectonic dislocations; (4) drained sea bottom; (5) islands.

confined groundwater equals 2.5 mln t. Hence, through 2005, 110 mln t of salt come to the drained bottom that causes salinization of the rocks of the drained bottom.

4.3.2.2   Water and Salt Regime of Aral Bottom Drying

Regression of the Aral Sea causes a growth of sea bottom outcrops, forming the hydrogeological system of modern marine deposits and the last salinization. Processes

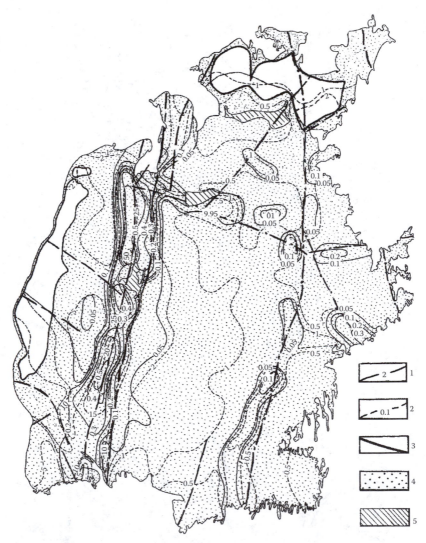

FIGURE 4.3.2.4 Forecast of groundwater and salt discharge to the Aral Sea on the basis of modeling (in 2020). Module of groundwater discharge: (1) salt, $g/sec \cdot km^2$; (2) water, $l/sec \cdot km^2$; (3) tectonic dislocations; (4) drained sea bottom; (5) islands.

of salt depositing on the drained bottom and its wind removal on other territories of the Priaral region negatively influence the total ecological situation in the area and the public health. Thus, discussion, analysis, and assessment of intensity are critical.

4.3.3 REGIME OF GROUNDWATER LEVEL OF THE DRYING SEA BOTTOM

The main aspect in assessment of atmogenic salt transfer is determining the dynamics of groundwater (silt water) level formation and its salinity and salt accumulation on the drained Aral Sea bottom. This problem can be solved using either mathematical modeling or decoding of aerospace information from remote sensing or land-based monitoring. The latter is the most reliable and unbiased and hence was used in this work.

The main mechanisms of groundwater discharge in the zone of sea bottom drainage are evaporation and transpiration from the groundwater level and through the aeration zone, spring and regional groundwater discharge, and local and regional interlayer overflow controlled by the permeability of tectonic fractures and impermeable layers.

In the first year after sea regression, the rocks of the drained bottom under the influence of tidal process are saturated by seawater up to the surface. The outcropped (exposed) surface is immediately covered by a salt crust that is dissolved frequently due to tidal processes. Groundwater is at a depth of 50–70 cm and its mineralization is closed to the seawater salinity due to rapid (quick) water and salt exchange. In the second year of draining, the zone of marsh solonchaks is formed on light soils. Groundwater level is at depths of 1–1.2 m and mineralization is 2 times more in comparison with seawater salinity. During the next 2–3 draining years, solonchaks transform into maritime grounds. Subsequently solonchaks change in typical desert groundwater of anthropogenic genesis under influence of exogenous (deflation-atmogenic) geological processes. In 3–4 years, the ground surface layer drained up to humidity <1% with the groundwater level deepening to 1.5 m. In the 4–5 draining years, the surface layer experience intensive deflation and fine basin sandy hilly relief is formed. After 9–10 years, the total natural complex experiences deflation, atmogenic accumulation of sandy material, and desert formation.

Evolution of hard rocks is different, because of high moisture capacity and low water yield. Thus, solonchaks are characterized by high salinization (no less than 1.5–2.5%) in profile to groundwater level. In hard rocks the groundwater level is at a depth of 3 m due to low water yield. Because of insignificant thickness and values of filtration coefficient of bottom sediments, water conductivity of the Holocene aquifer does not exceed several m²/day. This leads to the predominance of evaporative regime under the lateral filtration.

Considering that arid conditions are characterized by intensive evaporative concentration of groundwater and aeration zone salinization, a variegated combination of different ecological communities is observed. Some of them are typical to freshwater and slightly mineralized water, and others are typical of saltwater (Beideman 1983). Transpiration as well as evaporation accelerate salt accumulation in the groundwater and aeration zone. This process of biogenic salt accumulation (Beideman 1983) is multiphase and causes evolution of drained bottom ecosystems.

A study of spatio-temporal variations of the water and salt regime of groundwater of a drained bottom was carried out by determining the functional dependence of groundwater level (L) from the distance to the water area. The latter was considered as mediate temporal parameters of the bottom draining processes. On the basis of analyzing the regime data monitoring by several profiles (Sydykov et al. 1993), it was determined that the rate of aeration zone increasing (ω) is equivalent to the rate of groundwater level decrease and could be estimated by using the exponential function:

$$\omega L^{\alpha} = \beta \tag{4.3.2.7}$$

($\alpha = 0.82$, $\beta = 1.105$ [the Large Sea]; $\alpha = 1.94$, $\beta = 0.89$ [the Small Sea]). It shows that sea groundwater level decreases 10–100 times (Poryadin 1990, 1994).

Statistical analysis of calculated values of the rate of groundwater level decrease indicates the great variation in these values (0.1 to 1–2 m/yr). They reflect on lithology and morphology, as the role of evaporation and transpiration in the process of water and salt regime forms on the drained bottom. The last proves that evapo-transpiration predominates at the beginning of the draining process, and physical evaporation occurs only in the posterior years of desertification.

Aerovisual studying, land monitoring, and remote-sensing space monitoring, carried out in the 1980s and 1990s (Poryadin et al. 1997; Sydykov et al. 1993), showed that an important element of the hydrodynamics of the bottom draining is formation of a seepage zone on the boundary with the water area. This zone is formed due to filtration of seawater, enriching the rocks of the sea bottom, water-area regression, and also tidal effects and seiches. Hence, seepage zones continuously follow a regressing sea and provide uninterruptedly hydrodynamic connection between saltwater and the water area. The characteristic of the seepage zone forms under conditions where the rate of sea bottom draining (7 m/day) is 2–4 times higher than the rate of groundwater filtration and, moreover, than the rate of its movement (0.01/0.0001 m/day). These lead to disturbance of the hydrodynamic connection between sea area and the sea coast (within the boundaries in 1960) as the bottom drains.

The analysis of the equation of the rate of groundwater level decrease (Equation 4.3.2.7) shows that the width of the seepage zone (L_0) can be determined by the following equation (Poryadin 1990, 1994):

$$L_0 = \beta_0 / \omega_0 = \sqrt[a]{\beta / \omega_0} \tag{4.3.2.8}$$

where $\omega_0 = 2$ m/yr and is the limiting value of the rate of water inflowing to the seepage zone under evaporation equals 1 m/yr and rock porosity equals 50%; $\beta_0 = \beta L^{(1-\alpha)}$ is the averaged parameter of seepage zone. The width of the seepage zone varies from 40–360 m and greater (in the east and south parts of the draining sea bottom). Equation (3) makes it possible to estimate the filtration coefficient of the rocks of the sea bottom varying along the monitoring profile from 0.92–2.28 m/day.

It is necessary to note that the average value of the module of confined groundwater discharge along the total area of the sea is 10 times less than the module of the average values of evapotranspiration in the aeration zone of the drained bottom of the Aral Sea (0.0957 l/sec·km^2 for 1960 and 1.13–1.73 l/sec·km^2 for the present). Hence, it is possible to neglect the module of confined groundwater discharge in the balance of evaporation in the aeration zone of the drained sea bottom if assessing the module of moisture transferring in the aeration zone. Such an approach is acceptable if you account for the module of evapotranspiration within the water bottom sediments in the seepage area. Here it can be comparable with evaporation reaches in the defined water area of the Aral Sea, 900–1100 mm/yr (up to 35 l/sec·km²). The average value of the module of confined groundwater flow and evapotranspiration of the drained bottom of the Aral Sea differ 100 times and greater.

The quantity of sedimentation (residual) seawater in the bottom sediments evaporated from the aeration zone of the drained sea bottom for 40 years equals 34.5 km³, at the average thickness of the aeration zone of 1.5 m, rock porosity of 50%, and area of distribution of 46 ths km². Hence, the average value of the module of evapotranspiration equals 0.59 l/sec·km².

The quantity of silt water evaporated from the aeration zone for 44 years (to the beginning of 2005) equals 37.5 km³, at an average thickness of the aeration zone of 1.5 m, rock porosity of 50%, and area of distribution of 50,000 km². Hence, the average value of silt water discharge due to evapotranspiration equals 27 m³/sec at the module of evapotranspiration of 0.54 l/sec·km².

The quantity of seepage from bottom sediments sedimentation marine water due to sea bottom draining can be assessed in the following way. At the beginning of 2005, sea level decreased to 22 m. Hence, an average rate of level decrease for 44 years equals 0.5 m. Thus, the average value of gravitational water yield of sandy and silt bottom deposits equals 0.1 (range, 0.01–0.3); and for an area of drained bottom of 50,000 km², the quantity of seepage sedimentation marine water from the bottom sediments equals 2.5 km³. This value corresponds to an average value of module of seepage equal to 0.036 l/sec·km². Thus, the intensity of evapotranspiration exceeds approximately 15 times intensity of seepage.

Hydrodynamic assessment of silt water inflow to the drained sea bottom through the seepage zone as the sea regressed varies from 0.4 mln m³/yr (1985–1990) to 0.3 mln m³/yr (2005) at values of filtration coefficient of bottom sediments of 1.5 m/day, seepage zone thickness 0.5 m, and width, 500 m, and average value of sea bottom slope 0.001, depending on the water area. Thus the average value of the module of water inflow from the seepage zone equals 0.0325 l/sec·km².

The maximum modules of water loss in the seepage zone are typical of bottom sediments of the central part of the Arkhangelsky swell where oolitic sands are developed. The average values are typical of sandy deposits of the suburban parts of the basin for the first years of sea regression, and the minimal values are noted in the central shallow part of the Large Sea, where clay residuals of bottom sediments with paltry water loss predominate. Part of the seepage water is evaporated, forming a salt crust on the exposed sea bottom. The other flows into the water area. The ratio depends on bottom slope and rock water loss. According to carried-out calculations, the average minimal module of salt deposits is equal to 0.64 and 0.63 t/day·km²; the average maximum is 1.32 and 1.23 t/day·km²; the average is 1.23 and 1.21 t/day·km², respectively.

Maximum module and salinization of rocks of the aeration zone of the drained bottom are typical of oolitic sands of the Arkhangelsky swell in recent years. Minimal module and salinization are typical of fine-grained and sandy residuals of bottom sediments of suburban parts of the sea for the early years of draining. On the whole for light rocks inside salinization is typical. But for the heavy rocks surface and near surface salinization is noted. This leads to the formation of the crust and pump solonchaks.

Taking into account salt discharge from confined and unconfined groundwater into the drained bottom for the period 1960–2004, this equals 46.2 and 110 mln t, respectively; then salt mass resources of the drained bottom increase only 15.6% and equal 1141.6 mln t (maximum). Concerning the contribution of atmospheric precipitations to the aeration zone salinization, it is extremely insignificant (no more than 0.5 mln t/yr), i.e., it does not exceed 1–2% of the salt mass reserves of the aeration zone. If we consider the atmospheric component, potential salinization increases up to 1163.6 mln t. It corresponds to the aeration zone salinization of 0.8% and is equivalent to the salt layer of 9.3 mm. According to the above-mentioned assessments, about 25% of the potential contributes to the salinization of the light rocks in the aeration zone. The other 75% is localized on the surface of the drained bottom in the area of heavy rock distribution in the form of seepage zones, spatters, and seiches, as well as closed shallow pinched reservoirs with a clay bottom that is well noted on aerospace images (see Figure 4.3.1.1).

The total mass of salts accumulated in the soil-rock layer through 2005 and potentially involved in the salt and dust ejection equals 62.8 mln t at maximum values of the module of salt accumulation of 3.44 t/day·km². Hence, an average salt content from 1960 in the decimetric soil-rock layer equals 92.4 mln t. According to previous assessments, the annual possible salt ejection from the draining sea bottom in the first year of draining was 0.7 mln t/yr, with a range from 0 (1960) to 1.43 mln t/yr (2004) (Poryadin 1994).

Distribution of groundwater mineralization of the Holocene aquifers of the bottom draining of the Aral Sea is characterized by two opposite processes: salinization and desalinization. The first is connected with an increase of residual sea salinity (the source of silt water) and with an evaporative concentration of salt mass in the draining zone. The second is connected with a decrease of evapotranspiration influence and an increase of part of the atmospheric precipitation in water and salt balance of the shallow groundwater in areas of anthropogenic desertification as seas drain. Thus, the highest mineralization is noted in the coastal zone of the drained bottom (the zone of seepage in the first years of drainage), and the lowest mineralization is typical of the base coast. The intensity of soil-rock layer salinization of the drained bottom increases toward the water area where mineralization of the silt water increases due to sea regression and decreases toward the basin boundary due to infiltrative desalinization of the aeration zone by atmospheric precipitation with a norm of 8–10 mm/yr.

The total differentiation of the rate of rock salinization decreases from the water area to the basin boundary of the Aral Sea (1960) due to lower mineralization of silt water in the early years of sea regression and the posterior increase in infiltration desalinization of the rocks of the drained bottom. It is enough to note that under the norm of atmospheric precipitation of 135 mm/yr and part of the atmospheric alimentation of groundwater of 7.5% of this value (~10 mm/yr), the layer of the atmospheric moisture for the period from 1960 to 2005 was 455 mm. Numerous annual influence of infiltration lead to a decrease of mineralization of newly formed groundwater moving off the sea at several times and decrease of soil-rock layer salinization in the layer with a thickness of 0–30 cm from 6–10% to 1.5–3.5% or less.

4.3.3.1   Dust Storms and Salt Ejection in the Priaral Region

The exposed Aral Sea bottom, set by intensively saline marcs and coastal crust and pump solonchaks, as well as soil-rock sediments, represents a great potential source of salt removal by winds. Modern active solonchaks are connected through the capillary fringe with formation groundwater of the newest sediments of the draining bottom and are located at a depth of 0–3 m. Insignificant depth of occurrence, high soil-rock sediments hygroscopic properties, and evaporation in the first years of draining caused intensive salt removal from buried marine water to the surface. In the final stages of the Large Sea draining, uniform salt crusts form due to salt charging and the total drying of the residual sea brines will occur. Such processes were noted in the 1960s–1970s during the drying of the Aral Sea at the shallow Akpetkinsky Archipelago and were carefully observed during aerovisual hydrogeo-logical and engineering-geological investigations carried out in 1983.

According to the salt content, the soil-rock layer can be characterized as strongly saline, with a sandy cover in the form of dune ranges, and other atmogenic relief forms are slight and of medium salinity. Salt content in the dunes reaches 0.25–0.5%. Thus, in the area of the Aral Sea basin sandy storms produce ìsandy-salt.î Assessments of salt quantity removed by dust storms are different.

For the first time dust and salt storms in the Aral Sea region under initial drying processes were noted on May 22, 1975 from aerospace images received by the meteorological satellite *Meteor 18* (Chichasov 1990; Groundwater and Salt Discharge into the Aral Sea 1983). According to the data received from space and land actimometric investigations, the annual amount of aerosol equaled 15–75 mln t (salt content in the aerosol of 1% corresponds to the volume of salt removal equal to 0.15–0.75 mln t/yr). With the aerospace images it is possible to note that removal plumes from the strong storms have extended up to 200–400 km, i.e., the main mass of particles precipitates at a distance of several hundred kilometers, and only an insignificant part of the smallest particles can be carried for a long distance during the strongest storms.

In the paper by Rubanov and Bogdanova (1987), the mass of carried salt was estimated at 39 mln t/yr. Hence, for 27 years of drying, approximately 800 mln t of salt, i.e., 30 mln t/yr of salt, were carried away. This is 1000 times more than similar assessments carried out by soil scientists for the decimetric soil-rock layer of the Aral Sea bottom.

A model of sand storms was created. This model helped to estimate the mass of sand carried out by wind in the near-ground atmospheric layer (from several tens of meters to 150–200 meters) of the Aral Sea bottom. According to such assessments the wind transferred from 1560 t/km^2·yr (weather station "Barsakelmes") to 3845 t/km^2·yr (weather station "The Aral Sea"). The lowest volume of sand (330 t/km^2·yr) was carried out in the area of the weather station "Kazalinsk."

Estimations of salt amount in the aerosols of sand storms was done on the basis of special surface investigations and indicate that annual salt mass removal can reach

50–70 ths t/yr of salt content in the solid phase of a flow of 0.7–1%. Average perennial intensity of atmogenic denudation of the sea bottom surface is estimated at 2 mm/yr. Most of the removed salts are represented by calcium and magnesium sulfates. The main removal directions according to intensity decrease are west, southwest, and northeast. At present, the conditions of increasing the area of drained bottom estimations have to be significant. Data on sand and salt removal for 1966–1979 were 7.3 mln t/yr and 70 ths t/yr, respectively, with 5% probability. The total average perennial amount of removed aerosols of 50% probably from five sources of the Kazakh part of the drained sea bottom at a sea level drop of 15 m can be estimated at 1.175 mln t/yr. Forecast of future removal of aerosols for the two decades shows that due to an increase of the geometrical size of the removal sources, the average perennial volume of the removed aerosol will increase to 1.29 mln t/yr. In the case of the Small Sea filling up to a grade of 53 m, two sources of aerosol removal will be under the water. This will decrease the volume of sand-salt aerosol removal from the drained sea bottom approximately to 400 ths t/yr and may be ~900 ths t/yr to 1 mln t/yr. Appearance of new drained areas with heavy mechanical composition will lead to an increase of distance of salt aerosol removal due to a decrease of particle size of 10 μm or less. The behavior of such fine aerosol in the Aral region is not well studied. Its element and ion composition, content in the atmospheric flow, influence on diffusion, and change of spectral composition of the infrared balance of sun emission is not well known (Hydrometeorological Problems of the Priaral Region 1990).

4.3.4 ECOLOGICAL CONSEQUENCES OF DRYING THE ARAL SEA

Continuous regression of the Aral Sea leads to an increase in the bottom drained area and the formation of an anthropogenic desert. At the same time underground water flow into the basin, not exceeding 1%, does not play a significant role in the water balance of the Aral Sea. Salt flow, especially into the drained bottom, is quite significant in determining the behavior of the ecosystem with posterior deflation-atmogenic processes carrying out salts to adjacent territories. The predominance of chlorides in the salt composition of seawater and chlorides and sulfates in the salt composition of the Groundwater and the soil–rock layer negatively reflects on the situation and productivity of ecosystems of the Priaral area and causes degradation there.

Ecosystem degradation on large areas will promote further aridization of the climate, especially the decreasing humidity and increasing temperatures of the soil surface and near the ground layer. This will appear in drought-resisting plant development and thinning out of the vegetation layer, losses in bioavailability, decrease in productivity, decrease in resource potential, and finally decrease in substance and energy exchange in the biosphere.

The main reason for the continuous anthropogenic catastrophe in the Priaral area is the thoughtless, reckless disturbance of the natural ecological balance and the destruction of the ecosystem caused by exceeding the ecologically safe level of natural resources use, mainly water resources (Problems of the Aral Sea Basin 1999). It is difficult to overestimate the significance of water resources in ecosystem function and especially for human life. Water cover as a unified hydrosphere is represented by a triad of closely interconnected components — ground, underground, and over-head (atmosphere). Technologically regulated surface runoff disturbs the natural function of water systems and destroys the interconnection between surface and groundwater.

Meanwhile, the Aral Sea basin remains the main base for irrigated farming of Middle Asia and Kazakhstan, whose histories run several millenniums. During all times very moisture-loving crops (cotton and rice) were cultivated. Greater than 56% are irrigated lands, wherein are grown >95% cotton, >45% rice, and many other agricultural products. Thus, it is understood that having such significant value in agriculture, this region needs a great amount of water that is irreversibly removed from the Amudariya and Syrdariya Rivers for irrigation. At present, 2200 large and small channels with a total extension of 214 linear km are constructed. The extension of the drainage net is 73 ths linear km.

Irreversible retirement of surface runoff for hydroenergetic and irrigation pur-poses is particularly negative. On the one hand, this leads to the appearance of new man-made reservoirs, groundwater level rise in irrigated areas, and swamping and salinization; but on the other hand, this causes shallowing of surfacewater reservoirs, draining and desertification to adjacent territories due to groundwater level decrease, and transformations in the microclimate (Chichasov 1990). Thus, groundwater is a very important component of ecosystem functioning.

The negative ecological situation in the Priaral region varies. In the sea basin, including the water area and exposed bottom and also in the near-basin zone with a width of the first tens to hundreds of kilometers, the ecological situation has sharply worsened for the last one and one half to two decades. This is connected, first, with an intensive increase of bottom-drained areas becoming anthropogenic desert with increasingly high year-to-year potential dust and salt removal in the surrounding basin of irrigated lands and natural grazing areas that decreases productivity, includ-ing a sharp negative effect on people's health. The second, the decreasing sea level, significantly influences the hydrogeological situation in the Priaral area and is expressed in the deepening of the groundwater level as a component of the ecosys-tem, in increasing of desertification processes, and transformation and degradation of natural ecosystems. The third, transformations in the microclimate promote this negative process.

Intensive Aral Sea level decrease in the last three decades jointly with uncon-trolled increasing flow (2.2 m^3/sec) from year to year through exploitation of highly pressured artesian aquifers in the Priaral area, mainly in its eastern part (Veselov and Panichkin 2004), and the hydroeconomic measures and irrigation (Problems of the Aral Sea basin 1999) significantly change the hydrogeoecological conditions and interconnections of groundwater with river and seawater, and the water and salt runoff regime in the Aral Sea and its basin.

These changes are reflected mainly in changes of groundwater and confined water levels and hydrochemical regimes. During recent decades, their levels have decreased year after year.

During the last decades, close to the Aral Sea, the amplitude of groundwater level decrease reached 3–5 m. Toward the dividing plains the value of average annual groundwater level drop decreases, but is still noted at a distance of 7–10 km from the sea coastal zone of the North Priaral area to 60–80 km in the East Priaral area.

Such a significant extension of the area of regional groundwater level decrease in the East Priaral zone occurs not only under the influence of Aral Sea draining, but also due to climate aridity and decreasing infiltration of atmospheric precipitation. Significant decreases of piezometric levels of groundwater in the main aquifers of the Priaral area used for potable water supply of cities, settlements, agroindustrial complexes, and the spaceport "Baikonur" that lead to desertification of the surrounding territories are noted.

Perspective changes in hydrogeological conditions of water and salt flow of groundwater (as unconfined, as confined) of the Priaral area will in the future depend on not only the rate of the Large Aral Sea level decrease that will intensively appear in the nearby sea, but also on the intensity of changes in climate, surfacewater regime, hydroeconomic activity, and other elements of water balance. Decreasing groundwater levels in the future will influence the increase in thickness of the aeration zone and, hence, will intensify desertification of the territory.

On the whole, taking into account the increasing anthropogenic water deficit in the arid zone of the Aral Sea basin, it is possible to suppose that processes of landscape desertification will increase. Hence, it will lead to degradation of all organic forms, and finally to the decrease of natural and economic potential of the basin territory. Thus, the natural process of arid region desertification is intensified by anthropogenic impact. The desertification process both natural and anthropogenic can occur by two alternative and intercomplemented directions: salinization and desalination. Both trends are widely developed in the anthropogenic hydrogeoecological systems of the Aral Sea basin that is closely connected with irrational and uncontrollable use of water–land resources.

4.4 GROUNDWATER DISCHARGE TO THE BLACK SEA*

The Black Sea is a unique phenomena. On the one hand, it was separated from the ocean for a long time and was almost a freshwater reservoir. On the other, it is connected with the ocean through the Mediterranean Sea through shallow straits, distinguishing the Black Sea from others by the presence of a thin layer (200–220 m) of oxygenated mobile water (the "zone of life") occurring on the world's largest hydrogen sulfide almost immovable mass. According to bathymetric data, the Black Sea basin is subdivided into three parts: the shelf zone with depths of 100–200 m (~28% of the entire surface), the continental slope with a depth of 2000 m (30%), and the clap hollow with depths of 2000–2150 m (42%).

The origin of the Black Sea hollow is still under discussion. According to geophysical investigations it consists of two deep troughs (west and east) with a length >2000 m. The earth's crust of 20-km thickness has two layers. But on the continental slope the third layer, "granites," appears and hence the crust thickness

---

* Section 4.4 was written by Guram Buachidze (Institute of Hydrogeology and Engineering Geology, Georgia, bguram@gw.acnet.ge).

becomes 30–40 km. The characteristics of the Black Sea basin and main rivers of the region are given in Figure 4.4.1 and Table 4.4.1.

On the west the basin is bounded by the Alps folded system of Balkan and Carpathian, on the north by the Ukrainian plate and Crimean folded carbonate rocks, on the northeast by the Alps folded system of Major and Minor Caucasus, and on the south by the orogenic system of West and East Ponte (Figure 4.4.2).

4.4.1 NATURAL CONDITIONS OF THE COASTAL ZONE

Research of SGD of the Black Sea has a long history. Pliny the Elder (1st century) mentioned the Black Sea as "submarine springs bubbling freshwater as if from pipers." In the 20th century, sources of SGD were noted in the Crimea and Caucasus. In the 1970s, complex research was carried out along the Georgian coastal zone jointly by the Georgian Institute of Hydrogeology and the Russian Institute of Water Problems . The experiments proved the necessity of the following methods for assessing SGD: (a) submarine measurement using a diverging technique (for springs with a velocity of 1 m/sec) and seepage meter; (b) geophysical methods using the electronic resistivity of water for determining discharge boundaries (Ganthiadi, isolines of 1,2,3 m from the bottom); (c) geochemical (in assessing underground chemical flow, anomalies in Cl content, and mineralization values); (d) hydro-geochemical (sampling of the rocks composed of the bottom and analytical study of porous solution); (e) seismosounding (in detailed study of the geological structure of the first layer); (f) methods based on use of dissolved gases (mostly O_2, N_2, CH_4, CO_2); (g) geothermal measures (based on interaction between the temperature of the bottom water layer and the discharging groundwater). Moreover, use of the

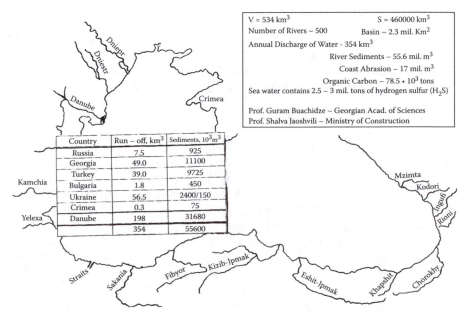

FIGURE 4.4.1 Survey scheme of the Black Sea basin.

TABLE 4.4.1
Main Rivers of the Black Sea Basin

No.	Rivers	Basin 10³ (km²)	Annual Runoff (km³)
1	Danube	817.0	198.0
2	Dnieper	503.0	43.3
3	Rioni	13.4	13.4
4	Dniester	72.1	10.2
5	Chorokhi	22.1	8.7
6	Kizil-Irmak	78.2	5.7
7	Eshil-Irmak	36.1	4.9
8	Sakharia	58.2	4.5
9	Kodori	2.03	4.2
10	Bziphi	1.5	3.8
11	Philios	13.0	2.9
12	South Boug	6.4	2.2
13	Khobi	1.3	1.6
14	Mzimtha	0.9	1.6
15	Inguri	4.1	1.3
16	Shache	0.6	1.2
17	Suphsa	1.1	1.2
18	Kharshithi	3.5	1.1
19	Gumista	0.6	1.1
	Total	1.6 10³.km²	310.9 km³ yr⁻¹

infrared radiometer seems to be very promising in the future, as it makes it possible to obtain a detailed cross-section of the shelf by measuring temperature and heat flow in bottom sediments. Using oxygen (O^{18}) and carbon (C^{13}, C^{14}) isotopes allows determining seawater origin and the age of some water elements.

4.4.1.1   Subdivision of Shelf Area

A clearly defined shelf is traced along the entire Black Sea perimeter, except some rocky coasts within the zone of sea conjugation with the East and West Ponte, the Adjaro-Trialety folded system, etc. (Table 4.4.1.1). The shelf is generally recognized down to the 100-m isobaths. However, in some areas the external edge of the shelf is pointed out within the 200-m isobaths (e.g., dispersion areas of drifts of the Danube and Dniper Rivers, rivers of the Turkish seacoast such as the Kizil-Irmack, Eshil-Irmac, etc., as well as those confined to individual structural blocks, under the influence of uplift during the neotectonic stage, such as the Dobrujla, Crimea, and Gudauta bank).

It is well known that shelf morphology depends on the geotectonic character-istics of the basement. In the areas of ancient continental plateaus and platforms (the Skythian and Mezy platforms and the Georgian Plate west end [Kolkhethi]),

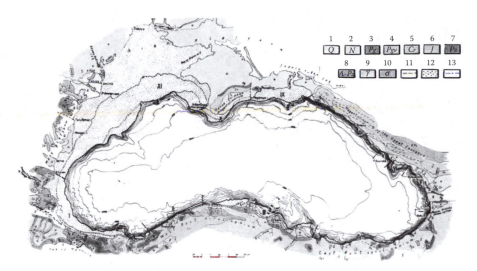

FIGURE 4.4.2 Map of the Black Sea basin geology: (1) Quaternary deposits: alluvial, marine, and bog deposits; (2) Neogene deposits: clays, sandstones, and conglomerates; (3) Paleogene sedimentary deposits: sandstones, clays, limestone, and flysch; (4) Paleogene vulcanogenous-sedimentary deposits: porphyrite, basalts, tuffs, tuff-breccias, tuff-sandstones, sandstones, and marls; (5) Cretaceous deposits: limestone, dolomite limestone, marls, and flysch; (6) Jurassic deposits: shale, sandstones, limestone, porphyrite, tuff-breccias, and tuffs; (7) Paleozoic deposits: limestone, marble, sandstones, and quartzite; (8) Archean–Paleozoic deposits: crystalline schist and granitoid; (9) Mezo-Cenozoic deposits: granites, diorites, and syenite; (10) serpentine, diorites, and diabase; (11) disjunctive dislocations; (12) shelf area; (13) shelf boundaries.

the shelf is mostly wide. A relatively wide shelf is typical of miogeosyncline zones (West Pontians and the Major Caucasus southern slope) and a quite narrow shelf is confined to the eugeosyncline zone of East Pontians and the Adjaro-Trialety folded system. Analyzing morphometric data, it is noted that a gentle submerging (0.001–0.002) is typical of a wide shelf (e.g., the Scythian and Mazy platforms). But the narrow shelf of eugeosyncline zones is characterized by a steep slope (0.03–0.08).

Regularity in distribution of the submarine canyons is observed. Thus, the submarine canyons are entirely absent in the zones of continental plateaus and platforms, while a vast development of canyons is recognized within the eugeosyncline and miogeosyncline zones of the Turkish and Caucasus coasts of the Black Sea.

According to this regularity the Black Sea shelf is subdivided into 11 parts: I Bosporus, II Zonguldak, III Sinop-Samsun, IV Trabson-Batumi, V Kolkhety, VI Gudauta, VII Gagra, VIII Sochi-Novorossiisk, IX Azovo-Kerch, X Mount Crimean, and XI Dniper-Danube (Figure 4.4.1.1).

Each listed part has some characteristics in geological composition, morphology, effect of wave regime, and processes of litho-dynamics on the coast. The combination of all those characteristics determines the SGD. The accepted principle makes it possible to determine the following areas.

TABLE 4.4.1.1
Shelf Classification of the Black Sea Basin

		Shelf Size			
No.	Area Name	Shelf Type	Length (km)	Width (km)	Depth (m)	Geology and Remarks

No.	Area Name	Shelf Type	Length (km)	Width (km)	Depth (m)	Geology and Remarks
I	Bosporus					Mezy Plate, wide shelf
	a. North part	Sh	250	35	100	West Pont–Folded System,
	b. South part	Sh	150	20	150	volcanogenic rocks ($Pg_2^2 + K_2$)
II	Zonguldag					West Pont-flysch, carbonic and
	a. Carbonate rock	K	170	7	100	volcanogenic rock
	b. Volcanic rock	V	75	10	150	
III	Sinop-Samsung					Same lithology, 4 big peninsulas
	a. Wide shelf	Sh	250	20	150	(2 stores each)
	b. Peninsulas	V	100	6		
IV	Trabzon-Batumi	V	400	3.5	100	East Point-West End of Adjara-Trialethi. The same lithology. The highest (3000 m) ranges in basin
V	Kolkhida					West end of Georgian Plate.
	a. Poti	Sh	100	25	100	Intensive sinking movements
	b. Ochamchire	Sh	40	20	200	
	c. Sokhumi	K	50	2.5	150	
VI	Gudautha	Sh	75	30	150	Q_1–20 m on uplift structure of Pont and Meotis age
VII	Gagra	K	60	0.5	50	Q_1–40 m, Pitsunda Peninsula has great depth; Gagra, the highest karstic system; Reprua, biggest submarine spring
VIII	Sochi-Novorossiisk	K	300	6	100	Same geology, many canyons and landslides (even on the beach)
IX	Azov-Kerch	Sh	200	250	100	Shelf is situated in Indol-Kuban Plate, mechanical material flow from Azov Sea
X	Maunt-Krimea					Carbonic rocks abrasia, the sea bottom sinks rapidly
	a. Carbonic rock	K	50	7	150	
	b. Wide shelf	Sh	75	25		
XI	Dnieper-Danube	Sh	600	150	100	End of Skiph Plate, Dobruja Massif, and Mazy Plate

Shelf Type: Sh, wide one —— stable area; V, volcanic rock —— folded system; K, carbonic rock —— karstic system.

The eugeosyncline type of shelf, comprising the West Pontians (Zongulak district), East Pontians, and the Adjaro-Trialety folded system (Trabson-Batumi district), has a relatively small width (average, 3 km), steep slope (0.07), and is rather poorly defined down to the 100-m isobaths. Morphologically the shelf base is poorly

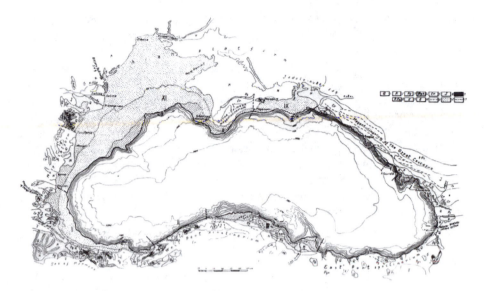

FIGURE 4.4.1.1 Division of the Black Sea basin shelf.

defined, being composed of volcanic-sedimentary rock of the Mezo-Cenozoic and locally by sandy-pebbly and volcanogenic flysch rocks of the Quaternary.

At present the shelf is characterized by a vast development of submarine canyons deeply penetrated into the bedrocks and in some cases with intensive SGD (springs and seeps).

The Sinop-Samsung district of the Turkish littoral, as well as the districts of the Major Caucasus southern slope (Gudauta, Gagra, and Sochi-Novorossiisk here are widely developed karstic processes) is of the *miogeosyncline-zone* shelf type. At present it is characterized by differential neotectonics. Some areas (e.g., the Sinop and Samsun capes, and the Gudauta bank) experience rising, which causes development of an extensive shelf, down to the 150-m isobaths.

The third, *platform* type of shelf, is confined to the Scythian and Mezy platforms and part of the Georgian Block-Kolkhethi. Within the Dnieper-Danube district it is characterized by a wide (200 km) and very gentle (0.001) shelf extended down to the 150-m isobaths. The shelf is composed only of Quaternary sands overlying the heterogeneous bedrocks and is characterized by an absence of submarine canyons. As a matter of fact, only the SGD for the coastal zone of Kolkhethi has been studied more or less in detail.

4.4.1.2 Coastal Part of Georgia

Along with other processes, general geotectonic structure influences SGD as well.

The coastal part of Georgia includes (from northwest to southeast) the southern slope of the Major Caucasus folded system (mostly carbonaceous rock), the Georgian block with thick Quaternary deposits (Kolkhethi Lowland), and the Adjara-Imereti folded zone of the Minor Caucasus (volcanogenic rock). According to the above zoning of the coastal part of Georgia, three types of SGD are noted: the first occurs mainly in karstic rocks, the second is typical of the phreatic aquifers and shallow subtidal zone, and the last prevails with deep offshore discharge from the fissures of volcanogenic rocks.

The karstic water recharge areas are located at an elevation of 2–2.5 km, and discharge is very intensive and equal to over hundred of liters per second (e.g., springs in the sea yielding, Gagra and Leselidze). The module of submarine water discharge for the limestone area of the Lower Paleogene and Upper Cretaceous is 30 l/sec·km^2 and resources equal 50 m^3/sec (Gantiadi, Tsintskaro springs). For the Gagra-area submarine spring Reprua, located 25 m from the coast with a depth of 6 m, has a debit equal to 1500–2500 m^3/sec. In total, resources for massive carbonate rock reach 90–199 m^3/sec^{-1}.

In the plane part of the Kolkhethi coastal zone, shallow groundwater resources of Lower Quaternary alluvial deposits of the ancient and modern terraces equal 9 m^3/sec. Also in the narrow strip of dune–sea–land interaction, area with values of submarine discharge equal to 1 m^3/sec is noted. Thus, the total resources here are no more than 10 m^3/sec.

In Adjara, SGD represents itself by a deep penetrating system of pressured fracture waters. Investigations here are considered to be the most difficult to do. But certainly discharge in this type of system is deepest and the common value of the module is quite high.

The active circulation zone in folded systems stretch out 600–1000 m and underground water salinity equals 100–1000 mg/l. Subzones of comparative active circulation, situated inside the layer of the slow circulated water. Its thickness is 1000 m, water salinity is ~1.5 g/l, and discharges mostly into the Black Sea. Due to the depth of circulation (2.5–3.0 km) these waters are thermal (t ≥ 100°C, Primorsky Termi). Depending on its location, waters of slow circulation zone have to discharge on the deep part of the continental slope. Underground flow of the active zone equals 360 m^3/sec or 13.3 l/sec·km^2. This value decreases toward the east with decreasing precipitation. The average value of modules equals 0.2 for Paleocene and 23.5 l/sec·km^2 for carbonated rocks of Pg + K + J complex. In general, underground water flow of West Georgia is 30.7% of river runoff and equals 95.4 m^3/sec. For Quaternary sediment, underground flow equals 45.7 m^3/sec, despite the fact that their module is only 9.6 l/sec·km^2.

Using the hydrogeochemical balance method it was determined that surface flow equals 188·10^3 g/sec, underground flow 77·10^3 g/sec, and mechanical flow 355·10^3 g/sec. Chemical flow and underground chemical flow were estimated at 0.5 and 0.2, respectively. The amount of dissolved hydrocarbons equals 166·10^3 g/sec and, hence, is close to that of full surface flow. The main result of such investigations is that the full chemical flow (i.e., interaction of underground water and rocks) is very similar to mechanical denudation.

4.4.1.3   Shelf Typology and Submarine Groundwater Discharge Estimation

Submarine groundwater discharge rate is determined by development of river system. Therefore, geology, precipitation, evapotranspiration, and topography are the main factors determining SGD. In our investigations it is possible to exclude recirculation of seawater in SGD. Thus, using the LOICZ methodology for SGD made it possible to determine the four main elements influencing SGD. They are precipitation, evapotranspiration, land elevation, and river runoff.

If we generalize these principles it is possible to determine the following types for an entire basin (Table 4.4.1.2).

1. Carbonate rocks (karsts intensive processes) with module and resources equal to 30 l/sec·km^2 and 50 m^3/sec, respectively. The type of discharge is concentrated (springs, Gagra, Ganthiadi) and dissipated (Leselidze).
2. Volcanogenic rocks of folded systems with fissured circulation. The module equals 1.5–2.0 times less than described above.
3. Plane part of coastal zone, so-called stable zone (or intensive sinking area) with wide shelf. The main aquifer is represented by Lower Quaternary alluvial deposits. In the whole module is 4–5 times less than previous, but even in some cases it is only 2–3 times less.

4.4.1.4   Submarine Groundwater Discharge and Pollution

The coastal zone is highly anthropogenically developed. Via submarine springs and seeps, SGD is considered to be the main pathway for pollution penetration into the coastal zone.

According to the results of work carried out in the Black Sea basin for over a century, it has been discovered that the hydrogen sulfide concentration profile has two maximums. The first one (13.5 ml/l) is near the sea bottom, due to sulfate reduction by activity of anaerobic bacteria in the upper part of the sedimentary layer. The other (7.0 ml/l) is situated below the interface, through which 20% of organic matter moves into the upper layer.

The existence and increase of the anoxic layer results in low velocities of water convection. This condition was noted in the Black Sea in the Holocene. Before, it was the New Euxine basin with almost freshwater, at a level 30–40 m below the modern level. The structure of the Bosporus Strait, where the shallowest depth on the south sill is 24 m, allowed the salty water of the Mediterranean Sea to flow into

---

TABLE 4.4.1.2
Summary of Submarine Groundwater Discharge (SDG) in the Black Sea Basin

No.	Shelf Type	Shelf Surface 10^3(km^2)	SDG (km^3·yr^{-1})	River Runoff (km^3·yr^{-1})	SGD Runoff (%)
I	Wide Shelf (stable area)	115.5	8.9	296	3.0
II	Shelf of carbonate rock (karstic systems)	3.3	4.5	13	34.6
III	Shelf of volcanogenic rock (folded systems)	2.8	3.0	14	21.4
	Total	121.6	16.4	323	5.9

the Black Sea as the ocean level approached contemporary values. As a result, the water stratification is controlled by density, e.g., the hydrogen sulfide zone, and now occupies 90% of the entire basin. The bathygraphic curve makes it possible to express the interface bedding as a logarithmic function of time. At the beginning this surface rose rapidly enough and the hydrogen sulfide zone covered half the surface of the sea bottom within 400 years. Thereafter the speed decreased over the last 2000 years; the contemporary period shows an asymptotic approach to the constant level (200–220m).

The aquatic basin adjoining the Georgian shore has been examined. The first systematic measurements were made by a joint expedition of the former USSR and the Georgian Academy of Sciences in 1973. At different distances from seashore, with different depths (300–1800 m), the thicknesses of the oxygenated water layer ("the layer of life") were 200–220 m for the basin. Over the next 15 years the depth interval remained the same. In 1998 on the Batumi traverse at 5 km from shore, it was discovered that the mean value of three measurements was just 150 m.

Systematic studies for the entire seacoast were planned, but only the Adjara zone has been measured. In this zone two distinct parts are noted. The boundary of the two large areas is situated near the Tsikhisdziri Cape. The mean depth of the interface (195 m) is near a large area, north of the Green Cape traverse (Figure 4.4.1.2.; Table 4.4.1.3). A statistically significant decrease on 5 m in such a short time (15–20 years) cannot be caused by natural regional changes. This value does not depend on the sea depth or structure of the shelf. The main reason of change in the interface has been an increase in organic matter inflow. This increase occurs due to intensive pollution of coastal land with agricultural and municipal wastes.

Abnormal interface depths are found on the traverses of Batumi and Makhinjauri, 155 m and 160 m, respectively. This area does not differ from the neighboring

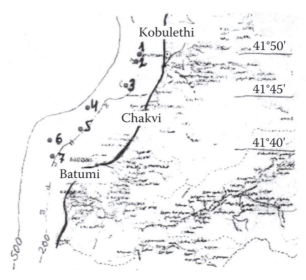

FIGURE 4.4.1.2 Marine stations in the Adjara coast.

TABLE 4.4.1.3
Depth of H$_2$S Surface

Station No.	Traverse Course, Gradient	Distance from Seashore (m)	Depth of the Sea (m)	Depth of Interface (m)
1	Kobuleti, 335°	2000	330	192–197
2	Kintrishi, 330°	3500	350	193–198
3	Tsikhisdziri, 335°	4200	400	192–197
4	Chakvi, 335°	4300	480	189–194
5	Green Cape, 330°	4600	380	194–199
6	Makhinjauri, 330⁰	4000	410	158–163
7	Batumi, 316°	5600	290	152–157

regions either morphologically or hydrologically. We assume it is a gross increase in the concentration of organic input. In fact, the source of pollution must be the large Batumi oil factory, where oil products are found at a depth of several dozen meters by drilling. Also there is a thick layer of black oil on the sea bottom 2–3 km from the shore.

It is possible to conclude that the change in the hydrogen sulfide interface depth reflects the pollution of the coast and adjacent waters. This can be considered as a universal criterion for assessing seashore ecological condition.

4.4.2 QUANTITATIVE ASSESSMENT OF SUBMARINE GROUNDWATER DISCHARGE

Estimation of SGD into the Black Sea has been made by extrapolating of data received from well-studied areas (e.g., Kolkhethi) of the entire basin. Dividing the shelf into three principal types makes it possible to evaluate (though roughly) a SGD value of 16.4 km³/yr or 5.9% of surface runoff. Submarine groundwater discharge of areas has values ranging in order (3% and >30%). A universal criterion of assessment of seashore ecological situation is change of H$_2$S surface in the littoral zone.

For future work different methods of measurements, balance, and estimation of recirculated seawater, and so on can be used.

In conclusion it is possible to use the quotation: "submarine groundwater discharge remains somewhat of a *mystery* because it's usually unseen and difficult to measure." Now and in the future it will be very interesting to study SGD.

4.5 GROUNDWATER DISCHARGE TO THE MEDITERRANEAN SEA

Submarine groundwater discharge to the Mediterranean Sea is formed within coastal zones of three continents. The position of the sea is such that SGD can be compared with other contributions to water balance. The basic role among them belongs to evaporation from water surface (3202 km³/yr) and the resultant water exchange through straits with the Atlantic Ocean and the Black Sea. Annual evaporation is considerably higher than precipitation (980 km³/yr) and river runoff (280 km³/yr), which leads to a deficit in the freshwater budget year-round (Ovchinnikov 1973).

Noncompensation of evaporation causes a decline of the Mediterranean Sea level, which, in turn, leads to constant seawater inflow from the neighboring basins. However, the existing water balance calculations did not take account of the role of subsurface runoff. At the same time it is known that SGD in the Mediterranean Sea is developed very intensively. There are numerous concentrated discharges of groundwater on the sea bottom, forming submarine springs with high yields. By the number of such springs the Mediterranean Sea is probably a unique sea basin (Zektser et al. 1972).

Total groundwater runoff to the Mediterranean Sea reaches almost 68 km³/yr, which is equal to ~24% of the river runoff. This figure is distributed among the continents as: discharge from the territory of Europe is ~49 km³/yr, Asia 8.3 km³/yr, Africa 5.1 km³/yr, and the largest islands ~6 km³/yr (see Table 3.2.1, Section 3.2 and Table 4.5.1, Section 4.5).

A major part of the groundwater discharge to the Mediterranean Sea is formed within the European continent. This is connected with favorable climate, orographic, and geologic-hydrogeological conditions. Precipitation is often >1000 mm/yr. Maximum precipitation falls in winter, favoring more active recharge of groundwater. Mountainous relief of the coasts exerts a screening effect on atmospheric circulation and causes a higher wetting. A basic factor causing intensive groundwater discharge is widely developed karst. Karst hollows absorb precipitation and surfacewaters and often conduct them into the sea. Yields of submarine karst springs reach

TABLE 4.5.1
Subsurface Runoff to the Mediterranean Sea

Mediterranean Sea	Water Runoff			Ion Runoff		
	Modulus (l/sec·km²)	Discharge (ths m³/day·km)	Total Value (km³/yr)	Modulus (t/yr·km²)	Discharge (ths t/yr·km)	Total Value (mln t/yr)
Africa	0.4	3.1	5.1	24.4	2.2	9.9
Asia	2.4	7.0	8.3	110.3	3.6	11.9
Europe	5.7	15.6	48.7	101.8	3.2	27.4
Major Islands	2.8	8.1	5.7	34.9	1.2	2.3
Total			67.8			51.5

10–15 m³/sec. Due to this, moduli of SGD achieve 5–6 l/sec·km², increasing in the area of the Dinara Mountains to almost 13 l/sec·km². In a structural-hydrogeological respect, the European coast of the Mediterranean Sea represents an alternation of hydrogeological formations and areas, associated with mountainous structures, and small artesian basins connected with tectonic depressions. Hydrogeological formations are composed, as a rule, of highly karstic carbonate rock of Mesozoic–Cenozoic age. In some areas, the groundwater is also associated with fissured volcanogenic formations. A degree of karst development and jointing of water-bearing rocks determine the intensity of groundwater discharge, the high moduli of which are evidence that the aquifers are well wetted. In this connection, mineralization of groundwater usually is not greater than 1 g/l and amount on average to 0.3–0.7 g/l (Aronis 1963; Custodio et al. 1977).

Artesian and coastal plains are composed, chiefly, of alluvial marine sediments of Neogene–Quaternary ages. Besides, their structure is considerably influenced by flysch strata of Cretaceous and Paleogene Periods. In some artesian structures on the Apennines and Pyrenean Peninsulas, some water-bearing rocks contain salt-bearing sediments. Water content in these rocks is lower than in karstic carbonate rocks. However, moduli of the SGD in these areas are also 3–4 l/sec·km². In general, groundwater mineralization is low (1 g/l), but in areas where evaporates are developed it can reach 5–7 g/l and greater.

Total ion runoff from the European continent to the Mediterranean Sea is equal to 27.4 mln t/yr. And the most significant ion runoff is typical of the coastal zones of the Balkan and Apennines Peninsulas. Areal moduli of ion runoff vary from 50–100 t/(yr·km²), and in the hydrogeological area of Dinar karst they reach 280 t/yr·km².

Favorable conditions for groundwater discharge into the Mediterranean Sea are observed in the Near East and Asia Minor where the moduli of submarine groundwater runoff reach 3 l/sec·km². There is a large amount of precipitation, with a maximum amount falling in the winter period when evaporation is low. In this region, there are distinguished Libyan-Sinaitic, Antal, Menderess, and Aden artesian basins composed, chiefly, of karstic limestone, dolomites, conglomerates, and sandstones of the Mesozoic and Neogene, with a thickness of from a few hundred to 3000 m (Hydrogeology of Asia 1974). High jointing and a large quantity of caverns in water-bearing rocks cooperate to form significant groundwater resources. However, their mineralization is varied and, as a rule, increases southward where sediments include gypsum-containing limestone and terrigenic formations. Fresh and low-brackish waters prevail in the northern and central parts; southward the mineralization reaches 12 g/l. Increased mineralization of groundwater in this region causes considerable ion runoff from the Asian continent to the Mediterranean Sea, the moduli of which gradually increase from north to south from 46–140 t/yr·km².

Basic groundwater discharge to the Mediterranean Sea from Africa is formed in the Atlas Mountains where three artesian basins are distinguished: the Muluinsky, Tell-Atlas, and Tunisian (Hydrogeology of Africa 1978; UNESCO 2004). Artesian structures consist of thick strata of Meso-Cenozoic rock. Basic aquifers are connected with limestone, conglomerates, and partly by effusives of the Neogene–Quaternary. Karstic limestone and conglomerates have a high water conductivity (2500 m²/day), which in combination with a relatively high precipitation (1000 mm/yr), mountainous relief, and tectonic fissuring create favorable conditions for groundwater recharge. Total submarine groundwater runoff reaches 3.5 km³/yr, and average areal modulus is 1.3 l/sec·km². In the upper part the groundwater has an increased mineralization (5 g/l) because of continental salinization processes. Confined aquifers contain a depth of 400–500 m of freshwater and low-brackish groundwater with a mineralization of 3 g/l. Total subsurface ion runoff is 3.5 mln t/yr with an average groundwater mineralization of 1 g/l for the entire Atlas hydrogeological region. Areal modulus of ion runoff is 41 t/yr·km², and linear discharge totals 1.8 ths t/yr·km (Table 4.5.1).

To the east of the Atlas hydrogeological area within the Libyan Desert, there is a complicated Libyan-Egyptian artesian basin. Conditions for groundwater discharge in the basin are extremely unfavorable, because it belongs to the driest zone on earth with an average annual precipitation of 1–2 mm (Hydrometeoizdat 1974). Basic aquifers are associated with limestone, conglomerates, sandstones, and sands of the Cretaceous, Paleogene, and Neogene. Generally, water content of sediments is slight; moduli of SGD are <0.2 l/sec·km². Near to the shoreline, the groundwater in all aquifers is gradually salinized (10 g/l), because of which the total ion runoff reaches 6.4 mln t/yr.

The Mediterranean subtropics belong to moderately wetted regions where groundwater runoff is not high. However, availability of karst in combination with other favorable factors lead to a considerable increase of groundwater discharge. Particular conditions for formation of SGD enable the Mediterranean Sea coast of Europe to be considered as an azonal region not corresponding to latitudinal zonality of groundwater discharges into seas. Moreover, evaluations of groundwater discharge into some lakes and seas of the former USSR show that it is equal to the initial percentages of the river runoff (Dzhamalov et al. 1977). At the same time the total groundwater discharge to the Mediterranean Sea amounts to 24% of the riverwater inflow. This outlines once more that the given sea basin is unique and that the subsurface water contribution is necessary to be taken into account in calculations of water balance.

4.6   GROUNDWATER DISCHARGE TO BISCAYNE BAY, FLORIDA*

Biscayne Bay is a coastal barrier island lagoon that relies on substantial quantities of freshwater to sustain its estuarine ecosystem. During the past century, field observations have suggested that Biscayne Bay changed from a system largely controlled by widespread and continuous SGD and overland sheetflow to one controlled by episodic discharge of surfacewater at the mouths of canals. Consequently, many of the organisms that require a narrow salinity envelope cannot tolerate the rapid salinity changes that occur in response to freshwater releases near the mouths of canals. Current ecosystem restoration efforts in southern Florida, which are focused on restoring the Everglades area to the west of Biscayne Bay, are examining alternative water management scenarios that could further change the quantity and

---

* Section 4.6 was written by Christian Langevin, Ph.D. (Center for Water and Restoration Studies, Miami, Florida).

timing of freshwater delivery to the bay. Ecosystem managers are concerned that these proposed modifications could adversely affect bay salinities by decreasing fresh groundwater or surfacewater flows. Currently, the two most important mechanisms for freshwater discharge to Biscayne Bay are thought to be canal discharges and SGD from the Biscayne aquifer. Canal discharges are routinely measured and recorded, but few studies have attempted to quantify rates of SGD. In 1996, the U.S. Geological Survey (USGS) initiated a project in cooperation with the U.S. Army Corps of Engineers to quantify the rates of SGD to Biscayne Bay. This project was accomplished through field investigation and the development of a numerical groundwater flow model that covers most of Miami-Dade County and parts of Broward and Monroe Counties, Florida (Figure 4.6.1).

For the study of SGD to Biscayne Bay, project objectives and the geometry of coastal hydrologic features required the development of a full 3-dimensional model. This section describes the model development process and the application of the variable-density SEAWAT code (Guo and Langevin 2002), a combined version of MODFLOW and MT3D, for the purpose of quantifying regional-scale SGD to a marine estuary. Detailed descriptions of the USGS study are given in Langevin (2001, 2003).

The physiographic features of southeastern Florida are relatively subtle, but because of the flat topography, small changes in land surface elevation can substantially affect surface and groundwater flow. The Atlantic Coastal Ridge separates the Everglades from the Atlantic Ocean and Biscayne Bay. The ridge, which is 5–15 km wide, roughly parallels the coast in the northern half of Miami-Dade County. In southern Miami-Dade County, the Atlantic Coastal Ridge is located farther inland, and low-lying coastal areas and mangrove swamps adjoin Biscayne Bay. Prior to development, high-standing surfacewater in the Everglades flowed through the transverse glades (low-lying areas that cut through the Atlantic Coastal Ridge) into Biscayne Bay.

Throughout much of the study area, a complex network of levees, canals, and control structures is used to manage local water resources. The major canals, operated and maintained by the South Florida Water Management District, are used to prevent low areas from flooding and to prevent saltwater from intruding into the Biscayne aquifer by maintaining the height of the water table. These water management canals are particularly effective in managing the height of the water table because they were dredged into the highly transmissive Biscayne aquifer. The sides of the canals are porous limestone, which means the canals are in direct hydraulic connection with the aquifer.

The hydrostratigraphy of southeastern Florida is characterized by the shallow surficial aquifer system and the deeper Floridan aquifer system. Previous studies suggest that groundwater discharging to Biscayne Bay originates from the Biscayne aquifer, which is part of the surficial aquifer system, and thus the numerical model does not extend beneath the Biscayne aquifer. The highly permeable Biscayne aquifer principally consists of porous limestone that ranges in age from Pliocene to Pleistocene (Fish and Stewart 1991). Hydraulic conductivities of the Biscayne aquifer range from 300 to >3000 m/day. The Biscayne aquifer is absent in much of western Miami-Dade County, but can be over 55-m thick near the coast.

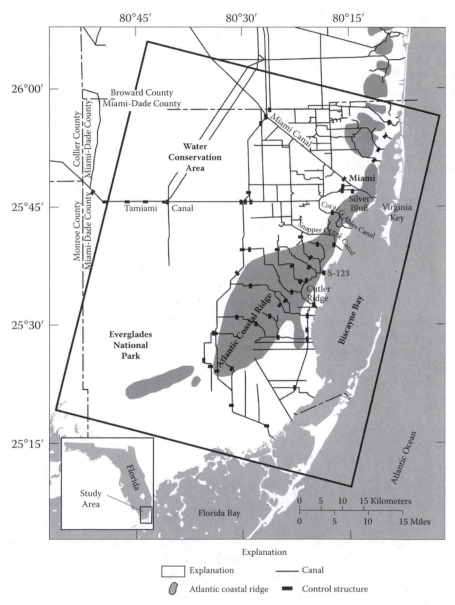

FIGURE 4.6.1 Map of southern Florida showing location of the study area, domain of the regional-scale model, and other hydrologic features.

The regional-scale model developed as part of this study simulates transient groundwater discharge to Biscayne Bay for an approximate 10-year period from January 1989 to September 1998. The model was developed using the conceptual hydrologic model shown in Figure 4.6.2. It is assumed that the Biscayne aquifer can be simulated with an equivalent porous medium (EPM). Historical observations suggest that submerged groundwater springs were once active within Biscayne Bay.

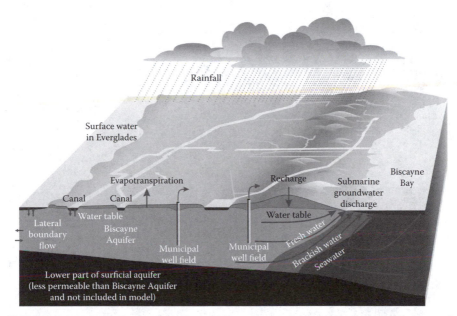

FIGURE 4.6.2 Conceptual hydrologic model used to develop the numerical model of groundwater discharge to Biscayne Bay.

By using the assumption of an EPM, individual springs and conduits are not explicitly simulated, but rather the properties of the conduits are averaged within model cells. This assumption limits the interpretation of model results at local scales, but is thought to be appropriate when conduits or fractures are much smaller than the scale of the model.

To simulate groundwater flow to Biscayne Bay, a regularly spaced, finite-difference model grid was constructed and rotated so that the y-axis would roughly parallel the coast (Figure 4.6.3). Each cell is 1000×1000 m^2 in the horizontal plane. The grid consists of 89 rows, 71 columns, and 10 layers, and the rotation angle from true North is clockwise 14°. The purpose for rotating the grid is to align model rows with the principal direction of groundwater flow, which is primarily toward Biscayne Bay. To maintain reasonable computer runtimes, the domain of the regional model was not extended south to Florida Bay. Future versions of the model, however, may extend into Florida Bay as a method for eliminating potential boundary effects. Boundary conditions for the uppermost model layer are shown in Figure 4.6.3, and the details for each boundary are described in Table 4.6.1.

The nearly 10-year simulation period is divided into 116 monthly stress periods. For each stress period, the average hydrologic conditions for that month are assumed to remain constant. This means that the model simulates seasonal and yearly variations rather than hourly or daily hydrologic variations. Further temporal discretization is introduced in the form of time steps within each stress period. Each stress period is divided into one or more time steps, the lengths of which are determined by SEAWAT to meet specified criteria associated with solving the solute transport equation. For the regional-scale model, about 20 time steps were required for each

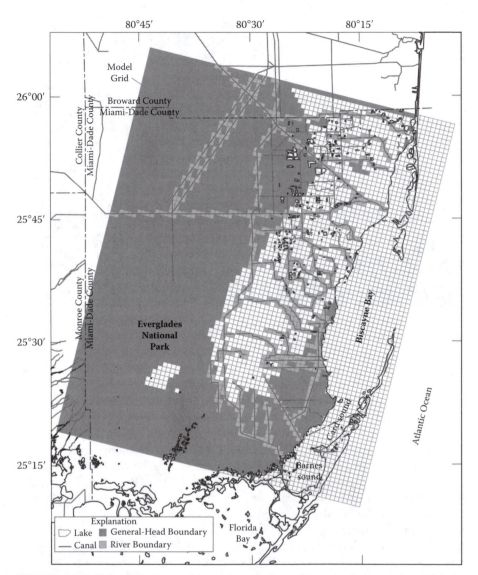

FIGURE 4.6.3 Finite-difference grid and boundary conditions in the upper layer for the regional-scale groundwater flow model in southern Florida. Heads and concentrations are specified for Biscayne Bay for all model cells in the upper layer.

stress period. The regional-scale model was calibrated using trial-and-error methods to match observed heads, groundwater exchange rates with canals, and the saltwater interface position. In general, the model was considered a reasonable approximation of field conditions. Details of the calibration are described by Langevin (2001).

The simulated groundwater flow from the active model cells into the coastal constant-head cells is assumed to represent groundwater discharge to Biscayne Bay. In addition to simulating the volumetric flow rate, the model also simulates the salt

TABLE 4.6.1
Boundary Conditions Used in Regional- Scale Groundwater Flow and Transport Model

Boundary Description	Boundary Type	Comments
Model Perimeter		
Active inland cells, layers 1–10	GHB	Time varying equivalent freshwater head value estimated from water table TINs[a] and salinity estimates.
Active offshore cells, layers 1–10	CHD	Time varying equivalent freshwater head calculated from Biscayne Bay stage and salinity.
Base of Model		
Base of Biscayne aquifer	No-flow	Model cells with center elevation below base of Biscayne aquifer were inactivated.
Biscayne Bay		
North Biscayne Bay	CHD	Time varying equivalent freshwater head calculated from Biscayne Bay stage and salinity value of 35 kg/m^3.
Central and South Biscayne Bay	CHD	Time varying equivalent freshwater head calculated from Biscayne Bay stage and salinity from hydrodynamic model.
Canals		
Primary and selected secondary water management canals (Figure 4.6.3)	RIV	Canal stages vary temporally based on field data; conductance determined through calibration.
Standing surface water		
Everglades and coastal wetlands (Figure 4.6.3)	GHB	Time varying heads calculated using water table TINs[a] and salt concentration of zero; conductance value of $1 \cdot 10^5$ m^2/day determined by calibration.
Recharge		
Uppermost active cell	RCH	Recharge concentration assigned value of zero; recharge rate estimated from rainfall and runoff.
Evapotranspiration		
Uppermost active cell	EVT	Evapotranspiration rate assigned based on linear function; extinction depths range from 0.15–1.8 m.
Groundwater withdrawals		
Municipal well fields	WEL	Monthly withdrawals for each well field were obtained from South Florida Water Management District and Miami-Dade County.

[a]For each month of the simulation, triangular irregular networks (TINs) were developed for the water table, using available surfacewater and shallow groundwater monitoring stations.

GHB, general head boundary; CHD, time variant constant head; RIV, river; RCH, recharge; EVT, evapotranspiration; WEL, well.

concentration of the groundwater flow into the constant-head cells. This salt concentration is assumed to represent the salinity of the groundwater that discharges to Biscayne Bay. The simulated salinity of groundwater discharge to Biscayne Bay ranges from nearly fresh at the shoreline to that of seawater some distance offshore. To simplify the results, simulated discharge estimates are presented as the freshwater portion of the total discharge. The freshwater portion of the groundwater discharge is calculated from the total discharge using the following Equation:

$$Q_f = Q_T \frac{(\rho_s - \rho)}{\rho_s - \rho_f} \tag{4.6.1}$$

where Q_f is simulated fresh groundwater discharge, in m³/day; Q_T is simulated total groundwater discharge, in m³/day; ρ_s is fluid density of seawater, in kg/m³; and ρ is simulated fluid density of groundwater discharging to Biscayne Bay, in kg/m³.

When the fluid density of the groundwater discharge is equal to 1000 kg/m³, the fresh groundwater discharge is equal to the total groundwater discharge. When fluid density of the groundwater discharge is equal to fluid density of seawater (1025 kg/m³), the fresh groundwater discharge is equal to zero. Destouni and Prieto (2003) used this type of relation to evaluate the freshwater component of the total discharge.

One potential problem with calculating fresh groundwater discharge estimates from simulated density is that the resulting freshwater discharge quantities are directly dependent on salt concentrations. (Density is a linear function of salt concentration.) As previously mentioned, groundwater salinities are considered calibrated when the simulated position of the saltwater toe matches with the observed position. This does not ensure that the simulated salt concentrations are calibrated. The average salt concentration of simulated discharge to Biscayne Bay is about half that of seawater, and thus half of the total discharge is freshwater. This suggests that if salt concentrations were in error by 100% (17.5 kg/m³), estimates of fresh groundwater discharge might be in error by about a factor of 2.

To show the relative contribution of freshwater to Biscayne Bay, a comparison was made between simulated fresh groundwater discharge and measured surfacewater discharge from the coastal control structures. The groundwater component was not known prior to this study, and thus the rates were estimated with the numerical model. Surfacewater discharges are routinely measured, however, and therefore the surfacewater component is known. Based on the results for the simulation period (1989–1998), fresh groundwater discharge seems to be an important mechanism of freshwater delivery to Biscayne Bay during some dry seasons (Figure 4.6.4). For the dry seasons of 1989, 1990, and 1991, model results suggest that fresh groundwater discharge exceeded the surfacewater discharge to Biscayne Bay. During the wet season, however, fresh groundwater discharge was about an order of magnitude less than the surfacewater discharge. For the total simulation period, groundwater discharge directly to Biscayne Bay was about 10% of surfacewater discharge.

Nearly 100% of the fresh groundwater discharge to Biscayne Bay was to the northern half of the bay; specifically, north of structure S-123 (Figure 4.6.1). This model result is also supported by salinity values at offshore monitoring wells.

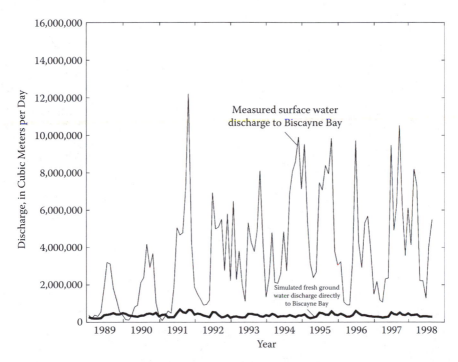

FIGURE 4.6.4 Simulated rates of fresh groundwater discharge compared with measured rates of surfacewater discharge to Biscayne Bay.

South of structure S-123 (Figure 4.6.1), where land surface elevations are <1 m above sea level, the water table was unable to develop enough head (due to runoff and high evapotranspiration rates) to drive groundwater discharge into the bay.

Results from the regional model presented here contain a high level of uncertainty. While the model seems adequately calibrated to heads and canal fluxes, there was no way to calibrate to SGD. This is particularly troublesome considering that the simulated estimate of SGD is only about 2% of the annual rainfall total. This means that the simulated discharge rate is well within the error range of the other water budget components (canal baseflow, evapotranspiration, runoff, and so forth). The reliability of the simulated SGD estimates will improve as estimates of recharge, surfacewater and groundwater interactions, and evapotranspiration improve.

CONTENTS

5.1 SUBMARINE GROUNDWATER DISCHARGE TO LAKE BAIKAL*

The submarine groundwater discharge (SGD) to Lake Baikal occurs in complicated geologic-geomorphological conditions, variable both along the perimeter of the shoreline and at the bottom of the Baikal depression. The calculation of lake water balance includes the groundwater that is discharged directly to the lake depression and not accounted for in the calculations of the river discharge.

Depending on the conditions of recharge, transient flow, and discharge, SGD consists of five basic components: spring, subchannel, interfluve, artesian, and deep discharges. The dynamics and regime of all these types of SGD are different and the methods used to evaluate them also differ considerably from each other. A relatively high reliability of calculations has been achieved in the evaluation of spring discharge. Comparatively speaking, the evaluation of subchannel discharge and discharge from interfluves through loose sediments is not as accurate. For the other discharge types, only tentative evaluation is possible with the use of indirect techniques and analogous methods. The description and assessment of each of the above-listed types of SGD are given below.

5.1.1 SPRING DISCHARGE

Springs located within the shore zone of Lake Baikal are represented by either concentrated jets discharged to the Earth's surface and adjoined to a single stream flowing into Lake Baikal or mochezhinas (permanently wet lands due to the outflow of underground water) and swamped areas formed due to the distributed stratum discharge of groundwater, as well as by submarine groundwater outbursts on the bottom of shallow creeks and bays.

The methodical procedure for the evaluation of spring yields is well known. A vital step is to reveal all the springs that exist in the shore zone. The basic indications for revealing open sources of SGD are the zones of tectonic disturbances, exposures of carbonates rocks, debris cones of river valleys and dry creak valleys, sharp slope discontinuities, areas with clear physical-geological phenomena, etc. The efficiency of spring mapping greatly increases with the preliminary decoding of aerocosmic photos, and special remote sensing or aero-visual flights, especially during winter when sources of SGD are easily registered by windows of open water and large icings of proparinas (glades) in the Lake Baikal ice cover. After the sources of SGD are revealed, they are monitored with follow-up examinations in February–March.

Using the above-mentioned methods, the thermal and cold-water springs and sources of distributed discharge, as well as a few submarine sources of cold water discharge, yielded a total of 2.53 m³/sec in the shoreline zone of Lake Baikal (Figure 5.1.1.1; Table 5.1.1.1). Thus, the results revealed seven thermal springs directly recharging the lake: one is on the western shore of North Baikal (Kotelnikovsky) with a water temperature of 81°C; and six springs are on the eastern shore

---

* Section 5.1 was written by Boris I. Pisarsky (Institute of Crystal Studies, Russia, pisarsky@irk.ru).

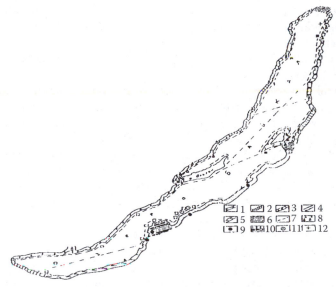

FIGURE 5.1.1.1 Geologic-geomorphological zoning of the shoreline; and the location of coastal and submarine discharge sources in the Lake Baikal water area. 1: Zone I (linear modulus of submarine groundwater discharge from interfluves, M = 0.2 l/secÞkm). 2: Zone II (M = 0.3 l/secÞkm). 3: Zone III (M = 0.4 l/secÞkm). 4: Zone IV (M = 0.5 l/secÞkm). 5: Zone V (M = 1 l/secÞkm). 6: Zone VI, delta areas of large rivers. 7: Cenozoic faults and those renewed in Cenozoic period. 8: Springs in the shore zone (a, thermal; b, cold). 9: Wells stripping thermal waters in the shore zone. 10: Proparinas (glades) in the ice cover of the lake, connected with cold (a) and thermal (b) waters and with emission of methane gases (c). 11: Wells that stripped thermal and subthermal waters within the lake area. 12: Hydrochemical anomalies in soil solutions of bottom sediments.

TABLE 5.1.1.1
Spring Discharge to Lake Baikal

Type of Spring	Spring Yield 1/sec	Total Yield 1/sec	Annual Average Discharge km³/yr
Thermal springs, 34–75°C	1–40	120	
Cold springs, 0.8–9°C	20–300	933	
Distributed discharge of cold waters (low areas of the shore), 0.5–4°C	50–200	1580	0.05
Total	—	2850	0.08

(Khakussky, Ayaisky, Davshinsky, Zmeiny, Goryachinsky) with one in the Kulinye swamps and their temperatures range from 34°–75°C. The total yield of hydrotherms is 0.10 m³/sec. Of particular note is the Kotelnikovsky spring. Testing in a well drilled in a hydrothermal field at a depth of 44 m showed that its yield, at 1-m lowerings, amounted to 11 l/sec, with a temperature of 80°C. This result enables estimation of the thermal water discharge to the lake (while taking into account natural onshore and submarine discharges) at not less than 20 l/sec. The multi- and intrayear thermal water regime is sufficiently stable. According to observation data at the Khakussky and Davshinsky springs, the changes in the yield in time do not exceed ±6%.

Sources of freshwaters are seldom met in the shore zone and (like thermal sources) are usually associated with the zones of tectonic disturbances (Figure 5.1.1.2) or with areas of actively manifested physical-geological phenomena (landslides, karst, permafrost, etc.).

On the western shore of Lake Baikal, the Zunduksky spring in the Priolkhonie area, located near the water line at the foot of exposed carbonate rocks, is especially distinguished. How much the spring opens depends entirely on the position of the lake level; therefore, the major part of the spring gryphon is submerged. In 1981 (when the water level in Lake Baikal was historically low), all the SGD appeared to be on land; its complete yield was measured and amounted to –0.3 m³/sec at 4.2°C. This powerfully concentrated discharge was associated with a large tectonically disturbed zone in the carbonate rocks. Other concentrated springs with large yields were revealed and examined on the cape of Elokhin, in the creek of Sosnovka, Chevyrkuisky Bay, and in South Baikal. In total, 29 fresh cold-water springs were registered with a temperatures of 4–9°C and yields of 5–300 l/sec. At Priolkhonie, where nearly half of all revealed springs are concentrated on the shore of the Maloye Sea, the groundwater discharges are associated with the bottom of bedrock slopes and debris cones in creek valleys.

Spring discharges may also be formed within sliding slopes. An example is the spring of the Kurminsky slope in Priolkhonie. There, within 2 km along the Maloye Sea on the exposed slope of a high terrace, there is a contact zone between thawed rocks (sand, gravel, pebbles) and frozen clays, where numbers of springs are located with a total yield of 2.4 l/sec. According to stationary observations, these springs have a seasonal character; in winter all their discharge is used for the formation of ice.

The total yield of all the cold-water springs is estimated at 0.93 m³/sec (Table 5.1.1.1). Intra-annual distribution of the discharge is relatively stable, though low-yield springs during the winter spend the major part of their discharge in the formation of ice. As stationary observations show, the lowest spring yields are in March and equal, on average, 70% of the annual average discharge.

Springs with distributed cold-water discharge are also detected in lowland areas of the shore. This type of spring discharge is widely met in the mouth areas of river valleys of the Middle and North Baikal as well as in some small intermontane depressions. The total distributed discharge in such areas is determined by direct measurements which amounts to 1.5 m³/l. In winter the discharge similarly decreases to the concentrated springs (70% of annual average value). The total spring discharge to Lake Baikal within its shore zone (land) amounts to 2.85 m³/sec (Table 5.1.1.1).

This type of SGD to Lake Baikal also involves the total fresh cold-water springs mapped at shallow depths (1–5 m) in the coastal zone of the Maloye Sea. Such submarine springs are usually registered in winter according to proparinas (glades) in the ice cover. Very often such proparinas are formed near river mouths because of a warming effect of the subchannel flow being discharged through areal seepage. Sometimes they are connected with submarine springs on the bottom and associated with the zones of tectonic disturbances. One such spring is found in the bay of Mukhor at the northwestern cape of the small island of Kharga (Maloye Sea) (Figure 5.1.1.1). This ascending submarine vent (facula) is discharged at a depth of 1 m with a yield of 0.5 l/sec at of 4°C (the lake water was 1°C). By composition, the water is of a hydrocarbonate calcium type, but its mineralization (0.4 g/l) is 4 times higher than the water in Lake Baikal.

5.1.2   SUBCHANNEL DISCHARGE

The major rivers in the Baikal area that have relatively small drainage areas and highly steep valley banks usually accumulate coarse-fragmented, well-washed alluvium with high filtration properties in the river mouths.

The subchannel discharge was estimated using data of detailed investigations on the Baikal Lake shore in 1985–2004. The most reliable estimation of subchannel discharge in the river valleys was carried out with the results of experimental work and calculations of the groundwater flow rate. For uninvestigated or poorly investigated rivers, the subchannel discharge was calculated taking the parametric data obtained by pumping results from the near located wells. Thus, the data of 292 hydrogeological wells were used, those having stripped the aquifers in loose Quaternary and Neogene sediments of the North Baikal and Priolkhonie coastal zones. In cases where the river drainage areas incurred losses of surface runoff for infiltration, the subchannel discharge was determined by a difference of the river flow rates between two measuring posts at the boundaries of the absorption zone. This method was used at 38 tributaries that were the least studied in the hydrogeological respect and where the shoreline relief was represented by low-dissected, relatively flat, piedmont plains formed by river debris cones. The upper parts of the debris cones, where coarse-fragmented sediments have the highest water permeability, serve as recharge areas, that is, zones of surface runoff absorption. In such areas the medium-sized rivers (with a drainage area of to 1000 km²) lose about 25%–40%, and small water streams up to 100% of their discharge.

Having analyzed the data for a number of North Baikal rivers, one established a direct relationship between annual average losses of the river discharge (equal to the subchannel discharge) and the size of the river mouth area, beginning from the debris cones to the Lake Baikal water line (Figure 5.1.1.2). The annual average subchannel discharge was determined on the basis of hydrometric works; the area of debris cones was calculated by the following simplified equation:

$$F_{con.} = \frac{a \cdot L}{2} \qquad (5.1.2.1)$$

where $F_{con.}$ is the area of the debris cone, a is the distance from the river mouth to a point where discharge losses start, and L is the length of the shoreline in the river mouth area.

The thickness of loose sediments composing the river valley debris cone, where a subchannel discharge is formed, was assumed to be within 40 m for all the rivers of the lake's northern shore.

The application of the given relationship to other concrete natural conditions can be specific and will require additional substantiation.

The western coast of South and North Baikal consists of continuous mountainous ridges composed of crystalline rocks and has a steep slope. There are only eight relatively large water streams and about 20 small ones in the mouth areas, of which loose and sufficiently thick sediments are developed with possible subchannel discharge. This discharge was estimated from the data of the hydrometric surveys. The investigations have shown that the hypsometric posts, installed below the boundary of water flow out of the mountainous frame to the lake terraces, miss (do not take into account) a part of the river discharge entering Baikal with the subchannel runoff. The overlooked discharge is sufficiently high so that in some water streams the surface runoff, 3–6 km from the lake, is entirely lost. Hydrometric data of subchannel discharge in the rivers of the western lake coast are presented in Table 5.1.2.1.

The least studied region in the hydrological and hydrogeological respect is the eastern coast of North Baikal. Systematic data on discharge in this area are available for only two rivers; for the other rivers only single characteristics were obtained during hydrometric surveying in summer and winter low-water periods. During the investigations of 1980–1981 when water levels in Baikal were historically low, the subchannel discharge from the Tompuda and Davsha Rivers amounted to 5.72 m³/sec, while the other streams were 2.74 m³/sec. The regime investigations during autumn-winter periods on the northeastern coast of Baikal made it possible to verify the surface runoff spent for infiltration in the coastal zone. The portion of subchannel discharge was 12%–23% of surface runoff, depending on the period of the year. The total subchannel discharge for the northeastern coast is estimated at 11.13 m³/sec (0.35 km³ /yr).

For the Turka and Kika Rivers on the eastern coast, there are no data on filtration properties of loose sediments widely spread in their valleys, which makes it impossible to determine their subchannel discharge to Lake Baikal by means of calculation.

TABLE 5.1.2.1

River	Drainage Area (km²)	Multiyear Average River Discharge (m³/sec)	Subchannel Discharge (m³/sec)	Subchannel Discharge of River Discharge (%)
Bughuldeika	1700	3.87	0.70	18
Sarma	768	5.93	2.37	40
Rel	567	13.4	2.68	20
Tya	2580	40.4	4.04	10
Kholodnaya	1050	6.10	0.60	10
Water streams in an area from Sarma to Slyudyanka River (North) (18 rivers)	1364	—	8.61	50–100
Total	8029	—	—	19.0

The most similar of these rivers by the valley structure and discharge type is the Rel River. By means of analogy, the subchannel discharge of Turka and Kika is taken as equal to 20% of the loss of groundwater recharge (19.2 m³/sec) and estimated at 3.84 m³/sec (0.11 km³/yr). Thus, the subchannel discharge to Lake Baikal at the current phase of knowledge is estimated at 48.1 m³/sec or 1.52 km³/yr.

5.1.3 SUBMARINE GROUNDWATER DISCHARGE FROM THE COASTAL ZONE IN INTERFLUVES

The total length of the shoreline perimeter of Lake Baikal is about 2000 km. Based on the decoded aerocosmic photos and detailed geologic-geomorphological and hydrogeological surveying, the coastal area was subdivided into zones according to the conditions of SGD formation (Pisarsky and Khaustov 1979; Pisarsky 1987). The major part of the shoreline is occupied by a near-lake zone, including the river deltas and interfluve spaces composed of loose and clastic sediments of the Baikal terraces and piedmont plains. In total, six zones with typical sediment profiles are distinguished within the shoreline (Table 5.1.3.1). Each zone has a discontinuous character because the distinguished zonal areas are often located alternately (Figure 5.1.1.1).

Zone I is represented by areas consisting of piedmont tails. It is the longest by perimeter. The profile is a sandy loam, i.e., a sand-pebble composition with equal proportions of boulders and pebble. The visible thickness of crystalline strata in the shore scarps is 15–20 m. Results of the experimental works in this area (Priolkhonie, eastern coast) show that filtration properties of the crystalline rocks on the shore slopes are exceptionally low (filtration coefficient is from 1/100th to 1 m/day). The total SGD along the entire zone length (600 km) is 120 l/sec; linear modulus of the SGD is 0.2 l/sec·km.

Zone II incorporates accumulative plains at the piedmonts of the Baikal and Khamar-Dabansky (partly) ridges and is composed of coarse-fragmented rock material. The thickness reaches tens and the first hundreds of meters. Alluvial sediments have a sandy-pebble composition. On the Baikal Holocene terraces, the linear modulus is registered at 2–3 l/sec·km (Solontsovy Cape, Large Spit) though, generally, the

---

TABLE 5.1.3.1
Submarine Groundwater Discharge (SDG) from Interfluves

Zone No.	Shoreline Length[a] (km)	Linear Modulus of SDG (l/sec·km)	Discharge Value (l/sec)
I	600	0.2	120
II	220	0.3	66
III	220	0.4	88
IV	200	0.5	100
V	140	1.0	140
Total			514

[a]Excluding river delta areas.

SGD for the zone as a whole is estimated at 66 l/sec and its modulus amounts to 0.3 l/sec·km.

Zone III includes areas of the northeastern coast and is composed of poorly graded aqueous-glacial sediments containing lenses of consolidated banded-layered sands and aleurolites of a lacustrine genesis. The thickness of Quaternary sediments reaches tens of meters. The island-type occurrence of permafrost has been observed (Bolshaya, Davsha), which forms, though not in large amount but sufficiently stable, slope SGD, estimated by the analogy method at 88 l/sec (modulus 0.4 l/sec·km).

Zone IV is an area of the Barghuzin and Selenga interfluve. At the foot of the shore terrace scarps, the Baikal alluvial, sometimes fluvio-glacial sediments predominantly of a sandy-pebble composition with a thickness of tens of meters, is stripped. Steep slopes, abundant forests, and relatively high rock transmissivity favor the formation of considerable SGD to the lake, which is estimated at 100 l/sec and with a modulus of 0.5 l/sec·km.

Zone V includes areas of the Khamar-Dabansky coast stretching from the mouth of the Khara-Murin River to Selenga. This shoreline area has the most complicated structure of Triassic loose sediments. The Neogene sediments include chiefly sandy-clayey rocks, often in the frozen state. From the surface the sediments are overlain with sandy-pebble rocks 5–10 m thick. The steeper shore slopes are mainly composed of Proterozoic metamorphic rocks consisting of gneiss and shale with a low hydraulic capacity and low water content. Groundwater flow is formed by precipitation; the discharge occurs in the form of seasonal springs with the contact of crystalline and loose rocks at a distance of 1–1.5 km from and directly within the lake. There are areas (near Tankhoi, eastward the from Mysovaya River) where the linear modulus of SGD reaches 6–8 l/sec·km. The total SGD to the lake along the entire southern coast amounts to approximately 140 l/sec and its average modulus is estimated at 1 l/sec·km.

Zone VI includes river delta areas where the above estimated subchannel discharge is formed.

The total SGD to the lake from the interfluves over the entire lake coast is estimated at 0.51 m³/sec (0.02 km³/yr).

5.1.4 ARTESIAN DISCHARGE

Conditions favorable for artesian discharge formation are revealed in the bottom sediments of the Selenga River low water zone and areas adjacent to it from the north and south. There, at a depth of 5–13 m, numerous proparinas (glades) were found and tested, which are constantly generated in the icy cover of Baikal near the Selenga River delta (Figure 5.1.1.1). Many proparinas are absent during winter cold periods and appear in March–April. The main feature of the Selenga River proparinas is the continuous and intensive gas emission (13 m³/day per one proparina). The emitted combustible gas contains 80–95% methane and 6–17% nitrogen. The gas remains even in ice in the form of blebs, reducing the cover strength of the ice. Examination of the proparinas showed that at the bottom where the gas is emitted there are no temperature anomalies. In some proparinas (in the area of Klyuchi-Stvolovaya) high-quality oil accumulates on the surface of water and ice.

It is acknowledged that proparinas are usually elongated as a chain, gravitating to zones of tectonic disturbances, i.e., large fissures, and to zones of small ruptures. One of the proparinas near Sukhaya Village has been long known by locals, who collect oil from it for their daily needs. Wells drilled through the lake ice cover stripped a 9-m-thick layer of sand and clay and gneiss saturated with oil. In one well, 550 m from the shore and 90 m deep from the bottom, drilling instruments repeatedly fell into the gaping fissures of the oil-saturated gneiss. In the same place, in the area of Stvolovaya Village in 1907, a well was drilled under a 10-m water layer (according to archive data) which passed through alternating sand and clay strata and, at a depth of 30 m from the bottom, stripped confined waters flowing for 7.4 m above lake level with a self-flowing yield of 0.4 l/sec at +3.4°C. The well emitted methane, and the water-bearing sands contained oil. And, in more recent times, a well near a large proparina, located opposite Kukui Bay at a depth of 35 m in Pleistocene loose sediments, stripped hydrocarbonate waters with a mixed composition of cations, mineralization of 3.2 g/l, and water temperature of 9°C (Borisenko and Borisenko 1981).

The availability of artesian confined waters in the Selenga River low-water area has been verified. It is evident that there is discharge of methane gas and oil in local source places of proparinas. The SGD from sedimentary sediments and metamorphic rocks cannot be denied. However, the quantitative assessment of such discharges is not yet feasible. Attempts to reveal a gryphon of confined waters by anomalies of mineralization, temperature, and composition in Lake Baikal waters failed. One can only speak now of small scales of such a discharge, because the permeability of the bottom sediments is very low and, due only to this, oil accumulations are met.

5.1.5 DEEP DISCHARGE

This type of SGD has been briefly described above and estimated as a part of the spring discharge in the coastal zone. The assessment of submarine deep discharge entering the water strata through the lake bottom is much more complicated. Extensive data exist which prove thermal water can discharge through the bottom. There are data from wells having stripped thermal waters both at large depths (3 km) where their temperature reaches 99°C and in tectonic zones at a relatively small depth. The well near Sukhaya Village, drilled at the lake water line at a depth of 272 m in Neogene sandy-clayey sediments around 6 m above the gneiss roof of the basement, stripped methane therms of a hydrocarbonate sodium type with a mineralization of 0.6 g/l. The well has been self-flowing for over 50 years with a stable water temperature of 35–37°C.

The hydrotherm discharge into the Baikal bottom is associated with proparinas being formed in ice and existing through the winter period. Such proparinas are usually adjacent to thermal springs on the shoreline (Khakussky, Kotelnikovsky, Zmeiny). The source of hydrotherm discharge is revealed according to a proparina near the peninsula "Saint Nose." There, near the bottom, the water temperature was +7°C at a depth of 7 m and +1°C on the surface. In summer, accumulations of thermophilic algae can be seen on the bottom.

Mizandrontsev (1975), who investigated the geochemical properties of ground solutions in the bottom sediments of Lake Baikal, revealed that in some places at depths of 200–900 m, there were anomalously high concentrations of sulfates (2–3 times), sodium (2–4 times), and chlorine (3–20 times) than the appropriate average values for Lake Baikal. Such anomalies can be observed along the western coast of North Baikal. Their locations are shown on Figure 5.1.1.1. In South Baikal significant geochemical anomalies were not observed. Explaining availability of the anomalies by physical-chemical processes in the system "water–ground solution–bottom sediments," Mizandrontsev proved, from the thermodynamic point of view, that the possible origination of the anomalies comes at the expense of the dissolution of a number of minerals. However, this does not preclude the possibility of a hydrothermal origin.

If we were to analyze the position of temperature and hydrochemical anomalies (many of which coincide or are adjacent to each other), it can be seen that the majority are located in or gravitate to Cenozoic or renewed (in Cenozoic) faults stretched on the Baikal bottom.

Thus, the above-stated data prove, with sufficient conviction, the presence of submarine thermal water discharge at large depths of Lake Baikal. For the tentative assessment of such a discharge, we adopt the value of the hydrothermal discharge modulus calculated for the Barghuzinskaya and Verkhne-Angarskaya depressions (0.005 l/sec·km^2) adjacent to Lake Baikal. Doing this, we obtain a value equal to 0.16 m^3/sec (~0.01 km^3/yr). If we were to adopt North Baikal as a large dug well working by means at its bottom, then the maximum possible discharge (calculated by the well-known formula) will be equal to 0.6 m^3/sec (0.02 km^3/yr). The thermal water discharge to the Lake Baikal bottom is estimated by calculations using the hydrogeothermal data. The calculations were made using 400 measurements of temperature from Baikal bottom sediments. This readout is sufficiently weighty and, in the future, the measuring grid can be made even denser.

In 1990 by means of a submerged apparatus of the Paisis type and using photorobotics, an attempt was jointly undertaken by a team of researchers from Russia and the United States to reveal hydrogeothermal anomalies directly on the bottom of Lake Baikal. However, hydrothermal springs have not yet been detected. At the point of the assumed Frolikhinsky submarine thermal spring, a powerful thermal anomaly in the bottom sediments was detected. According to the calculations of deep SGD to Lake Baikal by the methods described above, the average deep SGD is taken equal to 0.6 m^3/sec (0.02 km^3/yr).

5.1.6 CALCULATION OF TOTAL GROUNDWATER COMPONENT IN THE WATER BALANCE OF LAKE BAIKAL

Independent assessments of the groundwater component in the water balance of Lake Baikal were undertaken, each attempt obtaining new data, using new methods and techniques of calculation or satisfying demands of practice (Pisarsky 1973, 1984, 1987). Thus, calculations of the incoming part of the groundwater balance are based on a great amount of data and reflect the degree of knowledge about this unique water body. The calculation results are presented in Table 5.1.6.1.

TABLE 5.1.6.1
Components of Subsurface Discharge to Baikal Lake

	Annual Average Value		
	(m³/sec)	(km³/yr)	(%)[a]
Spring Discharge	2.53	0.08	5
Subchannel Discharge	48.1	1.52	93
Discharge from Interfluves	0.51	0.02	1
Submarine Water Discharge	0.60	0.02	1
Total Discharge	51.7	1.64	100

[a]Total subsurface inflow to the lake.

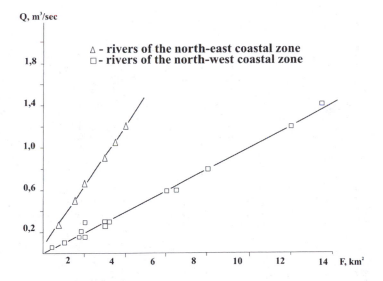

FIGURE 5.1.6.1 Losses of surface discharge (Q) versus area of debris cone (F).

As seen from Table 5.1.6.1, the determining portion of the subsurface inflow to the lake belongs to the subchannel discharge in the river valleys (93%), the second place belongs to the spring discharge (5%). Such are the annual average data. In the critical period of the year (February–March), the contribution of spring discharge increases to 13%, whereas the subchannel discharge decreases to 82%. Thus, the annual average subsurface component in the incoming part of the water balance is estimated at 51.7 m³/sec (1.64 km³/yr) and amounts, together with the total water balance residual of Lake Baikal, to 2.3 km³/yr.

As to groundwater outflow from Baikal, there are no data from direct measurements. It can be only noted that in the place where the Angara River originates (in the place of the most possible water outflow), there is a wedge of low-permeable crystalline rocks (Shaman Stone). A well drilled there to a depth of 1 km appeared to be dry.

TABLE 5.1.6.2
Components of Submarine Groundwater Discharge (SGD) to Lake Baikal

Component of SGD Type	Water-Bearing Rock	Rock Reservoir Type	Draining Zone Thickness	Filtration and Discharge		Conditions of Regulation
Springs	Loose and debris sediments, fissured consolidated rocks	Porous and fissured	Tens of meters	Concentrated flowing onto the surface and open discharge to the lake	Descending	Atmospheric regulation. Icing regulation up to water completely passes to a hard phase
	Karst carbonate rocks	Karst	Hundreds of meters		Descending and ascending	Icing regulation
	Zones of tectonic disturbances	Veined	Tens and the first hundreds of meters		Ascending	Weak icing regulation
Subchannel	Alluvial sediments of river valleys	Porous-stratum	Tens of meters	Descending, distributed submarine discharge in the coastal zone of the lake		Swamp regulation in coastal zone, hold up by lake waters in rivers mouths
Discharge from interfluves	Through loose sediments	Porous-stratum reservoir	Meters and the first tens of meters	Descending, distributed submarine discharge on a submerged coast slope		Atmospheric regulation
	Through igneous and metamorphic rocks	Fissured	The first meters	Descending, distributed submarine discharge on a submerged coast slope		Atmospheric regulation
Artesian	Consolidated metabolic, seldom sedimentary rocks	Porous and fissured-stratum	Hundreds of meters	Ascending, submarine, spring discharge along tectonic fissures and through hydrogeological windows		Filling of filtration paths by precipitating salts
Deep	Zones of open tectonic deep-seated faults	Veined	Thousands of meters	Submarine, spring and distributed discharge along zones (??)		Healing of deep faults, filling of filtration paths by precipitating salts

5.2   SUBMARINE GROUNDWATER DISCHARGE TO LAKE BALKHASH*

5.2.1   NATURAL CONDITIONS OF THE COASTAL ZONE

Lake Balkhash is located in the arid zone of Central Asia in the southeast of Kazakstan. The basin of the lake has an area of 413 ths km^2, over 27% (113 ths km^2) of which occupies the Xinjing Uigur Autonomous Region in the People's Republic of China. Also in China there is a zone of annually renewable water resources.

Various landscapes can be found in the drainage area. In the humid mountainous landscapes with highland-located glaciers are zones where water resources of the basin are formed. In the semiarid piedmont alluvial-proluvial detrital cones, the surface runoff formed in mountains is transformed into subsurface discharge. On the semiarid piedmont alluvial-proluvial, desert eolian and lacustrine-alluvial plains, the water discharge is distributed.

The lake represents a latitudinally elongated water body over 600 km long. The Saryesik Peninsula divides the lake into approximately two equal parts: a shallow and wide western part and a much deeper (in the upper east, to 26.5 m) and narrower eastern part. The Ile river, the main water source of the Balkhash region, provides 80% of the riverwater inflow to the lake and flows to the Western Balkhash. The other rivers (Karatal, Aksu, Lepsa) flow into the eastern part.

One of the specific features of Lake Balkhash is the distribution of mineralization in its water. In the western part, the water is fresh and low brackish (1.04–1.56 g/l), while in the eastern part it is brackish (3.65–4.42 g/l). Mineralization increases gradually from west to east. The water exchange in the Uzynaral Strait plays an important role in the water and salt regime of the lake.

The water level of Lake Balkhash is subject to large-scale perennial and age-long variations caused by cyclic climatic fluctuations typical of drainless water bodies in an arid zone. The annual variations of the water levels are determined by the annual changes in the elements of the water balance in the lake connected with an annual climatic cycle, as well as wind influences.

The water level in Lake Balkhash has been observed since the beginning of the 19th century, when three minimum and four maximum levels were observed. Three complete cycles (1840–1885, 1886–1946, 1947–1986) were observed. A rising phase started in 1987. The duration of a cycle lasts from 41–60 years. The level-rise phase of the 1947–1986 cycle was interrupted in 1970 due to the filling of the Kapshaghaiskoye storage reservoir on the Ile River, which caused an intensive decline in the water level of the lake. During the 1970–1986 period the water level declined by over 2 m (Figure 5.2.1.1).

Lake Balkhash is an important economic resource (e.g., technical water supply for mining-metallurgical manufacturing, fishing). The multifaceted use of water resources from the Balkhash basin is manageable when the water balance of the

---

* Section 5.2 was written by Oleg V. Podolny (Kazakh Hydrogeological Research and Production Project Company, Kazhydec@nursatkz).

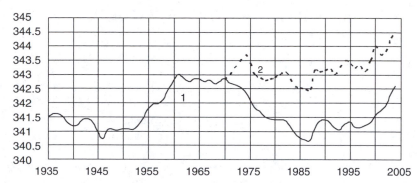

FIGURE 5.2.1.1 Change of real water level (1) and restored level due to water filling of the Kapshaghaiskoye storage reservoir (2) in Lake Balkhash.

lake is assessed. This requires reliable identification of the characteristics of the groundwater component in its water exchange.

5.2.2 QUANTITATIVE ASSESSMENT OF SUBMARINE GROUNDWATER DISCHARGE

The first estimates of groundwater components were obtained based on the general considerations of the character of water exchange. Shnitnikov (1936) presumed that the total inflow and outflow of groundwater are equal, and estimated the unbalanced inflow of the alluvial shallow groundwater to the lake at ~1.56–1.77 km³/yr. Values of the subsurface recharge, calculated by the residuals of water and salt balances, vary from 1.5 km³/yr of the inflow to the lake to 1.48 km³/yr of outflow from it.

Values of the subsurface recharge, obtained by hydrogeological methods, also differ considerably from each other at different steps in the study of the subsurface water-exchange processes, depending on the accumulation of factual materials. Based on the calculations of natural groundwater resources of the entire Balkhash basin and on hydrodynamic calculations, Akhmedsafin and Shapiro (1970) estimated the groundwater inflow to the lake at 0.8 km³/yr, and salt flux at 1.3 mln t/yr.

Further investigations by hydrogeologists, who used the latest data from mean-scaled geologic-hydrogeological surveying in the South Balkhash region and mathematical modeling of geofiltration to clarify the relationship between water in the lake and groundwater, enabled a more detailed assessment of the subsurface water-exchange formation in Lake Balkhash (Podolny and Shapiro 1992; Podolny et al. 1992; Sumarokova et al. 1992; Shapiro et al. 1982).

The lake is located at the interface of two hydrogeological structures: a hydro-geological massif of fissured waters in the north and a system of artesian basins in the south. The basic components of inflowing groundwater to the lake, which differ by their formation, hydrodynamic features, amounts, and water mineralization, are (a) subaqueous runoff of shallow and confined waters and (b) the runoff along local zones of tectonic disturbances. The formation of the SGD in the South Balkhash area was assessed with the use of mathematical modeling.

One step that is included in the construction of mathematical models of the hydrogeological systems of artesian basins is the comparison of natural resources or subsurface runoff of a basin, the "object and basin" model. Often, this is not a simple comparison of proportions of "model-object" equivalence. It is also a way to obtain new information on an object, which is difficult and sometimes even impossible to obtain in reality. Model-aided solutions enable obtaining a value of SGD. Also, it is possible to determine different sources of subsurface runoff formation with accuracy, depending on the scale of the model.

The system of artesian basins in the South Balkhash area consists of two artesian basins of the second order: Balkhashsky in the west and Lepsinsky in the east. The artesian basins are divided by an uplift of the Paleozoic basement. The water exchange takes part chiefly in continental Cenozoic aquifers, divided by a Pliocene Illinian aquiclude. The Pliocene aquiclude is overlain by a unified unconfined aquifer of the Upper Pliocene–Quaternary, and underlain by a Pliocene Illinian aquifer.

Such a hydrogeological cross-section is typical of the Balkhash artesian basin. A specific feature of the Lepsinsky artesian basin is its well-expressed hydrogeological zonality, which is typical of piedmont plains in the arid zone. The Pliocene Illinian aquiclude is absent in the piedmonts, where subsurface runoff is formed and which covers detrital cones. The Upper Pliocene–Quaternary unconfined aquifer is present in this zone. Apart from the mountains in the zone of groundwater wedging, the unified aquifer stratifies into several water-bearing layers divided by aquicludes. This schematic hydrogeological cross-section is similar to the cross-section of the Balkhash artesian basin.

A geofiltration process for this scheme is described by a system with the following equations:

$$\begin{cases} \nabla(T_1 \nabla H_1) + G_1(H_2 - H_1) + W_1 = \mu_1 \dfrac{\partial H_1}{\partial t} \\ \nabla(T_2 \nabla H_2) + G_1(H_1 - H_2) + W_2 = \mu_2 \dfrac{\partial H_2}{\partial t} \end{cases} \tag{5.2.2.1}$$

where ∇ is a 2-dimensional differential operator, H_1 and H_2 are levels of the upper and lower aquifers, G_1 is the filtration resistance of the dividing bed, T_1 and T_2 are water conductivity of the upper and lower aquifers, and W_1 and W_2 are the intensity of resultant recharge or discharge; all these terms depend on the coordinates x, y, and H_1, H_2, W_1, and W_2 also depend on time (t).

The equations were solved by the finite difference method on a computer using the TOPAZ-9 Code developed at Russia's VSEGINGEO. The grid area for the Balkhash artesian basin includes 1178 blocks (38×31) with a step of 10 km; for the Lepsinsky basin, 1248 blocks (48×26) with a step of 5 km, which corresponds in total to the model scale of 1:500,000. To confirm the reliability of the model, a number of reverse tasks were solved (Podolny et al. 1992). Detailed analysis of the information on the basin shows that at present the hydrogeological processes occurring in the basin and its geofiltration parameters, except in some parts, the reverse task cannot be formulated as a task of identification that would unambiguously assess coefficients and parameters of the initial equations. Therefore, the hydrogeological reliability of the model was checked by the water balance method or by geological regularization (Gavich 1989; Sokolov 1989).

Regional subsurface runoff of the Balkhash artesian basin is formed by the infiltration of precipitation, seepage from mountain rivers, filtration losses from the Ile and Karatal Rivers, as well as inflow from a small intermontane depression. The modeled solutions for the Balkhash artesian basin include the following contributions to the subsurface runoff: filtration from the Ile River (0.94 m³/sec) and the Karatal River (0.28 m³/sec); subsurface inflow and slope inflow from the bordering mountainous zone (2.61 m³/sec); inflow from a small intermontane depression (3.29 m³/sec); and infiltration of precipitation at different sand massifs (2.37 m³/sec). The total inflow is 9.49 m³/sec (Figure 5.2.2.2).

The subsurface runoff is mainly spent for recharge of the Karatal River (3.19 m³/sec) and for evapotranspiration in sands (6.32 m³/sec), totaling 9.51 m³/sec. Discharge of the confined aquifer (1.34 m³/sec) is fully spent for seepage into the upper aquifer.

In the Lepsinsky artesian basin, subsurface discharge is formed (22.7 m³/sec) the filtration losses from river beds (11.6 m³/sec), irrigated fields, irrigation network, and at the expense of infiltrating precipitation (8.27 m³/sec), and inflow from the mountains (2.84 m³/sec). In the wedging zone, the runoff is dispersed mainly through discharges into the river network and evapotranspiration (19.3 m³/sec). In the sandy plains, the subsurface discharge is increased at the expense of infiltrating precipitation by 1.68 m³/sec, and of filtrating river waters by 0.59 m³/sec. At the same time, the discharge is spent for evapotranspiration (3.8 m³/sec) and outcropping to the river network (1.95 m³/sec). The discharge of the confined aquifer (3.39 m³/sec) is also fully spent for seepage to the upper aquifer.

The regional model scale of 1:500,000 does not sufficiently reflect the actual conditions of the subsurface water exchange of Lake Balkhash. Nevertheless, the obtained solutions make it possible to assess the order of magnitude of confined filtration into the lake. Proceeding from the total SGD of the confined Pliocene Ilinian aquifer (4.73 m³/sec), possible underaccounting of subaqueous inflow of confined water to the lake amounts to ~0.5 m³/sec (maximum possible error of the model).

To make a quantitative assessment of shallow groundwater to Lake Balkhash, the analysis of hydrogeological conditions in the shore area was carried out. The zone within which groundwater inflow to the lake is formed covers a shore strip of 30 km wide, extending 100 km in the Ile River delta. Subaqueous discharge of the shallow groundwater occurs offshore within a narrow strip at a distance from the shore equal to 1–3 times the thickness of the aquifer, depending on the slope of the lake bottom.

Runoff of unconfined water is possible chiefly from the artesian basins of the South Balkhash area, and from the valleys of the North Balkhash area. Its subaqueous discharge disperses across the water area, vertically seeping through low-permeable bottom sediments. In fault zones, SGD is concentrated along tectonic zones in the north and possibly in the South Balkhash area, and is wedged out within the lake, forming anomalous mineralized areas. In the shore strip, shallow groundwater forms from inflow from hypsometrically overlying water-catching areas, filtration losses

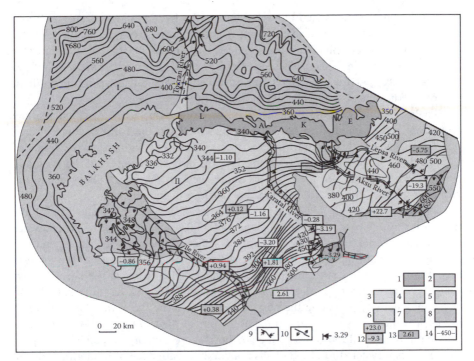

FIGURE 5.2.2.2 (Please see color insert following page 236.) Schematic map of the formation of submarine groundwater discharge (SGD) to Lake Balkhash. *Hydrogeological areas*: (I) hydrogeological massif of the North Balkhash zone; (II) system of artesian basins of the South Balkhash zone. Areas of basic formation of SGD in the system of artesian basins of the South Balkhash zone at the expense of: (1) losses of surface runoff, infiltrating precipitation, and irrigation waters; (2) losses from discharging temporary water streams and infiltrating precipitation; (3) water filtration from rivers; (4) infiltrating precipitation. *Distribution areas of SGD* by means of: (5) intensive discharge to rivers and evapotranspiration from shallow groundwater surface; (6) transpiration by desert communities; (7 and 8) areas with complicated conditions of interactions between surface and shallow groundwaters, the quantitative characteristics of which cannot be determined within the model; (10) hydrogeological massifs of fissured waters in Paleozoic rocks. *Interconnection of river and groundwaters*: (11) groundwater is wedged out to rivers; (12) rivers lose runoff for filtration. Quantitative characteristics of elements of SGD formation: (13) SGD from a small intermontane depression; (14) SGD in particular areas, m³/sec (+ - submarine groundwater recharge, - submarine groundwater discharge); (15) subsurface runoff from the mountainous rims of the Balkhash artesian basin; (16) hydrocontour line and its absolute elevation.

from rivers, and infiltrating precipitation. Also a partial discharge takes place there via evaporation and transpiration.

Processes of shallow SGD formation are different in different areas of the shore. These differences are explained by complicated hydrogeological, geological, geomorphologic, hydrochemical, and topographical factors. Among these factors, one of the main processes involves the draining ability of the area and the availability of nonephemeral water streams in conditions of a highly arid climate. The zoning of Lake Balkhash was carried out according to the hydrogeological cross-section, recharge and discharge, hydrogeochemical and hydrogeodynamic features, water level regime, and chemical composition. Several groups of shallow groundwater flows were thus distinguished. These are flows in the Ile River delta, the river valleys of Karatal, Aksu, Lepsa, and Tokrau, fissured waters, the fault zones in the northern part of Balkhash, and sandy massifs in the south.

Analysis of the shallow groundwater flow formation has shown that shallow groundwater is discharged to Lake Balkhash chiefly in the form of concentrated flows associated with old and modern river valleys, and with extensive zones of tectonic disturbances, as well as dispersed runoff of fissured waters of the West and North Balkhash areas. On the southern shore, beyond the river valleys where the influence zone is limited to 1–2 km, an outflow is formed from the lake. This phenomenon is typical of plain-type water bodies of the arid zone. The South Lake Balkhash shore is very steep with a slope of 0.0003–0.0008. Shallow groundwater flow reaches the lake only along active rivers. In conditions of arid climate and low drainage, the main regulating factor for discharge in the southern shore zone is evapotranspiration from the shallow groundwater table. To reach the lake, the discharging waters formed in the catchment area and not wedged out into rivers should have a productivity providing the possibility to overcome the "evapotranspirational barrier." At the values of geofiltration parameters of the aquifers and their recharge existing in reality, all the flows of the southern shore cannot overcome this barrier and bring their waters to Lake Balkhash beyond the influence of modern riverbeds. As a result, level depressions with a semistagnant geofiltration regime, where an outflow of the lake water is possible, are formed. Outflow from the lake cannot be high. It discharges via evapotranspiration and cannot fill the level depressions.

The type of shore determines the hydrodynamic features of subaqueous shallow SGD during a decrease in the lake water level. On the steep abraded western and northern shores, where shallow groundwater occurs at a considerable depth, hydraulic slopes of water flows increase with a decrease in the lake level, which leads to an increase of subsurface water inflow from these shores. On gently sloped shores, a decrease in the lake water level leads to the exposure of a considerable part of the lake bottom, thereby forming beaches.

In places where lake water outflows to the formed level depressions, recession of the lake leads to a decrease in shallow groundwater table, change in vegetation, and low displacement of the shallow groundwater level extremum toward a new shore. The size of the subsurface outflow can remain constant at slow lake decline when the shore vegetation keeps in step with the recessing shore, or can decrease with a sharp lake decline.

Quantitative assessment of SGD and lateral outflow of the lake water in the subsurface was carried out by the hydrodynamic method using a computer-constructed hydrodynamic grid of filtration (Figure 5.2.2.2). The basis of constructing the grid and calculating it was for the qualitative and quantitative treatment of hydrogeological information on the shore zone of the lake. The total SGD along the entire shoreline is determined by Darcy's law for specified flow intervals (see Chapter 1.3), without averaging hydrodynamic parameters over area.

In total, the subaqueous SGD into the lake is estimated at 11.6 mln m³/yr. Provided that its distribution is regular along the shoreline length of 1220 km, groundwater inflow will be 0.26 l/sec·km of the shoreline. However, groundwater is discharged into the lake as separate flows, the linear losses of which vary from 0.1–2.5 l/sec·km; 7.0 mln m³/yr SGD flows into the western freshwater part of Balkhash and 4.6 mln m³/yr into the eastern brackish part (Table 5.2.2.1).

Direct estimation of SGD to Balkhash was carried out in 1987–1988 by Meskheteli et al. (1991). Integrated hydrogeological investigations included seismo-acoustic, and thermo- and restivimetric profiling (VSEGINGEO 1987). The combination of these three independent determinations with thermal infrared surveying gives a sufficiently accurate picture of the possible places of SGD, and enables the quantitative assessment of groundwater yield on the bottom, discharges of which are not lower than the accuracy limit of the measuring equipment.

In the water area of the Balkhash, complex profiling of 25 sites (16 transversal and 9 longitudinal) were carried out, which, with an average step of about 22 km, covered the entire lake. Temperature and electrical conductivity were measured by means of a bottom-tow probe. Salt measurements by electrical conductivity ranged from 0–3, 0–6, and 3–6‰. The temperature ranges from 15–30°C (1987) and 10–20°C (1988). Temperature and electrical conductivity were checked by controlled measurements of samples taken at the start and finish of a profile. Areas

---

TABLE 5.2.2.1
Discharge of Shallow Groundwater to Lake Balkhash

Area of Flow	Discharge of Shallow Groundwater (mln m³/yr)	Average Mineralization of Shallow Groundwater (g/l)	Transfer of Dissolved Substances (ths t/yr)
West Balkhash			
Ile River delta	0.8	1.8	1.4
Fissured waters of North Balkhash area	5.2	2.0	10.4
Fault zones	1.0	1.0	10
Total	7.0		12.8
East Balkhash			
Valley of Karatal River	0.27	0.8	0.2
Valleys of Aksu and Lepsa Rivers	0.32	0.8	0.2
Karatal-Aksu interfluves	0.66	3.5	2.3
Sands	0.55	2.5	1.4
Fissured waters of North Balkhash area	0.56	2.0	1.1
Fault zones	2.2	1.0	2.2
Total	4.56		7.5
Lake Balkhash Total	11.6		20.3

with anomalous temperature and electrical conductivity were revealed through the scrupulous analysis of natural values of the measured parameters. Temperatures that were considered anomalous differed from natural levels by 0.5–0.6°C.

Thermal infrared surveying was carried out in the summer of 1987 with the aid of the thermal visor "Vulkan" installed onboard airplane IL-14 from a height of 1000 m, using a system of crossing routes. The cutting width of a route was 4 km. Temperature resolution of the surveying was within 0.25–0.5°C. The most representative thermal images were treated with a specialized optical-electronic measuring system. Thirty-seven profiles were made across the lake and along its coast. Thermal resistivimetry enabled us to reveal local anomalies in the geophysical fields of temperature and salt contents, which indicate possible discharge of deep groundwater to the lake. Calculations of this discharge is estimated at 0.3 mln m³/yr.

From the Lepsinsky artesian basin where the natural resources of SGD are estimated at 0.85 km³/yr (Podolny et al. 1992), only 0.3% of the groundwater is discharged to the lake. From the Balkhashsky artesian basin where the natural resources of SGD are estimated at 0.315 mln km³/yr (Podolny et al., 1992), only 7.0 mln m³/yr is discharged to the lake. From the entire South Balkhash area the lake receives about 2% of the natural groundwater resources, and from the North Balkhash area, 0.458 km³/yr (Shapiro 1974). Discharge from the latter to the lake amounts to only 2% of the total natural groundwater resources of the territory. The data show that Lake Balkhash is the main drainage area of groundwater in the basin, which is typical of many plain-type water bodies of the arid zone.

5.2.3 INTERACTION BETWEEN GROUNDWATER AND LAKE

Outflow of lake water in the subsurface was also estimated by the hydrodynamic method, using a geofiltration grid along flow strips beyond the influence zone of rivers. Discharge of subsurface runoff from the side of the lake within the southern shore is estimated at 24.8 mln m/yr (Table 5.2.3.1), including 19.1 mln m³/yr from the coastal part of the West Balkhash and 5.7 mln m³/yr from East Balkhash. If we compare these values with the subaqueous shallow groundwater discharge, we can see that the discharge from the lake is 2.1 times higher than the subaqueous discharge.

Dissolved substances transferred to the lake with groundwater were estimated on the basis of linear discharges of the shallow groundwater to the lake and generalized data on its chemical composition. According to the calculations, the total ion runoff to the lake amounts to 20.3 ths t/yr, including to West Balkhash (12.8 ths t/yr) and to East Balkhash (7.52 ths t/yr) (Sokolov 1989) (Table 5.2.3.1).

The salt transfer from the lake with the subsurface outflow was calculated earlier using the average mineralization of the lake water in different reaches. In areas where the groundwater table slope is directed from the lake, the groundwater has a mineralization of ≥20–30 g/l and a chloride and chloride-sulfate sodium composition. Mineralization increases rapidly from the lake toward the shore zone. The groundwater that flows from the Balkhash toward the level depressions carries great amounts of salt. Generally, the amount for the southern shore of Balkhash equals 586 ths t/yr, with 401 ths t/yr in the southwestern area and 185 ths t/yr in the

TABLE 5.2.3.1
Subsurface Water and Salt Exchange of Balkhash Lake

Area of Lake	Inflow of Shallow Groundwater	Discharge of Shallow Groundwater	Outflow of Coastal Zone (km³/yr ths t/yr) Water Losses from Lake for Providing Discharge of Shallow Groundwater	Losses of Shallow Groundwater for Evaporation and Transpiration	Balance
West Balkhash	0.007	0.019	0.378	0.359	−0.371
	12.8	401	401		−388
East Balkhash	0.003	0.006	0.068	0.062	−0.065
	7.52	185	185		−178
Balkhash Lake	0.010	0.025	0.447	0.422	−0.437
	20.3	586	586		−566

southeastern area (Table 5.2.3.1). The given values characterize the transfer of dissolved salts from the lake with the subsurface runoff.

Submarine groundwater discharge from the lake is formed as a result of the interaction among complicated processes of lake water infiltration into the shores and its intensive discharge in the coastal zone through transpiration by reeds and evaporation in conditions of very dry climate and recharge from water. Of considerable interest are the coastal areas adjacent to the lake, within which the basic transformation of surfacewater into groundwater takes place. The equation of groundwater balance in this zone has the following form:

$$V_{lake} - V_{evt} - V_{gw} = \Delta V \qquad (5.2.3.1)$$

where V_{lake} is water losses from the lake providing replenishment of the shallow groundwater; V_{evt} is evaporation and transpiration from the groundwater table; V_{gw} is outflow of groundwater into the shore beyond the zone of regulation, calculated by the hydrodynamic geofiltration grid; ΔV is a change in shallow groundwater reserves (with possible tendency of becoming equal to zero in perennial cross-section).

An equation of perennial hydrochemical balance of this zone can be written as:

$$V_{lake}M_{lake} - V_{gw}M_{gw} - L = 0 \qquad (5.2.3.2)$$

where M_{lake} is mineralization of the Lake Balkhash water; M_{gw} is mineralization of shallow groundwater; and L is salt losses for precipitation in soils and their eolian transfer.

The equation does not include internal sources of mineralization formation, because rocks of the aquifers and soils of the unsaturated zone in the coastal areas covered by reeds on swamp and meadow-swamp lands are not salted. An increase of the groundwater mineralization is explained only by concentration caused by evaporation typical to an arid climate.

As mentioned above, outflow of the shallow groundwater into the shore was calculated by the hydrodynamic method using a filtration grid. The obtained values of outflow and mineralization of groundwater were used to calculate the hydrochemical discharge toward the level depressions. The evapotranspirating processes of the shore zone were estimated via modeling of the geofiltration process of seasonal variations in the shallow groundwater level caused by vertical evapotranspiration-induced discharge at a constant (in time) water level in Lake Balkhash. Prevegetation maximum was taken at the initial level of the lake. The observed regime of the groundwater level was modeled with a temporal step of 1 month through the selection of a transpiration-induced discharge during the vegetative season, with the following geofiltration parameters: water yield (0.15–0.075), water conductivity (10–100 m^2/day), transpiration (200–1200 mm/yr). For a nonvegetative season, the transpiration was specified as equal to zero.

In this case, the recovered groundwater level was modeled. The main condition for the selection of a length of area and amount of transpiration discharge was complete recovery of the groundwater level up to the initial value along the entire profile occurring at the expense of groundwater inflow from the lake. It turns out that the zone of vegetation transpiration, provided by water inflow from Balkhash at its constant level, can be equal, depending on geofiltration parameters, to: from the first meters to a kilometer at a transpiration discharge of 200–900 mm/yr. The values of transpiration do not differ from the experimental ones (Sumarokova et al. 1992). Thus, losses of the lake water from transpiration of reeds at the constant water level in the lake can reach ≥ 0.3 km^3/yr.

However, the water level in the lake does not remain constant; it is subject to seasonal water variations. Due to the slight slope of the southern shore, the pileup of water substantially floods this part of the shore. At an average multiyear height of 0.2 m, a frequency of 50 times and a duration of 17 hours, the water pileups serve as a significant factor in the shallow groundwater recharge in the zone of forming outflow from the lake and expanded areas of transpiration-induced discharge by reeds. Under these circumstances, water losses from the lake can be not less than 0.3 km^3/yr.

Determination of the rest of the components of the balance equations — water losses from the lake providing groundwater flows to the level depressions, evaporation and transpiration from groundwater table — was made through solving a system of the above-given balance equations relative to these components, with a specified mineralization of lake water for different areas of the shore (salt losses for precipitation and mineralization of evaporating water were taken as equal to zero). Thus, water losses from the lake for the formation of SGD amounts to 0.447 km^3/yr, including from West Balkhash (0.38 km^3/yr) and East Balkhash (0.068 km^3/yr), and transpiration discharges at 0.42 km^3/yr, 0.36 km^3/yr, and 0.06 km^3/yr, respectively (Table 5.2.3.1). The presented values of outflow seem to be somewhat overestimated, because the mineralization of the lake waters was used in the calculations for the open part of Balkhash, whereas mineralization in the lake bays is much higher.

Thus, the discharge (though relatively small) of the subsurface flows toward the level depressions carries a great amount of dissolved salts. Whereas the subsurface outflow from the lake is twice as high as the SGD, discharge of salts from the lake is 30 times higher than the influx of dissolved substances with the shallow groundwater.

5.3 SUBMARINE GROUNDWATER DISCHARGE TO LAKE ISSYK- KUL

5.3.1 NATURAL CONDITIONS OF THE SHORE ZONE

The Lake Issyk-Kul basin represents an extensive oval-shaped trough elongated in a latitudinal direction. Its maximum dimensions are: from west to east 252 km, from north to south 146 km, with a total area of 22,080 km^2.

Issyk-Kul is a large lake in Kirghizia with physical-geographic, climatic, and hydrologic features distinguishing it from other water bodies of the arid zone. It is a lake of mountainous type with a trough of tectonic origin, located within the North Tian-Shan Mountains system on the bottom of an extensive intermontane depression bordered by the ridges of Kunghey and Terskey Ala-Tao. The average elevation of the highest ridge of Terskey over the lake level is 3600 m. Adjoining the lake is a plain shaped like a narrow belt, composed either of combined alluvial and proluvial cones of rivers or flattened surfaces of Issyk-Kul terraces. The plain is slightly sloped toward the lake. It is most developed on the western, northern, and eastern shores of the lake where its width reaches 15–20 km. On the southern shore, the plain is much narrower due to the close location of the Terskey front ridges, and has a width of >2 km. According to recent measurements, the morphometric characteristics of Lake Issyk-Kul are: length (178 km), width (60.1 km), shoreline length (688 km), average depth (278 m), and area (~6240 km^2). It is interesting to note that Lake Issyk-Kul is, on average, 3 times deeper than the average depth of the Baltic Sea.

Of considerable interest is the data on the isotope balance of the Lake Issyk-Kul, obtained by means of tritium measurements while studying the conditions of natural water formation in the Issyk-Kul depression (Romanov et al. 1989). Experimental tritium data obtained by the authors show that the groundwater is a Neogene aquifer.

The basic characteristics of the modern tectonic structure of the Lake Issyk-Kul basin is that its formation originated in the Late Paleogene intensity. Tectonic motions grew from the Paleogene to Neogene and reached a maximum in the yearly Quaternary period, when sagging of the Earth's crust changed into block displacements of the Paleozoic basement and the dislocations of loose Paleogene–Neogene sediments turned into faults.

The climate is formed under the influence of atmospheric circulation, radiation regime, and the type of underlying surface, which is determined, first of all, by the orographic closeness of the Issyk-Kul depression and the presence of a large deep lake. On the western shore, the amount of precipitation is ~100 mm/year and on the eastern shore 500–600 mm. The majority of precipitation (up to 96% in the west) falls in the year's wet season. The glaciers play a great role in river recharge.

The glaciers of the depression contain ~48 km^3 of freshwater, which is more than 10 times higher than the annual discharge of all the rivers in the lake basin. All the large rivers in the Lake Issyk-Kul basin flow from glaciers. Total melt water from the glaciers and, hence, water inflow to the rivers, depends on the duration of the glacial ablation. In years with longer ablation periods, a thicker ice layer melts and, hence, more water flows into Issyk-Kul. During warm years, the melting of a 300-cm-thick ice layer contributes ~3 mln m^3 of water to the rivers from a glacier area of 1 km^2. The rivers of the Issyk-Kul basin and the lake itself are recharged by precipitation in the high-mountainous zone. The river runoff is the basic incoming part of the water balance of the lake. Perennial and seasonal variations of the lake water level are connected to variations in precipitation, the regime of glacier melting, and with the extraction of riverwaters for irrigation and other purposes.

In the hydrogeological respect, the Issyk-Kul intermontane depression has a three-layered structure (Bergelson et al. 1986). The upper hydrogeological layer is associated with Quaternary sediments and composed of alluvial-proluvial rocks, sediments of detrital cones, and of lacustrine sediments of Quaternary detrital cones. Water conductivity of these rocks reaches 500–2000 m^2/day. In the upper and middle parts of the detrital cones, the depth of groundwater is 40–80 m, and the coefficient of filtration is ≥90 m/day. In the lower parts, the depth of groundwater is reduced to 5–20 m, the coefficient of filtration also decreases to 1 m/day, and the waters become confined (Romanov et al. 1989). Groundwater in the upper hydrogeological layer plays a central role in the formation of underground recharge of the lake.

The middle hydrogeological layer is associated with the rocks of Mesozoic age. Here, the groundwater is enclosed in sandy layers 50 m thick, alternating with low-permeable argillites and aleurolites. The waters are confined, with a pressure head on the shore reaching 20 m above the surface. Water mineralization in these sediments varies from 0.4–40 g/l, and its temperature reaches 500°C. Hydraulic conductivity of rocks ranges from 100–150 m^2/day.

Within the lower hydrogeological layer, where the rocks of the Paleozoic basement are present, the groundwater is enclosed in the crust of weathered crystalline rocks and fault zones. Recharge of fissure-veined waters occurring in the zones of faults and areas of open jointing is provided by the infiltrating precipitatin and the melting of glaciers and snow. Discharge of these waters occurs in mountainous gorges at the exit to the near-lake plain by the draining and seepage of river and spring water into the overlying aquifers in the artesian basin.

5.3.2   Quantitative Assessment of Submarine Groundwater Discharge to Lake Issyk-Kul

Groundwater flow in the Issyk-Kul basin is formed along the periphery of the depression due chiefly to filtration from the rivers, losses from irrigation systems, and infiltrating precipitation. The discharge is everywhere directed toward the lake as the regional basis of draining. As was mentioned above, the basic role in formation of SGD to the lake belongs to the upper (Quaternary) hydrogeological layer. The middle hydrogeological layer has a subordinated significance in subsurface recharge of the lake, but owing to its high mineralization it plays a significant role in the formation of ion discharge and the salt balance of the lake. The lower hydrogeological layer actually does not participate in formation of SGD to the lake.

To assess SGD to Issyk-Kul, complex studies were carried out for lake profiling, including seismo-acoustic profiles, thermometry, and conductometry (measurement of electrical conductivity). This enabled construction (by regular grid) of hydrogeological cross-sections (profiles) across the entire Issyk-Kul depression, with the purposes of revealing spatial features of distribution of the SGD and substantiating selection of hydrogeological areas on the shore for further detailed experimental and analytical investigations of subsurface inflow to the lake.

Submarine groundwater discharge to the lake was calculated on the basis of data on hydrogeological parameters of the aquifers, obtained by means of well pumpings in the shore zone and by means of determining of groundwater flow rate by Darcy's law. Assessment of the SGD directly through the lake bottom was made using data of the complex profiling, and by measuring pressure heads and velocities of groundwater filtration with the aid of filter meters and well points installed on the shore. From the filters, water samples were taken for chemical and isotopic analysis. These investigations were accompanied by vertical temperature and conductometric sounding (Meskheteli and Bergelson 1989).

The SGD was assessed by data of thermometry and conductometry using the following equation (Zektser et al. 1985):

$$f = \frac{D}{hv}(S_B - S_p)\left(1 - e^{\frac{vh}{D}}\right) + S_p \qquad (5.3.2.1)$$

where v is the velocity of groundwater seepage through bottom sediments, S_B is the salt content of water in the near-bottom layer, S_p is the salt content of groundwater, f is the salt content of pore waters saturating the upper layer of bottom sediments with a thickness h (measured by a probe), and D is the coefficient of convective diffusion.

In this case, the D value was determined in laboratory samples or taken from the literature. For quantitative assessment of SGD to the lake, filter meters were installed at 69 points to measure velocity of the groundwater filtration into the bottom sediments. At 16 points, vertical sounding of the sediments were carried out to determine temperature and conductivity. Vertical distribution of hydrostatic pressure in the sediments, saturated with groundwater, was studied at 9 points (Meskheteli and Bergelson 1986). From the filters, samples of discharging groundwater were taken for chemical and isotopic analysis. Additionally, seismo-acoustic profiling was carried out along 27 lines from onboard a specially equipped research ship, which enabled obtaining the geological cross-section of the bottom sediments and determining distribution of temperature and salt content in the near-bottom water bed of the lake.

The results provided the possibility of compiling a detailed scheme of SGD into the lake. The hydrogeological investigations established the modulus of SGD to Lake Issyk-Kul from all the water-bearing layers. Also, the SGD from the upper hydrogeological layer was assessed by the hydrodynamic method using Darcy's law.

Submarine groundwater discharge to the lake from the middle hydrogeological layer (so-called "deep groundwater discharge") was assessed for each area as a difference between the total groundwater inflow to the lake, determined by experiment, and the discharge estimated by hydrodynamic method for the upper layer (Table 5.3.2.1). The estimates of groundwater and ion discharges to the lake are presented in Table 5.3.2.2. According to these data, the SGD to Issyk-Kul amounts to 1.5 km³/year. The groundwater co-discharges to the lake ~750 ths t of dissolved salts per year. Compared with surface runoff, the groundwater co-discharges to the lake 60% of chlorides, 62% of sulfates, 44% of bicarbonates, 70% of sodium and potassium, 49% of calcium, 79% of magnesium, and a great amount of microelements.

It should be noted that groundwater also brings contaminants to the lake. This happens because the unsaturated zone and aquifers of the shore are chiefly composed of pebble and clastic sediments, which makes easy penetration of polluted surface-waters to water-bearing layers. Thus, intensive subsurface runoff can lead to pollution of the lake. Within the drainage area of the lake, there are located farms and plants that can be sources of pollution. The distributed character of pollutant penetration to the lake makes it difficult to control over-composition of the groundwater discharged to the lake, and the over-pollutants in it, which can negatively influence the water quality in Lake Issyk-Kul.

Annually, the groundwater co-discharges to the lake about 18 t of oil products, 10 t of synthetic surface-active substances, and 0.6 t of agricultural chemicals. The total penetration of soluble substances to the lake amounts to 19.4 kg/sec (of the total SGD ~1.5 km³/yr), including hydrocarbonates (9.4 kg/sec), calcium and magnesium (5.7 kg/sec), and sulfates (3.1 kg/sec).

Table 5.3.2.2 presents results of the tentative calculations of discharge of microelements and some pollutants with the subaqueous groundwater flow to Lake Issyk-Kul.

As stated earlier, despite the relatively low amount of groundwater discharged to the lake, its influence on salt composition and water quality in the lake is considerable and comparable with the influence exerted by surfacewaters. In total, the SGD equals 30–40% of the total water penetration to the lake with the river

---

TABLE 5.3.2.1

Estimates of Submarine Groundwater Discharge (SGD) (m³/sec) to Lake Issyk- Kul in Certain Areas

Area	Total SGD (Q_1)	SGD (Q_2) from Upper Hydrogeological Layer[a]	Deep Groundwater Discharge ($Q_3 = Q_1 - Q_2$)
Kurskoye	8.7	7.2	1.5
Chon-Uryukty	3.14	2.7	0.44
Karabulun	3.0	2.5	0.5
Tamgha	4.3	3.1	1.2

[a]Determined according to Darcy Equations.

TABLE 5.3.2.2
Transfer of Dissolved Salts by Groundwater to Lake Issyk- Kul

Hydrogeological Region	Groundwater Discharge (m³/sec)	Discharge of Dissolved Hard Substances (kg/sec)										
		Total	Cl	SO_4^{2-}	HCO_3^-	$(Na+K)^4$	Ca^{2+}	Mg^{2+}	NH_4^+	NO_3^-	NO_2^-	H_4SiO_4
Kurskoye	31.7	11.4	0.59	1.57	5.84	0.77	1.89	0.49	0.01	0.02	0.0003	0.40
Chon-Uryukty	8.1	2.9	0.14	0.40	0.73	0.13	0.25	0.09	0.001	0.006	0.0001	0.03
Karabulun	5.9	4.2	0.87	1.1	2.68	0.94	1.40	1.54	0.002	0.054	0.0002	1.62
Tamgha	1.6	0.9	0.028	0.062	0.17	0.057	0.05	0.01	0	0.005	0	0.02
Total	47.3	19.4	1.63	3.13	9.42	1.89	3.59	2.13	0.013	0.085	0.0005	2.07

TABLE 5.3.2.3
Microelements Penetrating with Groundwater to Lake Issyk- Kul

Hydrogeological region	Discharge of Dissolved Hard Substances (Ions), g/sec												
	Fe	Cu	Zn	Pb	As	Mo	Ni	Co	I	Br	F	Mn	P
I	3.8	0.36	0.78	0.028	0	0.014	0.005	0	0.19	0.19	5.8	0.034	0.35
II	0.21	4.5	13	0	—	—	—	—	—	—	0.39	—	0.24
III	0.90	0.031	0.065	—	0	0.023	0	0	0	0	5.9	0.04	0.10
Total													
(g/sec)	4.9	4.9	13.9	0.028	0	5.4	0.005	0	0.19	0.19	12.2	0.074	0.69
(t/yr)	155	151	438	0.88	0	170	170	—	6	6	350	350	22

runoff. The SGDs into the lake are over 50% of the total amount of dissolved substances.

5.4   GROUNDWATER DISCHARGE TO SOME LAKES OF KARELIA*

5.4.1   Natural Settings in Karelia

The Republic of Karelia lies in northwestern Russia. There are ~61,100 lakes in the republic, including Europe's largest ones: Ladoga and Onego. The lakes cover 12% of the territory, and if the Karelian parts of Lakes Onego and Ladoga are taken into account, the figure rises to 21% — one of the highest in the world. Most lakes have an area of >1 km^2. Only 1400 lakes are larger, and of those only 20 are >100 km^2 in area (Filatov and Litvinenko 2001). The hydrographic features of the region are determined by its natural settings, foremost the climate, geology, and topography.

Karelia's climate is moderately continental with maritime traits. It features long mild winters and short cool summers. The most significant factors shaping the hydrological regime are precipitation and evaporation. The republic is situated in an excessive moisture-supply zone, where precipitation is 550–750 mm/yr, with a north to south upward gradient. Owing to relatively low summer temperatures, high cloudiness, and humidity, Karelia is an area with comparatively low evaporation — from 310 mm in the north to 420 mm in the south.

Geologically speaking, Karelia is the eastern margin of the Fennoscandian (Baltic) Shield — a domain where Archaean–Proterozoic crystalline rocks predominate. They are covered by Quaternary deposits ranging from 0–130 m (south) in thickness. The terrain has been chiefly shaped by the ancient basement structures, tectonic movements, and Quaternary denudation and accretion processes. Owing to the major impact of the last glaciation, a highly broken hilly-ridge topography with relatively low dominant heights of 100–200 m was formed.

For the most part, Karelia falls into the Baltic fissure water basin, where the main aquifer, which occurs throughout, lies in the upper fractured crystalline rock zone. Only small areas in the south and southeast of the republic are situated in the margins of the Russian platform artesian basins. Nearly the entire republic has pore groundwater in loose Quaternary sediments.

The general features of the regional hydrogeology are as follows. Fissure waters are confined to the regional fractured rock zone 30–50 m thick. Rocks below this zone hardly contain any water. It is only in diastrophism zones that water-bearing fissures may reach 150–200 m in depth. The groundwater in overlying Quaternary sediments is in direct hydraulic connection with fissure waters. The groundwater is normally unconfined.

The groundwater is nearly totally recharged by seepage from precipitation, which is 70–100 mm/yr on average, as determined in natural groundwater resource estimates (Ieshina et al. 1987). The discharge mostly goes to rivers, lakes, and

---

* Section 5.4 was written by S. Borodulina and V.V. Trenin (Institute of Water Problems of the North, Karelian Scientific Center of RAS, Russia, borodulina@nwpi.krc.karelia.ru).

mires. The groundwater level in flatlands is rather close to the surface (within the upper 5 m), whereas in uplands the depth grows to 10 m, sometimes reaching 20 m. The groundwater table generally, although more smoothly, mirrors the surface topography.

The overall hydrogeologic setting of the open crystalline rock mass, which lacks regional confining layers, predetermines the fairly simple route for groundwater in all water-bearing rock complexes. The surface and groundwater catchments coincide, and the groundwater moves from drainage divides to the nearest surface watercourses and waterbodies.

Groundwater discharge was estimated for three lakes in southern Karelia: Syamozero, Kroshnozero, and Suojärvi (Table 5.4.1.1). Lake Syamozero is the largest lake of southern Karelia and the Shuja River watershed. Lakes Syamozero and Kroshnozero lie in the middle reaches of the Shuja River, where the Quaternary sediments are quite thick and composite. They are comprised of glacial sediments, which constitute the interlobate moraine, and pressure ridges, which are widespread along the southern shore of Lake Syamozero and southwest of Lake Kroshnozero. The interlobate moraine is made up of fine sands and sandy loams with abundant boulders. The sediments are 15–20 m thick. Where glacial meltwater broke through, the till was washed away, leaving coarse sand.

The glaciofluvial deposits include eskers, deltas, alluvial fans, and outwash plains. The largest esker massif stretches northward from Lake Kroshnozero to Lake Syamozero. Having a thickness of 15–30 m, it is 2–3 km wide and 30 km long. Several deltas, alluvial fans, and outwash plains adjoin it. Smaller esker ridges and glaciofluvial complexes are scattered around Lakes Syamozero and Kroshnozero. They are mainly composed of sands of varying particle size and composition, occasional sandy loams, and silts.

In the south of Lake Kroshnozero, a 4- to 30-m-thick intertill Onego aquifer made up of sands of varying particle size composition directly underlies till. The water in the aquifer is locally confined. The aquifer emerges as springs on the southern shore of Lake Kroshnozero. The high water abundance in the aquifer is evidenced by the discharge of one of Karelia's largest springs at ~100 l/sec.

Lake Suojärvi is situated in the Shuja River upper reaches. Quaternary sediments there are relatively shallow (5 m), with occasional bedrock outcrops (Archaean

---

TABLE 5.4.1.1
Morphological and Hydrological Characteristics of Lakes

Lake	Area (km²)			Depth (m)			Water Volume (mln m³)	Streamflow (mln m³)
	Catchment Area	Direct Groundwater Discharge	Water Surface	Mean	Max.			
Syamozero	1610	97	266	6.7	24.5		1789	326
Kroshnozero	187	21	9	5.7	12.6		50	53
Suojärvi	2087	78	61	4.7	26.0		285	505

From Lifshits (1970).

granites, granite-gneisses). Approximately 45% of the upper parts of the Shuja River watershed are paludified.

The degree of agricultural land use is the highest in the Syamozero and Kroshnozero lake catchments. The population density there is high and all the developed areas are spread along the lakes. The southern shore of Lake Suojärvi harbors the district center, the town of Suojärvi. The lakes are used for water supply, recreation, fisheries, and trout farming (Kroshnozero).

The water in the lakes is slightly mineralized. Total ion concentration ranges from 20–44 mg/l. The index is relatively invariable across the surface area and depth. The water composition is of the calcium-magnesium hydrocarbonate type.

The nutrient status of Lake Kitaev (1984) is mesotrophic, and Lake Kroshnozero is locally eutrophic. The main factor predetermining the nutrient status is the phosphorus content. The total phosphorus concentrations in the lakes range within 24–311 µg/l in Kroshnozero, 4–26 µg/l in Syamozero, and 15–33 µg/l in Suojärvi. A second component responsible for the lake productivity is nitrogen, with nitrates prevailing among its mineral forms. By the beginning of the phytoplankton growing season, the nitrate concentration in the lakes is 0.25 mg/l. Coupled with a mineral phosphorus concentration >20 µg/l and a water temperature >18°C, this results in water "blossoming" in summer (Karelian Research Centre 1998).

5.4.2 ESTIMATION OF SUBMARINE GROUNDWATER DISCHARGE TO SOME LAKES OF KARELIA

The SGD to the lakes under consideration was quantified using Darcy's law. To this end, published and archival data on the geological structure and hydrogeologic setting of the area were analyzed; exploratory drilling and test screening were also done. The data from the wells located about 1 km away from the shoreline were employed, and the assumption was made that the groundwater discharged into Lakes Syamozero and Kroshnozero formed chiefly in the glacial and aqueoglacial Quaternary sediments, whereas the groundwater discharged into Lake Suojärvi came from the upper fractured crystalline rock zone.

In 2002–2004, 25 hydrogeologic boreholes were drilled in sites with sandy Quaternary sediments along the shores of Lakes Syamozero and Kroshnozero, and 70 interval pumping tests were carried out. On the southern shore of Lake Syamozero, a 15-day multiwell pumping test was performed. The results indicate that the hydraulic parameters of the Quaternary sediments vary widely both across the area and across the profile, with differences observed at minor distances. The hydraulic conductivity factor ranged from 0.5–1 to 140 m/day, transmissivity from 5–800 m²/day. The values of the parameters are the highest at esker ridges.

The structural heterogeneity and uneven spatial distribution of individual types of Quaternary formations have made calculations for specific parts of the shore area rather challenging. Statistical processing yielded a hydraulic conductivity of 5 m/day for the Quaternary sediments. The crystalline rock parameters for Karelia were calculated using data from 600 wells; mean hydraulic conductivity for Archaean granite-gneisses was set at 0.2 m/day (Ieshina et al. 1987).

The calculated SGD from the shore areas composed of sandy sediments is much higher than that from areas where crystalline rock prevails (Table 5.4.2.1). Calculations show that direct discharge to the lakes (bypassing the river network) compared to the total streamflow is very low (<1%) for Lake Suojärvi, but reaches 13% for Syamozero and 18% for Kroshnozero.

5.4.3   GROUNDWATER CONTRIBUTION TO THE WATER AND SALT BALANCE OF THE LAKES OF KARELIA

The chemical composition of the groundwater discharged to the lakes was estimated using the results of aquifer sampling from wells and springs along the shore. A total of 82 groundwater samples were analyzed; Table 5.4.3.1 shows averaged results of the chemical analyses. To determine mean-weighted values of the chemical parameters of groundwater along lake shores, figures from repeated measurements at individual sampling sites were first averaged.

Under natural conditions, water mineralization in glacial and aqueoglacial sediments is quite low (0.1–0.2 g/l). The mineralization is the lowest (<0.1 g/l) in waters forming within esker ridges. In terms of the dominant components, the water is of the calcium and magnesium hydrocarbonate and sulfate type, the proportion of chlorides normally staying within 10 mmol-%, sodium 20 mmol-%, and potassium 5 mmol-%.

The groundwater in the intertill aquifer, which contributes ~30% to the total SGD to Lake Kroshnozero (Table 5.4.3.2), has a neutral pH (7.3 on average), calcium hydrocarbonate composition, anaerobic conditions, and high iron concentration (1–3.4 mg/l). The aquifer is noted for an elevated total phosphorus concentration (0.3 mg/l on average). Water mineralization ranges within 0.1–0.2 g/l.

The main agent shaping the groundwater chemical composition in the Lake Suojärvi catchment is crystalline bedrock. Groundwater with low acidity (pH 5.5) and low mineralization (0.04 g/l) may sometimes form in thin bands of till. In crystalline rocks, groundwater mineralization increases with depth from 0.1 to 0.5 g/l.

TABLE 5.4.2.1
Estimates of Submarine Groundwater Discharge (SGD) to the Lakes of Karelia

Lake	Front Width (km)	Hydraulic Conductivity (m/day)	Aquifer Thickness (m)	Hydraulic Gradient	SGD (mln m³/yr)
Syamozero	139.8	5.0	10	0.015	41.3 13[a]
Kroshnozero	22.9	5.0	15	0.01	9.4 18[a]
Suojärvi	92.9	0.2	40	0.01	2.7 0.5[a]

[a]Relative (%) to streamflow.

TABLE 5.4.3.1
Mean Indices of the Groundwater Chemical Composition along Lake Shores

Component (Parameter)	Syamozero (n = 47) I	Syamozero (n = 47) II	Kroshnozero (n = 15) I	Kroshnozero (n = 15) II	Kroshnozero (n = 15) III	Suojärvi (n = 20) I	Suojärvi (n = 20) II
pH	6.5	6.2	6.1	6.5	7.3	7.2	6.3
Na (mg/l)	8.6	14.0	3.8	22.7	6.5	15.5	15.6
K (mg/l)	2.2	12.8	1.0	24.0	1.5	2.5	7.9
Ca (mg/l)	11.5	14.7	14.0	21.7	28.6	17.0	21.0
Mg (mg/l)	5.2	4.6	5.0	7.6	16.8	4.5	6.1
HCO_3 (mg/l)	55.0	33.6	55.0	64.0	170	86	69.1
Cl (mg/l)	8.3	16.8	5.0	27.8	2.7	5.0	13.1
SO_4 (mg/l)	12.1	14.4	15.0	24.0	11.0	19.0	34.5
SIO_2 (mg/l)	10.2	10.8	9.5	10.0	8.4	13.9	8.4
Fe_{tot} (mg/l)	0.9	0.1	0.2	0.2	2.0	0.9	0.5
NO_3 (mg/l)	0.95	30.9	1.0	62.0	0.5	0.35	29.4
NO_2 (mg/l)	0.03	0.04	0.02	0.1	0.01	0.01	0.03
NH_4 (mg/l)	0.5	0.07	0.1	0.5	0.14	0.1	0.90
P_{tot} (mg/l)	0.06	1.3	0.05	0.9	0.3	0.04	0.04
Mineralization (g/l)	0.11	0.17	0.1	0.25	0.2	0.19	0.22

I, undisturbed areas; II, human impact areas; III, interfill aquifer discharge zone.

TABLE 5.4.3.2
Results of Calculations of Chemical Flow to Lakes with Groundwater

Lake	Discharge Zone	SGD (mln m³/yr)	Leakage with SGD (t/yr) Total Ions	Leakage with SGD (t/yr) NO_3	Leakage with SGD (t/yr) K	Leakage with SGD (t/yr) P	Rate (t/yr·km²)
Syamozero	I	35	3500	35	80	2	47
	II	6.3	1100	195	81	8	
	I	3.7	370	4	4	0.2	72
Kroshnozero	II	2.6	650	161	64	2	
	III	3.2	630	0.6	5	1	
Suojärvi	I	2.27	450	1	6	0.1	7
	II	0.44	100	13	3.5	0.02	

I, undisturbed areas; II, human impact areas; III, interfill aquifer discharge zone.

In compliance with vertical geochemical zonation, the oxidation conditions there change to anaerobic, pH increases from 6 to 7, and iron content rises to 38 mg/l.

It should be that the role of the chemical flow in shaping the salt regime of lakes may change significantly under human impact. Therefore, when estimating the chemical flow we distinguished human impact zones along the shore, including communities, farmland, and enterprise premises. Discharge from such areas may amount to 15–27% of the total SGD into the lakes (Table 5.4.3.2).

Becasue of the lack of water-impermeable deposits in the vadose zone, groundwater is easily prone to pollution from the surface, except for intertill aquifers. The crystalline bedrock lies close to the surface beneath sandy loams and sands, and is also highly vulnerable to pollution, so that both natural springs and wells and boreholes yield high concentrations as the main indicators of communal pollution.

Human activities cause transformations in the upper groundwater zone (Tyutyunova 1987; Samarina 2001). The most significant changes in groundwater mineralization and chemical composition in Karelia are caused by a sharp rise in nitrate and potassium concentrations (Borodulina and Perskiy 1998). Average nitrate concentrations in groundwater in undisturbed areas range from 0.1–0.5 mg/l, while all springs and wells within settlements that are contaminated with nitrates have concentrations reaching 100 mg/l. Furthermore, nitrates become a chief constituent in the anion composition (up to 50 mmol-%). At the same time, the concentrations of other nitrogen compounds stay within 0.08 mg/l (nitrites) and 0.1 mg/l (ammonium nitrogen), which is typical of aquifers with aerobic conditions and high redox potential values. Nitrate-contaminated waters usually contain high potassium concentrations as well, up to 20–50 mg/l (25 mmol-%), and their Na/K ratio, one of the main indicators of the hygienic water quality, is near zero, whereas the ratio in uncontaminated waters is ≥ 10 mg/l (Krainov and Zakutin 1993). The total phosphorus concentrations change from 0.03 to 4 mg/l. Concentrations of trace elements, including heavy metals, in groundwater rarely exceed regional background values, indicating a predominantly communal and agricultural nature of pollution. Groundwater in developed areas is generally more mineralized than that in the surrounding territories. The concentration of dissolved salts there reach 0.4–0.6 g/l. Results of the calculations of chemical seepage from shore areas with groundwater are shown in Table 5.4.3.2. The table provides also values for the influx of selected nutrients with groundwater flow. The groundwater discharged into Syamozero annually brings 4500 t of salts to the lake, including 230 t of nitrates, 160 t of potassium, and 10 t of phosphorus. The chemical flow to Lake Kroshnozero with groundwater from shore areas amounts to 1700 t, including 170 t of nitrates, 73 t of potassium, and 3 t of phosphorus. Salt influx to Lake Suojärvi is 550 t/year. The ion flow rate is the highest (72 t/yr·km²) in Lake Kroshnozero. It is 1.5 times higher than in Lake Syamozero (47 t/yr·km²), and an order of magnitude higher than in Lake Suojärvi (7 t/yr·km²).

A noteworthy fact is that the influx of major contaminants (nitrates, potassium) with groundwater to Lake Syamozero is 1.5–2 times greater than to Lake Kroshnozero and an order of magnitude greater than to Lake Suojärvi. Yet, the contaminated SGD to Lake Kroshnozero accounts for 40% of the total chemical flow

TABLE 5.4.3.3
Nutrient Influx to Lakes with Streamflow and Submarine Groundwater Discharge

Lake		Discharge (mln m³/yr)	Total Ions (1000 t/yr)	NO₃ (t/yr)	P (t/yr)	Si (t/yr)	Fe (t/yr)
Syamozero	Streamflow	326	8.2	82	12.6	1041	294
	Groundwater	41.3	4.6	230	10	206	41
Kroshnozero	Streamflow	53	2.5	16.4	5.5	170	49
	Groundwater	9.4	1.7	166	3	47	18
Suojärvi	Streamflow	505	7.5	126	13.4	1546	520
	Groundwater	2.7	0.6	14	0.1	14	3

(24% for Syamozero, 18% for Suojärvi), showing that anthropogenic pressure is the highest in the Kroshnozero area (Table 5.4.3.2).

Table 5.4.3.3 shows the ratio between salt leakage with groundwater and streamflow. Although direct discharge to the lakes (bypassing the river network) is rather low compared to the total streamflow (0.5–18%), SGD contribution to the chemical balance of the lakes is, however, more significant, reaching 50–70%.

The comparison of the nutrient mass the lakes receive from streamflow and from SGD has shown that in contrast to streamflow, where the main constituent is silicon (Surface waters…, 1991), SGD to the lakes contains as much (Syamozero) or more (3.5 times for Kroshnozero) nitrates than silicon. Also both streamflow and SGD are occupied by iron and phosphorus (Table 5.4.3.3). The groundwater discharged into Lakes Kroshnozero and Syamozero also carries about the same amount of phosphorus and 3–10 times as much nitrates as streamflow. One can thus conclude that groundwater flow plays a significant factor in the eutrophication of lakes.

5.5 ESTIMATION OF SUBMARINE GROUNDWATER DISCHARGE INTO LAKE LADOGA*

Lake Ladoga (Ladozhskoye Ozero) is one of the greatest lakes in Europe and is connected to the Baltic Sea via the River Neva. Lake Ladoga is located partially in Finland and Belarussia. Lake Ladoga is the main source of water supply to Petersburg. In this section we describe the problem of pollution via SGD in Lake Ladoga. Submarine groundwater discharge to the lake has not been previously studied. This subchapter is considered to be the first attempt to determine the role of SGD in the total water contamination of Lake Ladoga.

The basin of Lake Ladoga is a catchment system of Lakes Onega, Ilmen, and Saima. The total area is 28,000 km². The area of Lake Ladoga is 17,700 km² with

* Section 5.5 was written by Arkady Voronov, E. Viventsova, and M. Shabalina (Saint Petersburg State University, Russia).

an average depth of 51 m with a maximum depth of 230 m. The volume of Lake Ladoga is 908 km^3.

The climate in the Lake Ladoga area is humid and soils are well washed out. As a result, the lake water is slightly mineralized (~56 mg/l) and has a hydrocarbonate-calcium type composition.

A great number of islands are located in Lake Ladoga. Their total area is 457 km^2. There are 660 islands with an area >1 ha. They are distributed very irregularly; most (about 500) are situated near the northwest coast. They form a great number of groups separated from each other by narrow straights called skerries. At the center of the lake about 65 islands are noted. They belong to the Valaamsky and West archipelagos. About 80 are situated along the east coast, only 5 are noted along the west coast zone, and 16 in the south zone. They are represented by low sandy islands sowned by rubbles.

Approximately 32 rivers discharge to Lake Ladoga, but the water in the lake discharges to only one river — the Neva. Large rivers such as the Svir, Vuoksa, and Volkhov contribute 64.4 km^3/yr of water to Lake Ladoga. That is ~90% of total river runoff. Precipitation into Lake Ladoga is 8.8 km^3/yr. In general, the total volume of water going into Lake Ladoga is ~85.5 km^3/yr. Evaporation from Lake Ladoga is 6 km^3/yr. The SGD value is unknown. According to an article by Chernov (2000), SGD into Lake Ladoga is estimated at 1.3 km^3/yr, which equals 2% of the total water balance in the lake. The quantity of water coming into Lake Ladoga changes from year to year, thus varying from 55%–150% of the average yearly value.

Depth of the water table in Lake Ladoga is about 3 m. Two types of level fluctuation are noted. The first is connected with the total lake surface fluctuation that is connected with changes to the lake volume. The second is characterized by short-term rising of the level in one area and the similar decreasing in the other that is caused by wind regime.

Geologically the northern part of Lake Ladoga lies under the Baltic Shield that is represented by granites, gneisses, and other magmatic rocks of Archean age. Sometimes a thin layer (some several meters) of Quaternary sediments covers it. Fringed parts of the Russian Platform are noted in the south and west coasts of the lake. Bottom and coast relief is not similar in different parts of the lake. In the northern part of the lake, the bottom and coasts consist of dislocated crystalline rocks. Relief is highly disjointed and has a gradual structure. The southern shallow part of the lake is located on the Russian Platform and is characterized by plane relief of the bottom and coasts. Sedimentary cover is thin and consists of the rocks of Vendian and Cambrian age. Proterozoic sediments are represented by shale, quartzite, sandstone, conglomerate, crystalline, and dolomitized limestone. The Cambrian layer consists of sandstone, sand, and blue clays. The thickness of the sedimentary Quaternary layer reaches up to tens of meters (Kirillova and Raspopov 1971).

A crystalline basement, exposed on the Baltic Shield and constantly submerged in south and southeast directions under the sedimentary cover of the Russian Platform, is divided by fractures and fissures on separate ledges and depressions. Thus, the Lake Ladoga basin occurs in such tectonic graben.

Hydrogeologically Lake Ladoga is separated in two parts: the northern area of Quaternary fissured aquifers discharge and the southern area of Gdov and Cambrian–Ordovician aquifers discharge. Fissured aquifers have a regional character and are typical of the upper fissured zone of the crystalline massif in the northern part of the lake. The groundwater has a hydrocarbonate-calcium-natrium composition, and rare chloride-natrium composition. The thickness of the fissured zone is 20–30 to 70–80 m. Depth of the fissured zone of Archean rocks is less than in volcanic sediments. Maximum fissuring has been found by super-deep drilling at a depth of 30–40 m. In general, crystalline basements are rich-watering and the water table lies in the upper Quaternary layers. Only on some elevated areas can crystalline rocks be drained at depths of 1–5 m.

Filtration conditions in fractured rock are variable, though as a rule very low. Maximum filtration conditions are observed at depths of 40–50 m. Well yield which characterizes the water-bearing capacity of rock varies from 0.001–0.2 l/sec. This value is higher in fissured areas. Average values of transmissivity of different complexes are approximately in the same range and vary from 1–10 m^2/day. This allows characterizing crystalline rocks as slightly flooded. Spring yields are usually centesimal and 1/10, and rarely 1 l/sec.

A regime of fissured groundwater is more stable in Quaternary rocks. Groundwater temperature does not experience significant seasonal variations and equals 3–50°C. Also, groundwater is characterized by stable chemical composition. Groundwater in areas of a crystalline rock outlet has less stable regime. Spring yields vary significantly season to season, and some wells dry up in summer and freeze in winter.

All sediments of the Quaternary age are water bearing. Different genetic types of water-bearing rocks lay adjacent to each other, forming a unique unconfined aquifer. Groundwater of Quaternary sediments (except Intermorrenian aquifers) is hydraulically connected with the surfacewater, crystalline rocks, and each other. This occurs because of the absence in these rocks of a uniform aquitard.

There are several groundwater aquifers in the Lake Ladoga basin, in general of Quaternary age, and which play a significant role in the formation of groundwater flow and resources in this region. They are modern peat (organogenic), marine, fluvioglacial, glacial, and semiglacial sediments.

Groundwater runoff is estimated at 1.42 km^3/yr with an average module of SGD of 2.4 l/sec·km^2.

Lake Ladoga is currently under serious technogenic pressure as a result of incoming wastewater discharging from different industries. The total volume of wastewater going into Lake Ladoga was ~500 mln m^3 at the end of the 20th century. Annually 78.8 km^3 of pollutants flows from Lake Ladoga into the Neva River (Karelian Research Centre 1996). Biogenic substances inflowing into the lake cause anthropogenic eutrophication and contamination. As mentioned above, the pollution comes into Lake Ladoga with river runoff, SGD, precipitation, and waste water from industries along the coastal area. Approximately 500 industrial sites are in the coastal area of Lake Ladoga. Pollution is concentrated in the water, biota, and bottom sediments. The level of phosphorus concentration in the water has increased more than 3 times by the end of 20th century.

Among the main pollutants carried by river runoff into Lake Ladoga are organic substances, petroleum products, NO_3, Cu, and Zn. First, the pollution is concentrated at the mouth of rivers and in small rivers. The highest concentration of pollutants was discovered in the Volkhov, Svir, and Vuoksa Rivers. Wastewater from different industries in the coastal zone (for example, the SjasStroj cellulose-paper factory) presents a significant danger for the Lake Ladoga environment, especially taking into account the low water exchange rate of the lake (12 years). Less polluted areas are located near Valaam and the eastern part of the lake.

Some pollution could come into the lake with SGD. Among these pollutants are nitrates, petroleum products, heavy metals, etc. The natural composition of groundwater discharging into the lake can differ greatly. Fissured groundwater is radium, or radon-rich, while deep groundwater is highly mineralized. It is necessary to note that 20–25% of river flow is formed by SGD.

Lake Ladoga is a unique ecological locale deserving special attention. Thus, it is necessary for a detailed investigation of the SGD into the lake and the scale of groundwater pollution.

5.6 SPECIFIC FEATURES OF WATER EXCHANGE BETWEEN LAKES AND GROUNDWATER IN THE PERMAFROST OF THE PECHORA-URAL REGION *

In this region there are some quarter million lakes (not including the smallest ones) with a total area of over 12 ths km². The background lake-occupied area of its plain part is 1–5%, in some places 10–25%. In the Bolshezemelskaya tundra, thermokarst lakes prevail; while in the mountains, tectonic and glacial lakes predominate (Figure 5.6.1).

5.6.1 WATER EXCHANGE IN THE ZONE OF CONTINUOUS PERMAFROST

Ural and Pai-Hui. The tectonic Lake Large Shchuchie is the deepest (137 m) and most water abundant (0.78 km³) lake of the Ural and is one of the deepest mountainous lakes of the former USSR (Figure 5.6.1). Its shores are composed of very steep, partly cliffy slopes, rising to a height of 800–1000 m above sea level with an absolute elevation mark of 189.5 m. The lake has a narrow and long-shaped tub (1.35 × 12.72 km) with a flat bottom and steep slopes (Kemmerikh 1961). It represents a continental rift. The permafrost thickness reaches 900 m (Oberman 1998). At a depth of 136–137 m, the temperatures in the lake are usually equal to 2.3°C–2.6°C at the end of winter. These temperatures may indicate the existence of an open talik under the lake. The static levels of the subpermafrost waters lie at marks exceeding the largest depths of the lake by approximately 50–80 m. Thus, there is a favorable precondition for subpermafrost water discharge into it.

---

* Section 5.6 was written by Naum G. Oberman (Mining and Geological Company MIREKO, Russia, oberman@mireko.komi.ru).

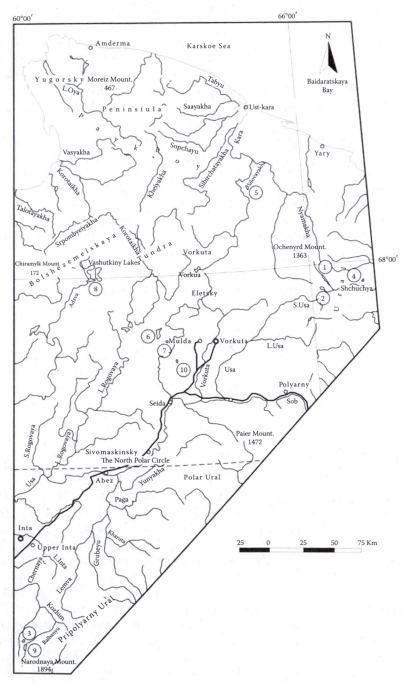

FIGURE 5.6.1 General map of the northeast part of the Pechora-Ural region. Lakes: (1) Large Shchuchie; (2) Small Shchuchie; (3) Grube-Pendi-Ty; (4) Baito; (5) Egor-Ty; (6) Small Kharbey-Ty; (7) unnamed lake in the Seida river basin; (8) Vashutkiny lakes; (9) Large Balban-Ty; (10) unnamed river in the Vorkuta river basin.

And this precondition is being implemented, if to judge from data on the microcomponent composition of the near-bottom waters sampled in the above-mentioned period. For example, the concentration of yttrium is on an order of magnitude higher than the clarke one; ytterbium by 5 times. The tin content exceeds its average concentration in the adjacent areas by 40 times and lead by 20 times. The water from one of the holes in the southern part of the lake, taken from a depth of 85 m, contained 1.5 g/l of free carbon dioxide with a 4.5 pH at a background value of 6.9 (Oberman 1984).

The summer chloride-hydrocarbonate, magnesium-calcium composition of the lake water is transformed by the end of the winter into low water of sulfate-hydrocarbonate, hydrocarbonate sodium-calcium types. The changes in water composition with its cryogenic metamorphism (Anisimova 1981) with the data on the composition of shallow-located groundwater in the vicinity of Large Shchuchie make it possible to interpret seasonal hydrochemical changes in its waters in the following way. These changes are caused by the discharge of above-permafrost waters into the lake from underchannel taliks by 12 temporary streams and a nonfreezing stream drained by the channel. The SGD to the lake is almost completely provided by the above-permafrost runoff.

The lake waters are discharged through the outflowing river of Large Shchuchie. It has been observed that the water discharge from the lake decreases with a decrease of the average annual air temperature, leading to a partial freezing of non-through taliks and the reduction of above-permafrost waters discharged from them into the lake. A large discrepancy between the registered amounts of water inflowing to the lake and outflowing from it has been evaluated. The discrepancy can be eliminated if it is assumed that the major part of the lake water is discharged through the underchannel talik of the Large Shchuchie River. The underchannel discharge at 1 km from the river outflow amounts to 3.92 m^3/sec (60% of the channel flow rate) during the summer low water; in winter, the icing is formed with a size of 0.5 mln m^3.

Located in the same area is Lake Small Shchuchie, which is similar to Large Shchuchie by origin and morphology, but is over 10 times smaller by water amount. The Gletcherny stream inflows into the Small Shchuchie and in the process drains three glaciers. The piezometric surface of under-permafrost waters lies deeper than the lake floor, therefore, their discharge to the lake is excluded.

The results of our work carried out in January 1980 (Table 5.6.1.1) only partly confirm the conclusions of predecessors' research. The dependence of ice thickness on the first of the factors is observed over the major lake area, and distinctly disturbed near the inflow of the Gletcherny stream and almost along the entire thalweg length. The anomalous decrease of ice thickness by 1.5–3 times (at the same thickness of snow) from Profiles VII[a], VII to Profiles I, A, and from Profiles V to II is accompanied by a decrease of water temperature by 0.9–1.4°C at a depth of 25 m and by 0.8°C at depths of 32–34 m. This occurs at equal temperatures within a depth interval of 5–20 m. This means that the anomaly of ice thickness is connected neither with specific features of snow accumulation nor with warming by shallow groundwater. It is impossible to correlate it and decreases of the near-bottom temperature with later freezing (due to discharge of the spring Gletcherny) of the water area is near the stream mouth. This disagrees with the absence of the analogous anomaly near

TABLE 5.6.1.1

Changes in Bottom Temperatures and Ice Thickness along Thalwegs of Lake Small Shchuchie

Numbers of Hole Profiles		VII	VII	VI	V	IV	III	II	I	A	River Outfall
Distance to Springs (km)	Bezimyanny	0.4	0.5	1.5	2.9	3.6	4.6	5.6	6.6	7.3	7.3
	Gletcherny	6.9	6.8	5.8	4.4	3.7	2.7	1.7	0.7	0.0	0.0
Thickness (m)	Snow	0.0	0.15	0.15	0.0	0.3	0.3	0.1	0.0	~0.0	0.0
	Ice	1.1	0.9	0.8	0.9	0.9	0.75	0.65	0.5	0.4	0.0
Water Temperature (°C) at Depth (m)	2.5		0.5	0.5	0.5	0.6	0.5	0.5	0.5		0.3[a]
	5		0.8	0.7	0.7	0.7	0.7	0.7	0.8		
	10	1.0	0.9	0.8	0.7		0.7	0.8	0.7		
	15	1.0	1.0	1.0	1.0	1.0	0.9	1.0	1.0		
	20		1.0	1.0		1.0		1.0	1.0		
	25	2.1[b]	2.6[b,c]	1.1	1.1	1.1	1.1	1.1	1.2[b]		
	(32–34)[b]			2.4	2.6	2.2	2.0	1.8			
Depth Interval of Homothermal Layer (m)		no d. 10–15	5–20	5–10 15–25	5–10 15–25	— 15–25	5–10 no d.	5–10 15–25	5–10 15–20	no d.	

[a] Near surface.
[b] On bottom.
[c] Depth of 27.5 m.

the area where another stream from the opposite side inflows into the lake. We speculate that the given anomalies are caused by cold waters discharged from melting glaciers along the underchannel talik of the Gletcherny stream. These waters not only decrease the near-bottom temperature in the adjacent part of the lake, but also form homothermal layers anomalous for the middle water mass during the cooling of the lake. The presence of such a layer can, counter to the stable density stratification, give evidence of induced-water convection. At turbulent heat exchange, the heat conductivity of water increases, forcing the heat flux to increase by 3.6 times toward the lower ice surface (Feldman 1984). This may be the most likely reason for the anomalous decrease of ice thickness in the southern part of the lake.

In the same area, opposite the Gletcherny stream mouth, the mineralization of lake waters was equal to 66 mg/l in January 1980 (at a background mineralization of 15–18 mg/l in the same period). The waters had anomalously high pH values and contained carbonates and silica acid. The summer chloride-hydrocarbonate, magnesium-sodium-calcium composition was preserved in only a quarter of the total water mass; in almost half the total water mass (16 samples of 36) its composition was of the hydrocarbonate magnesium-calcium type and in almost a third, its composition was of the hydrocarbonate calcium type. The highest calcium concentration was found opposite the Gletcherny stream mouth. The concentrations of the microelements in the Small Shchuchie water are higher than in its bottom sediments. The difference (compared with Large Shchuchie) in the proportions of microelements in these media may be caused by the absence of under-permafrost water discharge to the lake.

The frequent absence of surface runoff from the lake in winter, in addition to the availability of discharge from underchannel waters from the inflowing streams, points toward the probable infiltration of the lake waters through the bottom and, as a result, the recharge of the subpermafrost waters and underchannel waters of Small Shchuchie. The water discharge from the Small Shchuchie in winter during some years may be explained by the fact that the discharged above-permafrost waters into the lake exceed the filtration abilities of the lake bottom and underchannel talik. The winter water discharge from the Small Shchuchie, unlike the discharge from Large Shchuchie, is not correlated with the "warmest" year preceding the measurement. These observations point toward an above-permafrost water discharge to Small Shchuchie.

The Central Mountainous Area is also characterized by the availability of cirque lakes. One, Grube-Pendi-Ty, is located in the Pripolarny Ural, with an area of 0.26 km^2 and a maximum depth of 23.5 m. It is surrounded from three sides by steep slopes, rising above the lake for 300–350 m. On the fourth side it is dammed by moraine. The steep slopes are frozen to a depth of 700 m. The samples of the near-bottom lake waters showed abnormally high contents of microelements (in mg/l): beryllium 0.00087; ytterbium 0.002; germanium 0.00058; scandium 0.0058, and so on. The concentrations of these elements decrease toward the lake surface. These microelements generate geochemical anomalies also in the Proterozoic rocks of the lake bottom. Especially notable is the presence of oil in the lake waters, which has been registered by two laboratories (testing in February 1999 and April 2000). The oil content changes at the bottom from 6.5 mg/l (6 m depth) to 1.0 mg/l (1 m depth) (Oberman et al. 2004).

This indicates a subaqueous discharge of the subpermafrost water. Its discharge is not large, judging by the unchanged chemical (chloride-hydrocarbonate, sodium-magnesium-calcium) composition of water during the year. The only noticeable change is that the mineralization of the lake's near-bottom layers increases from 16–26 mg/l in summer to 44–49 mg/l in winter. The discharge occurs, probably, along the zone of regional thrust, with which the lake is associated.

The tectonic and thermokarst lakes are spread in the piedmont areas. They have the following features: shallow waters, permafrost thickness ~400–500 m, and a deep location of piezometric surface of subpermafrost waters. The area of the tectonic Lake Baito in the eastern piedmonts of the Polar Ural is 2.94 km^2 with a maximum depth of 22 m. Judging by the mineralization of the lake waters, 30–50 mg/l, the lake is mainly recharged by above-permafrost waters. The lake waters have an anomalous hydrocarbonate ammonium-calcium-magnesium composition in shallow waters (2 m depth) 400 m from shore. At 1300 m from shore (13 m depth), the near-bottom waters are of the hydrocarbonate sodium-magnesium-calcium type. It is acknowledged that the waters are being enriched with ammonium and magnesium because of the freezing of water-bearing rocks (Anisimova 1981). Therefore, the registered spatial changes in the lake water composition can be interpreted as a result of the cryogenic press-out of shallow groundwater into the lake from the frozen shallow waters and the seasonally thawed layer on the shore, as well as the water inflow from the subchannel talik.

The thermokarst Lake Egor-Ty with an area of ~0.95 km^2, a depth of 8.5 m, and a drainage area of 21 km^2 is located on Pai-Hui. Judging by the ratio of the levels of lake waters and groundwater, the lake may have water discharge only from the subchannel thin talik, the seasonally thawed layer on the shore, and from the seasonally thawed layer of the shallow waters. To assess the participation of each of these sources in the water discharge is possible (on the qualitative level) by data from Table 5.6.1.2. The analysis of this data shows that the seasonal changes of typical chemical components of the water are caused by the cryogenic concentration of lake waters with ice generation and inflow of groundwater subjected to cryogenesis. The sharp increase of magnesium and silica acid contents, as compared to their contents at the start of the ice over (sample 1), may mean that groundwater from the frozen sandy and clayey low-water areas is being discharged to the lake. The congealment of poorly "washed" (by precipitation due to the permafrost factor) icy-marine sediments of the seasonally thawed layer of the shore provide a leeway for the considerable penetration of sodium chloride into the deep part of the lake (samples 4, 3). The total SGD in the lake is not significant, taking into account that the lake depression is surrounded and underlain by clayey sediments and that the above-mentioned layer is thin (average, 0.8–0.9 m). Therefore, the lake water discharge is very low. It consists of icing ~700 m long and 5–15 m thick on the lake, and several tens of meters long and 40 sm thick in the valley of a stream outflowing from the lake.

Pechora Lowland. Typical of the Bolshezemelskaya Large Vanyuk-Ty are thermokarst lakes with depths of >1–40 m. In the eastern part of the lowland, predominantly Quaternary sediments of ~100–200 m thick are distributed. They are almost completely frozen. Judging by the ratio of absolute levels, the lake waters are recharged by subpermafrost waters of Paleozoic rocks through the Quaternary strata (Figure 5.6.1.1.A). The system of Vashutkiny lakes consists of a number of water bodies connected to each other by short ducts. The system is closed by the Lake "Large" Vanyuk-Ty which gives a start to the outflow of the Adzva River. The total area of the lakes is 88.3 km^2 with a drainage area of 560 km^2. Their maximum depths are 20–40 m. The lakes are located on a marine Middle Pleistocene plain composed chiefly of sands and pebbles; in deep-water areas, clays and silt underlie the lakes. The Quaternary cover is underlain by a carbonate strata of the Chernov uplift, complicated by a zone of regional disjunctive.

The discharge of the subpermafrost waters in this zone opens a path in gravel sediments for the penetration of chloride-sodium waters with a mineralization of 0.45 g/l to the Large Vanyuk-Ty. The discharge can influence the composition of the lake waters and, hence, of the water in the outflowing Adzva River, during anomalously warm and dry summers.

The chemical composition of waters in the lake and river of Adzva should be more distinct during the winter low-water period. However, this does not occur perhaps because of the discharge of subpermafrost waters mainly in the low-water areas. The disappearance of sulfates from the lake water is caused, probably, by the process of desulfatization, occurring in conditions of hydrogen deficiency under

TABLE 5.6.1.2
Hydrochemical Features of Lake Egor- Ty and Their Seasonal Variability

Sample No.	Testing[a] Date	Depth (m)	Mineralization (mg/l)	Oxidation Ability (mg/l O_2)	Content in Water (mg/l) Na+K	Mg	NH_4	Cl	H_4SiO_4	Chemical Type of Water
1	10/13/1991	0[b]	54	14.2	3	<1	0.6	6	6	$HCO^3_{74}Cl_{26}$-$Ca_{68}Na_{25}$
2	04/15/1991	1.76[c]	173	13.9	13	33	1.0	24	20	$HCO^3_{66}Cl_{32}$-$Ca_{44}Mg_{33}(Na+K)_{23}$
3	04/12/1991	4.7	71	8.0	7	2	0.8	15	6	$HCO^3_{54}Cl_{46}$-$Ca_{54}(Na+K)_{30}$
4	04/14/1991	6.5	57	3.7	9	<1	0.6	20	3	$Cl_{65}HCO^3_{35}$-$(Na+K)_{49}Ca_{46}$

[a]Near-bottom water.
[b]In spring outfall from the lake.
[c]Ice thickness 1.45 m.

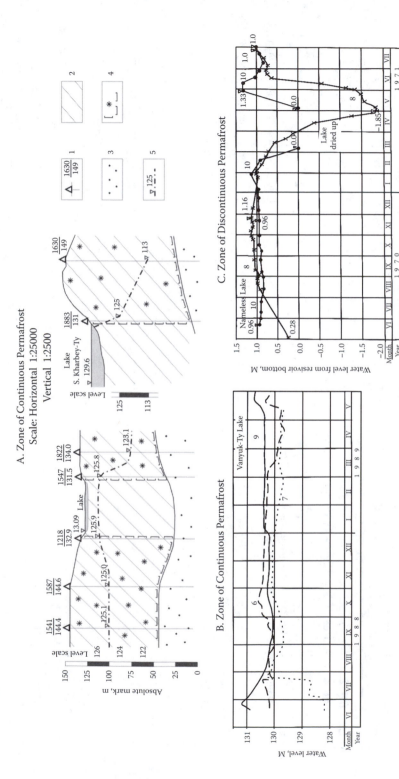

FIGURE 5.6.1.1 Correlation between groundwater and lake levels: (1) the well, in numerator [its number], in denominator [neck mark, m]; (2) the Quaternary deposits; (3) the Permian fractured rocks; (4) permafrost and its boundaries; (5) piezometric level of subpermafrost water and its marks; (6–8) the water of the Quaternary deposits — (6) flood plane lake taliks, (7–8) sublake open taliks; (9–10) water of nonfrozen and frozen lakes.

the ice. The significant enrichment of these waters by magnesium is correlated with the cryogenic press-out of shallow groundwater from freezing low-water areas: during immediate testing of the Lakes Large Vanyukta, Large Starikta, and Yurta, with average depths of 5.9, 2.9, and 1.5 m, respectively, the magnesium content in the near-bottom water amounted to 7, 14, and 46%-equiv., respectively. The concentration of this element near the same place in the first of these water bodies increased from 7% to 36%-equiv. with a decrease of air temperature during the year, preceding the testing, from –5.5°C to –6.9°C.

Taking into consideration the above-stated facts, the preferred method of confirmation of deep-water discharge in lakes would be the analysis of hydrochemical data during the summer dry period (for example, 1961). The considerable magnesium content in the lake waters of this time can be explained by the discharge of the shallow groundwater containing an increased concentration of this element. Thus, only the sodium chlorides and partly the sulfites can serve as indicators of water discharge from the zone of tectonic faults. The numerous springs on the same Chernov uplift have a similar sulfate-chloride, calcium-sodium composition, but are only present in the Vorkuta district.

Thus, the hydrochemical data enable establishing the availability of the ascending discharge of subpermafrost waters into the Vashutkiny lakes. The descending discharge of shallow groundwater of the lake flood plain is also observed. The comparison of the hypsometric marks of the shallow groundwater and lake waters indicates the seasonally variable interconnection of these waters (Figure 5.6.1.1B). Whereas during autumn and the first half of winter, the shallow SGD of the shore is directed toward the lakes with coming the critical water period, the discharge changes direction. In March, the level of these waters drops sharply due to their isolation by the frozen shallow water from the rest of the water area. The shallow groundwater surface of the water area itself is lower than the lake water line year-round. Through the bottoms of the water bodies, the ascending SGD from the Paleozoic strata and the infiltration recharge of the Quaternary shallow groundwater by the lake waters takes place at the same time. The "counter" movement of shallow and artesian waters through the "same talik gap" is typical of the permafrost area (Shvetsov 1968).

The hydrochemical data indirectly confirm the occurrence of groundwater inflow to the Vashutkiny lakes. However, the analysis of the mineralization levels and seasonal changes in the lake water composition, given the grounds to presume that the groundwater inflows into these lakes, is relatively low.

The first calculation of the water balance of the lakes was carried out by Goldina (1972). However, it did not include the subsurface portion of the water inflow into the lakes and their discharge, and, therefore, their drainage area was underreported by almost 1.5 times. Also, the surfacewater inflow to the lakes and discharge from them did not occur. Therefore, we made an attempt to calculate the water balance only for the period that was supported by the necessary initial data: since 19.10. (1988) to 18.03 (1989). Data of field investigations carried out by V. Ye. Karpovich and L.P. Goldina (1972) were used for assessing the water balance.

The equation of the water balance for the given case can be written in the following equation:

$$Yn + Ys + Yl + Oi = Yc + Ysc + Ysh + W \qquad (5.1.6.1)$$

where Yn is the water inflow to the closed talik and seasonally thawed layer from the drainage area before 02.01 (1989) (see Figure 5.6.1.1B), at a modulus of SGD of 1.72 l/sec km^2; Yn = $5.00 \cdot 10^6$ m^3; Ys is the inflow of subpermafrost waters to water areas, taking into account the data for the Vorkuta region, $0.00 \cdot 10^6$ m^3; Yl is the inflow of shallow groundwater from the zone of low water with a length of 108,000 m and width of 200 m, frozen through on average 0.7 m, with a rock water yield of 0.05, Yl = $1.50 \cdot 10^6$ m^3; Oi is the water amount pressed out by lake ice with an average thickness of 1.23 m, i = $9.90 \cdot 10^6$ m^3; Yc is the channel runoff of Adzva, $11.95 \cdot 10^6$ m^3, according to the hydrograph; Ysc is the subchannel runoff of Adzva (water transmissivity of sediments 45 m^2/day, width of subsurface flow 30 m, its slope ($4 \cdot 10^{-4}$) equals $0.00 \cdot 10^6$ m^3; Ysh is the discharge to other taliks of the shore, water transmissivity of rocks 161 m^2/day, width of subsurface flow 108,000 m, its slope $3.5 \cdot 10^{-4}$, discharge period since 02.01 (1989) equals $0.40 \cdot 10^6$ m^3; W is the recharge value of the shallow groundwater by the lake waters, determined after the parameters were inserted into the equation of water balance, and equals $4.05 \cdot 10^6$ m^3.

The seepage of the lake waters into the sublake talik amounted to 3.5 l/sec·km^2 of the water area. This is 2 times higher than the precipitation infiltration in similar geological, but subareal anthropogenic, conditions of the Vorkuta region. However, in the Vorkuta area the infiltration occurs sparsely due to the permafrost layers. To conclude, by taking into consideration the above-mentioned factors, the calculated bottom infiltration value from the Vashutkiny lakes can be viewed as sufficiently real.

5.6.2 WATER EXCHANGE IN THE ZONE OF DISCONTINUOUS PERMAFROST

Ural, Central Mountainous Area. Lake "Large" Balban-Ty is located in the axial zone of the Pripolyarny Ural. The ridges confining the lake rise above it for 600–650 m. The lake area is 0.66 km^2; the largest depth is 19 m; the water volume is equal to 0.0054 km^3; and the water level above sea level is 634 m (Kemmerikh 1961). The lake drains the metamorphic rocks of the Riphean to Lower Paleozoic periods and is located in the zone of the same tectonic fault where Lake Grube-Pendi-Ty is situated. The piezometric surface of the coastal subpermafrost waters lies above the lake level even during the end of winter low-water period.

The modulus of the through-channel groundwater inflow into the lake along the upper measuring range varied within 0.13–1.23 l/sec km^2. The modulus of the SGD ranges from 10.8–40.9 l/sec km^2. The high moduli of the subaqueous discharge and large range of their values indicate that their formation is due to the water discharge mainly from the zone of active water exchange and, hence, is subject to considerable influence from meteorological factors.

The waters of hampered water exchange also participate in the subaqueous discharge. As in Lake Grube-Pendi-Ty, 0.0029 mg/l yttrium and 0.05 mg/l natural oil have been detected in the lake water sampled from a polynia, from a depth of

3–4.5 m on 30.03 (2000). In the same sample, nitrogen gas probably emitted from more shallow depths has been registered.

Pechora Lowland. A rather typical thermokarst drainless lake is located on the icy-marine plain in the Vorkuta River basin. The lake area is 0.06 km^2, ~1 m deep, and the water line mark is 150 m. The through-sublake talik dissects the permafrost (50 m thick). The lake water level undergoes a rise of 0.1–0.75 m in the first half of winter due to the increasing ice pressure. As a result, the discharge of the Quaternary subaqueous shallow groundwater formed in the second half of summer changes into its recharge by the lake waters (Figure 5.6.11). The process is stopped only with the desiccation of the lake, registered according to formation of an under-ice air hollow 15–20 cm above the peaty-loam bottom. Infiltration of the lake waters begins again in springtime and stops by midsummer (Oberman and Kakunov 2002).

Conclusions. In the zone of continuous permafrost, the recharge and discharge of the tectonic and glacial lakes in the Central Mountainous Area is provided during winter mainly by waters from the subchannel closed taliks. In the piedmonts and on the plain areas, the lakes of the same and thermokarst genesis are recharged by cryogenically pressed-out waters from the seasonally thawed layer and frozen shallow waters. The discharge from the lakes is formed to a considerable degree by lake waters pressed out by thick ice cover. The discharge of nonfreezing lakes is used for recharge of shallow groundwater of the lake water area (year-round) and the shore (midwinter to midsummer). In draining water bodies, it forms an additional channel discharge.

In the zone of discontinuous permafrost, the winter recharge of lakes in the Central Mountainous Area and their discharge are supplied mainly by subpermafrost waters. The water exchange between lakes and groundwater in the thermokarst frozen water bodies differs from that in the nonfrozen ones by the shorter duration of winter infiltration of the lake waters.

6 Remediation in the Coastal Zone

CONTENTS

6.1 TECHNICAL REQUIREMENTS AND STANDARDS

The conceptual model for migration of contaminants from groundwater to coastal surfacewater is shown in Figure 6.1.1. The impact, both chemical and physical, may be heightened in smaller bodies of water such as embayments and lagoons due to their limited volume and restricted fluid exchange with the open ocean.

Submarine groundwater discharge (seepage) into coastal environments has been studied extensively using a variety of methods. The primary driver for seepage in nearshore environments is probably discharge from land to surfacewater induced by the hydraulic gradient in the terrestrial aquifer. However, significant contribution to seepage may also derive from groundwater circulation and oscillating flow induced by tidal stage. In coastal areas with strong tides, tidal mixing zones may be created by the movement of seawater into the aquifer. This tidally mixed zone may be important in controlling the exchange of groundwater due to a process referred to as tidal pumping. Tidal pumping occurs when seawater mixes with groundwater at

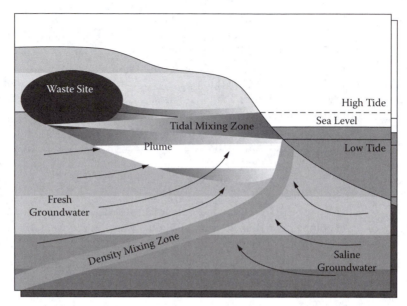

FIGURE 6.1.1 Conceptual model of coastal contamination migration process.

high tide and then, as the tide recedes, the mixture of seawater and groundwater is drawn out into the coastal waters. Because this process repeats every tidal cycle, appreciable volumes of groundwater can be extracted over time (Figure 6.1.2)

The notion of remediation in the coastal zone includes an appreciation of tidally influenced zones associated with ports, harbors, estuaries, and inland waterways. The remediation of contamination in coastal areas can range from navigation, sediment management, port and waterway development, coastal engineering, environmental services, and maritime security. Contaminants of concern can range from domestic sewage issues through contributions from industrial practices in coastal areas. Industrial contaminates can range from heavy metals, perhaps through boat repair and painting operations, through the discharge of petroleum and chlorinated hydrocarbon products.

The environmental concerns associated with coastal environments are primarily related to compliance, environmentalist investigation, design and remediation, natural resource management, environmental management and system consulting, environmental impact assessments, and solid waste engineering. Typical coastal environmental requirements are air quality permitting, compliance programs and monitoring; asbestos and lead-based paint surveys and management; emergency response; environmental management systems; health and safety programs, training, and assessment; erosion and sediment control; industrial hygiene support; natural resource management, including wetland and terrestrial resources; equipment operation and maintenance; pollution prevention and waste minimization; public outreach, involvement, and facilitation; regulatory compliance; remedial design and remedial action; risk assessment; site investigation, characterization, and remediation; solid waste management; spill prevention preparedness; water

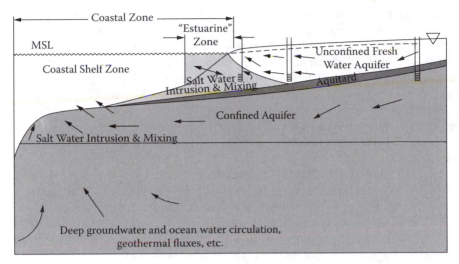

FIGURE 6.1.2 A hydrogeologic cross-section of the interaction between coastal groundwater aquifers and surfacewaters. (Source: Florida Sea Grant.)

quality permitting, compliance programs, and monitoring; wastewater treatment; wildlife and aquatic resources.

Navigation issues typically include planning, engineering, and designing dredging programs, including disposal options. Through the use of sophisticated state-of-the-art dredging and economic modeling programs, one can optimize the dredging process, predict production rights, and predict and minimize sediment plums. Activities typically associated with navigational requirements in coastal environments are dredging and disposal plans; disposal site alternative analysis (upland, nearshore, aquatic); technical plans and specifications; excavation/fill methods; predredging testing; permitting and regulatory compliance; numeric and economic cost modeling; operation cost estimates; constructability analysis; dredge equipment selection; discharge evaluations; dredge cut design.

Coastal engineering, on the other hand, requires analyzing dynamic processes and subsequently the design of shoreline protection measures, the development of sediment budgets and beach nourishment programs, and the recommendation of strategies to reduce or eliminate coastal flooding. State-of-the-art modeling techniques can be used to evaluate the fate and impact of sediment dredging and disposal operations, bottom scour by ship propellers, ship-to-ship pressure effects, beach erosion, literal drift, and sediment transport, including the design of confined disposal facilities and habitat enhancements. Typical coastal engineering requirements are computer modeling (coastal processes, hydraulics, dredging, and sediment fate and transport; confined disposal facilities and habitat enhancement; bottom scour by ship propellers); shoreline erosion prevention structures; continental shelf operations and offshore monitoring programs.

Sediment management offers several opportunities for developing cost-saving approaches. Typical services include strategic planning, sediment characterization, risk-based sediment removal and disposal of contaminated sediments, remedial investigations/feasibility studies, and remedial design. Comprehensive sediment management strategies often integrate facility planning, development, and habitat enhancement with remedial activities in order to optimize expenditures. Typical sediment management requirements are proposed remedial action plans (PRAPs) and records of decision (RODs) review, including evaluation, verification of cost projections, constructability analysis; site characterization/sampling; risk assessment; feasibility studies; remedial design; remedial alternative analysis; environmental restoration/contracting; long-term monitoring environments.

Port and waterway development typically require planners, scientists, economists, public involvement facilitators, engineers, and operation specialists who must work as a team to provide comprehensive and innovative approaches to port development and planning. In waterway development, it is important to integrate the requirements of intermodal transportation, marine facilities, structures, environmental standards, and migration requirements with community-driven concerns that are often a part of the planning process. Typical port and waterway requirements are strategic and master port plans; feasibility studies; market/forecast and cost-effectiveness analyses; operational analyses and plans; site, facility, and terminal development plans; environmental and permitting support.

The most accurate example of the scope of the problem can be seen in the large number of naval coastal landfills and hazardous waste sites in the United States. These cleanup sites with landfills/plumes are located adjacent to harbors, bays, estuaries, wetlands, and other coastal environments (NFESC, 2003). The total number of U.S. Naval landfills and hazardous waste sites by location and subject to tidal infiltration are given in Table 6.1.1.

TABLE 6.1.1
Scope of the U.S. Navy Problem for Groundwater–Surface Water Interaction at Coastal Landfills and Hazardous Waste Sites

Naval Area	Groundwater Contamination	Tidal Infiltration	Groundwater Infiltration
Atlantic Division	29	14	16
EFA Chesapeake	14	4	10
EFA Northeast	20	10	18
EFA West	29	14	31
Southwest Division	19	15	13
EFA Midwest	3	0	3
EFA Northwest	6	8	10
Pacific Division	5	10	8
Southern Division	27	26	50
Total	152	101	159

6.2   GUIDANCE FOR TIDAL STUDIES DESIGN, IMPLEMENTATION, AND ANALYSIS

This standard operating procedure (SOP) describes standards for tidal study design implementation and analysis, and discusses how such studies will be conducted and documented for projects executed. The SOP addresses technical requirements and required documentation. Responsibilities of individuals performing the work are also detailed. Additional project-specific requirements for design, implementation, and analysis of tidal studies may be developed, as necessary, to supplement this procedure and address project-specific conditions and/or objectives.

The following definitions are applicable to tidal studies and are used in this SOP:

high tide	Time and magnitude of maximum water level during a tidal cycle.
low tide	Time and magnitude of minimum water level during a tidal cycle.
flood tide	Period of water level rise from low tide to high tide.
diurnal	Daily tidal cycles.
semidiurnal	Twice daily tidal cycles.
ebb tide	Period of water level drop from high tide to low tide.
spring tide	Twice monthly highest tides occurring when sun and moon are in a straight line with Earth.
neap tide	Twice monthly lowest tidal range occurring when moon is at a right angle to the sun.
tidal range	The difference between low and high tide.
tidal amplitude	One half the tidal range.
tidal bore	A high tide-propagated wave that travels up an estuary.
tidal day	The period of time required for one tidal cycle equal to 24 hours and 50 minutes.
efficiency	The tidal range in the groundwater at a specific location divided by the tidal range at the adjacent shoreline. Tidal range is usually presented as a percentage.
lag time	The difference in time from one point in a tidal cycle (e.g., high tide) at one location to the corresponding point in the tidal cycle at another location.
priming the tides	The effect of acceleration of tide arrival time before the moon reaches local meridian.
lagging the tides	The effect of decrease of tide arrival time after the moon reaches local meridian.
barycenter	The common center of gravity of the Earth and the moon.

Design, implementation, and evaluation of tidal studies require an understanding of the tidal processes. The following is a brief description of the purpose and methodologies of tidal studies and the processes that are quantified during those investigations. More detailed discussion of these processes is presented in the reference documents listed above.

Tidal studies are conducted to describe and quantify the effect of the gravitational attraction of the moon and sun on surfacewater and groundwater. Tidal studies can be conducted to evaluate a number of issues affecting coastal environments (sites). Knowledge of the times, range, and extent of inflow and outflow of tidal waters is of importance in a wide range of practical applications such as the work on harbor engineering projects; construction of bridges, docks, breakwaters, and deep-water channels; the establishment of standard chart datums for hydrography; provision of information necessary for underwater demolition activities; and other military engineering uses.

Of particular interest in contaminant characterization investigations is the effect of tidal fluctuations on groundwater levels. Tides are the regular rise and fall of coastal water levels caused by gravity stimulated by the attraction of the moon and to a lesser degree, the sun. The tide-raising forces at the Earth's surface thus result from a combination of two basic forces: (1) the force of gravitation exerted by the moon (and sun) upon the Earth; and (2) centrifugal forces produced by the revolutions of the Earth and moon (and Earth and sun) around their common center-of-gravity (barycenter). The moon and Earth are held together by gravitational attraction, but are simultaneously kept apart by an equal and opposite centrifugal force produced by their individual revolutions around the center-of-mass of the Earth–moon system. In the hemisphere of the Earth turned toward the moon, a tide-producing force acts in the direction of the moon's gravitational attraction, or toward the center of the moon. The gravitational pull of the moon results in a bulge or high tide in the direction of the moon. On the side of the Earth directly opposite the moon, the net tide-producing force produced by the centrifugal force of the rotation around the Earth–moon center of gravity is away from the moon. Therefore, at the same time, the oceans on the opposite side of the Earth bulge in exactly the opposite direction from the moon.

Similar differential forces exist as the result of the revolution of the center-of-mass of the Earth around the center-of-mass of the Earth–sun system. The movement of water toward these high-tide bulges produces compensating low-tide troughs. Since it takes the moon a little more than a day to orbit the Earth there are two cycles of tides in roughly every 25 hours that defines the "tidal day." Tidal cycles generally occur twice per day and are defined as semidiurnal. At some locations, such as in low and high latitudes, tidal cycles may occur once per day and are called diurnal. From the low point in the tidal cycle the groundwater rises in a "flood tide" that lasts slightly more than 6 hours. After the maximum water level is reached the water level begins to drop in an "ebb tide," after which the minimum water level or "low tide" is reached. The difference in water level from low to high tide per given cycle is called the "tidal range." The tidal magnitude is one-half the tidal range.

In addition to these daily cycles there are cycles which occur twice a month, including the highest tides called "spring tides" and lowest tides called "neap tides." Spring tides occur when the moon and sun are in a straight line with the Earth such as during a full or new moon. Neap tides occur when the moon is at a right angle to the sun. There are other additional minor cycles that may affect tidal predictions. The type and magnitude of tidal study to be conducted is dependent on the goals of the investigator. Tidal fluctuations produce pressure waves that are propagated inland from large bodies of water such as the ocean. In surfacewater bodies such as in bays, rivers, or estuaries, the tidal signal is dampened as it travels away from the source of the tidal signal. The result is that a tidal wave decreases in range and size as the wave migrates away from the source.

Likewise, in a coastal aquifer, the aquifer matrix dampens the tidal signal as the wave is propagated through the aquifer. This dampening effect causes the tidal signal to decrease in magnitude away from the signal source. The ratio of the tidal range in the aquifer relative to the tidal range at the adjacent shoreline is referred to as the "tidal efficiency." The amount of time for the tidal signal to travel from one point to another is referred to as the "lag time." The dampening effect in the aquifer is controlled by the aquifer transmissivity and storativity. In unconfined aquifers the tidal response is the result of water moving in and out of the aquifer. In confined aquifers the tidal water level change is from a pressure wave migration through the aquifer. The difference in storativity of confined and unconfined aquifers causes tidal signals to propagate faster and further in confined aquifers than in unconfined aquifers. This is manifested as shorter lag times and greater efficiency in confined aquifers as compared to unconfined aquifers. Because of these characteristics, tidal fluctuations can be used to measure the transmissivity of an aquifer (Ferris 1951).

Groundwater investigations in coastal sites can be substantially affected by tidal fluctuations. The accurate determination of hydraulic gradient in aquifers adjacent to tidally fluctuating surfacewater bodies requires a detailed characterization of tidal fluctuation. Interpretation of aquifer tests such as slug tests and pumping tests can be substantially affected by groundwater tidal fluctuations. The effect of tidal fluctuation on coastlines also requires significant adjustment of dilution, dispersion, and transport effects in groundwater modeling efforts.

Tidal studies are conducted using a combination of field data collection and data evaluation techniques. The tidal study process consists of many individual steps, which include:

- Design
- Definition of problem
- Development of conceptual model
- Determination of data quality objectives
- Implementation
- Construction of data collection points
- Collection of data
- Analysis
- Evaluation of data
- Presentation of the data

These steps are common to all tidal studies regardless of their size or scope. All require documentation of their approaches and conclusions to complete the study process. Those individuals needing assistance in designing, implementing, or analyzing tidal studies may consult experts in hydrogeology or tidal studies/investigations.

The tidal study design consists of: definition of the problem; development of the conceptual model; and determination of data quality objectives. These components are necessary to collect the appropriate data and are described in the following text.

Tidal studies may be conducted to answer a number of problems. Tidal studies may be conducted to determine site hydrogeologic characteristics, including but not limited to the following:

- Determination of tidally influenced area(s)
- Correction of hydraulic gradient error based on tidal fluctuation
- Measurement of aquifer transmissivity using tidal-propagated waves
- Correction of pumping test results in tidally influenced aquifer
- Calibration of groundwater models
- Determining potential influence of tidal surfacewater bodies on the groundwater system

The determination of data quality objectives results in the identification of type, quantity/frequency, and quality of data to be collected for the tidal study. It also includes identification of specific instruments to be used to collect the data. This information comprises a data collection program as part of the tidal study design.

The data quality objectives are established based on the goals of the test as defined in the work plan. For instance, data requirements to determine tidal range at one location at a coastal site are substantially less than the data required for correction of tidal effects on pumping test results.

Important parameters to consider when determining data quality objectives include sample frequency, instrument detection limit, groundwater salinity and temperature, sources of interference including wind-driven waves, and anthropogenic sources such as pumping wells. Insufficient sampling intervals can result in an inability to accurately describe the tidal cycle characteristics. However, too many sampling intervals can create an unwieldy amount of data that require additional evaluation time and data management and create inefficiencies in data interpretations.

Care should be taken that the sampling instrument is appropriate for the project. For example, electric water level measuring tapes typically have a measurement precision of 1/100 of a foot while many automatic pressure transducers have a precision of 1/1000 of a foot. Automatic water level measuring devices such as pressure transducers are calibrated for use in a finite range of water depths and typically expressed as pounds per square inch (psi). Groundwater measurement using automatic pressure transducers is affected by water density and temperature. High salinity can produce substantial error in reported water levels if not corrected for density. The tidal study design, including the data collection program, should be incorporated into the work plan. The tidal study design should be reviewed and approved by the project hydrogeologist.

Data Collection

Tidal studies are conducted by measurement of changes in water level in surfacewater and groundwater. Surfacewater level measurement locations may include pits, lakes, rivers, bays, and oceans. Subsurface water level measurement devices may include wells and piezometers. Wells and piezometers used in tidal studies should be constructed in a manner that allows accurate measurement of actual groundwater fluctuations. Wells should be constructed in accordance with the appropriate SOP(s). Other locations may be useful for data collection including storm drains and channels. It is recommended that, if possible, the elevation of the surfacewater body be determined relative to mean sea level. Water levels may be determined using automatic recording devices such as data logging pressure transducers or stream gauges. These devices should be securely attached to a fixed object so that water level measurements are not compromised by instrument movement. A staff gauge may also be installed in the surfacewater body and surveyed to allow determination of surfacewater elevation. The water level in the surface body may be easily determined by visual observation.

When collecting water level data in large open bodies of water data can be substantially affected by the surface wind-driven waves. The effect of these waves can be minimized by data evaluation methods or by physical methods such as installation of stilling wells. Data evaluation methods for eliminating these effects are described below. A stilling well can be constructed of a vertical pipe open at both ends that is attached to a structure such as a dock and extending into the surfacewater body to a depth at which the effects of waves are minimal. Water level measurements inside the stilling well can be collected using manual or automatic data recording devices.

Tidal studies rely primarily on measurement of water level fluctuations. Water level data can be collected using a variety of instruments including staff gauges, stream gauges, pressure transducers with data loggers, ADLPTs, and water level indicators. An ADLPT is a self-contained pressure transducer–data logger system that can be secured inside the casing of a monitoring well. The appropriate water level measurement device used should be determined based on the data collection system, the data quality objectives, and ease of use in the field. A variety of suppliers provide these instruments. Data should be collected in accordance with applicable SOPs and/or project-specific requirements. It is recommended that ADLPTs be used for collection of water level data in tidal studies. These instruments allow collection of highly accurate, easily time-correlated data at the short time intervals necessary for characterization of tidal cycles. The setup of the automatic data logging pressure transducers varies depending on the instrument utilized. For each instrument the manufacturer's procedures should be followed. The general procedure for most ADLPT setup and data collection is as follows:

Identify locations where transducers are to be installed. It is recommended that one transducer be installed in the body of water from which the tidal signal is being propagated (e.g., ocean, bay, river, etc.). Utilize pressure transducers with the appropriate pressure range (usually 10–15 psi). If the aquifer is brackish or saline, determine salinity and temperature of the water. Correct water level readings for density recording on transducers, if possible. Synchronize clocks on all transducers.

If appropriate, synchronize the time to a published time chart source. For Instance, the U.S. Navy publishes its own data, which correspond to the official Navy clock.

Name the data file according to predetermined nomenclature in the work plan. The sampling interval should be from 1–10 minutes depending on data management requirements/limitations and the length of the test. Each transducer should be set to collect pressure data points at the same time for ease of data correlation. Install the transducer in the groundwater monitor well, piezometer, stilling well, or other sample location as required. If possible, a stilling well should be used to measure tidal fluctuations in surfacewater bodies. The transducer should be securely attached to the well or fixed object to prevent slippage during data collection. Utilize an additional transducer for measurement of barometric pressure in a monitoring well or piezometer located in an area not under tidal influence. Manually measure water levels in the well at periodic intervals; note the water level and measurement and time reported on the transducer.

Run the test for a minimum of 25 hours and preferably 2–3 days if possible. For evaluation of complete tidal cycles, tests may be run for a month or longer. Periodically collect manual measurements of the depth to water and the corresponding time for calibration of the transducers for quality assurance purposes. Remove the transducer from the well and record the time of removal. Stop the test after the pressure change resulting from removal from the well is recorded on the transducer. Download the data file from the transducer into a database. Back up data files. Record all manual measurements on the appropriate form(s) according to the work plan. Tidal studies can be conducted over large areas using multiple ADLPTs. If limited transducers are accessible and time is available, the study may be conducted by moving a portion of the ADLPT locations during the test. The movement of ADLPTs should be conducted in a manner that will allow an overlap of the data. The ADLPTs located in the wave propagation source should not be moved during the investigation. Each time ADLPTs are moved the test setup steps described above should be repeated for each ADLPT. Special attention should be paid to possible salinity differences in sample location at nearshore locations. Care should be taken to prevent cross-contamination of the well between sample events.

Short frequency and random water level fluctuations caused by wind-driven waves or anthropogenic sources can cause errors in determination of water level fluctuations resulting from tidal influence. In addition, these errors can substantially inhibit accurate description of tide characteristics. These errors can be minimized by the construction of suitable measurement structures (such as the stilling wells described above). Some ADLPTs provide a mechanism for automatic elimination of these errors. This is accomplished by averaging a group of water level readings collected over short periods of time to produce one average water level measurement. This method is recommended when short-period waves such as wind-driven waves affect water level measurement.

All data should be maintained in appropriate permanent records according to the work plan or other project-specific requirements. At a minimum the records should include:

- Names of field personnel
- Project name and project number
- Sample location name (e.g., well number)
- Date, time of start and finish of test
- Sample measurement interval, such as data averaging
- Adjustment to measurement devices (e.g., density calibration)
- Manual water level measurements and time of measurement
- Automatic data measurements and time of measurement
- Automatic data file name
- Tidal study data should be maintained in accordance with appropriate SOPs

6.2.1 Tides and Hydraulic Gradient Determination

An important feature of hydrogeologic investigations is the determination of the hydraulic gradient. In coastal aquifers the measurement and interpretation of average groundwater levels used to map potentiometric surfaces and determine hydraulic gradient is complicated by the groundwater fluctuations resulting from tides. Because the lag time and tidal efficiency vary based on distance from the wave source and site-specific hydrogeologic conditions, it is not possible to determine average groundwater level by measurement of water levels at one time. In tidally influenced aquifers, groundwater level measurements should be collected at regular intervals for one complete tidal cycle (25 hours). If the water level fluctuations produce a symmetrical wave, then the average water level can be determined by averaging all the water level data. In many coastal aquifers, however, the flood tide signal propagated in the aquifer is shorter than the ebb tide. If the water level fluctuations produce an asymmetrical wave then the average water level must be corrected for this asymmetry. A method for determination of average groundwater elevation in aquifers with asymmetrical tidal waves is presented by Serfes (1991).

6.2.2 Tides and Slug/Bail Tests

Groundwater tidal fluctuations can substantially affect determination of aquifer transmissivity determined by long-term slug/bail tests. If the transmissivity of the aquifer is high, these tests are generally of short duration usually lasting only a few minutes. In such cases the error caused by tidal fluctuation is small. If the transmissivity of the aquifer is low, however, the tidal fluctuation may substantially affect the slug/bail test interpretation. Conducting the test during the transition time from flood to ebb tide or from ebb to flood tide may minimize the effect. The test results may be corrected if tidal water level fluctuation is known. The data used for correction of the tidal effect should be those for which the tidal efficiency is similar to that at which time the slug/bail test was conducted. Tidal fluctuation data may be collected from a nearby well during the test if the lag time and tidal efficiency of the well are similar to those of the slug/bail test well. The slug/bail test should be conducted in accordance with the appropriate technical SOPs.

6.2.3 Tides and Aquifer Pumping Tests

Because pumping tests are of longer duration than slug/bail tests, the effects of tidal fluctuation are more possible. During pumping tests groundwater levels are drawn down around the pumping well. The amount of drawdown is minimized during flood tides and exaggerated during ebb tides.

Determining the water level fluctuations at the time of the test or using tidal studies for predicting the time and magnitude of the fluctuation can correct these tidal effects. This can be done by measuring water levels in wells outside the radius of influence of the pumping well. A significant problem with conducting and evaluating pumping tests in tidally influenced aquifers is that the delayed yield may be obscured for the test. This should be considered when evaluating the pumping test results. The pumping test should be conducted in accordance with the appropriate technical SOPs.

6.2.4 Tides and Measurement of Light Nonaqueous Phase Liquid (LNAPL)

The thickness of a LNAPL, such as petroleum, is often measured at contamination sites. It is often difficult to determine the actual thickness of LNAPL in most aquifers; however, it is particularly difficult to do so in tidally influenced aquifers. At some sites LNAPL may be present in wells at low tides and disappear at high tides. At other sites LNAPL may appear at high tides and disappear at low tides. LNAPL measurement at tidally influenced aquifers should be conducted in accordance with applicable technical SOPs. Additional free-phase measurements should be collected and evaluated to allow for determination of LNAPL thickness with respect to tidal cycle. Consideration should be given to variability in LNAPL thickness when evaluating and calculating water elevation beneath LNAPLs.

6.2.5 Tides and Groundwater Modeling

Tidal effects should be considered in groundwater modeling efforts. As discussed above, groundwater fluctuations in unconfined aquifers are the result of influx of water into the aquifer. The net effect of the tidal fluctuation is that the direction of groundwater movement in a tidally influenced area may be reversed on a semidiurnal basis. In addition, the hydraulic gradient and hence the rate of groundwater movement increase and decrease on a daily basis. These changes in groundwater level affect groundwater flow and contaminant transport characteristics. In tidally influenced areas, the reversal of hydraulic gradient effectively increases the groundwater transport distances. The increase in transport is often manifested in increased dispersion factors in areas of large groundwater fluctuations. These factors should be considered when preparing groundwater models in tidally influenced aquifers and should incorporate evaluation results from tidal studies. In addition to flow and transport, biological and chemical reactions in tidally influenced nearshore environments may be substantially different than in adjacent nontidally influenced aquifers.

When the groundwater chemistry of the surfacewater body is substantially different than the unconfined aquifer, the mixing of surfacewater and groundwater during changes in tidal cycle may alter the aquifer chemistry. One example of this is where an unconfined freshwater aquifer enters a saline environment. In such situations, the saline aquifer may provide a continual source of sulfate not present in the freshwater aquifer for enhancement of intrinsic biodegradation of organics. Inclusion of this source of sulfate has substantial impact on biodegradation modeling. Tidal effects on groundwater chemistry may need to be investigated and evaluated as part of the tidal study.

6.2.6   DATA PRESENTATION

Tidal study results may be presented in a variety of formats, including spreadsheets, graphs, 2- and 3-dimensional contour maps, movie (MPEG) files, and reports. The presentation style and format of the data and interpretation should be based on the reporting requirements. Contour maps may be prepared to graphically illustrate the spatial distribution of tidal efficiency and tidal lag time. The lag time and efficiency maps can be used for later correction of groundwater elevation data used for hydraulic gradient maps. Contour maps should be prepared in accordance with the appropriate technical SOPs.

A tidal study report should be generated and address the objectives of the study. The report should also include: main design parameters; description of data collected and general quality/usability of the data; results of the data evaluation; and conclusions and recommendations (optional). Raw data should generally be provided in the tidal study report unless otherwise requested. If data files are small this may be done in data tables included as appendices in reports. Often, however, data files from tidal studies contain large files. In such cases, the data files may be stored in an electronic format.

6.2.7   SUSPENDED PARTICULATES

Clay and iron oxide minerals are the main contributors to suspended particulates. These minerals have strong affinities for sorbing selected trace elements. Iron oxides strongly adsorb As, Sb, Se, and V, and clays strongly adsorb Ba, Cr, and Pb due to speciation and surface charge effects. Linear correlations of trace elements vs. Al (clays) or Fe in unfiltered samples indicate a natural origin for the trace elements. Plots of filtered/unfiltered ratios provide independent evidence (Hem 1985; Stumm and Morgan 1996; Brookins 1988; Electric Power Research Institute 1984; Pourbaix 1974; Bowell 1994).

6.2.8   SHALLOW AQUIFER

- Extends from 5–100 ft bgs [feet below ground surface].
- Unconsolidated coastal plain sediments (silty sands with some clay) with some fill material near surface.

- Groundwater is tidally influenced in areas adjacent to a tributary river of the Chesapeake Bay (Virginia and Maryland); the river is brackish-to-saline, thus the groundwater in coastal areas receives saline input.

6.2.9 SEAWATER INTRUSION

Simple Mixing: Seawater contains very high concentrations of major elements and some trace elements. The range of element concentrations are high in the intrusion zone, but trace-to-major element ratios are fairly constant in the absence of contamination. A simple mixing model is used to identify anomalous element concentrations. A natural seawater source is identified by a linear trend with major elements (Na, Mg, etc.).

Ion Exchange: Cyclical changes in the saline interface location induce adsorption-desorption reactions on mineral surfaces that can concentrate trace elements in groundwater.

Chloride Complexation: High-chloride environments allow the formation of metal-chloride complexes that can greatly enhance trace element solubility. Important complexes include:

Cadmium: $CdCl^+$, $CdCl_2^\circ$, $CdCl_3^-$
Copper: $CuCl_3^{-2}$, $CuCl_2^-$
Lead: $PbCl^+$, $PbCl_2^\circ$, $Pb\ Cl_3^-$

Natural chloride complexation effects are identified by correlations with chloride and via geochemical speciation modeling. Modeling performed using selected samples from this site's data set suggests that up to 97% of the Cd, 99% of the Cu, and 58% of the Pb concentrations are present as chloride complexes (Stumm and Morgan 1996; Drever 1997; Wolery 1992).

6.2.10 REDUCTIVE DISSOLUTION

Iron and manganese oxides concentrate specific trace elements on mineral surfaces. The release of organic contaminants (fuels, solvents, etc.) can establish local reducing environments caused by anaerobic microbial activity. Local reducing conditions can drive the dissolution of iron and manganese oxides, which become soluble as the redox potential drops below a threshold value. Dissolution of these oxide minerals can mobilize the trace elements that were adsorbed on the oxide surfaces. Reductive dissolution is identified by correlations of elevated trace elements (especially Fe, Mn, As, Se, and V) versus local depressions in redox indicators. Such indicators include lower dissolved oxygen, oxidation reduction potential, sulfate, and nitrate; and increases in soluble Fe and Mn (Electric Power Research Institute 1984; Sullivan and Aller 1996; Belzile et al. 2000).

6.2.11 CONCLUSIONS

The presence of a saltwater wedge in the aquifer underlying the coastal areas creates strong compositional gradients, which violate the assumptions required for valid statistical site-to-background comparisons.

Geochemical evaluations, based on elemental ratios rather than absolute concentrations, can distinguish between contaminations versus naturally high concentrations, even in the presence of a strong compositional gradient.

The majority of site samples contain element concentrations that are the result of natural processes. Some combination of seawater mixing, ion exchange, and/or suspended particulates is the likely cause of elevated concentrations of Al, Be, Ca, Cr, Mg, Hg, K, Se, Ag, Na, Tl, and V in all of the site samples evaluated in this study. Contamination is not indicated for these 12 elements.

Sb, As, Ba, Co, Cu, Fe, Pb, Ni, and Zn exhibit anomalous concentrations in at least one sample each, and these concentrations may reflect contamination. Anomalous Cd concentrations may be present in some samples, but such samples could not be clearly distinguished.

Redox effects are an important control on As, Fe, and Mn concentrations at some locations, and Ba concentrations may be indirectly controlled by redox effects. Local reducing conditions in groundwater at the site may be natural due to the presence of wetlands, or may be caused by fuel or solvent releases. In addition to the effects of low redox, elevated Mn concentrations may be caused by ion exchange and Cl– complexation in areas affected by seawater mixing.

This geochemical evaluation was performed using existing analytical data obtained during typical site investigations under the U.S. Comprehensive Environmental Response, Compensation, and Liability Act (CERCLA). This technique has also been successfully applied at other sites across the country, and the results of these evaluations are being used to eliminate natural metals from lists of chemicals of concern and to properly focus risk assessment, monitoring, and remediation efforts toward actual contamination.

6.2.12 RECOMMENDATIONS

Sediment site investigations should employ oceanographic sampling strategies in order to provide data that are ecologically significant and useful in understanding the relative risk of contaminants on Navy property compared to the rest of the industrialized watershed. As described above, several new methods are available that can be used in an overall sampling strategy that combines measures of transport and biodegradation with a seasonal and watershed-level sampling approach. More information on the above research is available in the Naval Facilities Engineering Command (NAVFAC) special publication titled, Accelerated Implementation of Harbor Processes Research, that was released in May 2003 (SP- 2135-ENV). This publication is also available at the Naval Facilities Engineering Service Center (NFESC) Website: http://enviro.nfesc.navy.mil/erb/erb_a/restoration/fcs_area/con_sed/sp-2135-harbor-proc.pdf.

6.3 LESSONS LEARNED FROM SELECTED
INVESTIGATIONS

Four brief case histories with lessons learned are presented that give insights into tidal issues encountered in remedial investigations in coastal environments.

Case History 1: In support of the $250 million, 5-year, Contract I of the Pacific Division Remedial Action Contracts (PACDIV RAC), OHM's Pacific base staff has executed approximately $90 million of remediation services at Pacific and Indian Ocean locations. Project sites have included the outer Hawaiian Islands of Kaho'olawe and Kauai, other Pacific Islands such as Midway, Johnston, and Wake, international locations such as Guam and Japan, and other remote locations such as Adak, Alaska, and Diego Garcia off the coast of Africa. Due in large part to OHM's successful execution of these projects, IT was recently awarded the follow-up RAC II Contract in the Pacific.

Through execution of approximately 30 remote projects under the PACDIV RAC, a variety of issues have been identified that are not typically experienced during execution of remediation projects located near major industrial cities. Such issues include technical challenges (cost-effective waste management, sample hold-times, communication with technical staff), operational challenges (mobilization, equipment operations and maintenance), and administrative challenges (staffing, safety, morale). Failure to address any one of these challenges can lead to rapid deterioration of project execution, resulting in problems with quality, safety, and/or client satisfaction. Projects on the islands of Midway, Wake, Johnston, Guam, and Diego Garcia best illustrate the challenges (and solutions) presented by these remote assignments. Projects on these islands encompass soil vapor extraction system installation, bio-slurper operation, stabilization of PCB-impacted soil, removal of underground storage tanks and underground pipelines, removal of offshore debris, and remediation of lead ash. At both Midway and Wake, on-site, NFESC-approved laboratories were established to ensure that all samples could be analyzed within sample hold-times, and real-time operational decisions could be made. On Guam, a field operations center has been established, and OHM has become the premier provider of environmental remediation service projects for both the Navy and the Air Force Center for Environmental Excellence.

Case History 2: Phytoremediation is an emerging remedial technology being applied at a variety of sites with contaminated groundwater. The technology has previously been shown to promote removal of contaminants from aquifers and degrade contaminants in place. This investigation was conducted to evaluate the potential for phytoremediation to control groundwater flow in a shallow, brackish aquifer.

The investigation was conducted at a former landfill site located adjacent to the San Francisco Bay at Hunters Point Shipyard in San Francisco, California. Soil and groundwater at the site had been impacted by a variety of organic and inorganic chemicals. An interim remedial action system consisting of a sheet pile wall and groundwater extraction system (GES) was previously installed to prevent PCBs in the unconfined shallow aquifer from entering San Francisco Bay.

This investigation included tidal studies and aquifer testing to refine the site conceptual model, and groundwater modeling to evaluate the effectiveness of phytoremediation as a passive remedial technology alternative to the GES. Baseline groundwater modeling indicated that the transport rate of PCBs is very slow and therefore not likely to reach San Francisco Bay. The GES effectiveness is limited because of low-permeability sediments comprising the unconfined shallow aquifer.

Surveys of water quality, substrate quality, and benthic macroinvertebrates were conducted in a variety of tidal creeks located in the vicinity of a municipal solid waste landfill prior to the construction of a leachate collection system. *In situ* water quality data indicated high water temperatures and low dissolved oxygen values along with high turbidities. Sediment chemistry data indicated that all sediment within the study area exceeded heavy metal criteria of the U.S. Environmental Protection Agency (EPA). Grain size and salinity data indicated that within similar concentrations with respect to indicators of leachate. The benthic macro-invertebrate community was consistently dominated by opportunistic *polychaete* and *oligochaete* worms. Both Shannon diversity and Rarefaction curves were used to evaluate trends in species diversity over time. The study included a comparison to data obtained by the EPA's Regional Environmental Monitoring and Assessments Programs. Large-scale biomonitoring baseline benthic communities within tidal creeks affected by multiple stressors included previous exposure and potential exposure to oil spills, continued point and nonpoint municipal and industrial wastewater discharges, and physical stressors such as elevated water temperatures, homogenous silt/clay substrate, and depressed dissolved oxygen values.

Case History 3: Considerable effort has been devoted recently in adapting enhanced oil recovery methods for remediation of dense, nonaqueous phase liquids (DNAPLs). *In situ* flushing of contaminated soil with DNAPL-solubilizing surfactants, coupled with partitioning tracer injection to estimate DNAPL volume, was pilot tested at a chemical plant on the gulf coast. The heterogeneity and low permeability of interbedded lenses of sandy silt and clay seriously challenged the feasibility of surfactant and tracer technologies.

Tests were conducted within a 15- × 25-ft test cell bounded by sealable-joint sheet piling. Vertical control of subsurface fluids was maintained by injecting water into a deep soil within the test cell to induce an upward vertical gradient. Solutions were flushed through a shallow unit using a three-row line drive configuration. A solution containing two nonpartitioning tracers, bromide and isopropyl alcohol, and two partitioning tracers, hexanol and heptanol, was injected for 4 days, followed by water injection. Flow rates varied greatly across the test cell, and tracer breakthrough curves were difficult to interpret because they reflected heterogeneity within the test cell.

Following completion of the tracer test, a 2% solution of sodium dodecylbenzene sulfonate, a surfactant used in household detergents, was injected continuously. Within 10 days, the injection rate had decreased by an order of magnitude, indicating permeability loss caused by interaction of the surfactant with clay minerals. With the decreased injection rate, surfactant concentrations were too low to significantly impact DNAPL mass or dissolved-phase concentrations.

Case History 4: Groundwater at the Air Operations Facility at Diego Garcia BIOT has been affected by a discharge of jet fuel. A groundwater investigation, including a tidal study and aquifer testing, was conducted to determine flow and transport characteristics of the aquifer affecting contaminant transport. From November 1998 to March 1999 groundwater elevation was measured at 27 locations and compared to measured lagoonal tidal fluctuation.

Groundwater responded substantially to ocean and lagoonal tidal fluctuations. Lag time and tidal efficiency in the aquifer were determined to be unrelated to distance from the ocean or lagoon. The results of the investigation indicate that groundwater tidal fluctuations are propagated vertically rather than laterally as at typical coastal environments. The vertical tidal propagation is considered to be controlled by geological conditions. The tidal fluctuations produce a vertical ground-water flow component.

6.4 COASTAL CONTAMINATION MONITORING TECHNOLOGIES

Landfills and hazardous waste sites located in coastal environments pose a potential environmental threat to surfacewater bodies through the exchange of groundwater-borne contaminants. It is estimated that one out of five Navy landfills are subject to groundwater exchange through tidal influence. Therefore, the ability to determine where groundwater is discharging, at what rate, and what contaminant concentrations are entering the surfacewater body is important to understanding these sites and determining the need for remediation.

Monitoring technologies focus on either flow detection or contaminant detection. The primary flow detection technologies include:

- Natural geochemical tracers
- Thermal infrared aerial imagery
- Seepage meters
- Thermal gradient flow meters
- Piezometers
- Tracer injection
- Colloidal borescope

The primary contaminant detection technologies include:

- Diffusion samplers
- Seepage meters
- Pore water probes
- Flow probe chemical analyzer

There are two major obstacles in studying groundwater exchange: (1) identifying the spatial location where exchange is likely to take place and (2) accurately measuring the groundwater seepage across the sediment–water interface. Two new techniques are described for identifying potential areas of groundwater impingement into surfacewater as well as quantifying flow rates and contaminant levels. These two new monitoring techniques, the Trident Probe and the UltraSeep Meter, were developed by Navy engineers in conjunction with scientists from Cornell University and are presented by Chadwick et al. (2003).

6.4.1 Trident® Probe

The Trident Probe is a flexible multisensor sampling probe for screening and mapping groundwater plumes at the surfacewater interface. It consists of a simple direct-push system equipped with temperature, conductivity, and pore water sampling probes. A schematic and photo of the Trident Probe are shown in Figure 6.4.1.1. Contrast in temperature and conductivity between surfacewater and groundwater are used to determine likely areas of groundwater impingement. The water sampler is used to collect samples for subsequent chemical analysis. The conductivity measurements can be used to detect contrast in salinity and/or clay content in unconsolidated sediments. The conductivity signal varies primarily as a function of clay content and porosity. Areas of likely groundwater seepage are generally associated with low conductivity, either as a result of low salinity, low clay content (high permeability), or both. Areas of groundwater seepage may appear either warmer or colder in contrast to the surfacewater depending on seasonal and site characteristics. The water sampling probe allows interstitial water to be extracted from the sediment at selected depths up to about 60 cm below the sediment–water interface. Pore water is collected by syringe or vacuum pump extraction through a small-diameter stainless steel probe.

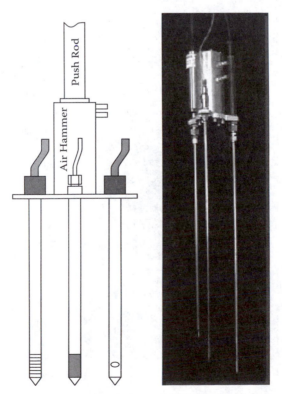

FIGURE 6.4.1.1 Schematic diagram and photo of the Trident Probe.

Recent trials show that the Trident Probe provided rapid spatial assessment of both groundwater exchange parameters (temperature contrast and conductivity and contaminant concentrations). Figure 6.4.1.2 is an example of temperature contours developed from Trident measurements taken at North Island Site 9 in San Diego, California. The contours mapped are the differences in temperature between surface and groundwater interface. The area having the greatest difference indicates a potential area of SGD.

6.4.2 ULTRASEEP® METER

The UltraSeep Meter is a modular state-of-the-art seepage meter for direct measurement of groundwater and contaminant discharges at the surface and groundwater interface. It features an ultrasonic flow meter that provides continuous, direct measurement of groundwater flow. The water sampler employs a low-flow peristaltic pump with sample selector valve and sample bag array. The onboard sensors measure

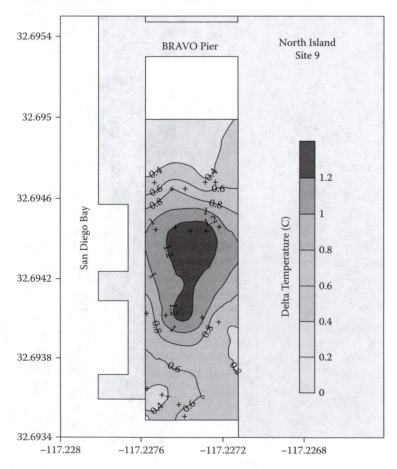

FIGURE 6.4.1.2 Temperature difference mapping at North Island Site 9.

temperature and conductivity and the controller stores data and controls sampling events. The feedback control system regulates water sampling to maximize sampling volume without restricting flow. The flow meter provides accurate detection of a specific discharge or recharge in the range of 0.1–150 m³/day. Figure 6.4.1.3 is a photograph of the meter.

The ability to collect a continuous seepage record is critical to understanding the dynamics of the exchange process, especially in areas with strong tidal influence. In addition, the flow sensing capability allows water samples to be collected in proportion to the seepage rate, enabling the direct quantification of the chemical loading associated with the SGD. At coastal sites, a typical deployment runs over a 12- to 18- hour period to capture an entire semidiurnal tidal cycle. During this time, the seepage rate is continuously monitored and up to six water samples can be collected for chemical analysis. At the end of deployment, the meter is recovered using either a lift line to the recovery boat or by diver assistance.

6.4.3 RECOMMENDATIONS

Cost avoidance from employing these new monitoring methods is potentially significant over conventional terrestrial investigations and fate and transport modeling. Improved site knowledge also leads to the selection of more appropriate, less expensive remedial alternatives. These technologies should be considered if one or more of the following conditions exist at a site:

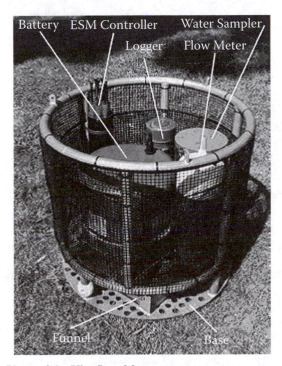

FIGURE 6.4.1.3 Photo of the UltraSeep Meter.

- A clear identification of a terrestrial contaminant plume migrating to the shoreward boundary of the surfacewater body.
- Applicable regulations or other compliance/clean-up drivers require identification of contaminant exposure levels in the surfacewater or at the interface.
- Hydrogeologic modeling results are ambiguous or require field validation.
- The area where the plume is impinging needs to be clearly delineated to address risk and/or remedial options (Trident Probe).
- The rate of discharge and associated contaminant loading requires delineation to address risk and/or remedial options (UltraSeep Meter).

Cornell University's development and evaluation of these new monitoring techniques were supported by the Space and Naval Warfare Systems Command (SPAWAR) and the Naval Facilities Engineering Service Center (NFESC).

6.5   COASTAL ZONE MODELING

With greater than 70% of the world population living in coastal areas, considerable emphasis has been placed on developing a quantitative understanding of freshwater and saltwater behavior. This chapter does not dwell on the development of modeling techniques since several recent books have been published which focus primarily on model development. One of the first books published on this subject, *Seawater Intrusion in Coastal Aquifers: Concepts, Methods and Practices* by Bear et al. (1999), presented the basic concepts, theories, and methodologies and can be used as a textbook. In 2001, the first International Conference on Salt Water Intrusion and Coastal Aquifers — Monitoring, Modeling and Management, was held in the coastal town of Essaouira, Morocco. From this conference and a second conference held in Merida, Mexico in 2003, a group of international experts who were practitioners in the field were assembled to write *Coastal Aquifer Management — Monitoring, Modeling, and Case Studies* (Cheng and Ouazar, 2004). This book has the added value of being co-published with a CD-ROM. The CD offers many advantages relative to data size, color graphics, computer programs, animations, etc., which support this as a fundamental document on which to develop a management understanding of coastal zone modeling. The readers are encouraged to review the above references relative to the subject of freshwater and saltwater modeling applications.

6.6   DETAILED CASE STUDY

6.6.1   FEASIBILITY ANALYSIS/INTERIM REMEDIAL ACTION PLAN FOR A LARGE CONFIDENTIAL PORT BULK FUEL TERMINAL IN CALIFORNIA

6.6.1.1   Site Location and Description

The site is located on portions of two port berths in California. It measures approximately 1500 × 1000 ft, and encompasses an area of about 34 acres.

The Outer Harbor bounds the site to the west, and other port berths bound the site to the north and south. The site is a marine container terminal, and is relatively flat and entirely paved with asphalt concrete. Most of the site consists of open paved areas dedicated to intermodal container operations and the storage of containers and container truck trailers. The most prominent structures are the container cranes for unloading marine cargo from container ships docked along the wharf on the western side of the site. A sheet pile wall extends beneath the approximate centerline of the crane rails along most reaches of the wharf and separates the Outer Harbor from land. An earthen embankment protected by rock rip-rap stands in lieu of the sheet pile wall beneath a 150-ft section of the wharf near the northern end of one of the berths. A network of underground utilities is present beneath the site, including storm drains, sanitary sewers, fire and potable water pipes, gas lines, electrical cables, and abandoned petroleum pipelines.

6.6.1.1.1   Surfacewater Flow

The site is entirely covered with concrete and/or asphalt pavement and surface water drainage is by way of overland flow to the storm drain system constructed beneath the site. Direct, vertical infiltration to the unconsolidated sand, silt, and clay underlying the site is likely negligible.

The Outer Harbor is part of a major California bay, and as such experiences tidal flow. Historical tidal data were obtained from the National Oceanic and Atmospheric Administration (NOAA) Website (http://co-ops.nos.noaa.gov/). The tidal data obtained from NOAA appeared to be delayed 1 hour relative to the onsite tidal data obtained. One tidal investigation measured tidal fluctuations relative to the port datum, which is reportedly 3.2 ft below the National Geodetic Vertical Datum of 1929 (NGVD29) sea level benchmark, and 0.41 ft below mean low low water (mllw). An elevation shift of 2.11 ft upward was required for the site tidal data to agree with the NOAA-obtained data.

At benchmark 9414750 (Tidal 8), which is operated by NOAA, the mllw is 2.93 ft below NGVD29 sea level. The following are reported at the Tidal 8 station, relative to mllw at 0.00 ft: the mean high high water is 6.60 ft, mean high water is 5.97 ft, mean tide level is 3.55 ft, mean low water is 1.13 ft, and mllw is 0.00 ft.

6.6.1.1.2   Stormwater Flow

Stormwater drainage at the site is controlled by a network of five main subsurface storm drains. The storm drain pipes range in diameter from 10–36 inches. The five main storm drains slope downward toward the outfalls at the Outer Harbor. The invert elevations of the storm drain pipes range from approximately 9 ft port datum (approximately 5 ft below ground surface [bgs]) to approximately 1.5–2.5 ft port data at the outfalls (approximately 11.5–12.5 ft bgs). The diameter of each main storm drain increases downstream, as smaller tributary drains connect with the main drains. These connections are typically made at concrete manholes with either round steel covers or rectangular steel grates. Inflow to the storm drains is through catch basins, manholes fitted with steel grates, and surficial trench drains located throughout the site. The storm drain system discharges directly to the Outer Harbor by way of five outfalls constructed through a sheet pile wall located beneath the wharf at one of the berths.

A previous inspection and survey of the storm drain system by others indicate that some of the pipes are cracked and/or have separated sections. Consequently, groundwater may be infiltrating the pipes and/or stormwater. Bay water may be discharging to groundwater, depending on the tide at those locations. In addition, water was observed seeping through the sheet pile wall immediately below storm drain outfalls SD-1 and SD-2.

6.6.1.2 Geology and Soil

The two berths are located along the eastern bay inargiil within an area generally referred to as the West Bay Plain groundwater basin. The geologic units of this basin consist of unconsolidated Pleistocene- and Holocene-age sediments overlying a Jurassic and Cretaceous rock melange, known collectively as the Franciscan Complex.

6.6.1.2.1 Soil Analytical Results

Soil Sample Analytical Data. Historical soil sample analytical data are: benzene concentrations are reported up to 53 mg/kg, total petroleum hydrocarbons as gasoline (TPHg) up to 18,000 mg/kg, total petroleum hydrocarbons as diesel (TPHd) up to 7800 mg/kg, and total petroleum hydrocarbons as oil (TPHo) up to 2100 mg/kg.

Soil Vapor Sample Analytical Results. The inferred historical extent is known of methane in soil vapor, with reported methane concentrations up to 560,000 parts per million by volume (ppmv). Benzene concentrations were reported up to 2600 ppmv; toluene up to 3700 ppmv, and xylenes up to 2300 ppmv.

6.6.1.3 Groundwater Occurrence and Flow

6.6.1.3.1 Hydrostratigraphic Units

The unconfined aquifer occurs within the artificial fill. The Young Bay Mud acts as an upper aquitard for the San Antonio Formation. Because the hydraulic conductivity of this unit is significantly lower than the units above and below it (i.e., about 1·cm/sec, as compared to about 1·10 cdsec), the Young Bay Mud forms an aquitard, impeding the downward flow of groundwater into deeper strata.

Groundwater in the San Antonio Formation is semiconfined, resulting from the presence of the overlying Young Bay Mud and underlying Old Bay Mud. The Old Bay Mud, consisting primarily of clay, forms a regionally continuous aquitard with vertical hydraulic conductivities of about 2·cdsec.

6.6.1.3.2 Tidal Influence

Tidal influence on the unconfined aquifer was evaluated by calculating tidal efficiencies at wells where tidal influence studies have been conducted. Tidal efficiency is defined as the change in the measured head in a well divided by the change in the tide level. Tidal data were obtained from the tidal investigations. The inferred extent of tidal influence on the unconfined aquifer is depicted as the 0.1% isopleth. Monitoring wells with an efficiency <0.1% would increase <0.01 ft (the typical error associated with water level measurements) during a tidal increase of 7 ft. The tidal influence appears to extend inland approximately 150–250 ft, except near storm drain SD-1 where tidal influence appears to extend up to 350 ft.

The time lag associated with the tidal influence (i.e., the time difference between the high tide and the peak of the water level in a well or low tide and the low point in the water level) varies directly with the tidal period. In general, time lags at the site were greater for low tides than for high tides. The average time lags associated with the wells completed in the artificial fill that demonstrate a tidal influence. The lag times ranged from 2.3–4.1 hours, except in the vicinity of storm drain SD-1, where lag times were a half hour or less. These data are consistent with storm drain SD-1 acting as a preferential pathway for tidal water. Other preferential pathways are not indicated by the data currently available on lag time and tidal efficiency.

The water table at monitoring wells TW-IBB, TW-IA, and SB-IC9 appears to be affected by a preferential pathway (storm drain SD-1) that facilitates movement of tidal water into and out of the formation at high tides but appeared to cease acting as a preferential pathway during low tide. The tidal efficiencies at these wells were calculated using high tide data.

Monitoring wells MW-37B and MW-37C, which are screened in the upper and lower portions of the San Antonio Formation, respectively, had estimated tidal efficiencies of 18 and 56%, respectively. The lag times associated with monitoring wells MW-37B and MW-37C were 3.3 and <0.3 hours, respectively. Historical data from 2000 indicate that vertical gradient reversals occur between the upper and lower portions of the San Antonio Formation in monitoring well clusters MW-42A-C and MW-46A-C. The reversals in the vertical gradients at these wells appear to be an indicator that these are tidally influenced.

6.6.1.3.3   Groundwater Flow

Historical quarterly monitoring reports provide indications of the direction and gradient of groundwater flow in the unconfined aquifer. The general groundwater flow direction in the unconfined aquifer is to the east, toward the Outer Harbor under an estimated hydraulic gradient of approximately 0.001, with localized areas where the gradient may range up to 0.003. At one berth, there is an apparent sink in the vicinity of monitoring well MW-13. The water table elevations reported for monitoring well MW-17 suggest that sometimes there is a mound in this vicinity, and at other times a sink. At the second berth, there is sometimes mounding near former monitoring well MW-23.

Tides can affect the water levels in wells located within 150–250 ft of the bay, and near preferential pathways. Wells influenced by tides, based on existing groundwater level data, were not used in the above assessments of flow direction and gradient. It should be noted that groundwater flow findings presented herein are provisional, pending completion of the tidal study proposed for the site.

6.6.1.4   Groundwater Quality

The Los Angeles Regional Water Quality Control Board (RWQCB) adopted a revised Water Quality Control Plan on June 21, 1995. The Basin Plan defines beneficial uses and water quality objectives for waters of the Bay area, including surfacewater and groundwater. The potential beneficial uses of groundwater underlying and adjacent to the site include: industrial process water supply, industrial service water supply, and agricultural water supply. At present, there is no known use of groundwater underlying the site for the above purposes, and groundwater in the Artificial Fill and San Antonio Formation is unsuitable for municipal/domestic uses because of brackish conditions. Saltwater intrusion has caused TDS concentrations in the San Antonio Formation to generally exceed 5000 mg/l. The Alameda Formation contains relatively freshwater with TDS concentrations generally below 1000 mg/l and is of suitable quality to be used for irrigation and municipal purposes. Benzene concentrations are reported up to 55,000 µg/l, TPHg up to 18,000,000 µg/l, TPHd up to 71,000 µg/l, and TPHo up to 28,000 µg/l.

6.6.1.4.1 Initial Vapor Extraction Pilot Testing

Objectives. The objectives of the initial vapor extraction pilot testing were to:

- Assess air permeability of the unsaturated zone for potential vapor extraction effectiveness based on the observed extraction flow/vacuum relationship.
- Assess the vapor extraction capture zone based on vacuum measurements at vapor extraction monitoring wells to specify extraction well locations and spacing for expanded vapor extraction testing.
- Estimate the initial extraction rate of VOCs and methane.

6.6.1.4.2 Initial Air Sparging Pilot Testing

Objectives. The objectives of the initial air sparging pilot testing were as follows:

- Assess air permeability of the saturated zone for potential sparging effectiveness, based on injection flow/pressure relationship.
- Observe the pressure response of saturated and unsaturated zones to air injection for gross assessment of sparging air flow pattern (i.e., stratified vs. nonstratified), based on monitoring well pressure measurements.
- Assess the potential extent of air sparging influence from an air sparging well.

Expanded pilot testing was also conducted for both vapor extraction and air sparging.

6.6.1.5 Hydrogeology

The water table surface at the site is located within the artificial fill. Depth to groundwater within the artificial fill in 2003 ranged from 4.35–10.40 ft bgs (7.90–3.95 ft port datum).

Groundwater elevation contour maps for the artificial fill were prepared for one berth using data collected on July 30, 2003, and for both berths using data collected on August 7, 2003. Groundwater at the site generally flows from west to east, toward the Outer Harbor. The average horizontal hydraulic gradient across the site is approximately 0.001. The presence of liquid phase hydrocarbons (LPH) in monitoring well MW-40 required that the groundwater elevation in this well be corrected for the affects of product density. A LPH density of 0.81 g/cm^3 was used for this correction.

Groundwater elevations obtained from monitoring well clusters MW-37A-C, MW-42A-C, MW-46A-C, and MW-48A-C were reviewed to estimate vertical gradients. Because tidal corrections have not been applied to the data, the calculated vertical gradients should be treated as qualitative indicators of groundwater flow direction. The data collected in 2003 indicate that a downward vertical gradient was present between monitoring wells screened in the artificial fill material ("A" wells) and those screened from 29 to 34 ft bgs in the upper portion of the San Antonio Formation ("B" wells). Downward vertical gradients in these wells ranged from approximately 0.05–0.6. At monitoring well clusters MW-46A-C and MW-48A-C, downward gradients of approximately 0.02–0.01 were noted between the "B" and "C" wells (the latter screened from 60–65 ft bgs, in the lower portion of the San Antonio Formation). Upward gradients were present between the "B" and "C" wells in monitoring well clusters MW-37A-C and MW-42A-C, ranging from approximately 0.4 to 0.1. These measurements are consistent with historical data from 2000, which indicate that a downward gradient generally exists between the artificial fill and the San Antonio Formation, that an upward gradient exists between the lower and upper portions of the San Antonio Formation at monitoring well cluster MW-37A-C, and that vertical gradient reversals occur between the upper and lower portions of the San Antonio Formation in well clusters MW-42A-C and MW-46A-C (SOMA Environmental Engineering 2000). The reversals in the vertical gradients at these wells appear to be an indicator that these are tidally influenced.

6.6.1.6   Chemical Analyses—Liquid Phase Hydrocarbons

6.6.1.6.1   Petroleum Constituents

Three-dimensional geostatistical models were developed for the constituents TPHg, TPHd, and TPHo in soil, utilizing software by Environmental Visualization. The data range for the models is 1997 to the present (i.e., August 1, 2003). This data range was chosen to yield a site-wide representation of the selected constituents and, in particular, to incorporate data for TPHd obtained in 1997. The data used in the models for TPHd and TPHo include samples analyzed in 2003 following silica gel cleanup and samples analyzed during previous investigations that did not include silica gel cleanup in the sample preparation. Results of the earlier samples not subject to silica gel cleanup may be considered conservative, in that the reported concentrations would be expected to be greater than the results would have been if silica gel cleanup had been performed.

The 3-dimensional models were created using the geostatistical method of kriging point data distributed about the site both horizontally and vertically. The kriging algorithm was fully 3-dimensional, in that the concentration value for each point on a grid was calculated from the measured values surrounding it in both the horizontal and vertical directions. Each data point used in the 3-dimensional models was assigned horizontal coordinates and an elevation in port datum. Boring locations were determined using recent survey data and, when historical elevation data were not available, the boring elevation was interpolated from a surface contour map derived from the recent survey data.

For the modeling, data reported as less than the reporting limit were assigned values equal to one-half the reporting limit. The reporting limit was that which corresponded to each sample result, and thus for samples run at high dilutions the assigned value of one-half the reporting limit may be biased high. Where duplicate samples were collected, the maximum value was used in the modeling.

Before kriging the concentration values were transformed to log base 10 values. Linear kriging was performed on the log-transformed concentration data assuming a horizontal to vertical anisotropy ratio of 10:1. For computing the value at each grid node, the nearest 20 data points were used. Because TPHo data in some regions of the site were limited, a blanking file of values set to one-half the reporting limit was used to represent the TPHo concentration at the ground surface at each horizontal sampling location to reduce apparent effects of interpolation of a data set that is less tightly constrained.

Other variables used to create the models were grid resolution and preprocessing and postprocessing clip values. The resolution was set to divide the site into a 100×100 node horizontal $\times$ 17 node vertical grid, resulting in approximately 20×20 ft^2 cells in the horizontal plane that were each 1 ft thick. The pre- and postprocessing clips are values used to allow the modeling software to recognize values less than the reporting limit during gridding calculations. The preprocessing clip was set to one-quarter of the method detection limit for each constituent, assuming a dilution factor of 1. The postprocessing clip was set equal to the reporting limit.

6.6.1.6.2 Bioremediation Indicator Parameters

Geochemical parameters can be used to ascertain whether aquifer conditions are conducive to bioremediation of petroleum hydrocarbons, establish if intrinsic bioremediation is ongoing at a site, predict potential modes of biodegradation, and to monitor the effectiveness of engineered remediation systems specifically designed to remove petroleum hydrocarbons from the subsurface. These geochemical parameters consist of naturally occurring chemicals used by indigenous microbial populations as electron acceptors in conjunction with metabolism of petroleum hydrocarbons. The metabolic process includes, in sequence of progressive reduction in redox intensity: aerobic respiration, denitrification, manganese (IV) reduction, iron (111) reduction, sulfate reduction, and methanogenesis. The geochemical data as a whole were used to evaluate the redox conditions at each well.

Upgradient Groundwater. Monitoring wells MW-16, MW-18, MW-26, MW-50, and MW-58 are the most hydraulically upgradient wells in the artificial fill at the site. Analysis of samples collected in 2003 from these wells reported TPHd concentrations ranging from 87–110 µg/l and TPHo ranging from less than 73–300 µg/l. These data are consistent with background water having some diesel and motor oil-range hydrocarbons either from an upgradient source, hydrodynamic dispersion from the site, or ambient conditions.

Aerobic Respiration. Aerobic respiration is the most efficient biochemical reaction whereby hydrocarbons are degraded to biomass, carbon dioxide, water, energy, and metabolites. In an aqueous environment, metabolism can switch from aerobic to anaerobic when dissolved oxygen concentrations decrease below approximately 1 mg/l (Tabak et al. 1981). Concentrations of dissolved oxygen measured in groundwater samples from both berths' monitoring wells in 2003 are generally less than 1.0 mg/l. These data indicate that under ambient conditions, aerobic respiration of the petroleum hydrocarbons is not likely to be significantly decreasing dissolved hydrocarbon concentrations.

Denitrification. Some microbes can utilize nitrate in groundwater as an electron acceptor by anaerobic respiration. Decreased nitrate concentrations indicate bioremediation through denitrification may be occurring. Approximately 90% (37/41) of groundwater samples recently collected and analyzed from the port did not contain detectable concentrations of nitrate. Of the four monitoring wells (MW-16, MW-34, MW-44, and MW-50) in which groundwater samples did contain measurable concentrations of nitrate (>0.5 mg/l), three ranged from 1.2–2.1 mg/l. Monitoring well MW-44 with 8.5 mg/l contained the greatest concentration. With the exception of monitoring well MW-50, denitrification appears to be the primary redox reaction in the wells with reportable nitrate. Petroleum hydrocarbon concentrations reported for samples from the wells where denitrification appears to be occurring are generally within the range of upgradient "background" wells.

Manganese Reduction. Anaerobic bioremediation by manganese-reducing microbes converts solid manganese (IV) oxides into soluble manganese (11) thereby increasing dissolved manganese concentrations. The geochemical parameters reported for samples collected from monitoring well MW-35 are consistent with manganese reduction being the dominant redox reaction. However, the petroleum hydrocarbon concentrations reported for samples from monitoring well MW-35 are within the range reported for samples from upgradient background wells.

The groundwater beneath most of the site appears to have already undergone manganese reduction. As such, it is unlikely that significant bioremediation by manganese-reducing bacteria is currently occurring at the site.

Iron Reduction. Ferric hydroxide minerals may be used as the terminal electron acceptor by anaerobic iron-reducing microbes degrading petroleum hydrocarbons. The reduction of solid ferric iron to soluble ferrous iron increases dissolved iron concentrations.

Iron reduction appears to be the predominant redox process occurring at monitoring wells MW-26, MW-45, MW-47, MW-50, MW-53, and MW-56. Iron reduction is not typically an efficient method of degrading petroleum hydrocarbons. Stoichiometrically, approximately 0.045 mg/l of benzene may be degraded for every 1 mg/l of ferrous iron produced. Significant petroleum degradation by iron-reducing bacteria is not anticipated to be occurring at the site.

Sulfate Reduction. Sulfate-reducing microbes can utilize sulfate in groundwater as a terminal electron acceptor during bioremediation processes by performing anaerobic respiration. Elevated sulfate concentrations suggest bioremediation is not occurring, whereas depressed concentrations or the presence of sulfide, which is highly reactive, indicates that it is.

Of the wells screened in the artificial fill, 27/37 appear to have sulfate reduction as the predominant redox reaction. Of these wells, sulfate concentrations ranged from 0.53–270 mg/l, with a median of 53 mg/l. In contrast, the sulfate concentrations reported for samples collected from areas where sulfate reduction does not appear to be occurring ranged from 150–1100 mg/l, with a median of 610 mg/l. Approximately 0.22 mg/l of benzene can be degraded per 1.0 mg/l of sulfate consumed. It appears that sulfate reduction is a significant biodegradation process at the site and has the capacity to further reduce hydrocarbon concentrations.

Methane. Methane-producing microbes (methanogens) are anaerobic bacteria that use carbon sources including petroleum hydrocarbons as a source of energy, and produce methane as a by-product.

The geochemical data from at least 19 monitoring wells are consistent with methanogenesis occurring. The distribution of wells where methanogenesis appears to be occurring generally coincides with the extent of methane in soil vapor. There is overlap between the range of redox conditions in which iron reduction, sulfate reduction, and methanogenesis can occur (Strumm and Morgan 1996). The geochemical data from monitoring wells MW-45, MW-53, and MW-56 are consistent with methanogenesis occurring alongside iron reduction. The data from approximately 16/27 wells where sulfate reduction appears to be occurring suggests methanogenesis is also occurring.

Methane concentrations in groundwater greater than 1 mg/l, and in soil vapor at concentrations of up to 56% by volume indicate that a robust population of methanogens is present in the subsurface, and that these bacteria are actively biodegrading petroleum hydrocarbons in the soil and groundwater. Approximately 1.3 mg/l of benzene are degraded per 1 mg/l of methane produced.

Summary of Bioremediation Indicator Parameters. Bioremediation indicator parameters show that subsurface conditions at the site are anoxic and provide a suitable environment for anaerobic microbial degradation of petroleum hydrocarbons. Analysis of ferrous iron, manganese, and nitrate data indicates that bacteria capable of using these compounds for metabolism or respiration may be locally present, but are not degrading hydrocarbons to a significant degree at the site.

Sulfate reduction is the primary redox reaction occurring at the site and is often coupled with methanogenesis. There is often sufficient sulfate remaining in the vicinity of the wells undergoing sulfate reduction to further degrade petroleum hydrocarbons. Methane concentrations in groundwater and soil vapor generally coincide with the distribution of petroleum hydrocarbons in the subsurface at the site. Concentrations of methane in groundwater and soil vapor indicate that methanogenic microbes are actively biodegrading petroleum hydrocarbons in the subsurface at the site.

6.6.1.7　Results of Initial Vapor Extraction Pilot Test

6.6.1.7.1　*Vapor Flow vs. Applied Vacuum Relationship*

The measured soil vapor flow rate was observed to increase from 6.2 scfm at the lowest vacuum level of 12 in. H$_2$O to 27.1 scfm at the highest applied vacuum level of 62 in. H_2O. The vacuum vs. flow rate graph demonstrates subsurface air permeability and is also a primary tool for vapor extraction (VE) system design.

6.6.1.7.2   Radius of Influence

Vacuum influence was observed during the initial vapor extraction pilot test at all vapor monitoring wells, including at the farthest monitoring point (MW-53) 146 ft distant from the extraction well VE-1. Exponential curves (straight lines on semilog plots) were found to fit the data with correlation coefficients ranging from 0.92–0.96. This degree of fit indicates relatively homogeneous soil permeability with respect to radial orientation in the subsurface.

The recording of absolute pressure in the headspace of groundwater monitoring well MW-5 1, located approximately 142 ft from extraction well VE-1, was intended to provide data on background pressure fluctuations to facilitate interpretation if induced vacuum at a monitoring well was of the same magnitude as background pressure fluctuations.

Radius of vacuum influence was estimated graphically by using an induced vacuum of 0.1 in. H_2O as the defining point for limit of influence. The exponential curves fit to the monitoring well vacuum measurement data in Figure 42 were extrapolated to 0.1 in. H_2O vacuum. The radius of vacuum influence predicted in this manner ranges from 153–313 ft for the applied vacuum levels ranging from 12 in. H_2O to 62 in. H_2O.

Empirical calculations based on a geometric model (Johnson et al. 1990) were performed as an alternate method for estimating radius of vacuum influence. The calculations were performed for each combination of extraction well applied vacuum and monitoring well observed vacuum. The estimated radius of influence from the empirical calculations ranges from 121–966 ft.

6.6.1.7.3   Extracted Soil Vapor Analytical Results

Evidence of biodegradation is also supported by observation of depleted oxygen (as low as 1.9 vol%) and the presence of carbon dioxide (to 8.7 vol%) and hydrogen sulfide (to 26 ppbv). Total gasoline range petroleum hydrocarbons were reported to 8200 ppmv and benzene to 22 ppmv. During the 6-hour steady-state extraction period, the methane content decreased by one-third, (from 38 vol% to 26 vol%) while the nonmethane content remained relatively constant. Concentrations of methyl tert butyl ether (MTBE) in soil vapor were reported up to 2 1 ppmv.

6.6.1.7.4   Hydrocarbon Mass Removal Rate

Mass removal rates were calculated for each step test vacuum level and for the 6-hour steady-state extraction period based on the vapor flow rate data and the vapor composition as reported by the analytical laboratory. The total mass removal rate from vapor extraction well VE-1 increased from approximately 5.8–29.8 lb/hr as vacuum was increased from 12–62 in. H_2O. During the 6-hour, steady-state extraction period, the mass removal rate decreased by about 30% to approximately 21.0 lb/hr because methane content decreased during the test. The initial vapor extraction pilot test data indicate that the methane concentrations in soil vapor will decrease more rapidly than nonmethane concentrations. The total methane and nonmethane hydrocarbon mass removal during the initial vapor extraction pilot test was approximately 182 lb.

6.6.1.7.5   Mass Removal Calculations

The thermal oxidizer includes an extracted vapor flow instrument (pilot tube) at the discharge of the vacuum blower and an hour meter recording operating time. The mass of hydrocarbons removed by the thermal oxidizer system can be estimated by combining vapor concentration data, the total flow rate, and hours of operation. Over the period July 9 to August 12, 2003, an estimated 3139 lb of hydrocarbons were removed. Of this total, 2002 lb were methane and 1137 lb were nonmethane hydrocarbons.

6.6.1.7.6   Vacuum Influence, Permeability, and Recommended
Design Radius of Influence

The initial vapor extraction pilot test consisted primarily of a radius of vacuum influence determination based on extraction from a single well (VE-1). To assess the permeability and vacuum influence over the entire Treatment Cell 1 area and refine design parameters for treatment cells at Berth 24, additional data were collected during the expanded vapor extraction pilot test.

6.6.1.7.7   Air Flow vs. Sparging Well Pressure

Pressure was applied in three steps to derive a flow versus pressure relationship. Air flow increased from 3 scfin at an applied well head sparging pressure of 4.5 psig to 7.1 scfin at an applied pressure of 10.3 psig.

6.6.1.7.8   Pressure Response

The 25-hour initial air sparging test was started at the maximum pressure step of 10.3 psig with a flow of approximately 7–8 scfnl. The flow remained relatively constant for the duration of the test while the well head pressure steadily decreased to a final value of 6.1 psig. A positive pressure response was observed at all air sparging monitoring wells. The pressures generally remain significantly elevated compared to pretest conditions. This type of pressure response is indicative of stratified flow; that is, sparging flow is preferentially outward from the sparging well instead of upward. Stratified flow is expected given the extensive asphalt paving overlying the site.

6.6.1.7.9   Dissolved Oxygen Concentration

Dissolved oxygen was monitored continuously with *in situ* probes placed in monitoring wells AM-1 through AM-4, and by flow through a dissolved oxygen measurement cell during groundwater sample collection from monitoring wells MW-15 and MW-52. Dissolved oxygen concentrations increased from presparging levels in monitoring wells AM-1 through AM-4 and MW-15, to a distance of approximately 20 ft from air sparging well AS-1. Dissolved oxygen did not increase measurably in the sample collected from monitoring well MW-52, located 40 ft from the sparging well. However, data for other groundwater parameters suggest a significant oxygen demand in the subsurface that could account for the lack of a dissolved oxygen concentration increase in monitoring well MW-52 over the test duration of 25 hours.

6.6.1.7.10 Ferrous Iron Concentrations

Concentrations of ferrous iron were measured before and after the 25-hour air sparging period. Groundwater samples were collected from monitoring wells AM-1 through AM-4, MW-15, and MW-52 and analyzed utilizing a chemical reagent/colorimetric test kit. A decrease in ferrous iron concentration indicates air sparging influence as the oxygen content of the sparged air oxidizes ferrous iron (soluble) to ferric oxide (insoluble).

Table 44 shows the initial ferrous iron concentrations for the six monitoring wells and the postsparging ferrous iron concentrations for four of the six wells. Monitoring wells AM-1 and AM-3, the two closest monitoring wells to air sparging well AS-1, went dry during the test. Table 44 shows ferrous iron concentrations increased in monitoring wells AM-2 and AM-4, which is contradictory to the increased dissolved oxygen concentrations observed. Monitoring well MW-15 had a decreased ferrous iron concentration consistent with the increased dissolved oxygen concentration observed. Monitoring well MW-52 also had a decreased ferrous iron concentration indicative of air sparging influence 40 ft from air sparging well AS-1. The ferrous iron results for monitoring well MW-52 are not inconsistent with the basically unchanged dissolved oxygen concentration observed for this monitoring well, given that ferrous iron and other groundwater parameters represent an oxygen demand to be satisfied before observing increases in dissolved oxygen concentration.

6.6.1.8 Estimated Radius of Influence

Changes in groundwater sample concentrations of dissolved oxygen, ferrous iron, and VOCs were assessed to estimate radius of air sparging influence at conditions of 7–8 scfm flow rate over a 25-hour period. Influence is estimated as follows:

6.6.1.9 Method Observed Radius of Influence

- Dissolved oxygen in groundwater samples 20–40 ft
- Ferrous iron in groundwater samples >40 ft
- Groundwater sample VOC concentrations >40 ft

The majority of data collected in the initial air sparging pilot test suggest a radius of influence of at least 40 ft from a sparging well operated with an air flow rate of 7–8 scfm. An air sparging well grid spacing based on a 40-ft radius of influence and a 15% overlap for complete aerial coverage would be 68 ft ($40 \times 2 \times 0.85$). Approximately 50% of the 6.0-acre Treatment Cell 1 area overlays groundwater impacted by hydrocarbons in excess of the anticipated remediation goals. Applying a well grid spacing of 68 ft would result in approximately 32 air sparging wells installed to cover the impacted area.

Following the initial air sparging pilot test, 58 air sparging wells were installed in Treatment Cell 1 to conduct expanded air sparging pilot testing and interim remediation. The 58 air sparging wells are on a grid spacing of 50 ft, corresponding to a design radius of influence of approximately 29 ft. The conservative grid spacing provides for the possibility of less permeable soils in the expanded pilot test area, the use of a lower air flow rate, and expedited remediation. Additional data were obtained in the expanded air sparging pilot test to refine air sparging design parameters for one berth.

6.6.1.9.1 Measurement of Relative Permeability

The relative permeability to air flow for the air sparging wells was evaluated by measuring the flow vs. applied pressure relationship, and comparing those data with air sparging well AS-1, which was utilized in the initial air sparging pilot test. The data show that 24/54 (44%) air sparging wells appear to be installed in sediments with a lower permeability than AS-1, while 30 wells are installed in sediments of higher permeability than AS-I. The wells demonstrating lower permeability are generally found in the southeast portion of Treatment Cell 1.

6.6.1.10 Risk Assessment under Current Conditions

6.6.1.10.1 Site Conceptual Model

Complete exposure pathways include inhalation of volatile emissions from subsurface soil and groundwater, inhalation of dust released during subsurface construction activities, and incidental ingestion of soil and dermal contact with soil during construction excavation activities. Consumption of groundwater is not a complete exposure pathway because domestic use is not an existing or potential beneficial use of groundwater beneath the site due to excessive salinity.

The potential is for contaminants of potential concern (COPC) dissolved in groundwater to impact ecological receptors in the Outer Harbor by way of direct groundwater flow and indirect flow pathways (preferential pathways). The conceptual model includes groundwater flow, stormwater flow, tidal flow, and the interaction among the three. The figure is a schematic cross-sectional view oriented generally west to east along a typical storm drain, and thus does not represent the configuration of a particular storm drain. The figure includes the range of tidal flow in the storm drains and the net direction of groundwater flow.

6.6.1.11 Summary of Exposure Assessment

For acute hazards, the EPA's Human Health Risk Assessment/Material Handling Equipment (HHRA/MHE) used 10% of the lowest effect level (LEL) for methane as a benchmark, and focused on a short-term exposure scenario for the evaluation of potential explosion hazards to a hypothetical trench worker. The previous risk assessment by SOMA (2000) had applied the LEL as a benchmark to a probabilistic evaluation of potential explosion hazards under a similar scenario. Potential acute risks to indoor and outdoor commercial/industrial workers were also estimated.

For chronic hazards, the HHRA/MHE estimated exposure point concentrations based on the 95%-th upper confidence limit of the arithmetic mean from concentration data drawn from a defined "zone of contamination," thereby excluding data from areas of relatively lesser concentrations. For the outdoor worker exposure scenario, a 25-year occupational period was considered representative of long-term dock workers or other outdoor workers at the site not involved in subsurface construction activities. For this scenario, the calculations of vapor emissions conservatively assumed that the porosity of the ground surface is equal to that of soil, and thus assumed that the asphalt pavement, which would have a much lower porosity than soil, was absent for the entire exposure duration.

The HHRA/MHE employed a different methodology for estimating the exposure point concentration for the evaluation of indoor air exposure from emission of VOCs into a building onsite. The risk assessment assumed that the building was located at the greatest concentration of benzene detected at the site and applied an EPA implementation of the Johnson and Ettinger model of vapor intrusion to the maximum concentration. The HHRA/MHE characterized this approach as a maximum worst-case indoor air exposure scenario.

6.6.1.12   Human Health Risk Characterization

For acute risks, the HHRA/MHE identified approximately 24 acres of the site beneath which soil vapor concentrations exceeded 10% of the LEL. Identified acute risks were associated with intrusive construction and maintenance activities, such as trenching. The HHRA/MHE also asserts that the act of cutting or breaking the pavement alone would result in potentially explosive atmospheres, but does not include a quantitative analysis of these risks. The risk assessment also identifies an area where benzene concentrations in soil vapor exceed the 500 ppmv as a potential acute risk, as set forth by the National Institute for Occupational Safety and Health's Immediately Dangerous to Life or Health Concentrations.

For chronic risks, the HHRA/MHE reports that the estimated excess cancer risk exceeds the 1E-05 acceptable risk for commercial and industrial workers for three exposure scenarios: trench worker, indoor worker, and outdoor worker. The trench worker risks are based on results of the 1997 ICF Kaiser Risk Assessment. As stated previously, the indoor commercial/industrial worker exposure scenario is based on the maximum observed benzene concentration in soil vapor and represents a maximum worst case. The estimated risk to the outdoor commercial/industrial worker is also conservative, as the risk calculations assume that the land surface is bare soil and thus assume that the asphalt pavement is absent. The risks estimated for benzene account for over 90% of the calculated risks for both the indoor and outdoor commercial/industrial worker.

6.6.1.13   Ecological Screening Evaluation

The evaluation included an ecological screening evaluation that addressed the potential impacts of COPC in groundwater to the Outer Harbor. The ecological screening evaluation identified benzene/xylenes (BTEX) as COPC in groundwater and applied groundwater flow and chemical transport modeling, which was used to predict the concentration of BTEX in groundwater that would discharge to the Outer Harbor.

The groundwater flow and transport model incorporated flow through the sub-surface and the effect of the sheet pile wall; it did not address a 150-ft section where a rip-rap embankment stands in place of the sheet pile wall or preferential flow pathways in the storm drains. The predicted BTEX concentrations in ground-water discharging to the Outer Harbor were less than water quality criteria for bays and estuaries as set forth by federal and state regulatory agencies. The exclusion of the 150-ft section of rip-rap embankment and preferential flow pathways from the model may have resulted in an underestimation of the flux of COPCs to the Outer Harbor. The parallel PPITS investigation described in the next section will include an evaluation and verification of the groundwater flow and chemical trans-port modeling.

6.6.1.14   Parallel Preferential Pathway/Tidal Study Investigation and Risk Assessment

Proposed remedial activities at the site will also include an investigation of potential preferential pathways for the migration of petroleum constituents in groundwater to the Outer Harbor through the storm drains beneath the site and through the 150-ft section where a rip-rap embankment stands in place of the sheet pile wall. The EPA's Pollution Prevention Information Tracking System (PPITS) investigation will consti-tute a parallel investigation and feasibility study. The primary objective of the PPITS investigation is to assess the potential for COPC dissolved in groundwater to impact ecological receptors in the Outer Harbor through groundwater flow and indirect preferential pathways. The significance of the risk, if any, to ecological receptors in the Outer Harbor would depend on the actual flux of COPC into the Outer Harbor. The PPITS Work Plan will include a detailed discussion of the proposed work.

6.6.2   REMEDIAL ACTION OBJECTIVES — MITIGATION OF HUMAN HEALTH RISKS

The primary objective of the recommended remedial actions is to mitigate potential acute and chronic human health risks posed by COPC in soil, soil vapor, and groundwater. In the short term, implementation of the recommended remedial actions will focus on the reduction of soil vapor concentrations of flammable gases and VOCs to facilitate planned renovation activities at the site.

Remedial actions will also be implemented to mitigate potential human health risks under long-term exposure scenarios (e.g., indoor commercial/industrial worker). Long-term remedial activities will be conducted to reduce COPC concen-trations remaining in soil and groundwater.

6.6.2.1   Control of Long-Term Methane Rebound

Another objective of implementing the recommended remedial actions is to miti-gate the potential for long-term rebound of methane and other flammable gas concentrations in soil vapor. Implementation of these remedial actions considers the factors contributing to subsurface methane generation, including total petroleum hydrocarbons concentrations in soil and groundwater and anaerobic conditions within the artificial fill. Residual LPH will also be addressed when observed to minimize this potential source for methane generation.

6.6.2.2 Preliminary Action Levels

Sources for PALs include:

- 10% of the LEL as defined by the Occupational Safety and Health Administration.
- The RWQCB September 4, 2003 Screening for Environmental Concerns at Sites with Contaminated Soil and Groundwater, which presents environmental screening levels (ESLs).

The human health risk-based action levels for soil and soil vapor are also derived from estimates presented in the HHRA/MHE and ICF Kaiser Risk Assessment. The human health risk-based action levels are calculated COPC concentrations that are equivalent to risk levels accepted for commercial or industrial land use. These accepted risk levels are equal to a 1E-05 excess cancer risk or 1.0 noncancer hazard (i.e., 1/100,000). The human health risk-based action levels are provided in part as a comparison against the other action levels. The calculations begin with the maximum estimated excess cancer risk or the maximum calculated noncancer hazard reported in the HHRA/MHE or ICF Kaiser Risk Assessment. For COPC that present cancer and noncancer hazards, the calculations use the greatest risk relative to the 1E-05 excess cancer risk level or 1.0 noncancer hazard. The exposure point and exposure point concentration corresponding to the maximum estimated human health risk are also provided. Next, the table lists the soil vapor or soil COPC concentration from which the exposure point concentration was derived. In the case of soil, the two values are equal. The COPC concentration equivalent to 1E-05 cancer risk or 1.0 noncancer hazard is calculated from the ratio of these risk levels to the appropriate maximum estimated risk level.

6.6.2.3 Soil Vapor

For the primary short-term objective of reducing risks related to the presence of flammable gases and VOCs in soil vapor, the PALs are set at 10% of the LEL. The soil vapor PALs for the other COPC are set at the ESLs. Values for both flammability and the ESL are provided for benzene. The ESLs for soil vapor are based on conservative assumptions; for example, a continuous source of COPC in soil vapor is assumed to be present directly beneath the concrete floor slab of a hypothetical building. The assumptions incorporated into the ESLs explain the differences between the human health risk-based action level for benzene (180 ppmv) vs. the ESL (0.088 ppmv). The benzene PAL is set at the human health risk-based action level of 180 ppmv.

6.6.2.4 Soil

The soil PALs for TPHg, TPHd, and TPHo are those specified in the order for the site. The PALs for BTEX and MTBE are set at the ESLs because the orders do not address these COPC.

The soil criteria for lead and 6 of the 7 PAHs are based on ESLs because the ESLs are judged to be more relevant. The numerical criteria for soil in the order served a rather specific purpose, in that they were limits for reuse of excavated soil and not remediation of *in situ* soils. Soil criteria were established for three zones: a 300-ft wide Shoreline Protection Zone, a 700-ft wide Buffer Zone, and an Upland Zone located at least 1000 ft from the Middle Harbor. The soil criteria in the order draw from a variety of sources. In some instances, a site-specific soil and groundwater transport model was used to estimate attenuation of COPC concentrations in soil. In other cases where observed COPC concentrations in soil were less than the concentration criteria calculated with the soil and groundwater transport model, the soil limits were set equal to water quality criteria for sediment. For lead and 6 of the 7 PAHs references, either NOAA aquatic sediment criteria or disposal criteria for dredge sediment for the Puget Sound.

6.6.2.5 Groundwater

Groundwater criteria were established for the Shoreline Protection Zone and were based in part on the Basin Plan's Shallow Water Effluent Limitations for marine water and the U.K.'s Sea Fish Industry Authority (SFIA) Tentative Order. The groundwater PALs for TPHd and the seven PAHs are based on the SFIA Tentative Order while the PAL for lead references the Basin Plan. The groundwater PALs for TPHg, BTEX, and MTBE are set at the ESLs.

6.6.2.6 Risk Assessment for Cessation of Active Remediation

Based on remedial progress monitoring, a human health risk assessment will be performed to support the decision to cease active remediation. The results from the human health risk assessment will be combined with the results from the ecological screening evaluation or risk assessment to assess the adequacy of PALs and to establish final remedial goals.

6.6.2.7 General Response Actions

General response actions will be implemented so that the remedial action objectives can be satisfied. Separate lists of general response actions have been developed for each of the affected environmental media at the site (Table 6.6.2.1).

6.6.3 TECHNOLOGY IDENTIFICATION—SOIL AND SOIL VAPOR

6.6.3.1 Natural Attenuation

Remediation of soil and soil vapor by natual attenuation relies on naturally occurring biological or physical processes to destroy, degrade, immobilize, and/or attenuate

TABLE 6.6.2.1
General Response Actions for Remedial Action Objectives

Environmental Media	General Response Actions	Representative Technologies/Methods
Soil and soil vapor	Natural attenuation	Natural attenuation
	Extraction or removal and *ex situ* treatment or disposal	Passive venting
		Vapor extraction
		Soil excavation and offsite disposal
Groundwater	*In situ* treatment	Bioventing
	Natural attenuation	Natural attenuation
	Extraction and *ex situ* treatment	Pump and treat
	In situ treatment	Enhanced bioremediation
		Air sparging
		Air-ozone sparging

COPCs. can occur if the COPCs are amenable to biodegradation, if suitable degrading microbes are present, and if other geochemical conditions (e.g., the presence of nutrients and oxygen) are met. Aerobic biological processes utilize oxygen as the oxidizer or electron acceptor. In the presence of sufficient oxygen, aerobic microbes (aerobes) metabolize (degrade) hydrocarbons into carbon dioxide and water. In the absence of oxygen or on the depletion of available oxygen, aerobes cease to function and anaerobic microbes (anaerobes) predominate. Anaerobes utilize chemical species other than oxygen as the electron acceptor to degrade hydrocarbons. As environmental conditions become more reducing, the electron acceptor order of preference for anaerobes is nitrate, manganese, ferric iron, and sulfate. In the absence of these anaerobic electron acceptors, methanogenic anaerobes degrade hydrocarbons using carbon dioxide as an electron acceptor and produce methane as a by-product.

Settings amenable to natural attenuation remediation are those with biodegradable COPCs and favorable distributions of COPC in the environment (i.e., lack or minimal presence of LPH, low threat to drinking water, and absence of toxic or hazardous by-products of biodegradation). Investigation data reported herein indicate that anaerobic microbes are actively degrading petroleum hydrocarbons at the site. The biodegradable COPC should include the diesel and motor oil petroleum fractions (Ruthermich et. al. 2002; Baedecker et. al. 1993). Monitoring and reporting are the only costs to implement natural attenuation, but this relatively inexpensive alternative is slow compared to other remedial actions.

6.6.3.2 Passive Venting

Passive venting involves the exchange of soil vapor with atmospheric air using vent wells installed in the vadose zone. Subsurface pressure differentials with respect to atmospheric pressure are the driving force for gas flow. Such pressure differentials develop due to barometric pressure variations, temperature differences, and groundwater elevation changes. When atmospheric air is introduced to the subsurface, the added oxygen can support aerobic biodegradation. When soil vapor is vented to the atmosphere, VOCs are removed from the subsurface. In some applications, gas flow may be desired in only one direction (i.e., soil vapor venting only or air inlet only); this venting can be accomplished by incorporating a low differential pressure check valve at the well head or vent manifold piping. Vented soil vapor generally requires treatment in accordance with air quality regulations.

6.6.3.3 Vapor Extraction

Vapor extraction involves the installation of wells in the vadose zone and withdrawing soil vapor with a vacuum blower. Volatile compounds equilibrate between soil vapor and sediments, so VOCs are present in the extracted vapor. Replacement of the extracted soil vapor with flow of pore-volume air from outside the zone of contamination drives the diffusion of VOCs adsorbed on sediments to the soil vapor. In addition, the replacement air will have a greater oxygen content and can contribute to aerobic biodegradation processes. Emissions of extracted soil vapor typically require treatment such as carbon adsorption, biofilters, or thermal/catalytic oxidizers in accordance with air quality regulations.

Settings favorable for vapor extraction include those with VOCs, those with COPC amenable to aerobic biodegradation, and those with permeable sediments. Pilot testing has demonstrated favorable soil permeability and mass extraction rates and effective reduction of methane and VOC concentrations in soil vapor.

6.6.3.4 Bioventing

Bioventing involves the introduction of atmospheric air to the subsurface by either withdrawal of soil vapor or the injection of fresh air. Indigenous aerobic biological processes are enabled or enhanced by the oxygen content of the added air. Bioventing can stimulate the aerobic degradation of both volatile (gasoline) and nonvolatile (diesel fuel and motor oil) hydrocarbons. Bioventing processes can be designed with sufficiently low ventilation or injection rates that emissions treatment is either unnecessary or less costly than vapor extraction. The biological processes utilized in bioventing results in less expedient remediation of volatile hydrocarbons than more aggressive alternatives such as vapor extraction.

6.6.3.5 Soil Excavation and Offsite Disposal

This remedial option involves the physical removal of impacted soil from the site. Excavation is often the most time-expedient remedial alternative when feasible. Buildings and other site improvements can limit feasibility by restricting access to impacted soil, and by posing structural safety hazards from soil undermining. Fugitive emission of vapor and dust is a potential problem. The area required for excavation can disrupt facility operations.

6.6.4 GROUNDWATER

6.6.4.1 Natural Attenuation

Remediation of groundwater by natural attenuation involves the same processes and principles for soil and soil vapor; namely, naturally occurring biological and/or physical processes resulting in the degradation, retardation, or attenuation of COPC. Biological processes can be generally classified as either aerobic or anaerobic, depending on the presence or absence, respectively, of dissolved oxygen. The success of natural attenuation depends on the biodegradability of the contaminant, the presence of suitable degrading microbes in the subsurface, and geochemical factors such as the presence of nutrients or dissolved oxygen.

Settings favorable for natural attenuation remediation are those with biodegradable COPC and with a favorable distribution of the COPC (i.e., lack or minimal presence of LPH, low threat to drinking water, and absence of toxic or hazardous by-products of biodegradation). Investigation data indicate that anaerobic microbes are actively degrading petroleum hydrocarbons at the site. Natural attenuation is relatively inexpensive but slow compared to active remedial alternatives.

6.6.4.2 Enhanced Bioremediation

Enhanced bioremediation can accelerate the processes discussed for natural attenuation. Suitable contaminant-degrading microbes can be introduced if they are not present in sufficient quantities (bioaugmentation). Nutrients and air (oxygen) can be injected into the groundwater (bioenhancement). Oxygen can also be introduced with placement of oxygen-release compound (ORC) in groundwater wells. Air sparging is an enhanced bioremediation process that increases dissolved oxygen but is also a physical removal process. Enhanced bioremediation has lesser implementation costs compared to physical remediation processes such as air sparging, as microbes accomplish the contaminant removal instead of positive air flow provided by mechanical equipment.

Settings favorable for enhanced bioremediation include those with biodegradable COPC and favorable contaminant distributions (i.e., lack or minimal presence of LPH, low threat to drinking water, and absence of toxic or hazardous by-products of biodegradation). Historical aquifer testing and the pilot testing indicate acceptable hydraulic characteristics and subsurface permeability for feasibility of this alternative. Although accelerated compared to natural attenuation, enhanced bioremediation is less expedient than more aggressive alternatives such as air sparging.

6.6.4.3 Air Sparging with Concurrent Vapor Extraction

In the air sparging process, wells are installed with screened intervals below the water table. Compressed air from an oil-free compressor is injected through the sparging wells into the groundwater. The injected air will volatilize or strip VOCs from the dissolved phase and transfer them to the soil vapor phase, where typically the VOCs are captured by a concurrently operated soil vapor extraction system.

Air sparging can also be effective in removing volatile LPH when present on the groundwater table. In addition to the physical stripping process, air sparging increases the dissolved oxygen content of the impacted groundwater and thus enhances or stimulates aerobic biodegradation. Thus, less volatile but biodegradable COPC such as weathered gasoline or diesel can be remediated. Air sparging systems are frequently operated in a pulse mode in order to avoid the creation of stable, preferential flow passages in the subsurface.

6.6.4.4   Air–Ozone Sparging

Air–ozone sparging is an enhanced version of air sparging. Instead of ambient air, an ozone generator is employed to inject a mixture of air and ozone (typically 100–300 ppmv ozone in air) into the saturated zone. Ozone oxidizes many organic chemicals including petroleum hydrocarbons. Because the oxidized COPCs are destroyed rather than transferred to the vapor phase as in air sparging, a concurrently operating vapor extraction system may not be required and the corresponding implementation costs are less than for air sparging. Another favorable aspect is that excess ozone will decompose to oxygen within hours, leaving no residual ozone in the subsurface and providing dissolved oxygen for aerobic degradation. A potential disadvantage of air–ozone sparging is that ozone generating equipment is likely to have less operating reliability than air sparging equipment.

The settings favorable for air–ozone sparging are the same as discussed for air sparging in the preceding section, and the alternative is also feasible for the same reasons.

6.6.4.5   Groundwater Pump and Treat

In a pump and treat process, groundwater is extracted, treated above ground if necessary, and discharged to a suitable location or reinjected. Remediation of the aquifer occurs as adsorbed COPC diffuse from sediments and into the pore water. Aboveground treatment processes employed with pump and treat remediation include air stripping, carbon adsorption, biological processes, and advanced oxidation. Pump and treat as a remedial process offers the advantage of hydraulic containment that may be desired if drinking water or other sensitive receptors are impacted or threatened. Pump and treat is typically not time-expedient compared to other remedial alternatives because of the diffusion-limited mass transfer in water. Short-term costs for pump and treat may compare favorably to other active groundwater remediation alternatives but long-term costs can be the greatest of any alternative because of the extended operating time.

6.6.5   Analysis of Remedial Alternatives

Tentatively identified compounds compare the remedial technologies to select interim remedial measures to mitigate potential acute and chronic human health risks posed by COPCs in soil, soil vapor, and groundwater. The remedial technologies are evaluated against the three general criteria of effectiveness, implementability, and cost. The particulars of these criteria are derived from the characteristics of the

site and attendant operational constraints. Specific elements evaluated within each
of these three general criteria include the following:

Effectiveness
- Protection of human health and the environment
- Compliance with laws and regulations
- Reduction of toxicity, mobility, and volume
- Long-term effectiveness
- Short-term effectiveness

Implementability
- Compatibility with site conditions and constraints
- State and community acceptance
- Timeliness

6.6.5.1 Technologies for Remediation of Soil and Soil Vapor

Technologies considered for remediation of soil and soil vapor include natural
attenuation, passive venting, vapor extraction, bioventing, and excavation.

6.6.5.2 Natural Attenuation

Effectiveness. Current site conditions and historical investigation data suggest that
natural attenuation generates excessive methane as a product of the biodegradation
of COPC. The continued generation of methane directly conflicts with the primary
short-term remedial objective of mitigating the acute risk of flammable gases in the
subsurface. Although engineering controls could be employed to mitigate acute risks
where warranted by site construction, the presence of toxic COPC in the soil vapor
during the natural attenuation remedial timeframe conflicts with the remedial objective of mitigating chronic health risks. The effectiveness of natural attenuation for
soil and soil vapor is judged unacceptable and this remedial alternative is not
evaluated further.

6.6.5.3 Passive Venting

Effectiveness. The subsurface permeability demonstrated by vapor extraction
pilot testing is conducive to the implementation of passive venting. Compared to
the more aggressive vapor extraction alternative, the relatively low soil vapor
venting rate anticipated for passive venting correlates to a lesser degree of short-
term effectiveness. With a sufficient number of venting wells long-term effectiveness of passive venting may be comparable to vapor extraction, especially
with respect to remediation of nonvolatile COPC through the mechanism of
aerobic biodegradation supported by the introduction of atmospheric air to the
subsurface.

    Implementability. Construction of venting wells with completion depths of 5 ft
is a favorable factor, but a greater number of venting wells are likely necessary to
implement passive venting in comparison to the number of extraction wells necessary

to implement vapor extraction. Additional pilot testing would be necessary to determine final design parameters for a passive venting system.

6.6.5.4 Vapor Extraction

Effectiveness. Pilot testing indicates that vapor extraction can effectively reduce concentrations of methane and volatile COPC to achieve short-term remedial goals, and that ongoing operations would control concentrations to achieve long-term goals. The major limitation of vapor extraction effectiveness is the time to remediate nonvolatile COPC. Extended operation of a vapor extraction system may be necessary to maintain acceptable methane and COPC concentration reductions during the time necessary for aerobic biodegradation processes to sufficiently remediate the nonvolatile COPC.

    Implementability. Favorable conditions for implementation of vapor extraction include construction of vapor extraction wells with completion depths of 5 ft, and permeability as demonstrated by pilot testing that allows relatively few wells to remediate relatively large areas.

6.6.5.5 Bioventing

Effectiveness. The subsurface permeability demonstrated by vapor extraction pilot testing is conducive to the implementation of bioventing. Although not suitable for methane in soil vapor without emission controls, the remedial expediency of bioventing is probably second only to excavation for remediation of nonvolatile COPC in soil. With the primary remedial mechanism of aerobic biodegradation, bioventing is not the most time-effective short-term alternative, but should have long-term effectiveness comparable to any other soil remediation alternative.

    Implementability. Construction of venting wells with completion depths of 5 ft is a favorable factor. Successful implementation of bioventing may require no greater number of wells than necessary to implement vapor extraction. Additional pilot testing would be necessary to determine final design parameters for a bioventing system.

6.6.5.6 Soil Excavation and Offsite Disposal

Effectiveness. Physically removing and disposing impacted soil from the site is the most effective short-term remedial alternative. Because of the potential for impacted groundwater to generate and release methane to soil vapor, active remediation of groundwater may be required for long-term effectiveness of the soil excavation alternative.

    Implementability. Except for the thickness of paving installed to withstand the traffic of cargo container operations, the lack of aboveground site improvements is conducive to excavation. However, numerous underground utilities are present including gas, electric, communication, sewers, storm drains, fire water mains, and potable water lines. The underground utilities present a safety challenge to conducting extensive excavation. Because the area of impacted soil covers approximately 20 acres and the depth to water is generally 8 ft bgs, the volume of soil to be removed would exceed 250,000 yd^3. The California Environmental Quality Act (CEQA) process would likely delay or even preclude implementation of this alternative.

6.6.6 TECHNOLOGIES FOR REMEDIATION OF GROUNDWATER

Technologies considered for remediation of groundwater include natural attenuation, enhanced bioremediation, air sparging with concurrent vapor extraction, air–ozone sparging, and pump and treat.

6.6.6.1 Natural Attenuation

Effectiveness. Current site conditions and historical investigation data suggest that natural attenuation processes are causing degradation of COPCs, but these processes are not time expedient and continued generation of methane is a concern. Although not time expedient, natural attenuation processes are theoretically capable of eventually achieving groundwater PALs with implementation of appropriate soil and soil vapor remediation. The acute health risks posed by COPCs in groundwater generating and releasing methane to soil vapor can also be mitigated with the appropriate soil and soil vapor remediation.

   Implementability. The natural attenuation alternative creates no implementation issues.

6.6.6.2 Enhanced Bioremediation

Effectiveness. Hydraulic characteristics reported from historical aquifer testing are conducive to successful introduction of agents to enhance or accelerate natural attenuation processes. Whether from introduction of oxygen by an ORC, direct injection of air or oxygen, addition of nutrients, and/or inoculation of specific microbes designed to degrade the COPC, enhanced bioremediation is likely capable of acceptable long-term remedial effectiveness. Bioremediation probably offers minimal short-term effectiveness in mitigating acute risks from generation of methane compared to some other groundwater remedial alternatives, but the appropriate soil and soil vapor remediation can mitigate these acute risks.

   Implementability. Construction of remedial wells with completion depths on the order of 10 ft is a favorable factor, but a greater number of wells are likely necessary to implement enhanced bioremediation in comparison to the number of wells necessary to implement air sparging or air–ozone sparging. Additional pilot testing is necessary to determine final design parameters for an enhanced bioremediation system, possibly including microbial studies and bench scale tests.

6.6.6.3 Air Sparging with Concurrent Vapor Extraction

Effectiveness. Pilot testing indicates air sparging to be a feasible alternative for short- and long-term effectiveness to reduce concentrations of VOCs in groundwater. Air sparging has the additional benefits of promoting aerobic biodegradation of nonvolatile COPC and the removal of residual volatile LPH that may be present on the water table.

Implementability. Potential emissions of VOCs and methane require concurrent vapor extraction with implementation of air sparging described the favorable conditions for implementation of the vapor extraction component of this alternative. The air sparging component has similar favorable implementation factors, including relative ease of well installation and subsurface permeability that allows relatively few wells to remediate relatively large areas.

6.6.6.4 Air–Ozone Sparging

Effectiveness. The feasibility of air–ozone sparging is similar to air sparging for potential shorthand long-term effectiveness in reducing concentrations of VOCs in groundwater, promoting aerobic biodegradation of nonvolatile COPC, and the removal of residual volatile LPH that may be present on the water table. However, this alternative has an advantage with the remedial mechanism of direct oxidation and destruction of COPC, compared to the air sparging remedial mechanisms of medium (phase) transfer for VOCs and aerobic biodegradation for nonvolatile COPC. The direct oxidation of COPCs potentially makes air–ozone sparging the most time-expedient remedial alternative for groundwater. Air–ozone sparging is not a mature technology compared to air sparging, and long-term reliability of ozone generating equipment is less certain.

Implementability. Air–ozone sparging has similarly favorable implementation factors such as those discussed for air sparging, including relative ease of well installation and subsurface permeability that allows relatively few wells to remediate relatively large areas.

6.6.6.5 Groundwater Pump and Treat

Effectiveness. Historical aquifer testing indicates hydraulic characteristics conducive to groundwater pumping. The theoretical feasibility of groundwater pumping is attractive for hydraulic control of COPC potentially migrating toward the bay and depression of the water table in the vicinity of potential preferential pathways. However, the thickness of the saturated zone within the artificial fill (generally <10 ft) limits the achievable drawdown of pumping wells and the ultimate effectiveness for hydraulic control. Groundwater pumping has limited shorthand long-term effectiveness in reducing COPC concentrations to PALs.

Implementability. The limited depth of groundwater pumping wells at approximately 15 ft maximum is favorable for installation. Potential fouling of groundwater pumping wells is a concern. The CEQA process would likely delay implementation of this alternative.

6.6.7 RECOMMENDED ALTERNATIVES

6.6.7.1 Soil and Soil Vapor Remediation

Vapor extraction is recommended for soil and soil vapor remediation. Vapor extraction is a mature and reliable technology that offers the best combination of short-term effectiveness and cost. Excavation is judged comparable in short-term effectiveness but is costly and difficult to implement compared to vapor extraction. Other alternatives are less costly than vapor extraction but do not offer the short-term effectiveness in mitigation of acute risks posed by flammable gases and VOCs in soil vapor.

The major limitation of vapor extraction is that remediation of nonvolatile COPC is not time expedient, as the biodegradation mechanism for COPC remediation is slow compared to the extraction process for VOCs. As remediation progresses and concentrations of VOCs and methane become significantly less than nonvolatile COPC, operation of vapor extraction can be pulsed (e.g., operation every other week) to provide oxygen for aerobic biodegradation in a more cost-effective manner. Depending on the length of time estimated to achieve PALs or final remedial goals for COPC, the vapor extraction system can be converted to a bioventing system to reduce operating costs further while maintaining remedial effectiveness.

6.6.7.2 Groundwater Remediation

Air sparging with concurrent vapor extraction is recommended for remediation of groundwater. Air sparging is a mature and reliable technology to remediate VOCs in groundwater. Air sparging also provides oxygen to support aerobic biodegradation of COPCs and accomplishes removal of volatile LPH where present on the water table. With implementation of vapor extraction for soil and soil vapor remediation, the incremental cost for air sparging is favorable. Air sparging shares the limitation of vapor extraction in the lack of expediency for remediation of nonvolatile COPCs. If acceleration of the remedial timeframe to achieve PALs or remedial goals for nonvolatile COPC is necessary, the air sparging system can be converted to air–ozone sparging. Air–ozone sparging is not the first choice for remediation of groundwater because the technology is relatively immature compared to air sparging and the reliability of ozone generating equipment is less certain.

6.6.8 REMEDIAL ACTION PLAN

6.6.8.1 Summary Overview

Historical investigations and the additional investigations performed in 2003 have characterized impacts to soil, soil vapor, and groundwater related to the Bulk Fuel Terminals. The investigations have identified a variety of COPC, including volatile petroleum hydrocarbon constituents related to gasoline and nonvolatile and relatively immobile petroleum hydrocarbon constituents related to diesel fuel and motor oil. Risk assessments have identified acute and chronic risks associated with the COPC, most notably related to the presence of the flammable gas methane and the toxic VOC benzene.

A set of PALs are proposed as remedial objectives to address the identified acute and chronic risks. Potential remedial actions were evaluated for effectiveness, implementability, and cost to achieve the PALs. Based on the evaluation of remedial alternatives, vapor extraction for remediation of soil and soil vapor and air sparging for remediation of groundwater are recommended. Vapor extraction and air sparging are not the least expensive remedial alternatives but appear to present the best combination of short-term effectiveness and cost. Passive skimming will be employed to mitigate the residual nonvolatile LPH observed in groundwater monitoring well MW-39.

The vapor extraction operations consist of monitoring and optimizing the hydrocarbon mass removal, managing supplemental hel usage, and tracking equipment operation and performing preventive maintenance. The vapor extraction remediation systems are designed to operate continuously.

The Treatment Cell 1 vapor extraction system will have one oxidizer with a nominal capacity of 500 scfm extracting from the 29 vapor extraction wells. The design extraction vacuum is 30 in. H_2O. The predicted vapor extraction well flow at 30 in. H_2O vacuum is approximately 15 scfm. The typical combined extraction flow rate for the 29 vapor extraction wells observed during pilot testing and interim remediation is approximately 430 s c h.

The conceptual vapor extraction design for Treatment Cells 2, 3, and 4 includes three oxidizers with a nominal capacity of 500 scfm each. At a design extraction vacuum of 30 in. H_2O a combined flow of approximately 1000 scfm would be anticipated for the 66 vapor extraction wells, suggesting two oxidizers would have sufficient capacity. The third oxidizer is included in the conceptual design to accommodate the greater concentrations of methane and VOCs anticipated at start-up. Long-term operation will likely require operation of two oxidizers for the vapor extraction wells installed in Treatment Cells 2, 3, and 4.

Air sparging operations include tracking equipment operation and performing preventive maintenance. The air sparging systems are designed to operate in a pulse (cycling) mode, with the anticipated operation to be a cycle of 1 hour on followed by 1 hour off.

6.6.9 LONG-TERM MONITORING AND ASSESSMENT OF REMEDIAL ACTIONS — REMEDIAL PROGRESS MONITORING

Assessment of soil vapor and groundwater sampling results and remedial system performance will occur on a semiannual or more frequent basis.

Assessment of vapor extraction performance will include review of cumulative mass extraction and mass extraction rates, and evaluation of soil vapor monitoring and sampling results. Methane and VOC concentrations of influent soil vapor to oxidizers will be measured as part of weekly operation monitoring and maintenance. Soil vapor samples will be obtained from accessible vapor extraction wells on a semiannual basis.

Assessment of air sparging performance will include evaluation of groundwater monitoring and sampling results for both COPC and geochemical indicator parameters. Groundwater monitoring and sampling will be performed quarterly through 2004 and semiannually thereafter. Groundwater samples will be collected following a low-flow purge/stabilization technique. Collected samples will be analyzed for dissolved oxygen and ferrous iron with a flow-through cell and a colorimetric test kit, respectively.

Groundwater samples will be analyzed for the following analytes by Sequoia Analytical:

- Total petroleum hydrocarbons as diesel and TPHo by EPA Method 8015M (with silica gel cleanup by EPA Method 3630)
- Total petroleum hydrocarbons as gasoline by EPA Method 8015M
- Benzene, toluene, ethylbenzene, xylenes, MTBE, oxygenates, and VOCs by EPA Method 8260B

6.6.10 Methane Rebound Testing and Evaluation

The remedial actions have the objective of mitigating the potential for long-term rebound of methane and other flammable gas concentrations in soil vapor. Testing for rebound of methane concentrations may occur when mass extraction decreases to asymptotic rates or when soil vapor sampling results indicate methane has been reduced to asymptotic concentrations.

Rebound testing will be performed by shutting the remedial systems down for a predetermined period and performing soil vapor sampling to assess increases in methane concentrations compared to concentrations before shutdown. If methane concentrations rebound during the shutdown period remedial systems will be restarted and operated until asymptotic conditions are again indicated, at which time the rebound testing will be repeated. If repeated methane rebound testing indicates consistent methane rebound, the site conditions will be further investigated to assess the continuing source of methane generation. The effectiveness of remedial actions will be evaluated to address identified sources of methane rebound.

6.6.11 Risk Assessment for Cessation of Active Remediation

Based on remedial progress monitoring, a risk assessment will be performed to support the decision to cease active remediation. Achievement or nonachievement of PALs will be reviewed in relation to the long-term remedial objectives. The risk assessment will consider both human health and ecological risks. Evaluation of methane rebound testing will be a critical component of the risk assessment.

Conclusion

After analyzing and generalizing the results of investigations carried out by many scientists from many countries, it can be concluded with certainty that at present the interface of two adjacent sciences have formed a new branch — *marine hydrogeology*.

Marine hydrogeology is the science of studying submarine groundwater, its properties, circulation, and distribution. The study subject of marine hydrogeology is subsurface water exchange between land and sea and the interaction processes of submarine waters with seawaters, water-bearing rocks, and the biosphere.

Marine hydrogeology as a science has its own object of study: submarine groundwater, which possesses a number of independent methods of investigations.

The investigations, the results of which are briefly discussed in this book, considerably widen the understanding of general water circulation. For the first time, quantitative estimates are obtained for groundwater discharge to seas and oceans and the Earth as a whole, as well as many large lakes. The role of groundwater in the water and salt balances of coastal zones and seas and oceans is shown. The basic regional regularities are established in the formation and distribution of submarine groundwater in different natural and anthropogenically disturbed conditions of coastal zones. It has been determined that though the role of groundwater discharge in the water balance of seas is relatively low (compared with river water discharge), groundwater inflow is one of the important factors of formation of the salt balance of seas, as well as of temperature and hydrobiological regimes in coastal regions.

The results presented indicate that in studying the present and expected water and salt balances of seas (especially inland ones) and large lakes, it is necessary to carry out, based on accurate hydrological and hydrogeological data, the quantitative assessment of subsurface water exchange between land and sea and to predict its changes under influences of natural factors and human activity in drainage areas.

In certain cases, quantitative assessment of groundwater discharge to seas can possibly reveal additional fresh groundwater resources for water supply and irrigation uses.

The basic tasks of further investigations within the problem under consideration can be formulated as follows:

Work out new methods and improve existing methods of quantitative assessment and prediction for changes in groundwater discharge to seas within different natural and anthropogenically disturbed conditions, including (in the first turn) the influence of climatic changes.

Create mathematical models of interaction between groundwater and seawaters under different geologic-hydrogeological conditions of coastal zones, as well as methods for prediction of seawater intrusion to aquifers in conditions of intensive groundwater extraction by coastal wells.

Assess the influence of groundwater on the formation of chemical compositions, including microcomponents and gaseous compositions of seawaters, as well as the thermal regime in near-bottom water layers and bottom sediments.

Study the role of groundwater discharge to seas and oceans in geological processes, foremost in the formation of mineral deposits.

Work out techniques to assess submarine groundwater safe yield and to determine prospects of its practical use.

References

Abraham, D.M., Charette, M.A., Allen, M.C., et al. 2003. Radiochemical estimates of submarine groundwater discharge to Waquoit Bay, Massachusetts. *Biological Bulletin* 205(2):246–247.

Acton Mickelson Environmental, Inc. 2003. Work Plan for Additional Soil and Ground Water Investigation, Former Mobil Bulk Fuel Terminal, Port of Oakland, Berth 23, 909 Ferry Street, Oakland, California, unpublished report prepared for ExxonMobil Refining and Supply Company, Global Remediation, and submitted to the California Regional Water Quality Control Board, San Francisco Bay Region.

Acton Mickelson Environmental, Inc. 2003. Work Plan for Rehabilitation and Sampling of Existing Ground Water Monitoring Wells, Former Mobil Bulk Fuel Terminal, Port of Oakland, Berth 24, 909 Ferry Street, Oakland, California, unpublished report prepared for ExxonMobil Refining and Supply Company, Global Remediation.

Acton Mickelson Environmental, Inc. 2003. Work Plan for Vapor Extraction and Air Sparging Pilot Testing/Interim Remediation, Port of Oakland, Oakland, California.

Acton Mickelson Environmental, Inc. 2003. Work Plan for Additional Soil and Ground Water Investigation at Treatment Cell #2, Former Mobil and Ashland Bulk Fuel Terminals, Port of Oakland, Oakland, California.

Acton Mickelson Environmental, Inc. 2003. Work Plan Addendum for Additional Investigation at Treatment Cell #2, Berth 23, Former Mobil and Ashland Bulk Fuel Terminals, Port of Oakland, Oakland, California.

Aeschbach-Hertig, W., Kipfer, R., Hofer, M., et al. 1996. Quantification of gas fluxes from the subcontinental mantle: The example of Laacher See, a maar lake in Germany. *Geochimica et Cosmochimica Acta* 60(1):31–41.

Afanasev, A.N and Didenko, A.A. 1976. Calculation of groundwater discharge to Baikal Lake. Proceedings of IV All-Union Hydrology Forum, VIII, Leningrad, Hydrometeoizdat, 190–195.

Agosta, K. 1985. The effect of tidally induced changes in creekbank water table on pore water chemistry. *Estuarine, Coastal and Shelf Science* 21(3):389–400.

AGPS. 1975. *Groundwater Resources of Australia.* Canberra: Australia Government Public Service, 143.

Akhmedsafin, U.M., Sydykov, Zh.S., and Dalian, I.B. 1961. Artesian water of the East Priaral region and conditions of its forming. *Izv. AS KazSSR. Ser. Geology*, 2, p. 86–95.

Akhmedsafin, U.M. and Shapiro, S.M. 1970. Groundwater discharge to Balkhash Lake. *Vestnik AN KazSSR*, 5(30):44–53.

Alekin, O.A. and Brazhnikova, L.V. 1964. *Discharge of Dissolved Solids from the Territory of the USSR.* Moscow, Nauka, 144.

Alekin, O.A. and Pyakhin, Yu.I. 1984. *Chemistry of Ocean.* Leningrad: Hydrometeoizdat, 343.

Alisto Engineering Group. 1993. Supplemental Site Investigation Report, Mobil Oil Corporation, Former Bulk Terminal, Port of Oakland, Oakland, California.

Alisto Engineering Group. 1993. Groundwater Monitoring and Sampling Report, Former Mobil Oil Bulk Terminal, 909 Ferry Street, Oakland, California. November 1993.

Alisto Engineering Group. 1994. Groundwater Monitoring and Sampling Report, Former Mobil Oil Bulk Terminal, 909 Ferry Street, Oakland, California. November 1994.

Alisto Engineering Group. 1995. Groundwater Monitoring and Sampling Report, Former Mobil Oil Bulk Terminal, 909 Ferry Street, Oakland, California. April 1995.

Alisto Engineering Group. 1995. Groundwater Monitoring and Sampling Report, Former Mobil Oil Bulk Terminal, 909 Ferry Street, Oakland, California. November 1995.

Alisto Engineering Group. 1996. Groundwater Monitoring and Sampling Report, Former Mobil Oil Bulk Terminal, 909 Ferry Street, Oakland, California. February 1996.

Alisto Engineering Group. 1996. Groundwater Monitoring and Sampling Report, Former Mobil Oil Bulk Terminal, 909 Ferry Street, Oakland, California. October 1996.

Alisto Engineering Group. 1997. Groundwater Monitoring and Sampling Report, Former Mobil Oil Bulk Terminal, 909 Ferry Street, Oakland, California. June 1997.

Alisto Engineering Group. 2000. Remedial Investigation and Methane Hazard Evaluation Report, Former Mobil Terminal Facility, Oakland, California.

Aller, R. 1980. Quantifying solute distributions in the bioturbated zone of marine sediments by defining an average micro-environment. *Geochimica et Cosmochimica Acta* 44(12):1955–1965.

Aller, R. 2001. Transport and reactions in the bioirrigated zone. In: *The Benthic Boundary Layer Transport Processes and Biogeochemistry.* B.P. Boudreau and B.B. Jorgensen, Eds. New York: Oxford University Press, 269–301.

Aller, R. and Aller, J. 1992. Meiofauna and solute transport in marine muds. *Limnology and Oceanography* 37(5):1018–1033.

Anderson, M.P. and Woessner, W.W. 1992. *Applied Groundwater Modeling, Simulation of Flow and Advective Transport.* San Diego, CA: Academic Press, 381.

Anisimova, N.P. 1981. *Cryohydrochemical Features of the Permafrost Zone.* Novosibirsk: Nauka, 153.

Anthoney, H. 1998. Combining a System Dynamics Model with a GIS: the Mono Lake Basin. USGS Water Resource Investigations. Washington, DC.

ArcNews. 2003. Monitoring port of Rotterdam with GIS. *ArcNews.* 25(3):1. Redlands, CA: Environmental Systems Research Institute.

Aronis, G.A. 1963. Special case of karst hydrology. *AIH Memories* 4:61–64.

Atlas of Kazakhskaya SSR. 1982. 1. Natural Conditions and Resources. Moscow.

AWRC. 1975. Australian Water Resources Council. *Review of Australia's Water Resources, 1975.* Canberra: Australian Government Public Service, 170.

Baedecker, M.J., Cozzarelli, I.M., and Eganhouse, R.P. 1993. Crude oil in a shallow sand and gravel aquifer—III. Biogeochemical reactions and mass balance modeling in anoxic groundwater. *Applied Geochemistry* 6:569–586.

Babinets, A.E., Mitropol'skiy, A.Y., and Ol'shanskiy, S.P. 1973. *Hydrogeological and Hydrochemical Peculiarities of Deep-sea Sediments of the Black Sea.* Kiev. 159.

Bakker, M. 2003. A Dupuit formulation for modeling seawater intrusion in regional aquifer systems. *Water Resources Research* 39(5):1131.

Bakker, M., Oude Essink, G.H.P., and Langevin, C.D. 2004. The rotating movement of three immiscible fluids—a benchmark problem: *Journal of Hydrology* 287(1–4): 270–278.

Bar, Ya., Zaslavski, D., and Irmey, S. 1971. *Physical-Mathematical Principles of Water Filtration.* Moscow: Mir, 451

Barnett, W.C. 1999. Offshore springs and seeps are focus of working group. EOS, 80, 13–15. Baseline. 2002. Draft Remedial Investigation Report, Former McGuire Chemical Company Leasehold, Port of Oakland, Outer Harbor Terminal, Oakland, California, prepared for the Port of Oakland.

Baydon-Ghyben, W. 1888–1889. Nota in verband met de voorgenomen putboring nabij Amsterdam. Koninklyk Instituut Ingenieurs Tijdschrift (The Hague), 8–22.

de Beer, D., Wenxhofer, F., Ferdelman, T., Boehme, S., et al. 2005. Transport and mineralization rates in North Sea sandy intertidal sediments, Sylt-Romo Basin, Wadden Sea. *Limnology and Oceanography* 50(1):113–127.

Bear, J., Cheng, A., Sorek, S., et al. (Eds.) 1999. *Seawater Intrusion in Coastal Aquifers: Concepts, Methods and Practices*. New York: Springer-Verlag.

Beideman, I.N. 1983. *Manual on Water Use by Vegetation in the Natural Zones of the USSR (Geobotanic and Ecological Characteristics)*. Novosibirsk: Nauka.

Belanger, T.V. and Mikutel, D.F. 1985. On the use of seepage meters to estimate groundwater nutrient loading to lakes. *Water Resources Bulletin* 21:265–272.

Belanger, T.V. and Montgomery, M.T. 1992. Seepage meter errors. *Limnology and Oceanography* 37(8):1787–1795.

Belanger, T.V. and Walker, R.B. 1990. Ground water seepage in the Indian River Lagoon, Florida. In: *International Symposium on Tropical Hydrology and Fourth Caribbean Islands Water Resources Congress*. J.H. Krishna, V. Quinones-Aponte, F. Gomez-Gomez, and G.L. Morris, Eds. American Water Resources Association, San Juan, Puerto Rico, 367–375.

Belzile, N., Chen, Y.W., and Xu, R., 2000. Early diagenetic behavior of selenium in freshwater sediments. *Applied Geochemistry* 15:1439–1454.

Berelson, W., Hammond, D.E., and Fuller, C. 1982. Radon-222 as a tracer for mixing in the water column and benthic exchange in the southern California borderland. *Earth and Planetary Science Letters* 61(1):41–54.

Bergelson, G.M., Drushchits, V.L., Kuznetsov, D.V., and Meskheteli, A.V. 1986. Study of groundwater flow to the lake Issyk-Kul. *Geologiya Morey I Okeanov* 2.

Beyer, R., Schlosser, P., Bönisch, G., et al. 1989. Performance and blank components of a mass spectrometric system for routine measurements of helium isotopes and tritium by the 3He in-growth method. In: Sitzungsberichte der Heidelberer Akademie der Wissenschaften. *Mathematisch-nauturwissenschaftliche Klasse* 5:241–279.

Bhadha, J., Martin, J., Jaeger, J., Lindenberg, M., and Cable, J. 2007. Re-circulation of shallow lagoon water and its significance on chemical fluxes in the Banana River Lagoon, Florida, *Journal of Coastal Research* (in press).

Bokuniewicz, H. 1980. Groundwater seepage into Great South Bay, New York. *Estuarine and Coastal Marine Science* 10:437–444.

Bokuniewicz, H. 1992. Analytical descriptions of subaqueous groundwater seepage. *Estuaries* 15(4):458–464.

Bokunievicz, H. and Buddemeier, R.W. 2002. IHP-OHP, Koblenz, Germany.

Borisenko, I.M. and Borisenko, L.V. 1981. Mineral water discharge to Baikal Lake [Abstract]. The V All-Union Meeting of Limnologists, Irkutsk, V: 274–279.

Borodulina, G.S. and Perskiy, N.E. 1998. Engineering and ecological problems of groundwater utilization for water supply. *Russian Journal of Engineering Ecology* [Russian] 5:57–62.

Bottomley, D.J., Ross, J.D., and Clarke, W.B. 1984. Helium and neon isotope geochemistry of some ground waters from the Canadian Precambrian shield. *Geochimica et Cosmochimica Acta* 48(10):1973–1985.

Boudreau, B. 1997. *Diagenetic Models and Their Implementation Modelling Transport and Reactions in Aquatic Sediments*. Berlin: Springer-Verlag.

Boudreau, B. and Jorgensen, B. 2001. *The Benthic Boundary Layer: Transport Processes and Biogeochemistry*. New York: Oxford University Press, 404.

Bowell, R.J., 1994. Sorption of arsenic by iron oxides and oxyhydroxides in soils. *Applied Geochemistry* 9:279–286.

Brookins, D.G. 1988. *Eh-pH Diagrams for Geochemistry*. New York: Springer-Verlag.

Brown, R.H. and Parker, G.G. 1945. Salt water encroachment in limestone at Silver Bluff Miami, Florida. *Economic Geology* 40(4):235–262.

Brusilovskiy, S.A. 1971. On possibility of assessment of submarine discharge by its geo-chemical looses. *Complex Investigations of the Caspian Sea*, 2, 68–74.

Buachidze, G. 1977. The Problem of the Black Sea Hollow Origin. *Geotectonica*. 2 (in Russian).

Buachidze, G. 2002. SGD in Kolkhida and universal criterion IHP-OHP, Koblenz, Germany.

Buachidze, G. and Buachidze, I. 1972. *Proc. Acad. Sci. of Georgia*, Tbilisi (in Georgian).

Buachidze, G. and Meskheteli, A. 1996. Different Types of Submarine Discharge from the Georgian Coast to the Black Sea. *LOICZ Report*, 8 (in English).

Buachidze, I. and Meliva, A. 1972. Discharge of Ground Waters in Black Sea in Gagra. *Proc. Acad. Sci. Res. Lab.*, Tbilisi (in Russian).

Buachidze, I.M. and Meliva, A.M. 1967. To the groundwater discharge into the Black Sea in the Gagry region. Proceedings of the Scientific-Research Lab of Hydrogeological and Engineering Problems of Georgia Institute. 3:17–24.

Buachidze, G., Buachidze, I., and Zverev, V. 1976. Underground Chemical Outflow on West Georgia. *Proc. Acad. Sci. of Georgia*, Tbilisi (in Georgian).

Bugna, G., Chanton, J., Cable, J., et al. 1996. The importance of groundwater discharge to the methane budgets of nearshore and continental shelf waters of the northeastern Gulf of Mexico. *Geochimica et Cosmochimica Acta* 60(23):4735–4746.

Burnett, W., Bokuniewicz, H., Huettel, M., et al. 2003a. Groundwater and pore water inputs to the coastal zone. *Biogeochemistry* 66(1–2):3–33.

Burnett, W., Chanton, J., Christoff, J., et al. 2002. Assessing methodologies for measuring groundwater discharge to the ocean. *EOS: Transactions. American Geophysical Union* 83:117–123.

Burnett, W.C. 1999. Offshore springs and seeps are focus of working group. *EOS: Transactions. American Geophysical Union* 80:13–15.

Burnett, W.C. and Dulaiova, H. 2003. Estimating the dynamics of groundwater input into the coastal zone via continuous radon-222 measurements. *Journal of Environmental Radioactivity* 69:21–35.

Burnett, W.C., Chanton, J.P. and Kontar, E. 2003b. Special issue: Submarine groundwater discharge. *Biogeochemistry,* 66(1–2):202.

Burroughs, P. 1986. *Geographic Information Systems for Land Resource Assessment*. London: Oxford University Press.

Bussmann, I., Dando, P.R., Niven, S.J., and Suess, E. 1999. Groundwater seepage in the marine environment: Role for mass flux and bacterial activity. *Marine Ecology Progress* Series 178:169–177.

Cable, J., Bugna, G., Burnett, W., and Chanton, J. 1996a. Application of 222Rn and CH4 for assessment of ground water discharge to the coastal ocean. *Limnology and Oceanography* 41(6):1347–1353.

Cable, J.E., Burnett, W.C. Chanton, J.P., and Weatherly, G.L. 1996b. Estimating groundwater discharge into the northeastern Gulf of Mexico using radon-222. *Earth and Planetary Science Letters* 144(3–4):591–604.

Cable, J.E., Burnett, W.C., Chanton, J.P., et al. 1997a. Field evaluation of seepage meters in the coastal marine environment. *Estuarine Coastal and Shelf Science* 45(3):367–375.

Cable, J.E., Burnett, W.C. and Chanton, J.P. 1997b. Magnitude and variations of groundwater seepage along a Florida marine shoreline. *Biogeochemistry* 38(2):189–205.

Cable, J.E., Martin, J.B., Swarzenski, P.W., et al. 2004. Advection within shallow pore waters of a coastal lagoon, Florida. *Ground Water* 42(7):1011–1020.

California Regional Water Quality Control Board, San Francisco Bay Region, Groundwater Committee. 1999. East Bay Plain Groundwater Basin Beneficial Use Evaluation Report, Alameda and Contra Costa Counties. June 1999.

California Regional Water Quality Control Board, San Francisco Bay Region. 1999. Order No. 99-055, Waste Discharge Requirements for Port of Oakland, Berth 55–58 Project, Oakland, Alameda County.

California Regional Water Quality Control Board, San Francisco Bay Region. 1999. Order No. 99-063, Adoption of Site Cleanup Requirements for: Mobil Oil Corporation and Port of Oakland for the Property located at the Former Mobil Bulk Fuel Terminal at the Port of Oakland, Alameda County. July 1999.

California Regional Water Quality Control Board, San Francisco Bay Region. 1999. Tentative Order - Adoption of Revised Site Cleanup Requirements and Rescission of Order Nos. 95-1 36, 95-1 52, 92-1 40 for the Property at San Francisco International Airport, San Mateo County.

California Regional Water Quality Control Board, San Francisco Bay Region. 2003. Screening for Environmental Concerns at Sites with Contaminated Soil and Ground Water.

Capone, D. and Bautista, M. 1985. A groundwater source of nitrate in nearshore marine sediments. *Nature* 313(5999):214–216.

Capone, D. and Slater, J. 1990. Interannual patterns of water table height and groundwater derived nitrate in nearshore sediments. *Biogeochemistry* 10:277–288.

Cecil, L.D. and Green, J.R. 2000. Randon-222. In: *Environmental Tracers in Subsurface Hydrology*. P. Cook and A.L. Herczeg, Eds., Boston: Kluwer, 175–194.

Chan, S.Y. and Mohsen, M.F.N. 1992. Simulation of tidal effects on contaminant transport in porous media. *Ground Water* 30(1):78–86.

Chanton, J.P. and Burnett, W.C. 1995. Delivery of nutrients to Florida Bay via submarine groundwater discharge. *Florida Department of Environmental Protection Final Report* 66.

Chanton, J.P., Burnett, W.C., Dulaiova, H., et al. 2003. Seepage rate variability in Florida Bay driven by Atlantic tidal height. *Biogeochemistry* 66(1–2):187–202.

Chanton, J.P., Burnett, W.C., Young, J., and Bugna, G. 1995. Submarine ground water discharge. *Proceedings of the Florida Bay Science Conference* 41–43.

Charette, M. and Buesseler, K. 2004. Submarine groundwater discharge of nutrients and copper to an urban subestuary of Chesapeake Bay (Elizabeth River). *Limnology and Oceanography* 49(2):376–385

Charette, M. and Mulligan, A. 2004. Water flowing underground. *Oceanus Magazine* 43(1):29–33.

Charette, M. and Sholkovitz, E. 2002. Oxidative precipitation of groundwater-derived ferrous iron in the subterranean estuary of a coastal bay. *Geophysical Research Letters* 29(10):1444, DOI 10.1029/2001GL014512.

Charette, M. and Sholkovitz, E. 2004. Development of an automated chemical sampler/analyzer for submarine groundwater discharge in estuaries. NOAA/UNH Cooperative Institute for Coastal and Estuarine Environmental Technology, NOAA Grant Number(s) NA07OR0351, NA17OZ2507 Final Report.

Charette, M., Sholkovitz, E. and Hansel, C. 2005. Trace element cycling in a subterranean estuary: Part 1. Geochemistry of the permeable sediments. *Geochimica et Cosmochimica Acta* 69(8):2095–2109.

Charette, M.A., Splivallo, R., Herbold, C., et al. 2003. Salt marsh submarine groundwater discharge as traced by radium isotopes. *Marine Chemistry* 84(1–2):113–121.

Cheng, A.H-D. and Ouazar, D. (Eds.) 2004. *Coastal Aquifer Management — Monitoring, Modeling, and Case Studies*. Boca Raton, FL: CRC Press.

Cherkauer, D. and Hensel, B. 1986. Ground Water Flow into Lake Michigan from Wisconsin. *Journal of Hydrology* 84:261–271.

Cherkauer, D.S. and Nader, D.C. 1989. Distribution of groundwater seepage to large surface-water bodies—the effect of hydraulic heterogeneities. *Journal of Hydrology* 109(1–2):151–165.

Chernenko, I.M. 1965. Groundwater discharge in the Priaral region and dependence of its level from the Aral Sea level. *Izv. Dnepropenrovsk Institute of Mines*, issue. 46, 318–320.

Chernenko, I.M. 1970. Groundwater inflow into the Aral Sea and its role in the solving of Aral problems. *Problems of the Desert Development*, 4:28–38.

Chernov, V. (Ed.). 2000. Water of Russia. In: *Water-Resource Potential*. Ekaterinaburg: Akva-Press.

Chichasov, G.N. (Ed.). 1990. *Hydrometeorological Problems of the Priaral Region*. Leningrad: Gidrometeorizdat, 278.

Christophorou, L.G., Olthoff, J.K. and Van Brunt, R.J. 1997. Sulfur hexafluoride and the electric power industry. *IEEE Electrical Insulation Magazine* 13(5):20–23.

Church, T.M. 1996. An underground route for the water cycle. *Nature* 380(575):579–580.

Clark, J.F. and Hudson, G.B. 2001. Quantifying the flux of hydrothermal fluids into Mono Lake by use of helium isotopes. *Limnology and Oceanography* 46(1):189–196.

Clark, J.F., Hudson, G.B., Davisson, M.L., et al. 2004. Geochemical imaging of flow near an artificial recharge facility, Orange County, CA. *Ground Water* 42(2):167–174.

Clark, J.F., Stute, M., Schlosser, P., et al. 1997. A tracer study of the Floridan aquifer in southeastern Georgia: implications for groundwater flow and paleoclimate. *Water Resources Research* 33(2):281–289.

Cochran, J.K. 1992. The oceanic chemistry of uranium- and thorium-series nuclides. In: *Uranium-Series Disequilibrium: Applications to Earth, Marine and Environmental Sciences*. M. Ivanovich and R.S. Harmon, Eds. Oxford: Clarendon Press, 334–395.

Collins, M.A. and Gelhar, L.W. 1971. Sea water intrusion in layered aquifers. *Water Resources Research* 7(4):971–979.

Connor, J.N. and Belanger, T.V. 1981. Groundwater seepage in Lake Washington and the Upper St. Johns River Basin, Florida. *Water Resources Bulletin* 17(5):799–805.

Cook, P.G., Favreau, G., Dighton, J.C. and Tickell, S. 2003. Determining natural groundwater influx to a tropical river using radon, chlorofluorocarbons and ionic environmental tracers. *Journal of Hydrology* 277(1–2):74–88.

Cooper, M., Crenshaw, L., Penman, T., and Saade, E. 2002. Application of GIS in the search for the German U-559 submarine. In: *Undersea with GIS*. Dawn Wright, Ed. Redlands, CA: ESRI Press.

Corbett, D.R. and Cable, J.E. 2003. Seepage meters and advective transport in coastal environments: comments on seepage meters and Bernoulli's revenge by Shinn, Reich, and Hickey. *Estuaries* 26(5):1383–1389.

Corbett, D.R., Chanton, J.P., Burnett, W.C., et al. 1999. Patterns of groundwater discharge into Florida Bay. *Limnology and Oceanography* 44(4):1045–1055.

Corbett, D.R., Burnett, W., Cable, P.H., and Clark, S.B. 1997. Radon tracing of groundwater input into Par Pond, Savannah River Site. *Journal of Hydrology* 203(1–4):209–227.

Corbett, R., Cable, J., and Martin, J. 2005. Direct measurements of submarine ground water discharge using seepage meters. In *Submarine Groundwater*, Zekster, I.S., Dzhamalov, R.G., and Everett, L.G., Eds. (in press).

Corbett, D.R., Dillon, K., Burnett, W., and Chanton, J. 2000a. Estimating the groundwater contribution into Florida Bay via natural tracers 222Rn and CH4. *Limnology and Oceanography* 45(7):1546–1557.

Corbett, D.R., Dillon, K., and Burnett, W. 2000b. Tracing groundwater flow on a barrier island in the north-east Gulf of Mexico. *Estuarine, Coastal and Shelf Science* 51(2):227–242.

Corbett, R., Chanton, J., Burnett, W., Dillon, K., Rutkowski, C. and Fourqrean, J. 1999. Patterns of Groundwater Discharge into Florida Bay. *Limnology and Oceanography* 44(4):973–1185.

Corbett, R., Dillon, K., Burnett, W., and Schaefer, G. 2002. The spatial variability of nitrogen and phosphorus concentration in a sand aquifer influenced by onsite sewage treatment and disposal systems: a case study on St. George Island, Florida. *Environmental Pollution* 117(2):337–345.

Corbett, R., Kump, L., Dillon, K., Burnett, W., and Chanton, J. 2000. Fate of wastewater-borne nutrients under low discharge conditions in the subsurface of the Florida Keys, USA. *Marine Chemistry* 69(1–2):99–115.

Corliss, J.B., Baross, J.A., and Hoffman, S.E. 1981. An hypothesis concerning the relationship between submarine hot springs and the origin of life on Earth. *Oceanologica Acta* NoSP: 59–69.

Correl, D.L., Jordan, T.E., and Weller, D.E. 1992. Nutrient flux in a landscape: effects of coastal land use and terrestrial community mosaic on nutrient transport to coastal waters. *Estuaries* 15(4): 431–442.

COSODII. 1987. Fluid circulation in the crust and the global geochemical budget. Conference on Scientific Ocean Drilling, Strasbourg, France.

Custodio, E., Bayo, A., and Batista, E. 1977. Sea water encroachment in Catalonian coastal aquifers. *AIH Memories* 13:F1–F14.

Demeritt, D. 2001. The construction of global warming and the politics of Science. *Annals of the Association of American Geographers* 91(2):307–337.

De Wist, R. 1969. *Hydrogeology with Principles of Land Hydrology.* Moscow: Mir, 312.

Destouni, G. and Prieto, C. 2003. On the possibility for generic modeling of submarine groundwater discharge: *Biogeochemistry* 66(1–2):171–186.

Dhzamalov, R.G., Zektser, I.S., and Meskheteli, A.V. 1977. *Groundwater Discharge to Seas and The World Ocean.* Moscow: Nauka, 94.

Diersch, H.G. 1996. Interactive, graphics-based finite element simulation system FEFLOW for modeling groundwater flow, contaminant mass and heat transport. Berlin, Germany: WASY Institute for Water Resource Planning and System Research Ltd.

Diersch, H.-J. 1988. Finite element modelling of recirculating density-driven saltwater intrusion processes in groundwater. *Advances in Water Res.* 11(1):25–43.

Dillon, K.S., Corbett, D.R., Chanton, J.P., et al. 1999. The use of sulfur hexafluoride (SF6) as a tracer of septic tank effluent in the Florida Keys. *Journal of Hydrology* 220(3–4):129–140.

Dillon, K.S., Corbett, D.R., Chanton, J.P., et al. 2000. Bimodal transport of a wastewater plume injected into saline ground water of the Florida Keys. *Ground Water* 38(4):624–634.

Dobrentsov, N.L. and Kirdyashkin, A.G. 1994. Abyssal Geodynamic. Novosibirsk, SO RAS Publ.

Drever, J.I., 1997. *The Geochemistry of Natural Waters: Surface and Groundwater Environments,* 3rd ed. Upper Saddle River, NJ: Prentice Hall.

Dyunin, V.I. 2000. *Hydrodynamics of Deep Layers of Oil- and Gas-bearing Basins.* Moscow: Nauchny Mir, 471.

Dzhamalov, R.G. 1973. *Groundwater Discharge in the Tersko-Kumskiy Artesian Basin.* Moscow, Nauka, 95.

Dzhamalov, R.G., Zektser, I.S., and Obyedkov, Yu.L. 1977. Interpretation of aero- and cosmic images at studying groundwater flow. *J. Razvedka i Okhrana nedr.* (1):42–52.

Dzhamalov, R.G., Zektser, I.S., and Ivanov, V.A. 1978. Studying groundwater from the space. *Zemlya i Vselennaya* (1):120–132.

Dzhamalov, R.G., Obyedkov, Yu.L., and Safronova, T.I. 1979. Peculiarities in using space images for hydrogeological investigations. *Water Resources* (1):120–132.

Dzhamalov, R.G. 1996. A conceptual model of subsurface water exchange between the continent and the sea. *Water Resources* 23(2):124–128.

Dzhamalov, R.G. and Safronova, T.I. 1999. Influence of submarine sedimentation waters on water and salt balance of oceans. *Water Resources* 26(6):722.

Electric Power Research Institute. 1984. *Chemical Attenuation Rates, Coefficients, and Constants in Leachate Migration, Volume 1: A Critical Review.* EPRI EA-3356. Palo Alto, CA: Electric Power Research Institute.

Ellins, K.K., Roman-Mas, A., and Lee, R. 1990. Using ^{222}Rn to examine groundwater/surface discharge interactions in the Rio Grande De Manati, Puerto Rico. *Journal of Hydrology* 115:319–341.

Emerson, S., Jahnke, R., and Heggie, D. 1984. Sediment-water exchange in shallow water estuarine sediments. *Journal of Marine Research* 42:709–730.

Enright, R.V. 1990. Relating the effects of oceanic tidal loading of a confined aquifer in Sarasota County, Florida, to fluctuations in well-water levels. M. Sc. thesis, Florida State University.

EPA. 1992. Guidelines for Use of the DRASTIC Method in Aquifer Vulnerability Assessment. Washington DC: U.S. Environmental Protection Agency.

EPA. 1999. Salt water intrusion barrier wells. U.S. Environmental Protection Agency Technical Report, Office of Drinking Water. Washington DC: U.S. Environmental Protection Agency.

ERM-West, Inc., 1994. Progress Report Former Ashland Oil Site, unpublished report prepared for the Port.

EPA. United States Environmental Protection Agency. 1993. Presumptive Remedies: Site Characterization and Technology Selection for CERCLA Sites with Volatile Organic Compounds in Soils. Quick Reference Fact Sheet, EPA 540-F-93-048.

EPA. United States Environmental Protection Agency. 1994. Test Methods for Evaluating Soil Waste Physical/Chemical Method. (SW-846), Update 111, Office of Solid Waste.

EPA. United States Environmental Protection Agency. 1998. Guidance for Conducting Remedial Investigations and Feasibility Studies under CERCLA, Interim Final. Office of Solid Waste and Emergency Response, EPA-540/R-941012.

Essaid, H.I. 1990. The computer model SHARP, a quasi-three-dimensional finite-difference model to simulate freshwater and saltwater flow in layered coastal aquifer systems. *U.S. Geological Survey Water-Resources Investigations Report* 90-4131: 181.

Fanning, K.A., Byrne, R.H., Breland, J.A. and Betzer, P.R. 1981. Geothermal springs of the west Florida continental shelf: evidence for dolomitization and radionuclide enrichment. *Earth and Planetary Sciences Letters*, 52(2):345–354.

Feldman, G.M. 1984. *Thermokarst and Permafrost.* Novosibirsk: Nauka, 261.

Fellows, C.R. and Brezonik, P.I. 1980. Seepage flow into Florida lakes. *Water Resources Bulletin* 16(4):635–641.

Ferris, J.G. 1951. Cyclic fluctuation of water level as a basis for determining aquifer transmissibility. *International Association of Scientific Hydrology* 33:148–155.

Ferronsky, V.I., Polyakov, V.A., Kuprin, P.N. and Lobov, A.L. 1999. Nature of water level variations in the Caspian Sea (by results of studying bottom sediments). *Water Resources* 26(6):652–666.

Fetter, C. 1994. *Applied Hydrogeology.* Prentice Hall, Upper Saddle River, NJ.

Fetter, C. 2001. *Applied hydrogeology.* Upper Saddle River, NJ: Prentice-Hall.

Filatov, N.N. and Litvinenko, A.V. (Eds.). 2001. *Index of Lakes and Rivers of Karelia* [Russian]. Petrozavodsk: Karelia Publishers, 290.

Fish, J.E. and Stewart, M.T. 1991. Hydrogeology of the surficial aquifer system, Dade County, Florida. *U.S. Geological Survey Water-Resources Investigations Report* 90-4108.

Fisher, T., Carlson, P., and Barber, R. 1982. Sediment nutrient regeneration in three North Carolina estuaries, *Estuarine, Coastal and Shelf Science* 14:101–116.

Flipse, W., Katz, B., Lindner, J., and Markel, R. 1984. Sources of nitrate in ground water in a sewered housing development, central Long Island, New York, *Ground Water* 22(4):418–426.

Foresman, T. 1993. *A History of Geographic Information Systems*. London: Taylor & Francis.

Forest, C.E., Stone, H.P., Sokolov, A.P., Allen, M.R., and Webster, M.D. 2002. Quantifying uncertainties in climate system properties with the use of recent climate observations. *Science* 295:113–116.

Freeze, A.R. and Cherry, J.A. 1979. *Groundwater.* Englewood Cliffs, NJ: Prentice Hall, 604.

Frimpter, M. and Gay, F. 1979. Chemical quality of groundwater of Cape Cod, Massachusetts. U.S. Geological Survey Water Resources Investigation. Boston, MA: U.S. Department of the Interior.

Frolov, A.P. 1991. Intrusion of sea water into fresh water nonartesian strata. *Water Resources* 18(4):364–370.

Frolov, A.P. and Yumashev, I.O. 1998. Interaction of fresh groundwater on land with sea water on the Baltic Sea coast. *Water Resources* [Russian]. 1:1–3.

Fu, J.-M. and Winchester, J. 1994. Sources of nitrogen in three watersheds of northern Florida, USA: Mainly atmospheric deposition, *Geochimica et Cosmochimica Acta* 58(6):1581–1590.

Fukuo, Y. and Kaihotsu, I. 1988. A theoretical analysis of seepage flow of the confined groundwater into the lake bottom with a gentle slope. *Water Resources Research* 24:1949–1953.

Furukawa, Y., Bentley, S., and Lavoie, D. 2001. Bioirrigation modeling in experimental benthic mesocosms. *Journal of Marine Research* 59:417–452.

Furukawa, Y., Bentley, S., Shiller, A., et al. 2000. The role of biologically-enhanced pore water transport in early diagenesis: An example from carbonate sediments in the vicinity of North Key Harbor, Dry Tortugas National Park, Florida. *Journal of Marine Research* 58:493–522.

Gaines, A., Giblin, A., and Mlodzinska-Kijowski, Z. 1983. Freshwater discharge and nitrate input into Town Cove. In: *The Coastal Impact of Groundwater Discharge: An Assessment of Anthropogenic Nitrogen Loading in Town Cove, Orleans, MA*. Teal, J.M. Ed. Woods Hole, MA: Woods Hole Oceanographic Institution, 13–37.

Gallagher, D., Dietrich, A., Reay, W., et al. 1996. Ground water discharge of agricultural pesticides and nutrients to estuarine surface water. *Ground Water, Ground Water Monitoring and Remediation* (Winter):118–129.

Gamlin, J.D., Clark, J.F., Woodside, G., and Herndon, R. 2001. Large-scale tracing of ground water with sulfur hexafluoride. *Journal of Environmental Engineering ASCE* 121:171–174.

Garrels, R.M. and Mackenzie, F.T. 1971. *Evolution of Sedimentary Rocks.* New York: Norton, 397.

Gascoyne, M. 1977. Hydrogeology and solution chemistry of north Venezuelan karst. *AIH Memoirs* 12:553–566.

Gavich, I.K. 1980. *Theory and Practice of Modeling in Hydrogeology.* Moscow: Nedra, 385.

Gatalsky, M.A. 1954. Underground water and gases of the Northern part of Russian platform. *Proc. VNIIGRI* 9:28–41.

Geological Survey Water-Resources Investigation Report 98-4005, 90.

Geraghty & Miller, Inc. 1998. Results of Surface and Subsurface Vapor Characterization Activities, prepared for the Port of Oakland.

Geraghity, J.J., Miller, D.M., Van der Leeden, and Troise, F.L., Eds. 1973. *Water Atlas of the United States*. Washington, DC: Public Water Information Center, 200.

Giblin, A. and Gaines, A. 1990. Nitrogen inputs to a marine embayment: the importance of groundwater. *Biogeochemistry* 10:309–328.

Gieskes, J.M. and Lawrence, J.R. 1981. Geochemical significance of geagenetic reaction in ocean sediments: an evaluation of interstitial water data. *Oceanologica Acta* NSP:111–113.

Gilboa, Y. 1971. Replenishment sources of the Peruvian coast. *Ground Water* 9(4):39–46.

Gilliam, J., Daniels, R., and Lutz, J. 1974. Nitrogen content of shallow ground water in the Northern Carolina coastal plain. *Journal of Environmental Quality* 3(2):147–151.

Giorgi, F., Marinucci, M.R., and Bates, G.T. 1993a. Development of a second-generation regional climate model (RegCM2). Part I: Boundary layer and radioactive transfer processes. *Monthly Weather Review* 121:2794–2816.

Giorgi, F., Marinucci, M,R., and Bates, G.T. 1993b. Development of a second-generation regional climate model (RegCM2). Part II: Convective processes and lateral boundary conditions. *Monthly Weather Review* 121:2814–2832.

Giorgi, F., Shields-Brodeur, C., and Bates, G.T. 1994. Regional climate change scenarios over the United States produced with a nested regional climate model. *Journal of Climate* 7:375–399.

Glazovsky, N.F., Batoyan, V.V., and Brusilovsky, S.A. 1976. Integrated investigations of the Caspian Sea. Moscow: MSU, 5:189.

Gleick, P. 1989. Climate change, hydrology, and water resources. *Reviews of Geophysics* 27(3):329–344.

Goldberg, V.M. 1982. Intrusion of sea water into fresh groundwater aquifers. In: *Hydrogeological Investigations Abroad*. Moscow: Nedra. 74–88.

Goldberg, V.M., Kovalevsky, Yu.V., Baranov, A.P., and Serebryakova, L.A. 1989. Analysis of dynamics of groundwater contamination area under influence of protective well field. *Vodnye Resursy* 3, 58–63 (in Russian).

Goldina, L.P. 1972. *Geography of Lakes in the Bolshezemelskaya undra*. Leningrad: Nauka, 102.

Gordeyev, V.V. 1983. *River Runoff to Ocean and Its Geochemical Features*. Moscow: Nauka, 160.

Gordon, D.C., Boudreau, P.R., Mann, K.M., et al. 1996. LOICZ geochemical modeling guidelines. *LOICZ Reports and Studies* 5. LOICZ Core Project, Netherlands Institute for Sea Research. The Netherlands: Texel.

Gradsteyn, I.S. and Ryzhik, I.M. 1994. *Table of Integrals, Series, and Products*, 5th ed., Academic Press, San Diego, CA.

Grady, S. and Mullaney, J. 1998. Natural and human factors affecting shallow water quality in surficial aquifers in the Connecticut, Housatonic, and Thames River basins. Washington, DC: U.S. Geological Survey.

Gregorauskas, M.M., Mokrik, R.V., and Iokshas, K.K. 1986. Hydrochemical aspects of studying fresh groundwater discharge into the Baltic Sea. *Water Resources* 13(4):313–322.

Gregorauskas, M.M., Mokrik, R.V., and Meyeris, S.A. 1987. On modeling of sea water intrusion into aquifers. *Water Resources* 14(4):315–323.

Grilikhes, M.C. and Filanovskiy, B.K. 1980. *Contact Conductometry.* Leningrad: Chemistry, 175.

Groffman, P., Howard, G., Gold, A., and Nelson, W. 1996. Microbial nitrate processing in shallow groundwater in a riparian forest. *Journal of Environmental Quality* 25:1309–1316.

Groundwater discharge in the coastal zone. *Proceedings of and International Symposium.* 1996. Buddemeierm, R.W. (Ed.). LOICZ Report 8, LOICZ, Texel, the Netherlands, 179.

Groundwater and salt discharge into the Aral Sea. — Alma-Ata: 1983. 160

GUGK, 1974. Map of groundwater flow of the USSR. Scale: 1: 2 500 000. Moscow, GUGK.

GUGK, 1983. Map of groundwater flow of Central and East Europe. Scale: 1:1 500 000. Moscow, GUGK.

Guo, W. and Langevin, C.D. 2002. User's guide to SEAWAT: A computer program for simulation of three-dimensional variable-density ground-water flow: U.S. Geological Survey Techniques of Water-Resources Investigations, Book 6, Chapter A7, 77. Washington, DC: U.S. Geological Survey.

Halbert, W.E. and Jenson, R.E. 1996. Influence of tidal fluctuations on coastal aquifers, general principals and case studies. Proceedings of the Tenth National Outdoor Action Conference and Exposition, May 13–15, Las Vegas, Nevada, National Ground Water Association.

Hallie, A. 1998. Combining a System Dynamics Model with a GIS: the Mono Lake Basin. U.S.G.S. Water Resource Investigations. Washington, DC.

Hammond, D.E., Simpson, H.J., and Mathieu, G.P. 1977. 222Rn distribution and transport across the sediment–water interface in the Hudson River estuary. *Journal of Geophysical Research* 82:3913–3920.

Hancock, G.J. and Martin, P. 1991. The determination of radium in environmental samples by alpha-particle spectrometry. *Applied Radiation and Isotopes* 42:63–69.

Hancock, G.J. and Murray, A.S. 1996. Sources and distribution of dissolved radium in the Bega River estuary, Southeastern Australia. *Earth and Planetary Science Letters* 138(1–4):145–155.

Hancock, P., Boulton, A., and Humpreys, W. 2005. Aquifers and hyporheic zones: towards an ecological understanding of groundwater. *Hydrogeology Journal* 13(1):98–111.

Hantush, M.S. 1968. Unsteady movement of fresh water in thick unconfined saline aquifers. *Bulletin of the International Association of Scientific Hydrology.* 13(2):40–60.

Harbaugh, A.W., Banta, E.R., Hill, M.C., and McDonald, M.G. 2000. MODFLOW-2000, the U.S. Geological Survey modular ground-water model—user guide to modularization concepts and the ground-water flow process: *U.S. Geological Survey Open-File Report* 00-92, 121.

Hartman, B. and Hammond, D.E. 1984. Gas exchange rates across the sediment-water and air-water interfaces in South San Francisco Bay. *Journal of Geophysical Research* 89:3593–3603.

Harvey, J. and Odum, W. 1990. The influence of tidal marshes on upland groundwater discharge to estuaries. *Biogeochemistry* 10:217–236.

Harvey, J., Germann, P., and Odum, W. 1987. Geomorphological control of subsurface hydrology in the Creekbank zone of tidal marshes. *Estuarine, Coastal and Shelf Science* 25(6):677–691.

Heath, R. 1989. *Basic Groundwater Hydrology.* U.S. Geological Survey, Water Resource Investigation. Denver, CO: Earth Science Information Center.

Heinle, D. and Flemer, D. 1976. Flows of materials between poorly flooded tidal marshes and an estuary. *Marine Biology* 35:359–373.

Hem, J.D. 1985. *Study and Interpretation of the Chemical Characteristics of Natural Water,* 3rd ed. Water Supply Paper 2254. Washington, DC: U.S. Geological Survey.

Henderson-Sellers, A. and Pitman, A.J. 1992. Land-surface schemes for future climate models: specification, aggregation, and heterogeneity. *Journal of Geophysical Research* 97(D3):2687–2696.

Hendricks, A. 1992. *Implementing a Wellhead Protection Program Using GIS.* Washington, DC: U.S. EPA

Herrera-Silveira, J. 1994. Spatial heterogeneity and seasonal patterns, in a tropical coastal lagoon. *Journal of Coastal Research* 10(3):738–746.

Herrera-Silveira, J. and Comin, F. 1995. Nutrient fluxes in a tropical coastal lagoon. *Ophelia: International Journal of Marine Biology* 42:127–146.

Herzberg, A. 1901. Die wasserversorgung einiger Nordseebader. *Journal Gasbeleuchtung und Wasserversorgung* (Munich) 44:815–819, 842–844.

Hesseldahl, A. 2004. Europe reinvents GPS. *Forbes Magazine,* January 25, 2004.

Hill, M.C. 1998. Methods and guidelines for effective model calibration. U.S. Geological Survey Water Resources Investigations Report 98-4005.

Hinchee, R.E., Johnson, P.C., Johnson, R.L. and Leeson, A. 1999. Air Sparging for Site Remediation, seminar presentation sponsored by the Department of Defense, Environmental Security Technology Certification Program, in conjunction with the Fifth International *In Situ* and On-Site Bioreclamation Symposium in San Diego, California.

Ho, D.T., Schlosser, P., and Caplow, T. 2002. Determination of longitudinal dispersion coefficient and net advection in the tidal Hudson River with a large-scale, high resolution SF6 tracer release experiment. *Environmental Science and Technology.* 36:3234–3241.

Houghton, J.T., Meira Filho, L.G., Callander, B.A., Harris, N., Kattenberg, A., and Maskell, K. 1996. Climate change 1995: *The Science of Climate Change.* Cambridge University Press, Cambridge, UK.

Huettel, M. and Gust, G. 1992. Impact of bioroughness on interfacial solute exchange in permeable sediments. *Marine Ecology Progress Series* 89:253–267.

Huettel, M. and Webster, I. 2001. Pore water flow in permeable sediments. In: *The Benthic Boundary Layer Transport Processes and Biogeochemistry.* B.P. Boudreau and B.B. Jorgensen, Eds. New York: Oxford University Press, 144–179.

Huettel, M., Ziebis W., and Forster, S. 1996. Flow-induced uptake of particulate matter in permeable sediments. *Limnology and Oceanography* 41(2):309–322.

Huettel, M., Ziebis, W., Forster, S. and Luther, G. 1998. Advective transport affecting metal and nutrient distributions and interfacial fluxes in permeable sediments. *Geochimica et Cosmochimica Acta* 62(4):613–631.

Hussain, N., Church, T.M. and Kim, G. 1999. Use of 222Rn and 226Ra to trace groundwater discharge into Chesapeake Bay. *Marine Chemistry* 65:127–134.

Huyakorn, P.S., Andersen, P.F., Mercer, J.W., and White, H.O. 1987. *Water Resour. Res.* 23(2):293.

Hydrogeology of Africa. 1978. Moscow: Nedra, 370.

Hydrogeology of Asia. 1974. Moscow: Nedra, 576.

Hydrogeology of the USSR. 1966–1972. Vols. 3, 9, 12, 13, 24, 30–33, 45. Moscow: Nedra.

Hydrological Atlas of Canada. 1975. Bedrock hydrogeology. Canada: Department of Fisheries and the Environment Inland Water Directorate.

Hydrometeoizdat. 1974. *World Water Balance and Water Resources of the Earth.* Leningrad: Hydrometeoizdat, 638.

Hydroscience Press. 1999. World Map of Hydrogeological Conditions and Groundwater Flow. St. Paul, MN: Hydroscience Press.

ICF Kaiser. 1997. Final Construction Worker Risk Assessment, Port of Oakland, Berth 24. Submitted to Port of Oakland, Environmental Health and Safety Compliance Department.

Ieshina, A.V., Polenov, I.K. et al. 1987. Resources and geochemistry of groundwater resources of Karelia. 151.

Illades, J.M.L. 1976. Estudio hidrogeologico e hydrogeoquimico de la peninsula de Yucatan. Direccion de geohydrologia y de zonas aridas. *Proyecto Conacym* NSF 704:1–64.

Intergovernmental Panel on Climate Change (IPCC). 2001. *Climate Change 2001: Impacts, Adaptation, and Vulnerability.* Cambridge University Press, Cambridge, UK.

International Geological Congress. 1984. *Origin and History of Marginal and In-Land Seas.* Leiden, Netherlands.

Iris-Canbria Environmental, Joint Venture. 2002. Preferential Pathway and Related Tidal Influence Investigation, Former Mobil Bulk Fuel Terminal, 909 Ferry Street (Berth 24). Prepared for the Port of Oakland.

Isiorho, S.A. and Meyer, J.H. 1999. The effects of bag type and meter size on seepage meter measurements. *Ground Water* 37(3):411–413.

Israelson, O. and Reeve, R. 1944. Canal lining experiments in the delta area, Utah. *Utah Agricultural Experiment Station Technical Bulletin* 13:52.

Ivanovich, M. and Harmon, R.S. 1992. *Uranium-Series Disequilibrium: Applications to Earth, Marine and Environmental Sciences,* 2nd ed. Oxford: Clarendon Press.

Jacinthe, P., Groffman, P., Gold, A., and Mosier, A. 1998. Patchiness in microbial nitrogen transformations in ground water in a riparian forest. *Journal of Environmental Quality* 27:156–164.

Jacobs, T. and Gilliam, J. 1985. Riparian losses of nitrate from agricultural drainage waters, *Journal of Environmental Quality* 14:472–478.

Johannes, R. and Hearn, C. 1985. The effect of submarine groundwater discharge on nutrient and salinity regimes in a coastal lagoon off Perth, Western Australia. *Estuarine, Coastal and Shelf Science* 21(6):789–800.

Johannes, R.E. 1980. The ecological significance of the submarine discharge of groundwater. *Marine Ecology* Progress Series 3:365–373.

Johnson, P.C., Kemblowski, M.W., Colthart, J.D., et al. 1990. A practical approach to the design, operation, and monitoring of in-situ soil venting systems. *Ground Water Monitoring Review* lO(2):150–178.

Johnson, P. 2003. *Hydrologic Mapping and Assessment of New Mexico Aquifers.* Soccoro, NM: New Mexico Bureau of Geology.

Jordan, T. and Correll, D. 1985. Nutrient chemistry and hydrology of interstitial water in brackish tidal marshes of Chesapeake Bay. *Estuarine, Coastal and Shelf Science* 21:45–55.

Jordan, T., Correll, D., and Weller, D. 1993. Nutrient interception by a riparian forest receiving inputs from adjacent cropland. *Journal of Environmental Quality* 22:467–473.

Kaleris, V., Lagas, G., Marczinek, S. and Piotrowski, J.A. 2002. Modelling submarine ground-water discharge: An example from the western Baltic Sea: *Journal of Hydrology* 265(1–4):76–99.

Kalinin, A.V., Kalinin, V.V., and Pivovarov, B.A. 1975. Potential efficiency of electro-spark sources grouping. *Applied Geophysics* 82:106–114.

Kalinin, G.P. and Kuznetsova, L.P. 1972. Calculation of elements of the water balance in atmosphere and hydrosphere [Russian]. *Water Resources* 1:15–31.

Karelian Research Centre. 1996. Change in groundwater composition in cities anthropogenic systems. Procedings of the Conference *50 Years of Karelian Scientific Centre of Russian Academy of Science*. Petrozavodsk: Karelian Research Centre of RAS.

Karelian Research Centre. 1998. Present-day status of water objects in Republic of Karelia. 1992–1997 monitoring results. Petrozavodsk: Karelian Research Centre of RAS, 188.

Kasmarek, M. and Robinson, J. 2003. Hydrogeology and Simulation of Ground-Water Flow and Land-Surface Subsidence in the Northern Part of the Gulf Coast Aquifer System, Texas. Texas Water Development Board, Austin, Texas, USA.

Katz, B., Hornsby, H., and Bohlke, J. 1999. Sources of nitrate in water from springs and the Upper Floridan aquifer, Suwannee River basin, Florida. In: *Impact of Land-Use Change on Nutrient Loads from Diffuse Sources*. Proceedings of IUGG 99 Symposium HS3. L. Heathwaite, Ed. Birmingham, UK: International Association of Hydrological Sciences, 117–124.

Kazakhstan. 1969. *Natural Conditions and Resources of the USSR*. Moscow: 482.

Keller, R. 1965. *Waters and Water Balance of the Land*. Moscow: Progress, 435.

Kemmerikh, A.O. 1961. *Hydrography of the Northern, Near-Arctic and Arctic Ural*. Moscow: Academy of Sciences of the USSR, 138.

Khodzhibaev, N.N. and Miraliev, D.U. 1971. On quantity of groundwater recharging the Aral Sea. *Mathematical modeling for solving of the hydrogeological problems*. Tashkent.

Khublaryan, M.G. 1995. Phenomenon of the Caspian Sea. *Vestnik RAN*, 65(7)616–621.

Khublaryan, M.G. 2000. Water level variations of the Caspian Sea and its environmental and economic consequences. Proceedings of the International Workshop on Environmental Problems of the Caspian Sea Region. Moscow, 5–11.

Khublaryan, M.G. and Frolov, A.P. 1988. Modeling of intrusion processes in water areas of estuaries and underground aquifers. Moscow: Nauka, 144.

Khublaryan, M.G. and Frolov, A.P. 1994. Some problems of interaction of natural waters in different media. In: *Waters of Land: Problems and Solutions*. Moscow: Water Problems Institute of Russian Academy of Sciences, 232–249.

Khublaryan, M.G., Churmaev, O.M., and Yushmanov, I.O. 1984. Investigation of hydrodynamic problem of seepage and convective diffusion in inhomogenous and anisotropic porous media. *Water Resources* 11(3):185–191.

Khublaryan, M.G., Frolov, A.P., and Yumashev, I.O. 1996. Sea water intrusion in inhomogenous layers. *Water Resources* 23(3):262–266.

Khublaryan, M.G., Frolov, A.P., and Yumashev, I.O. 2002. Modeling of salt accumulation processes in shallow groundwater in areas with insufficient wetting. *Water Resources* 29(6):632–637.

Kim, G., Burnett, W.C., Dulaiova, H., et al. 2001. Measurement of Ra-224 and Ra-226 activities in natural waters using a radon-in-air monitor. *Environmental Science and Technology* 35:4680–4683.

King, P.T., Michel, J., and Moore, W.S. 1982. Ground water geochemistry of 228Ra, 226Ra and 222Rn. *Geochimica et Cosmochimica Acta* 46(7):1173–1182.

Kiriakov, P., Lisichenko, G.V., and Emel'yanov, V.A. 1982. Radon survey detection of submarine groundwater discharge zones. *Water Resources* 9(5):557–561.

Kirillova, V. and Raspopov, I. 1971. *Lakes of St. Petersburg Region*. St. Petersburg.

Kiryukhin, V.A. and Tolstikhin, N.I. 1988. *Hydrogeology of the World Ocean Bottom*. Leningrad: 104.

Kitaev, S.P. 1984. Ecological Basics of Bioproductivity of Lakes in Different Natural Zones. Moscow: 207.

Klenova, M.V. and Solovieva, V.F. 1962. Geological structure of the submarine slope of the Caspian Sea. *Izd-vo AN SSSR* 638.

Kohout, F. 1966. Submarine springs: A neglected phenomenon of coastal hydrology. *Hydrology* 26:391–413.

Kohout, F. and Kolipinski, M. 1967. Biological zonation related to ground water discharge along the shore of Biscayne Bay, Miami, Florida. In: *Estuaries*. G. Lauff, Ed. AAAS No. 83, Washington, DC: American Association for the Advancement of Science, 488–499.

Kohout, F.A. 1960a. Cyclic flow of saltwater in the Biscayne aquifer of southeastern Florida: *Journal of Geophysical Research* 65(7):2133–2141.

Kohout, F.A. 1960b. Flow pattern of fresh water and salt water in the Biscayne aquifer of the Miami area, Florida. *International Association of Scientific Hydrology* 52:440–448.

Kolditz, O., Ratke, R., Diersch, H.G., and Zielke, W. 1998. Coupled groundwater flow and transport: 1. Verification of variable density flow and transport models: *Advances in Water Resources* 21(1):27–46.

Kononov, V.I. 1983. *Geochemistry of Thermal Water of Areas of the Modern Volcanism.* Moscow: Nauka, 215.

Kontar, A.E. and Zekster, I.S. 1999. Submarine discharge and its effect on oceanographic processes in the coastal zone. *Water Resources* 26:512.

Kontar, E. and Burnett, W. 1998. Development of submarine monitors for investigations of the coastal zones. Ocean Sciences Meeting, American Geophysical Union (AGU), San Diego, California, February 9–13, 1998.

Kontar, E. and Zektser, I.S. 1999. Submarine discharge and its effect on oceanic processes in the coastal zone. *Water Resources* 26, 512.

Korotaev, S.M., Shneyer, V.S., and Gurevich, V.I. 1980. Assessment of rates of upward water movement in submarine springs by filtration electrical field. *Soviet Geology* 66:106–111.

Korotkov, A.I., Pavlov, A.I., and Yurovskiy, Yu.G. 1980. *Hydrogeology of the Shelf Areas.* Leningrad: Nedra, 220.

Korzh, V.D. 1991. *Geochemistry of Elementary Composition of the Hydrosphere.* Moscow: Nauka, 244.

Krainov, S.R. and Shvets, V.M. 1980. *Principles of Geochemistry of Groundwater.* Moscow: Nauka, 286.

Krainov, S.R. and Zakutin, V.P. 1993. Groundwater pollution in agricultural regions. *Hydrogeology, Engineering Geology: Review.* Moscow: Geoinformmark Ltd., 86.

Krainov, S.R., Ryzhenko, B.N., and Shvets, V.M. 2004. *Geochemistry of Groundwater.* Moscow: Nauka, 677.

Krest, J., Moore, W., Gardner, L., and Morris, J. 2000. Marsh nutrient export supplied by groundwater discharge: Evidence from radium measurements. *Global Biogeochemical Cycles* 14(1):167–176.

Krishnaswami, S., Graustein, W.C., Turekian, K.K., and Dowd, J.W. 1982. Radium, thorium and radioactive lead isotopes in groundwater: Application to the *in situ* determination of adsorption-desorption rate constants and retardation factors. *Water Resources Research* 18:1663–1675.

Krupa, S.L., Belanger, T.V., Heck, H.H., et al. 1998. Krupaseep — the next generation seepage meter. *Journal of Coastal Research* 25:210–213.

Kudel'skiy, A.V. 1976. Hydrogeology, hydrochemistry of iodine. *Science and Technology.* [Minsk] 214.

Kudelin, B.I. 1960. *Principles of Regional Estimation of Natural Groundwater Resources.* Moscow: Izd-vo MGU, 344.

Kulakova, L.S. and Lebedev, L.I. 1983. *Geologic-Geomorphologic Investigations of the Caspian Sea.* Moscow: Nauka, 65.

L'vovich, M.I. 1974. *Water Resources of the World and Their Future.* Moscow: Nauka, 448.

Lal, P.B. 1990. Groundwater capturing by dams [Russian]. *Vodnye Resursy* 2:185–187.

Lambert, M. and Burnett, W.C. 2004. Submarine groundwater discharge estimates at a Florida coastal site based on continuous radon measurements. *Biogeochemistry* 66(1–2):55–73.

Landon, M.K., Rus, D.L., and Harvey, F.E. 2001. Comparison of in-stream methods for measuring hydraulic conductivity in sandy streambeds. *Ground Water* 39(6), 870–885.

Lane-Smith, D.R., Burnett, W.C., and Dulaiova, H. 2002. Continuous radon-222 measurements in the coastal zone. *Sea Technology* 43:37–45.

Langevin, C.D. 2001, Simulation of ground-water discharge to Biscayne Bay, southeastern Florida: U.S. Geological Survey Water-Resources Investigations, Report 00-4251, 127.

Langevin, C.D. 2003, Simulation of submarine ground water discharge to a marine estuary: Biscayne Bay, Florida: *Ground Water* 41(6):758–771.

Langevin, C.D., Oude Essink, G.H.P., Panday, S., et al. 2004. MODFLOW-based tools for simulation of variable-density groundwater flow. In: *Coastal Aquifer Management—Monitoring, Modeling, and Case Studies.* A.H-D. Cheng and D. Ouazar, Eds. Boca Raton, FL: CRC Press, 49–76.

Langevin, C.D., Wexing, G. 2006. MODFLOW/MT3DMS-based simulation of variable density groundwater flow and transport. *Ground Water,* 44(4), 339–351.

Lanyon, J.A., Eliot, I.G., and Clarke, D.J. 1982. Groundwater level variation during semidiurnal spring tidal cycles on a sandy beach. *Australian Journal of Marine and Freshwater Research* 33:377–400.

Lapointe, B. and Matzie, W. 1996. Effects of stormwater nutrient discharges on eutrophication processes in nearshore waters on the Florida keys. *Estuaries* 19(2B):422–435.

Lapointe, B.E. and O'Connell, J. 1989. Nutrient-enhanced growth of Cladophora-Prolifera in Harrington Sound, Bermuda — eutrophication of a confined, phosphorus-limited marine ecosystem. *Estuarine Coastal and Shelf Science* 28(4):347–360.

Lapointe, B., O'Connell, J., and Garrett, G. 1990. Nutrient couplings between on-site sewage disposal systems, groundwater and nearshore surface waters of the Florida Keys. *Biogeochemistry* 10:289–307.

Law, C.S., Watson, A.J., and Liddicoat, M.I. 1994. Automated vacuum analysis of sulphur hexafluoride in seawater: derivation of the atmospheric trend (1970–1993) and potential as a transient tracer. *Marine Chemistry* 48:57–69.

Ledwell, J.R., Watson, A.J., and Law, C.S. 1993. Evidence for slow mixing across the pycnocline from an open-ocean tracer-release experiment. *Nature* 364:701–703.

Lee, C.-H. and Cheng, R.T.-S. 1974. On seawater encroachment in coastal aquifers. *Water Resources Research* 10(5):1039–1043.

Lee, D.R. 1977. A device for measuring seepage flux in lake and estuaries. *Limnology and Oceanography* 22(1):140–147.

Lee, V. and Olsen, S. 1985. Eutrophication and management initiatives for the control of nutrient inputs to Rhode Island coastal lagoons. *Estuaries* 8(2B):191–202.

Leibo, A.B. 1977. Electrotechnical effects connected with the sea waves. *Geomagnetism and Aeronomy* 17(3):502–506.

Leick, A. 2004. *GPS: Satellite Surveying,* 2nd ed. New York: John Wiley & Sons.

Leigh, D.P., Gaudlitz, J., and Hayes, D. 2000. Vertically propagated tidal fluctuation at Atoll Island Diego Garcia BIOT. Proceedings of the Pacific Environmental Restoration Conference, Honolulu, HI, April 4–7, 2000.

Leipnik, M., Zektser, I.S., and Loaiciga, H. 1994. The use of geographic information systems for studying ground water. *Water Resources* 21(6):608–611.

Leipnik, M. and Zektser, I.S. 1989 Groundwater GIS. *Journal of Water Problems,* Russian Academy of Sciences, Moscow.

Lester, D. and Greenberg, L.A. 1950. The toxicity of sulfur hexafluoride. *Archives of Industrial Hygiene and Occupational Medicine* 2:348–349.

Li, H.L. and Jiao, J.J. 2003. Tide-induced seawater-groundwater circulation in a multi-layered coastal leaky aquifer system. *Journal of Hydrology* 274(1–4):211–224.

Li, L., Barry, D.A., Stagnitti, F., and Parlange, J.Y. 1999. Submarine groundwater discharge and associated chemical input to a coastal sea: *Water Resources Research* 35(11):3253–3259.

Libelo, E.L. and MacIntyre, W.G. 1994. Effects of surface-water movement on seepage-meter measurements of flow through the sediment-water interface. *Applied Hydrogeology* 2(4):49–54.

Lifshits, V.H. 1970. Physiographic characteristics of the Shuja river (Lake Onego tributary) watershed and discharge regime. *Water resources of Karelia and their uses.* Petrozavodsk, P. 235–276. (in Russian).

Lindzen, R.S. 1990. Some coolness concerning global warming. *Bulletin American Meteorological Society* 71, 288–299.

Linsley, K., Koller, M.A., and Paulus, D.L.H. 1962. *Applied Hydrology.* Leningrad: Hydrometeoizdat, 756.

Lisitsyn, A.P. 1974. *Sedimentation in the Oceans.* Moscow: Nauka.

Lisitsyn, A.P. 1978. *Processes of Ocean Sedimentation.* Moscow: Nauka, 392.

Loáiciga, H.A., Haston, L., Michaelsen, J. 1993. Dendrohydrology and long-term hydrologic phenomena. *Reviews of Geophysics* 31(2), 151–171.

Loáiciga, H.A., Valdes, J.B., Vogel, R., Garvey, J., Schwarz, H.H. 1996. Global warming and the hydrologic cycle. *Journal of Hydrology* 174(1–2), 83–128.

Loáiciga, H.A. 1997. Runoff scaling in large rivers of the world. *The Professional Geographer* 49(3), 356–363.

Loaiciga, H.A., Maidment, D.R., and Valdes, J.B. 2000. Climate-change impact in a regional karst aquifer, Texas, USA. *Journal of Hydrology* 227:173–194.

Loaiciga, H.A. and Zektser, I.S. 2001. Methods to estimate groundwater discharge to the ocean. *Journal of King Abdulaziz University: Marine Sciences.* 12:24–32.

Loaiciga, H.A. 2003. Climate change and groundwater. *Annals of the Association of American Geographers* 93(1):30–41.

Lorenz, E. 1963. Deterministic non-periodic flow. *Journal of Atmospheric Sciences* 20:130–141.

Lorenz, E. 1967. The predictability of a flow which possesses many scales of motion. *Tellus* 21:289–307.

Low, E. and Riemersma, G. 2004. *Offshore GIS.* Edinburgh, UK: ESRI.

Lowrance, R. 1992. Groundwater nitrate and denitrification in a coastal riparian forest. *Journal of Environmental Quality* 21:401–405.

Lowrance, R., Todd, R., and Asmussen, L. 1983. Nutrient budgets for the riparian zone of an agricultural watershed. *Agriculture, Ecosystems, and the Environment* 10:371–384.

Lowrance, R., Todd, R., and Asmussen, L. 1984. Nutrient cycling in an agricultural watershed: 1. Phreatic movement. *Journal of Environmental Quality* 13:22–27.

Lusczynski, N.J. and Swarzenski, W.V. 1962. Fresh and salty groundwater in Long Island, New York. Proceeding of the American Society of Civil Engineers. *Journal of Hydrology* 88(4):173–194.

Lvovich, M.I. 1974. *World Water Resources and Their Future*. Moscow: Nauka, 448.

Lyalko, V.I., Mitnik, M.M., and Wulfson, L.D. 1978. Study of submarine sources by geothermal methods. *Geological Journal* 38:46–52.

Lyubimova, E.A. et al. 1974. Data on studying of heat fluids through the bottom of the Caspian Sea. *Izv. AS USSR, ser. Earth Physics* 4:98–103.

Makarenko, F.A. and Zverev, V.P. 1970a. Subsurface chemical denudation on the territory of the USSR. *DAN* 192(2): 424.

Makarenko, F.A. and Zverev, V.P. 1970b. Underground chemical outflow on the territory of USSR [Russian]. *Lithology and Fossil*.

Magara, N. 1982. Rocks consolidation and fluids migration. (translated from English), Moscow, Nedra.

Manheim, F.T. and Paull, C.K. 1981. Patterns of groundwater salinity changes in a deep continental-oceanic transect off the southeastern Atlantic coast of the USA. *Journal of Hydrology* 54(1–3):95–105.

Manheim, F.T. 1967. Evidence for submarine discharge of water on the Atlantic continental slope of the southern United States and suggestion for further search. *Transactions of the New York Academy of Sciences* 29(7):838–853.

Marinelli, R. 1994. Effects of burrow ventilation on activities of a terebellid plychaete and silicate removal from sediment pore waters. *Limnology and Oceanography* 39:303–317.

Marquis, S.A., Jr. and Smith, E.A. 1994. Assessment of ground-water and chemical transport in a tidally influenced aquifer using geostatistical filtering and hydrocarbon fingerprinting groundwater. 32(2):190–199.

Martin, J., Cable, J., and Jaeger, J. 2005. Quantification of advective benthic fluxes to the Indian river lagoon, Florida, final report (216). 32178–1429; Contract No. SG458AA. Palatka, FL: St. Johns River Water Management District.

Martin, J., Cable, J., and Swarzenski, P. 2002. Quantification of ground water discharge and nutrient loading to the Indian River Lagoon. Palatka, FL: St. Johns River Water Management District, 244.

Martin, J., Cable, J., Swarzenski, P., and Lindenberg, M. 2004. Enhanced submarine ground water discharge from mixing of pore water and estuarine water. *Ground Water (Special Oceans Issue)* 42:1001–1010.

Martin, J.B., Cable, J.E., and Swarzenski, P.W. 2000. Quantification of ground water discharge and nutrient loading to the Indian River lagoon, Florida. Final report for phase one. 32178–1429; Contract No. 99G245. Palatka, FL: St. Johns River Water Management District, 263.

Masterson, J.P. and Walter, D.A. 2001. Hydrologic analysis of the sources of water to coastal embayments, western Cape Cod, Massachusetts. Proceedings of the Geological Society of America Annual Meeting, November 5–8, 2001. Boulder, CO: GSA.

Mathieu, G.P., Biscaye, P., Lupton, R., and Hammond, R. 1988. System from measurement of 222Rn at low levels in natural water. *Health Physics* 55:989–992.

Matson, E. 1993. Nutrient flux through soils and aquifers to the coastal zone of Guam (Mariana Islands). *Limnology and Oceanography* 38(2):361–371.

McBride, M.S. and Pfannkuch, H.O. 1975. The distribution of seepage within lakebeds. *U.S. Geological Survey Journal of Research* 3:505–512.

McCaffrey, R., Myers, A., Davey, E., et al. 1980. The relation between pore water chemistry and benthic fluxes of nutrients and manganese in Narragansett Bay, Rhode Island, *Limnology and Oceanography* 25(1):31–44.

McDonald, M.G. and Harbaugh, A.W. 1988. A modular three-dimensional finite-difference ground-water flow model: *U.S. Geological Survey Techniques of Water Resources Investigations* 6:586

McDuff, R.E. and Gieskes, J.M. 1976. Calcium and magnesium profiles in DSDP interstitial water: diffusion or reaction? *Earth Planet Sciences Letter* 33:1–10

Meile, C., Koretsky, C., and Cappellen, P. 2001. Quantifying bioirrigation in aquatic sediments: An inverse modeling approach. *Limnology and Oceanography.* 46:164–177.

Mercer, J.W., Larson, S.P., and Faust, C.R. 1990. *Groundwater* 18(4):374.

Meskheteli, A.V., Drushchits, V.A., and Podolny, O.V. 1991. Components of Lake Balkhash groundwater recharge — hydrogeological aspects. *Water Resources* 18(3):242–252.

Miller, D. and Ullman, W. 2004. Ecological consequences of estuarine ground water discharge at Cape Henlopen, Delaware Bay, USA. *Ground Water* 42(7):959–970.

Millham, N. and Howes, B. 1994. Freshwater flow into a coastal embayment: groundwater and surface water inputs. *Limnology and Oceanography* 39(8):1928–1944.

Milnes, E. and Renard, P. 2004. The problem of salt recycling and seawater intrusion in coastal irrigated plains: an example from the Kiti aquifer (Southern Cyprus). *Journal of Hydrology*, 288, 327–343.

Mitchell, J.F.B., Jones, T.C., Gregory, J.M., and Tett, S.F.B. 1995. Climate response to increasing levels of greenhouse gases and sulphate aerosols. *Nature* 376:501–504.

Mizadrontsev, I.B. 1975. To geochemistry of shallow groundwater fluids. In: *Dynamic of the Baikal Depression*. Novosibirsk, 203–230.

Moon, D.S., Burnett, W.C., Nour, S., et al. 2003. Preconcentration of radium isotopes from natural waters using MnO_2 Resin. *Applied Radiation and Isotopes* 59:255–262.

Moore, W. 1996. Large ground-water inputs to coastal waters revealed by 226Ra enrichment. *Nature* 380(575):612–614.

Moore, W. 1999. The subterranean estuary: a reaction zone of ground water and sea water. *Marine Chemistry* 65:111–125.

Moore, W.S. 1976. Sampling radium-228 in the deep ocean. *Deep-Sea Research* 23:647–651.

Moore, W.S. 1984. Radium isotope measurements using germanium detectors. *Nuclear Instruments and Methods* 227:407–411.

Moore, W.S. 2000a. Ages of continental shelf waters determined from 223Ra and 224Ra. *Journal of Geophysical Research* 105:22117–22122.

Moore, W.S. 2000b. Determining coastal mixing rates using radium isotopes. *Continental Shelf Research* 20:1993–2007.

Moore, W.S. 2003. Sources and fluxes of submarine groundwater discharge delineated by radium isotopes. *Biogeochemistry* 66(1–2):75–93.

Moore, W.S. and Arnold, R. 1996. Measurement of 223Ra and 224Ra in coastal waters using a delayed coincidence counter. *Journal of Geophysical Research* 101:1321–1329.

Moore, W.S. and Church, T.M. 1996. Submarine groundwater discharge. *Nature* 382:121–122.

Moore, W.S. and Shaw, T.J. 1998. Chemical signals from submarine fluid advection onto the continental shelf. *Journal of Geophysical Research — Oceans* 103:21543–21552.

Morris, J. 1995. The mass balance of salt and water in intertidal sediments: Results from North Inlet, South Carolina. *Estuaries* 18(4):556–567.

Mu, Y., Cheng, A., Badiey, M., and Bennett, R. 1999. Water wave driven seepage in sediment and parameter inversion based on pore pressure data. *International Journal for Numerical and Analytical Methods in Geomechanics* 23:1655–1674.

Muller, J.L., Botha, J.F., and Tonder, G.J. 1987. Modeling sea-water intrusion in the Atlantis aquifer. *Water SA*, 13, N 3, 171–174.

Mulligan, A. 2005. Submarine groundwater discharge: identification and quantification via remote sensing, hydrologic sampling, and geochemical tracers. Funded project summaries 2002. Coastal Ocean Institute and Rinehart Coastal research Center, Woods Hole Oceanographic Institute.

Murdoch, L.C. and Kelly, S.E. 2003. Factors affecting the performance of conventional seepage meters. *Water Resources Research.* 39(6):1163.

Nace, R.L. 1967. Are we running out of water? U.S. Geological Survey, Circ. 536, 7. Washington, DC: U.S. Department of the Interior.

Nace, R.L. 1970. World hydrology status and prospects. Proceedings of the IASH–UNESCO–WHO Symposium on World Water Balance: Reading, UK: AIHS, Louvain, I, 92: 1–10. Retrieved April 24, 2003 at the National Oceanic and Atmospheric Administration Website http://co-ops.nos.noaa.gov/.

Napolitano, P. 2002. Assessing Aquifer Vulnerability in Italy. International Institute for Geo-Information Science and Earth Observation. Naples, Italy.

Nelson, N.T. and Brusseau, M.L. 1996. Field study of the partitioning tracer method for detection of dense nonaqueous phase liquid in a trichloroethene-contaminated aquifer. *Environmental Science and Technology* 30:2859–2863.

Nelson, W., Gold, A., and Groffman, P. 1995. Spatial and temporal variation in groundwater nitrate in a riparian forest. *Journal of Environmental Quality* 24:691–699.

NFESC. 2003. RITS: Remediation Innovative Technology Seminar. Washington.

Nickson, R.T., McArthur, J.M., Ravenscroft, P., Burgess, W.G., and Ahmed, K.M. 2000, Mechanism of arsenic release to groundwater, Bangladesh and West Bengal. *Applied Geochemistry*, 15:403–413.

Nielsen, P. 1990. Tidal dynamics of the water table in beaches. *Water Resources Research* 26(9):2127–2134.

Oberdorfer, J.A., Hogan, P.J., and Buddemeier, R.W. 1990. Atoll Island hydrogeology: flow and freshwater occurrence in a tidally dominated system. *Journal of Hydrology* 120:327–340.

Oberdorfer, J.A. 2003. Hydrogeologic modeling of submarine groundwater discharge: Comparison to other quantitative methods. *Biogeochemistry.* 66(1–2):159–169.

Oberman, N.G. 1984. Groundwater of the Arctic and Near-Arctic Ural. In: *Rational Use and Protection of Groundwater.* Krasnoyarsk: SibNIIGiM, 21–35.

Oberman, N.G. 1998. European North-East. Ural. In: *Principles of Geocryology. Part 3. Regional and Historical Geocryology of the World.* E.D.Yershov, Ed. Moscow: MSU, 228–237.

Oberman, N.G. and Kakunov, N.B. 2002. Monitoring of groundwater in cryolithozone of northeastern European Russia and Ural region. In: *Monitoring of Groundwater in Cryolithozone.* Yakutsk: Permafrostology Institute of SORAN, 18–43.

Oberman, N.G., Shesler, I.G., and Rubtsov, A.I. 2004. *Ecogeology of Republic of Komi and Eastern Part of the Nenets Autonomous Area.* Syktyvkar: PrologPlus, 256.

Ogilvi, N.A. 1959. The theory of geotemperature fields in geothermal methods of groundwater exploration. In: *The Problems of Geothermic Studying and Practical Use of the Earth Heat.* Moscow:1, 53–85.

Osmond, J.K. and Cowart, J.B. 1992. Ground water. In: *Uranium-Series Disequilibrium: Applications to Earth, Marine and Environmental Sciences.* M. Ivanovich and R.S. Harmon, Eds. Oxford: Clarendon Press, 290–333.

Ovchinnikov, I.M. 1973. Water balance of the Mediterranean Sea. Proceedings of the All-Union Symposium on the Study of the Black and Mediterranean Seas, Use and Protection of their Resources. Part 1. Oceanography. Kiev, 3–18.

Panagoulia, D. 1992. Impact of GIS-modeled climate changes in catchment hydrology. *Hydrologic Sciences Journal* 37(2):141–163.

Pandit, A. and El-Khazen, C. 1990. Groundwater seepage into the Indian River Lagoon at Port St. Lucie. *Florida Scientist* 53:169–179.

Panin, G.N., Mamedov, R.M. and Mitrofanov, I.V. 2005. *The Current State of the Caspian Sea.* Moscow: Nauka.

Park, S. 2004. Visualization for oil exploration. Center for Subsurface Modeling. Austin, TX: University of Texas.

Pashkovskiy, I.S. 1969. Groundwater discharge into the Aral Sea, its present and future. *Bulletin of MOIP. Section Geology.* 64(4):110–119.

Paull, C.K., Hecker, B., Commeau, R., et al. 1984. Biological communities at the Florida escarpment resemble hydrothermal vent taxa. *Science* 226:965–967.

Paulsen, R.J., Smith, C.F., O'Rourke, D., and Wong, T.-F. 2001. Development and evaluation of an ultrasonic ground water seepage meter. *Ground Water* 39(6):904–911.

Pavlov, A.N. 1977. *Geological Water Circulation on the Earth.* Leningrad: Nedra, 142.

Peck, A.A. 1968. *Dynamics of Juvenile Solutions.* Moscow, Nauka.

Pederson, T.A. and Curtis, J.T. 1991. Soil Vapor Extraction Technology Reference Handbook, reference prepared for the Risk Reduction Engineering Laboratory, Office of Research and Development, U.S. Environmental Protection Agency.

Peter Kaldveer and Associates. 1975. Soil and Foundation Engineering Services for Container Yard Improvement, Berths 2, 3, and 4, Port of Oakland, California.

Peltonen, K. 2002. *Direct Groundwater Inflow to the Baltic Sea.* Temanord: Nordic Council, 76.

Perelman, A.I. and Kasimov, N.S. 1999. Geochemistry of landscape. Moscow, 1999, 763.

Peterjohn, W. and Correll, D. 1984. Nutrient dynamics in an agricultural watershed: Observations on the role of a riparian forest. *Ecology* 65:1466–1475.

Pfannkuch, H.O. and Winter, T.C. 1984. Effect of anisotropy and groundwater system geometry on seepage through lakebeds. 1. Analog and dimensional analysis. *Journal of Hydrology* 75:213–237.

Phillips, P., Denver, J., Shedlock, R., and Hamilton, P. 1993. Effect of forested wetlands on nitrate concentrations in ground water and surface water on the Delmarva Peninsula. *Wetlands* 13(2):75–83.

Pisarsky, B.I. 1987. Regularities in formation of groundwater discharge in Baikal Lake basin. Novosibirsk: Nauka, 155.

Pisarsky, B.I. and Khaustov, A.P. 1979. Subsurface chemical discharge in Baikal Lake basin. In: *Groundwater Discharge on the Siberian Territory and Methods of Its Study.* Novosibirsk: Nauka, 24–26.

Podolny, O.V. and Shapiro, S.M. 1992. Subsurface water exchange of Balkhash Lake. In: *Balkhash Segment. Groundwater.* Alma-Ata: Ghylym, 182–190.

Podolny, O.V., Smolyar, V.A., and Darishev, K.M. 1992. Formation and evaluation of natural groundwater resources: South Balkhash Area. In: *Balkhash Segment. Groundwater.* Alma-Ata: Ghylym, 154–182.

Podsechin, V.P. and Frolov, A.P. 1989. Unsteady intrusion of sea water into coastal confined aquifers. *Water Resources* 16(2):156–163.

Popov, O.V. 1968. *Subsurface Recharge of Rivers.* Leningrad: Hydrometeoizdat, 291.

Portnoy, J., Nowicki, B., et al. 1998. The discharge of nitrate-contaminated groundwater from developed shoreline to marsh-fringed estuary. *Water Resources Research* 34 (11):3095–3104.

Poryadin, V.I., Podolny, O.V., Veselov, V.V., et al. 1997. Geoecological map (Priaral region and the Aral Sea basin). Scale 1:500 000. Tashkent: Natural Resources of the Kazakhstan Republic.

Poryadin, V.I. 1990. Hydrodynamic regime of the open sea area and drauned bottom of the Aral Sea and problems of salts inflowing. *Problems of Desert Development* 4:12–22.

Poryadin, V.I. 1994. Ecological aspects of studying of the draining Aral Sea bottom regime. *Geoecology* 1:76–87.

Potie, L. 1973. Investigations and capture of submarine fresh-water springs. Submarine springs at Port-Miou, Cassis, France. 2nd Intern. Symp. Ground Water. Palermo, 30.

Pourbaix, M. 1974. *Atlas of Electrochemical Equilibria in Solutions*. Houston, TX: National Association of Corrosion Engineers.

Power, H. 1990. The propagation of a wave packet on a free surface within a saturated porous layer. *Journal of Hydrology* 119:263–270. Amsterdam: Elsevier Science Publishers B.V.

Prieto, C. 2001. Modeling Freshwater-Seawater Interactions in Coastal Aquifers: Long-term Trends and Temporal Variability Effects: Stockholm, Sweden: Licentiate Thesis, Royal Institute of Technology.

Problems of the Aral Sea Basin. 1999. NAN RK UNESCO, 58

Problems of the Aral Sea. State of the open sea area and drained bottom of the Aral Sea. Alma-Ata: 1983. 236

Ragone, S., Katz, B., Lindner, J., and Flipse, W. 1976. Chemical quality of ground water in Nassau and Suffolk Counties, Long Island, New York: 1952–1976. U.S. Geological Survey Open File Report 76-845.

Rakhmanov, R.R. 1987. *Mud Volcanoes and Their Importance in Prediction of Subsurface Oil and Gas-bearing Potential*. Moscow: Nedra, 182.

Rama, L. and Moore, W. 1996. Using the radium quartet for evaluating groundwater input and water exchange in salt marshes. *Geochimica et Cosmochimica Acta* 60(23):4245–4252.

Ramanathan, V., Collins, W. 1992. Thermostat and global warming. *Nature* 357, 649–653.

Raper, J. 1989. The 3-dimensional geoscientific mapping and modeling system: a conceptual design. In: *Three Dimensional Applications in Geographic Information Systems*. Jonathan Raper, Ed. London: Taylor & Francis.

Rasmussen, L. 1998. *Groundwater Flow, Tidal Mixing and Haline Convection in Coastal Sediments. Oceanography*. Tallahassee: Florida State University, 119.

Reay, W., Gallagher, D., and Simmons, G. 1992. Groundwater discharge and its impact on surface water quality in a Chesapeake Bay Inlet. *AWRA Water Resources Bulletin* 28(6):1121–1134.

Redlands Institute for Environmental Design, Analysis and Policy. 2002. *Salton Sea Atlas*. Redlands CA: ESRI Press, 132.

Reeves, M., Ward, D.S., Johns, N.D., and Cranwell, R.M. 1986. Theory and implementation of SWIFT II, the Sandia waste-isolation flow and transport model for fractured media. Report SAND83-1159. Albuquerque, NM: Sandia National Laboratory.

Reilly, T.E. 1990. Simulation of dispersion in layered coastal aquifer systems. *Journal of Hydrology* 114(3–4):211–228.

Reilly, T.E. 2001. System and boundary conceptualization in ground-water flow simulation. *U.S. Geological Survey Techniques of Water-Resources Investigations,* Book 3, Chapter B8, 30. Washington, DC: U.S. Geological Survey.

Remedial Action Corporation. 1992. Revised Work Plan Petroleum Hydrocarbon Investigation, Port of Oakland, Oakland, California.

Reilly, T.E. and Goodman, A.S. 1985. Quantitative analysis of saltwater-freshwater relationships in groundwater systems — a historical perspective. *Journal of Hydrology* 80(1–2):125–160.

Rengarajan, R., Sarin, M.M., Somayajulu, B.L.K., and Suhasini, R. 2002. Mixing in the surface waters of the western Bay of Bengal using Ra-228 and Ra-226. *Journal of Marine Research* 60:255–279.

Rennie, G. 2005. Monitoring the Earth's subsurface from space. In: *Science and Technology Bulletin*. University of California, Lawrence Livermore National Lab.

Ressl, R. and Ptichnikon, A. 2001. Aral Sea GIS. Project sponsored by USAID and Institute of Geography, Russian Academy of Sciences, Moscow, Russia.

Reynolds, D.H.B. 1970. Under draining of the Lower Chalk of South East Kent. *Journal of the Institution of Water and Environmental Management (London)* 24(8):471–480.

Riedl, R., Huang, N., and Machan, R. 1972. The subtidal pump: A mechanism of interstitial water exchange by wave action. *Marine Biology* 13:210–221.

Robinson, M.A. 1996. A finite element model of submarine ground water discharge to tidal estuarine waters. PhD dissertation, Virginia Polytechnic Institute.

Robinson, M.A. and Gallagher, D.L. 1999. A model of ground water discharge from an unconfined coastal aquifer. *Ground Water* 37(1):80–87.

Ronov, A.B. 1993. *Stratisphere or Sedimentary Cover of the Earth*. Moscow: Nauka, 144.

Rosenberry, D.O. 1990. Inexpensive groundwater monitoring methods for determining hydrologic budgets of lakes and wetlands. National Conference on Enhancing the States' Lake and Wetland Management Programs. J. Taggart, Ed. U.S. Environmental Protection Agency, North American Lake Management Society. Madison: University of Wisconsin Press, 123–131.

Rosenberry, D.O. 2000. Unsaturated-zone wedge beneath a large, natural lake. *Water Resources Research* 36(12):3401–3409.

Rosenberry, D.O. and Morin, R.H. 2004. Use of an electromagnetic seepage meter to investigate temporal variability in lake seepage. *Ground Water* 42(1):68–77.

Rubanov, I.V. and Bogdanova, N.M. 1987. Quantitative assessment of the salt deflation of the drained bottom of the Aral Sea. *Problems of the Desert Development* 3:9–16.

Ruthermich, M., Hayes, L., and Lorley, D. 2002. Anaerobic sulfate-dependent degradation of polycyclic aromatic hydrocarbons in petroleum-contaminated harbor sediments. *Environmental Science and Technology* 36(22).

Rutkowski, C.M., Burrnet, W.C., Iverson, R.L., and Chanton, J.P. 1999. The effect of groundwater seepage on nutrient delivery and seagrass distribution in the northeastern Gulf of Mexico. *Estuaries* 22:1033–1040.

Sadov, A.V. and Krasnikov, V.V. 1987. Finding of the sources of submarine groundwater discharge into the Aral Sea using remote sensing methods. *Problems of the Desert Development* 1:28–36.

Samarina, V.S., Gaiev, A.Ya., et al. 1999. Technogenic metamorphisation of the chemical composition of natural waters (example of eco-hydrochemical mapping of the Ural River watershed, Orenburg Region). Ekaterinburg: RAS Ural Branch Publishers, 444 (in Russian).

Sandnes, J., Forbes, T., Hansen, R., et al. 2000. Bioturbation and irrigation in natural sediments, described by animal-community parameters. *Marine Ecology* Progress Series 197:169–179.

Sano, Y., Kusakabe, M., Hirabayashi, J.-I., et al. 1990. Helium and carbon fluxes in Lake Nyos, Cameroon: Constraint on next gas burst. *Earth and Planetary Science Letters* 99(4):303–314.

Schafran, G.C. and Driscoll, C.T. 1993. Flow path-composition relationships for groundwater entering an acidic lake. *Water Resources Research* 29(1):145–154.

Schlosser, P., Stute, M., Sonntag, C., and Munnich, K.O. 1989. Tritiogenic 3He in shallow groundwater. *Earth and Planetary Science Letters* 94(3–4):245–256.

Schluter, M., Sauter, E., Hansen, H.-P., and Suess, E. 2000. Seasonal variations of bioirrigation in coastal sediments: Modeling of field data. *Geochimica et Cosmochimica Acta* 64:821–834.

Schmorak, S. and Mercado, A. 1969. Upconing of fresh water-sea water interface below pumping wells, field study. *Water Resources Research* 5(6):1290–1311.

Scientific Software Group. 2003. Manual for MODFLOW (USGS Developed Groundwater Flow Model). Sandy, UT.

SCOR/LOICZ Working Group 112. 1997. Magnitude of submarine groundwater discharge and its influence on coastal oceanographic processes. Retrieved March 2006 from the Scientific Committee on Ocean Research Website at http://www.jhu.edu/~scor/wg112.htm.

Scripps Institution of Oceanography. Initial Reports of the Deep Sea Drilling Project. University of California, San Diego, 1–50.

Senger, R.K. and Fogg, G.E. 1990. Stream functions and equivalent fresh-water heads for modeling regional flow of variable-density groundwater (1) review of theory and verification. *Water Resources Research* 26(9):2089–2096.

Serfes, M.E. 1991. Determining the mean hydraulic gradient of ground water affected by tidal fluctuations. *Groundwater* 29(4):549–555.

Serfes, M.S. 1987. Interpretation of Tidally Affected Ground-Water Flow Systems in Pollution Studies, *Proceedings Petroleum Hydrocarbons and Organic Chemicals in Ground Water; Prevention, Detection and Restoration – A Conference and Exposition*, National Water Well Assoc., Dublin OH, pp. 55–73.

Sewell, P. 1982. Urban groundwater as a possible nutrient source for an estuarine benthic algal bloom. *Estuarine, Coastal and Shelf Science* 15(5):569–576.

Shamir, U. and Dagan, G. 1971. Motion of the seawater interface in coastal aquifers: numerical solution. *Water Resources Research* 7(3):644–657.

Shapiro, S.M. 1974. *Groundwater in the Southeast of Central Kazakhstan*. Alma-Ata: Nauka, 184.

Shapiro, S.M. 1988. Groundwater in water and salt balance of Balkhash Lake. *Water Resources*, 5: 180–183.

Shapiro, S.M. and Podolny, O.V. 1990. Problems of formation of subsurface water exchange in plain water bodies of arid zone (on the example of Balkhash Lake). *Proceedings of V All-Union Hydrol. Forum*, 8. Leningrad, Hydrometeoizdat, 155–161.

Shapiro, S.M., Pavlichenko, L.M., and Podolny, O.V. 1982. *Hydrogeological Predictions of Groundwater Discharge to Balkhash Lake*. Alma-Ata: Nauka, 128.

Shaw, R.D. and Prepas, E.E. 1989. Anomalous, short-term influx of water into seepage meters. *Limnology and Oceanography* 34(7):1343–1351.

Shaw, R.D. and Prepas, E.E. 1990. Groundwater-lake interactions: I. Accuracy of seepage meter estimates of lake seepage. *Journal of Hydrology* 119(4):105–120.

Shestopalov, V.M. 1981. *Natural Groundwater Resources in Platform-type Artesian Basins of Ukraine*. Kiev: Dumka, 196.

Shih, D.C.F., Lee, C.D., Chiou, K.F. and Tsai, S.M. 2000. Spectral analysis of tidal fluctuations in ground-water level. *Journal of the American Water Resources Association*, 36(5):1087–1100.

Shinn, E.A., Reich, C.D., and Hickey, T.D. 2002. Seepage meters and Bernoulli's revenge. *Estuaries* 25:126–132.

Shishkina, O.V. 1972. *Geochemistry of Sea and Ocean Water*. Moscow: Nedra, 228.

Shishkina, O.V. 1980. *Methods of Studying Organic Matter Content in the Oceans*. Moscow: Nauka, 343.

Shnitnikov, A.V. 1936. Elements of water and salt balance of the Balkhash Lake. Proceedings of Salt Laboratory, All-Union Institute on Mineral-Salt Production. 11:5–82.

Sholkovitz, E., Herbold, G., and Charette, M. 2003. An automated dye-dilution based seepage meter for the time-series measurement of submarine groundwater discharge. *Limnology and Oceanography: Methods* 1:16–28.

Shum, K. 1992. Wave-induced advective transport below a rippled water-sediment interface. *Journal of Geophysical Research* 97:789–808.

Shum, K. 1993. The effects of wave-induced pore water circulation on the transport of reactive solutes below a rippled sediment bed. *Journal of Geophysical Research* 98:10,289–10,301.

Shvartsev, S.L. 1996. *Hydrogeology*. Moscow: Nedra, 423.

Shvartsev, S.L. 1999. *Hydrogeochemistry of the Hypergenesis Zone*. Moscow: Nedra, 366.

Shvetsov, P.F. 1968. *Regularities of Hydrogeothermal Processes in the Far North and Northeast of the USSR*. Moscow: Nauka, 111.

Silin-Bekchurin, A.I. 1958. Hydrodynamic and Hydrochemical Regularities in the Baltic Region. *Proceedings of the Lab of Hydrogeologic Problems of the USSR Academy of Sciences* 20:3–28.

Simmons, G. and Netherton, J. 1986. Groundwater discharge in a deep coral reef habitat: Evidence for a new biogeochemical cycle? American Academy of Underwater Sciences Sixth Annual Scientific Diving Symposium. Tallahassee, Florida, American Academy of Underwater Sciences.

Simmons, G., Jr. 1992. Importance of submarine groundwater discharge (SGWD) and seawater cycling to material flux across sediment/water interfaces in marine environments. *Marine Ecology* Progress Series 84:173–184.

Simmons, G., Miles, E., Reay, W., and Gallagher, D. 1992. Submarine groundwater discharge quality in relation to land use patterns in the southern Chesapeake Bay. Proceedings of the First International Conference on Ground Water Ecology. U.S. EPA and the American Water Research Association. April, 341–350.

Slater, J. and Capone, D. 1987. Denitrification in aquifer soil and nearshore marine sediments influenced by groundwater nitrate. *Applied and Environmental Microbiology* 53(June):1292–1297.

Slomp, C. and Van Cappellen, P. 2004. Nutrient inputs to the coastal ocean through submarine groundwater discharge: controls and potential impact. *Journal of Hydrology* 295(1–4):64–86.

Smethie, W., Nittrouer, C., and Self, R. 1981. The use of Radon-222 as a tracer of sediment irrigation and mixing on the Washington continental shelf. *Marine Geology* 42:173–200.

Smirnov, S.I. 1974. *Introduction in Studying of Geochemical History of Groundwater of Sedimentation Basins*. Moscow: Nedra, 263.

Smith, A.J. and Nield, S.P. 2003. Groundwater discharge from the superficial aquifer into Cockburn Sound Western Australia: Estimation by inshore water balance: *Biogeochemistry* 66(1–2):125–144.

Smith, L. and Zawadzki, W. 2003. A hydrogeologic model of submarine groundwater discharge: Florida intercomparison experiment: *Biogeochemistry* 66(1–2):95–110.

Smith, R., Howes, B., and Duff, J. 1991. Denitrification in nitrate-contaminated groundwater: occurrence in steep vertical geochemical gradients. *Geochimica et Cosmochimica Acta* 55(7):1815–1825.

Snyder, M., Taillefert, M., and Ruppel, C. 2004. Redox zonation at the saline-influenced boundaries of a permeable surficial aquifer: Effects of physical forcing on the biogeochemical cycling of iron and manganese. *Journal of Hydrology* 296(1–4):164–178.

Sokolov, A.A. (Ed.). 1989. *Hydrological and Water-Management Aspects of the Ile-Balkhash Problem.* Leningrad: Hydrometeoizdat, 310.

Solomon, D.K., Hunt, A., and Poreda, R.J. 1996. Source of radiogenic helium-4 in shallow aquifers: Implications for dating young groundwater. *Water Resources Research* 32(6):1805–1813.

SOMA Environmental Engineering, Inc. 1996. Additional Characterization Activities, Former Mobil Terminal Facility, Oakland, California.

SOMA Environmental Engineering, Inc. 1997a. Groundwater Flow and Contaminant Transport Modeling Report, Former Mobil Oil Terminal. Prepared for Mobil Oil Corporation.

SOMA Environmental Engineering, Inc. 2000. Evaluation of final cleanup objectives and risk management plan at former Mobil terminal facility. 909 Ferry Street, Oakland, California.

Sorek, S. and Pinder, G.F. 1999. Survey of computer codes and case histories. In: *Seawater Intrusion in Coastal Aquifers—Concepts, Methods and Practices*: J. Bear et al, Eds. Dordrecht, Netherlands: Kluwer Academic Publishers, 625.

Sorokhtin, O.G. 1974. *The Global Evolution of the Earth.* Moscow, Nedra.

Souza, W.R. and Voss, C.I. 1987. Analysis of an anisotropic coastal aquifer system using variable-density flow and solute transport simulation. *Journal of Hydrology.* 92(1–2):17–41.

Starr, J., Sadeghi, A., Parkin, T., and Meisinger, J. 1996. A tracer test to determine the fate of nitrate in shallow groundwater. *Journal of Environmental Quality* 25:917–923.

Staver, K. and Brinsfield, R. 1996. Seepage of groundwater nitrate from a riparian agroecosystem into the Wye River Estuary. *Estuaries* 19(2B):359–370.

Stieglitz, T.C. 2004. Submarine groundwater discharge into the near-shore zone of the Great Barrier Reef, Australia. *Marine Pollution Bulletin*, submitted.

Stieglitz, T., Ridd, P., and Muller, P. 2000. Passive irrigation and functional morphology of crustacean burrows in a tropical mangrove swamp. *Hydrobiologia* 421:69–76.

Strakhov, N.M. 1962. *Basis of the Kithogenesis Theory.* Volume III. Lows of composition and distribution of the arid sediments. Moscow.

Strozzi, T., Wegmüller, U., Werner, C., et al. 2003. JERS SAR Interferometry for land subsidence monitoring. *IEEE Transactions on Geoscience and Remote Sensing.* 41(7):1702–1708.

Stumm, W. and Morgan, J. 1996. *Aquatic Chemistry*, 3rd ed. New York: Wiley-Interscience.

Stute, M., Sonntag, C., Deák, J., and Schlosser, P. 1992. Helium in deep circulating groundwater in the Great Hungarian Plain: Flow dynamics and crustal and mantle helium fluxes. *Geochimica et Cosmochimica Acta* 56(5):2051–2067.

Subsurface Consultants and Todd Engineering. 1999. Hydrogeologic Investigation, Oakland Harbor Navigation Improvement (-50 foot) Project, Port of Oakland, Oakland and Alameda, California. Prepared for the Port of Oakland.

Sudarikov, S.M. 1992. *Description of Subaquatic Hydrotherms–Hydrothermal Sulfide Ores.* St. Petersburg: Nedra, 39–57.

Suess, E. 1888–1909. Das Antlitz der Erde, 2 Aufl., Bd 1–3, Prag.

Sugio, S. and Nakada, K. 1992. *Hydrocomp '92. Int. Conf. on Interaction of Computational Methods and Measurements in Hydraulics and Hydrology.* Budapest. P.81.

Sullivan, K.A. and Aller, R.C. 1996. Diagenetic cycling of arsenic in Amazon shelf sediments. *Geochimica et Cosmochimica Acta.* 60(9):1465–1477.

Sumarokova, V.V., Tsytsenko, K.V., and Podolny, O.V. 1992. *Aerocosmic Investigations and Water Balance of the Ile River Delta.* St. Petersburg: Hydrometeoizdat, 128.

Sun, Y. and Torgersen, T. 1998. The effects of water content and Mn-fiber surface conditions on Ra-224 measurement by Rn-220 emanation. *Marine Chemistry* 62:299–306.

Surface Waters of the Shuja Lake-River System under Human Impact. Petrozavodsk: Karelia Publishers, 1991. 212 (in Russian).

Swarzenski, P.W., Reich, C.D., Spechler, R.M., et al. 2001. Using multiple geochemical and radionuclide tracers to characterize the hydrogeology of the submarine spring off Crescent Beach, Florida. *Chemical Geology* 179:187–202.

Sweet, R. 2003. *GPS for Mariners.* Camden, ME: Ragged Mountain Press.

Sydykov, Zh.S., Veselov, V.V., et al. 1993. Experience and results of aerospace monitoring and mapping of hydrogeological processes in the areas of ecological crisis — basins of the Aral, Caspian Seas and Lake Balkhash. Proceedings of the International Scientific Seminar on Aerospace Monitoring of the Earth Cover and Atmosphere. Kiev, 139–151.

Szikszay, M. and Teissedre, I.M. 1978. Aspects hydrogeochimiques des sources limitrophes du bassin de Sao Paulo, Brasil. *AIH Memoires* II:651–662.

Szikszay, M., Teissedre, I.M., Barner, U., and Matsiu, E. 1981. Geochemical and isotopic characteristics of spring and groundwater in the state of Sao Paulo, Brasil. *Journal of Hydrology* 54(1–3):23–32.

Tabak, H.H., Quave, S.A., Mashni, C.I., and Barth, E.F. 1981. Biodegradability studies with organic priority pollutant compounds. *Journal of the Water Pollution Control Federation* 53:1503–1518.

Talbot, J., Kroeger, K., Rago, A., et al. 2003. Nitrogen flux and speciation through the subterranean estuary of Waquoit Bay, Massachusetts. *Biological Bulletin* 205(2):244–245.

Taniguchi, M. 2002. Tidal effects on submarine groundwater discharge into the ocean. *Geophysical Research Letters* 29(12).

Taniguchi, M. and Fukuo, Y. 1993. Seepage using an automated seepage meter. *Ground Water* 31(4):675–679.

Taniguchi, M. and Tase, N. 1999. Nutrient discharge by groundwater and rivers into Lake Biwa, Japan. In: *Impact of Land-Use Change on Nutrient Loads from Diffuse Sources.* L. Heathwaite, Ed. Proceedings of IUGG 99 Symposium HS3. Birmingham, UK: International Association of Hydrological Sciences, 67–74.

Taniguchi, M., Burnett, W., Cable, J., and Turner, J. 2002. Investigation of submarine groundwater discharge. *Hydrological Processes* 16(11):2115–2129.

Taniguchi, M., Burnett, W.C., Cable, J.E., and Turner, J.V. 2003. Assessment methodologies for submarine groundwater discharge. In: *Land and Marine Hydrogeology,* M. Taniguchi, K.Wang, and T.Gamo (Eds.). Elsevier Publications, Oxford, 208.

Testa, J., Charette, M., Sholkovitz, E., et al. 2002. Dissolved iron cycling in the subterranean estuary of a coastal bay: Waquoit Bay, Massachusetts. *Biological Bulletin* 203(2):255–256.

Theodorsson, P. and Gudjonsson, G.I. 2003. Increased radon detection sensitivity: Extraction from 200 mL of water and liquid scintillation counting. *Health Physics* 85:610–612.

Timmermann, K., Christensen, J., and Banta, G. 2002. Modeling of advective solute transport in sandy sediments inhabited by the lugworm Arenicola marina. *Journal of Marine Research* 60(1):151–169.

Timofeev, P.P., Kholodov, V.N., and Zverev, V.P. 1988. *Izv. An. SSSR. Geology Series* 6:3.

Tobias, C., Harvey, J., and Anderson, I. 2001. Quantifying groundwater discharge through fringing wetlands to estuaries: Seasonal variability, methods comparison, and implications for wetland-estuary exchange. *Limnology and Oceanography* 46(3):604–615.

Todd, D.K. and Huisman, L. 1959. *Journal of the Hydraulics Division, Proceedings of the ASCE* 85(7):63.

Top, Z., Brand, L., Corbett, R., et al. 2001. Helium as a tracer of ground water input into Florida Bay. *Journal of Coastal Research* 17:859–868.

Treadwell and Rollo. 2002. Berth 23 Soil Gas Sampling Report, Former Mobil Bulk Fuel Terminal, Port of Oakland Berths 23 and 24, Oakland, California, unpublished report prepared for the Port and submitted to the California Regional Water Quality Control Board, San Francisco Bay Region.

Treadwell and Rollo. 2002. Revised Human Health Risk Assessment and Methane Hazard Evaluation, Former Mobil Bulk Fuel Terminal, Port of Oakland Berths 23 and 24, Oakland, California, unpublished report prepared for the Port and submitted to the California Regional Water Quality Control Board, San Francisco Bay Region.

Tricca, A., Wasserburb, G.J., Porcelli, D., and Baskaran, M. 2001. The transport of U- and Th-series nuclides in a sandy unconfined aquifer. *Geochimica et Cosmochimica Acta* 65(8):1187–1210.

Troitsky, L.S., Khodakov, V.G., Mikhalyev, V.I., Guskov, A.S., et al. 1966. Glaciation of Ural. Glaciology, 16, Moscow, Nauka, 150.

Tyutyunova, F.I. 1987. *Hydrochemistry of Technogenesis* [Russian]. Moscow: Nauka, 335.

Uchiyama, Y., Nadaoka, K., Rolke, P., et al. 2000. Submarine groundwater discharge into the sea and associated nutrient transport in a sandy beach: *Water Resources Research* 36(6):1467–1479.

UNESCO. 1987. *Groundwater Problems in Coastal Areas*. Paris: UNESCO, 596

UNESCO. 2004. *Groundwater Resources of the World and their Use*. Paris: UNESCO, 346.

URL: http://www.whoi.edu/institutes/coi/research_02mulligan.htm

Vaccaro, J.J. 1992. Sensitivity of ground water recharge estimates to climate variability and change, Columbia Plateau, Washington. *Journal of Geophysical Research* 97(D3): 2821–2833.

Vacher, H.L. 1978. Hydrology of small ocean islands — influence of atmospheric pressure on the water table. *Ground Water* 16:417–423.

Valdes, A.A. 1985. Case of clay layer surface runoff recharged from groundwater. *Habana* 7.

Valiela, I. and Costa, J. 1988. Eutrophication of Buttermilk Bay, a Cape Cod coastal embayment: concentrations of nutrients and watershed nutrient budgets. *Environmental Management* 12(4):539–553.

Valiela, I. and Teal, J.M. 1979. The nitrogen budget of a salt marsh ecosystem. *Nature* 280:652–656.

Valiela, I., Costa, J., Foreman, K., et al. 1990. Transport of groundwater-borne nutrients from watersheds and their effects on coastal waters. *Biogeochemistry* 10:177–197.

Valiela, I., Foreman, K., LaMontagne, M., et al. 1992. Couplings of watersheds and coastal waters: sources and consequences of nutrient enrichment in Waquoit Bay, Massachusetts. *Estuaries* 15(4):443–457.

Valiela, I., Teal, J., Volkman, S., et al. 1978. Nutrient and particulate fluxes in a salt marsh ecosystem: tidal exchanges and inputs by precipitation and groundwater. *Limnology and Oceanography* 23(4):798–812.

Valyashko, M.G. Evolution of ocean water chemical composition. In *The History of the World Ocean*. Academy of Sciences of the USSR, 97–103.

Vanek, V. 1993. Groundwater regime of a tidally influenced coastal pond. *Journal of Hydrology* 151:317–342.

Vassoevich, N.B. 1986. Geochemistry of the organic matter and oil formation. Selec. Proc., Moscow, Nauka.

Verigin, N.N. 1963. Estimation of river water discharge. *Proceedings of VNII VODGEO*, 4:16–17.

Verkhovsky, A.B., Yurgina, E.K., and Shukolyukov, Yu.A. 1985. *Degassing of the Earth and Geotectonics.* Moscow: Nauka, 46.

Veselov, V.V. and Panichkin, V.Yu. 2004. Geoinformatic and mathematical modeling of hydrogeological conditions of the Priaral region. Almaty: Gylym, 428.

Vigdorchek, M.E. 1980. *Arctic Pleistocene History and Development of Submarine Permaforst.* Boulder, CO: Westview Press, 286.

Vinogradov, A.P. 1967. *Introduction to Chemistry of the Ocean.* Moscow: Nauka, 242.

Visocky, A.P. 1970. Estimating the groundwater contribution to storm runoff by the electrical conductance method. *Ground Water* 8(2):5–10.

Vörösmarty, C.J., Green, P., Salisbury, J., and Lammers, R.B. 2000. Global water resources: vulnerability from climate change and population growth. *Science*, 289:284–288.

Volker, R.E. and Rushton, K.R. 1982. An assessment of the importance of some parameters for seawater intrusion in aquifers and a comparison of dispersive and sharp interface modeling approaches. *Journal of Hydrology* 56:293.

Voss, C.I. 1984. A finite-element simulation model for saturated-unsaturated, fluid-density-dependent ground-water flow with energy transport or chemically-reactive single-species solute transport: U.S. Geological Survey Water-Resources Investigation, Report 84-4369. Washington, DC: U.S. Geological Survey, 409.

VSEGINGEO. 1987. *Methodical Recommendations on Hydrogeological Study of Aqueous Areas of Seas and Large Lakes.* All-Russian Scientific and Research Institute for Hydrogeology and Engineering Geology. Moscow: VSEGINGEO, 66.

Vsevolozhsky, V.A. 1983. *Subsurface Runoff and Water Balance of Platform-like Types.* Moscow: Nedra, 167.

Walton, W.C. 1970. *Groundwater Resources Evaluation.* New York: McGraw-Hill, 664.

Wanninkhof, R., Ledwell, J.R., Broecker, W.S., and Hamilton, M. 1987. Gas exchange on Mono Lake and Crowley Lake, California. *Journal of Geophysical Research* 92:14567–14580.

Warnken, K., Gill, G., Santschi, P., and Griffen, L. 2000. Benthic exchange of nutrients in Galveston Bay, Texas. *Estuaries* 23:647–661.

Waterloo Hydrogeologic Inc. 2000. *Visual Modflow User's Manual.* Waterloo, Canada.

Wayne, C., Shanks, I.I., and Callender, E. 1992. Thermal springs in Lake Baikal. *Geology*, 20:495–497.

Webb, J.E. and Theodor, J. 1968. Irrigation of submerged marine sands through wave action. *Nature*, 220:682–683.

Webb, J.E. and Theodor, J.L. 1972. Wave-induced circulation in submerged sands. *Journal of Marine Biological Association* 52:903–914.

Weil, R., Weismiller, R., and Turner, R. 1990. Nitrate contamination of groundwater under irrigated coastal plain soils. *Journal of Environmental Quality* 19:441–448.

Weiskel, P. and Howes, B. 1991. Quantifying dissolved nitrogen flux through a coastal watershed. *Water Resources Research* 27(11):2929–2939.

Weiskel, P. and Howes, B. 1992. Differential transport of sewage-derived nitrogen and phosphorus through a coastal watershed. *Environmental Science and Technology* 26(2):352–360.

Weiss Associates and ENTRIX, Inc. 2003. Draft Initial Study/Mitigated Negative Declaration for Soil and Groundwater Investigation Vapor Extraction and Air Sparging Pilot Testing/Interim Remediation at Berths 23 and 24, Port of Oakland, Maritime Street, Oakland, California.

Weiss, R.F. 1971. Solubility of helium and neon in water and seawater. *Journal of Chemical Engineering Data* 16:235–241.

Westbrook, S.J., Rayner, J.L., Davis, G.B. et al. 2005. Interaction between shallow ground-water, saline surface water and contaminant discharge at a seasonally and tidally forced estuarine boundary. *Journal of Hydrology*, 302:255–269.

Whiting, G. and Childers, D. 1989. Subtidal advective water flux as a potentially important nutrient input to southeastern U.S.A. saltmarsh estuaries. *Estuarine, Coastal and Shelf Science* 28:417–431.

Wilkins, B., Dominick, T.F., and Roberts, H. 1971. Mathematical model for beach groundwater fluctuations. *Water Resources Research* 7(6):1626–1635.

Willebrand, J. 1993. Forcing the ocean by heat and water fluxes. In: *Energy and Water Cycles in the Climate System*. Raschke, E. and Jacob, D. (Eds.). NATO ASI Series I, vol. 5, Springer-Verlag, Berlin, 215–233.

Wilson, A.M. 2003. The occurrence and chemical implications of geothermal convection of seawater in continental shelves. *Geophysical Research Letters* 30(21):1649.

Wilson, R.D. and Mackay, D.M. 1993. The use of sulfur hexafluoride as a conservative tracer in saturated sandy media. *Ground Water* 31:719–724.

Winter, T.C. and Pfannkuch, H.O. 1984. Effect of anisotropy and groundwater system geometry on seepage through lakebeds. 2. Numerical simulation analysis. *Journal of Hydrology* 75:239–253.

Woessner, W.W. and Sullivan, K.E. 1984. Results of seepage meter and mini-piezometer study, Lake Mead, Nevada. *Groundwater* 22:561–568.

Wolery, T.J. 1992 EQ3NR, A Computer Program for Geochemical Aqueous Speciation-Solubility Calculations: Theoretical Manual, Users Guide, and Related Documentation (Version 7.0). University of California, Lawrence Livermore National Laboratory, UCRL-MA-110662 PT

Wong, F., Eittreim, S.L., Degnan, C.H. and Lee, W.C. 1999. USGS Seafloor GIS for Monterey Sanctuary—Selected Data Types. ESRI International Users Conference, San Diego, California.

Woodward Clyde Consultants, 1980. Project File Notes, Project No. 14567A, November and December 1980. Provided by the Port of Oakland, September 2003.

Worts, G.F. 1963. A brief appraisal of ground water conditions in the coastal artesian basin of British Guiana, South America. Geological Survey Water Supply, Paper No. 1663-B:1–44.

Wright, D. 2002. *Undersea with GIS*. Redlands, CA: ESRI Press.

Wylie, F.E. 1979. *Tides and the Pull of the Moon*. Brattleboro, VT: Stephen Green Press, 241.

Yakubov, A.A., Dadashev, R.G. and Mekhtiev, A.K. 1983. *Geologic-Geomorphologic Investigations of the Caspian Sea*. Moscow: Nauka, 70.

Yohe, G. and Andronova, N. 2004. To hedge or not against an uncertain climate future? *Science*, 306:416–417.

Yurovskiy, Yu.G. 1993. Peculiarities of natural processes in the zone of groundwater submarine discharge. Abstract of DSc thesis, Kiev, 44

Zatenatskaya, N.P. 1965. Experimental data on salt diffusion in the clay rocks—Postsedimentation changes of the Quaternary and Miocene clay deposits of Bakinsk archipelago. Moscow: Nedra, 143–159.

Zektser, I.S. 1968. *Natural Fresh Groundwater Resources of the Baltic Region*. Moscow, Nedra, 105.

Zekster, I.S. 1973. On the groundwater discharge into the Baltic Sea and methods of estimating. *Nordic Hydrology* 4(2):105–118.

Zekster, I.S. 1977. *Regularities in Formation of Groundwater Flow and Scientific-Methodical Principles of Its Study*. Moscow: Nauka, 173.

Zekster, I.S. 1983. Present-day status of investigations of interaction between groundwater and sea waters. *Water Resources* 10(6):537–543.

Zekster, I.S. 1996. Groundwater discharge into the seas and oceans: State of art. Groundwater Discharge in Coastal Zone. In *Proceeding of an International symposium, LOICH Reports & Studies* #8, Moscow.

Zektser, I.S. 2000. *Groundwater and the Environment*. Boca Raton, FL: Lewis Publishers.

Zektser, I.S. 2001. *Groundwater as a Component of the Environment*. Scientific World, Moscow.

Zektser, I.S. and Dzhamalov, R.G. 1981a. Groundwater discharge to the Pacific Ocean. *Hydrological Sciences Bulletin* 26(3):271–279.

Zektser, I.S. and Dzhamalov, R.G. 1981b. Groundwater discharge to the world's oceans. *Nature and Resources* 17(3):18–20.

Zektser, I.S. and Dhzamalov, R.G. 1989. *Groundwater in Water Balance of Large Regions*. Moscow: Nauka, 124.

Zektser, I.S. and Loaiciga, H.A. 1993. Ground-water fluxes in the global hydrology cycle: past, present, and future. *Journal of Hydrology* 144(1–4):405–427.

Zektser, I.S. Dzhamalov, R.G. and Meskheteli, A.V. 1984. *Subsurface Water Exchange between Land and Sea*. Leningrad: Hydrometizdat, 207.

Zektser, I.S. Ivanov, V.A., and Meskheteli, A.V. 1972. Groundwater discharge to seas [Russian]. *Vodnye Resursy* 3:125–146.

Zektser, I.S. Ivanov, V.A., and Meskheteli, A.V. 1973. The problem of direct groundwater discharge to the seas. *Journal of Hydrology* 20(1):1–36.

Zektser, I.S., Plemennov, V.A., and Kasianova, N.A. 1994. Role of modern tectonics and mud volcanism in water and salt balance of the Caspian Sea. *Water Resources* 21(4):437.

Zheng, C. and Bennett, G.D. 2002. *Applied Contaminant Transport Modeling*, 2nd ed. New York: John Wiley & Sons, 621.

Zimmermann, C. Montgomery, J., and Carlson, P. 1985. Variability of dissolved reactive phosphate flux rates in nearshore estuarine sediments: Effects of groundwater flow. *Estuaries* 8(2B):228–236.

Zimmermann, C.F. 1991. Submarine groundwater discharge to the Patuxent River and Chesapeake Bay. In: *New Perspectives in the Chesapeake Bay System: A Research and Management Partnership*. J.A. Mihursky and A. Chaney, Eds., Baltimore, MD. Chesapeake Research Consortium Publication 137, 663–668. Solomons, MD: Chesapeake Research Consortium.

Zimmermann, C.F., Montgomery, J.R. and Carlson, P.R. 1985. Variability of dissolved reactive phosphate flux rates in nearshore estuarine sediments: Effects of groundwater flow. *Estuaries*, 8(2B):228–236.

Zolotarev, V.P. 1990. Modeling of groundwater and salt exchange in the Aral Sea basin on the modern stage and perspectives of its changing. Abstract on Ph.D. competition in technical sciences. Tashkent.

Zverev, V.P. 1968. The role of atmospheric precipitations in chemical elements circulation between atmosphere and lithosphere. *Far East Academy of Sciences* 181(3):716–719.

Zverev, V.P. 1993. *Hydrogeochemistry of Sedimentary Process*. Moscow: Nauka, 176.

//Ieshina A.V., Polenov, I.K. et al. 1987. *Resources and Geochemistry of Subsurface Waters in Karelia* [Russian]. Petrozavodsk: Karelia Publishers, 151.

Index